Food Packaging and Shelf Life

A Practical Guide

Food Packaging and Shelf Life

A Practical Guide

Edited by

Gordon L. Robertson

CRC Press
Taylor & Francis Group
Boca Raton London New York

CRC Press is an imprint of the
Taylor & Francis Group, an **informa** business

CRC Press
Taylor & Francis Group
6000 Broken Sound Parkway NW, Suite 300
Boca Raton, FL 33487-2742

Printed in the United States of America on acid-free paper
10 9 8 7 6 5 4 3 2 1

International Standard Book Number: 978-1-4200-7844-2 (Hardback)

Library of Congress Cataloging-in-Publication Data

Food packaging and shelf life : a practical guide / [edited by] Gordon L. Robertson.
p. cm.
Includes bibliographical references and index.
ISBN 978-1-4200-7844-2 (alk. paper)
1. Food--Packaging. I. Robertson, Gordon L., 1946- II. Title.

TP374.F653 2009
664'.09--dc22 2009013021

Visit the Taylor & Francis Web site at
http://www.taylorandfrancis.com

and the CRC Press Web site at
http://www.crcpress.com

Contents

Preface

Food packaging is an area with which, sooner or later, every practicing food scientist and technologist becomes involved. The importance of packaging hardly needs stressing, as only a handful of foods are sold in an unpackaged state. Furthermore, the fact that, on average, around 25% of the exfactory cost of consumer foods is for their packaging provides the incentive and the challenge for food packaging technologists to design and develop functional packages at minimum cost.

There is an old saying that any fool can do for $10 what a good engineer can do for $1. This saying also applies to food packaging technologists. Anyone can overpackage a food, but to provide just enough protection to ensure that the food maintains its acceptability until the end of its shelf life requires detailed knowledge and understanding of both food and packaging and how together they combine to deliver the desired shelf life. Although there are several books on shelf life, they tend to treat packaging in a superficial and unsatisfactory way. It is my hope that this book, by clearly demonstrating the nexus between packaging and shelf life, will provide valuable insights lacking in other texts.

This book introduces for the first time in print the concept of indices of failure (IoFs), first introduced to me when I was an undergraduate student at Massey University, New Zealand, by the late H.A.L. Morris, then a reader in food processing. IoFs are discussed in Chapter 2, and the chapter authors have all adopted this approach in discussing the shelf life of specific foods. I am confident that readers will find it a useful concept. It is hoped that this book will lead to the informed development of food packages that provide just the required amount of protection—no more and no less. With an increasing focus on sustainability, responsible companies no longer want to overpackage their food products and yet many remain unsure just where reductions can effectively be made. This book should help them in their endeavors.

It would obviously not have been possible to complete this book without the active participation of the authors, and I here place on record my appreciation for their willingness to contribute. It is a largely thankless, unpaid task to write a book chapter, and we should all be grateful that busy people are prepared to give of their time in this way. A special mention must also go to Steve Zollo, Senior Editor, Food Science and Technology at CRC Press/Taylor & Francis Group, who first suggested a book on this subject. His encouragement and support have been very much appreciated, as has the efficient attention to administrative details by Kari Budyk, Senior Production Coordinator, Editorial Project Development.

Gordon L. Robertson

Editor

Gordon L. Robertson is a food packaging consultant, author, trainer, and an adjunct professor in the School of Land, Crop, and Food Sciences at the University of Queensland in Brisbane, Australia. Previously he was vice president for environmental and external affairs for Tetra Pak in their regional headquarters in Asia. Before that he was foundation professor of packaging technology at Massey University, New Zealand, where he taught courses on food packaging for 21 years.

A member of several editorial boards, he is a fellow of the International Academy of Food Science and Technology, a fellow of the U.S. Institute of Food Technologists, a fellow of the Australian Institute of Packaging, and a fellow and former president of the New Zealand Institute of Food Science and Technology. Dr. Robertson received the BTech, MTech, and PhD degrees in food technology from Massey University.

The second edition of his book, *Food Packaging Principles and Practice*, was published by CRC Press in the United States in 2006 and is widely used in universities and by industry around the world. He offers workshops and training courses on food packaging in many countries (see www.gordonlrobertson.com for further details).

Contributors

Philip R. Ashurst
Dr P.R. Ashurst & Associates
Ludlow
Shropshire, United Kingdom

Charles W. Bamforth
Department of Food Science and Technology
University of California Davis
Davis, California

Yulissa Y. Belisario-Sánchez
Food Engineering and Agricultural Equipment Department
Technical University of Cartagena
Cartagena, Spain

Sonia Calligaris
Department of Food Science
University of Udine
Udine, Italy

Kay Cooksey
Department of Packaging Science
Clemson University
Poole Agricultural Center
Clemson, South Carolina

Alex O. Gill
Bureau of Microbial Hazards
Health Canada
Sir F.G. Banting Research Centre
Ottawa, Ontario, Canada

Colin O. Gill
Agriculture and Agri-Food Canada
Lacombe Research Centre
Lacombe, Alberta, Canada

Nathalie Gontard
Agropolymers Engineering and Emerging Technologies
University of Montpellier II
Montpellier, France

Carole Guillaume
Agropolymers Engineering and Emerging Technologies
University of Montpellier II
Montpellier, France

Michel Guillet
Creascience
Montreal, Quebec, Canada

Jung H. Han
PepsiCo Fruit and Vegetable Research Center
Frito-Lay Inc.
Plano, Texas

Vibeke Kistrup Holm
Danish Technological Institute
Kolding, Denmark

Young T. Kim
Department of Packaging Science
Clemson University
Clemson, South Carolina

Michael G. Kontominas
Laboratory of Food Chemistry and Technology
Department of Chemistry
University of Ioannina
Ioannina, Greece

John M. Krochta
Department of Food Science and Technology
University of California Davis
Davis, California

Dong Sun Lee
Department of Food Science and Biotechnology
Kyungnam University
Masan, South Korea

Sara Limbo
Department of Food Science and Microbiology
University of Milan
Milan, Italy

Antonio López-Gómez
Food Engineering and Agricultural Equipment Department
Technical University of Cartagena
Cartagena, Spain

Roger D. MacBean
Parmalat Australia Ltd.
South Brisbane, Australia

Lara Manzocco
Department of Food Science
University of Udine
Udine, Italy

Sea C. Min
Division of Food Science
Seoul Women's University
Seoul, Korea

Maria Cristina Nicoli
Department of Food Science
University of Udine
Udine, Italy

Luciano Piergiovanni
Department of Food Science and Microbiology
University of Milan
Milan, Italy

Manuela Pintado
Food Packaging Department
Biotechnology College
Portuguese Catholic University
Porto, Portugal

Maria de Fátima Poças
Food Packaging Department
Biotechnology College
Portuguese Catholic University
Porto, Portugal

Malcolm J. Reeves
Faculty of Science and Technology
Eastern Institute of Technology
Taradale, New Zealand

Gordon L. Robertson
University of Queensland and
Food•Packaging•Environment
Brisbane, Australia

Natalie Rodrigue
Creascience
Montreal, Quebec, Canada

María Ros-Chumillas
Food Engineering and Agricultural Equipment
Department
Technical University of Cartagena
Cartagena, Spain

Steve Slattery
Innovative Food Technologies
Emerging Technologies
Primary Industries and Fisheries
Queensland Department of Employment,
Economic Development and Innovation
Hamilton, Australia

Kees Sonneveld
Packaging and Polymer Research Unit
Victoria University
Melbourne, Australia

Elmira Arab Tehrany
Nancy-Université
Laboratoire de Science & Génie
Alimentaires
Vandoeuvre lés Nancy, France

Abbreviations, Acronyms, and Symbols

θ	time
θ_s	shelf life
°C	degrees Celsius
A	area (m^2); frequency factor in Arrhenius equation
A_{420}	absorbance at 420 nm
AA	acrylic acid; arachidonic acid; acetaldehyde
ABS	acrylonitrile-butadiene-styrene
Ac	acrylic
ACTIS	amorphous carbon treatment on internal surface
AITC	allyl isothiocyanate
alufoil	aluminum foil
AM	antimicrobial
AmPA	amorphous polyamide
AN	acrylonitrile
ANMA	acrylonitrile-methyl acrylate
ANOVA	analysis of variance
ANS	acrylonitrile/styrene
AnV	anisidine value
APC	aerobic plate count
APET	amorphous poly(ethylene terephthalate)
Ar	argon
ASLT	accelerated shelf life testing
atm	atmosphere
ATP	adenosine triphosphate
a_w	water activity
BET	Brunauer–Emmett–Teller
BHT	butylated hydroxytoluene
BIB	bag-in-box
BO	biaxially oriented
BON	biaxially oriented nylon
BOPP	biaxially oriented polypropylene
BPA	bisphenol A
BPAA	bis-anthracene-trimethylphenylammonium dichloride
BTW	by the way
c	concentration of permeant in polymer
ca.	approximately
cfu	colony-forming unit
CLCs	calcium lactate crystals
cm Hg	centimeters of mercury
CO	carbon monoxide
CO_2	carbon dioxide
cP	centipoise (10^{-3} Pa sec)
CPET	crystalline poly(ethylene terephthalate)

D	diffusion coefficient (cm^{-2} sec^{-1})
Da	dalton = mass of a single hydrogen atom (1.66 x 10^{-24} gram)
DAL	defect action level
DEHA	di-(2-ethylhexyl) adipate
DEHP	di-(2-ethylhexyl) phthalate
DFD	dark, firm, and dry (meat)
DHA	docosahexanoic acid
DLC	diamond-like coating
DO	dissolved oxygen
DMA	dimethylamine
DWI	drawn and wall ironed
E_a	activation energy (J mol^{-1})
EAA	ethylene-acrylic acid copolymer
ECCS	electrolytically chromium-coated steel
E_d	activation energy for diffusion (kJ mol^{-1})
ECTA	ethylenediaminetetraacetic acid
EFA	essential fatty acid
EFSA	European Food Safety Authority
E_h	oxidation-reduction potential
EO	essential oil
Ep	activation energy for permeation (kJ mol–1)
ESC	environmental stress cracking
ESE	easy serving expresso
EPS	expanded polystyrene
EU	European Union
EVA	ethylene-vinyl acetate copolymer
EVOH	ethylene-vinyl alcohol copolymer
EVOO	extra-virgin olive oil
FA	formaldehyde
FAST	fluorescence of advanced Maillard products and soluble tryptophan
FCOJ	frozen concentrated orange juice
FDA	Food and Drug Administration
FMP	filled milk powder
FSOJ	freshly squeezed orange juice
GAB	Guggenheim–Anderson–de Boer
GMP	good manufacturing practice
GRAS	generally recognized as safe
H3O+	hydronium ion
HACCP	hazard analysis critical control point
HDL	high density lipoprotein
HDPE	high density polyethylene
HEPA	high-efficiency particulate air
HHRS	highly heat-resistant spore
HIPS	high-impact polystyrene
HPC	hydroxypropyl cellulose
HPMC	hydroxypropyl methylcellulose
HPP	high pressure processing
HQL	high-quality shelf life
HRS	heat-resistant spore
HTST	high-temperature short-time
IBMP	iso-butyl-methoxypyrazine

IoFs	indices of failure
IU	international unit
J	flux of permeant in polymer
LAB	lactic acid bacteria
LBM	Le Bouchage Mecanique
LDPE	low density polyethylene
LLDPE	linear low density polyethylene
LSD	least significant difference
LTLT	low-temperature long-time
k	rate constant
kGy	kiloGray
k_0	Arrhenius pre-exponential factor
K	Kelvin
KM	Kaplan–Meier
LC-PUFAs	long-chain polyunsaturated fatty acids
LDL	low density lipoprotein
LTLT	low-temperature-long-time
MA	methyl acrylate; also modified atmosphere
MAP	modified atmosphere packaging
Mb	myoglobin
MC	methylcellulose
MCP	methylcyclopropene
Met	Metalized
MetMb	metmyoglobin
mg	milligram (1×10^{-3} gram)
MH	mercaptohexanol
MHA	mercaptohexyl acetate
MJ	megajoule
mL	milliliter (1×10^{-3} liters) = cc = cm^3
MMT	montmorillonite; million metric tons
MPa	megapascal
mPET	metalized PET
MPFVs	minimally processed fruits and vegetables
MPN	most probable number
MPPO	modified phenylene oxide
MRA	metmyoglobin reducing activity
MXD6	meta-xylylene diamine/adipic acid nylon
NADH	nicotinamide adenine dinucleotide
NDM	nonfat dry milk
NEB	nonenzymic browning
NFCOJ	not-from-concentrate orange juice
nm	nanometer (1×10^{-9} meter)
NSLAB	nonstarter lactic acid bacteria
OCR	oxygen consumption rate
O_2Mb	oxymyoglobin
OPA	oriented polyamide
OPET	oriented polyester
OPP	oriented polypropylene
ORP	oxidation-reduction potential
OS	oxygen scavenger
OTR	oxygen transmission rate

p	partial pressure (cm Hg)
P	permeability coefficient (mL cm cm^{-2} sec^{-1} (cm Hg^{-1}))
P/X	permeance
PA	polyamide
PAN	polyacrylonitrile
PBT	poly(butylene terephthalate)
PC	polycarbonate
PDLA	poly-D-lactic acid
PDLLA	mixture of poly-D-lactic acid and poly-L-lactic acid
PE	polyethylene
PECVD	plasma-enhanced chemical vapor deposition
PEN	poly(ethylene naphthalate)
PET	poly(ethylene terephthalate) = polyester
PFB	printed fibreboard
PFO	polyfuryloxirane
pg	picogram (1 x 10^{-12} gram)
PGA	propylene-glycol alginate
PHA	polyhydroxyalkanoate
PHB	polyhydroxybutyrate
PLA	polylactic acid or polylactate or polylactide
PLLA	poly-L-lactic acid
PME	pectinmethylesterase
POD	peroxidase
PP	polypropylene
ppb	parts per billion (1×10^{-9})
ppm	parts per million (1×10^{-6})
PPO	polyphenoloxidase
ppt	parts per trillion (1 x 10^{-12})
PS	polystyrene
PUFA	polyunsaturated fatty acid
PV	peroxide value
PVC	poly(vinyl chloride)
PVD	physical vapor deposition
PVdC	poly(vinylidene chloride)
Q	total amount of permeant passing through polymer
Q_{10}	temperature quotient = ratio of reaction rates for 10°C temperature difference
QIM	quality index method
R	ideal gas constant (= 8.314 J K^{-1} mol^{-1} = 1.987 cal K^{-1} mol^{-1})
R&G	roasted and ground
RCF	regenerated cellulose film
RFID	radio frequency identification
RH	relative humidity
ROPP	roll-on pilfer-proof
ROTE	roll-on tamper-evident
RTE	ready-to-eat
S	solubility coefficient of permeant in polymer (mL cm^{-3} (cm Hg^{-1}))
SiO_2	silicon dioxide
SiO_x	oxides of silicon
SLO	sulfur-like odor
SLOs	sulfur-like odors
SMP	skim milk powder

SO	sunflower oil
SO_2	sulfur dioxide
SSO	specific spoilage organism
SSSP	Seafood Spoilage and Safety Predictor
STP	standard temperature and pressure
TBA	thiobarbituric acid
TBHQ	tert-butylhydroquinone
TBP	tribromophenol
TCA	trichloroanisole
TCP	trichlorophenol
TDI	tolerable daily intake
TDN	trimethyldihydronaphthalene
TEBO	tail-end blow-off
TeCA	tetrachloroanisole
Tg	glass transition temperature
TiO2	titanium dioxide
Tm	crystalline melting temperature
TMA	trimethylamine
TMAO	trimethylamine oxide
TNF	thickness normalized flux
TPA	terephthalic acid
TPB	trimethylphenylbutadiene
TPE	thermoplastic elastomer
TQM	total quality management
TR	transmission rate
TTI	time-temperature indicator
UHT	ultra-heat-treated or ultra-high-temperature
ULDPE	ultra-low density polyethylene
UP	ultrapasteurized
UPC	universal product code
UV	ultraviolet
UVA	380–320 nm
UVB	320–280 nm
UVC	280–100 nm
VCM	vinyl chloride monomer
VP	vacuum packaging
VSP	vacuum skin packaging
WMP	whole milk powder
WPNI	whey protein nitrogen index
WVP	water vapor permeability
WVTR	water vapor transmission rate
X	thickness of polymeric material
μg	microgram (1×10^{-6} gram)
μm	micrometer (1×10^{-6} meter)
μmax	maximum specific growth rate

1 Food Packaging and Shelf Life

Gordon L. Robertson
University of Queensland and
Food•Packaging•Environment
Brisbane, Australia

CONTENTS

1.1 INTRODUCTION

1.1.1 Role of Food Packaging

Food packaging is essential and pervasive: essential because without packaging the safety and quality of food would be compromised, and pervasive because almost all food is packaged in some way. Food packaging performs a number of disparate tasks: it protects the food from contamination and spoilage; it makes it easier to transport and store foods; and it provides uniform measurement of contents. By allowing brands to be created and standardized, it makes advertising meaningful and large-scale distribution and mass merchandising possible. Food packages with dispensing caps, sprays, reclosable openings, and other features make products more usable and convenient.

A distinction is usually made between the various "levels" of packaging. A primary package is one that is in direct contact with the contained product. It provides the initial, and usually the major, protective barrier. Examples of primary packages include metal cans, paperboard cartons, glass bottles, and plastic pouches. It is frequently only the primary package that the consumer purchases at retail outlets. This book will confine itself to the primary package.

A secondary package contains a number of primary packages, for example, a corrugated case. It is the physical distribution carrier and is sometimes designed so that it can be used in retail outlets for the display of primary packages. A tertiary package is made up of a number of secondary packages, the most common example being a stretch-wrapped pallet of corrugated cases. In interstate and international trade, a quaternary package is frequently used to facilitate the handling of tertiary packages. This is generally a metal container up to 40 m in length that can hold many pallets and is intermodal in nature.

Four primary and interconnected functions of packaging have been identified: containment, protection, convenience, and communication (Robertson, 2006).

1.1.1.1 Containment

This function of packaging is so obvious as to be overlooked by many, but it is the most basic function of packaging. Food products must be contained before they can be moved from one place to another. The containment function of packaging makes a huge contribution to protecting the environment from the myriad of products that are moved from one place to another on numerous occasions each day.

1.1.1.2 Protection

This is often regarded as the primary function of the package: to protect its contents from the outside environmental effects of water, water vapor, gases, odors, microorganisms, dust, shocks, vibrations, compressive forces, and so on.

For the majority of food products, the protection afforded by the package is an essential part of the preservation process. For example, aseptically packaged milk in paperboard laminate cartons remains aseptic only for as long as the package provides protection; vacuum-packaged meat will not achieve its desired shelf life if the package permits O_2 to enter. In general, once the integrity of the package is breached, the product is no longer preserved.

Freedom from harmful microbial contaminants at the time of consumption can also be influenced by the package. First, if the packaging material does not provide a suitable barrier around the food, microorganisms can contaminate the food and make it unsafe. Microbial contamination can also arise if the packaging material permits the transfer of, for example, moisture or O_2 from the atmosphere into the package. In this situation, microorganisms present in the food but posing no risk because of the initial absence of moisture or O_2 may then be able to grow and present a risk to the consumer.

Effective packaging reduces food waste, and in doing so protects or conserves much of the energy expended during the production and processing of the product. For example, to produce, transport, sell, and store 1 kg of bread requires 15.8 MJ (megajoules) of energy. This energy is required in the form of transport fuel, heat, power, refrigeration in farming and milling the wheat, baking and retailing the bread, and distributing both the raw materials and the finished product. To produce the polyethylene bag to package a 1 kg loaf of bread requires 1.4 MJ of energy. This means that each unit of energy in the packaging protects 11 units of energy in the product. Although eliminating the packaging might save 1.4 MJ of energy, it would also lead to spoilage of the bread and a consequent waste of 15.8 MJ of energy (Robertson, 2006).

1.1.1.3 Convenience

Modern, industrialized societies have seen tremendous changes in lifestyle, and the packaging industry has had to respond to those changes, which have created a demand for greater convenience in household products: foods that are pre-prepared and can be cooked or reheated in a very short time, preferably without removing them from the package; condiments that can be applied simply or by means of pump-action packages; dispensers for sauces or dressings that minimize mess; reclosable openings on drink bottles to permit consumption on the go; and so on. Thus, packaging plays an important role in allowing products to be used conveniently.

Two other aspects of convenience are important in package design. One of these can best be described as the apportionment function of packaging. In this context, the package functions to reduce the output from industrial production to a manageable, desirable "consumer" size. An associated aspect is the shape (relative proportions) of the primary package in relation to convenience of use by consumers (e.g., easy to hold, open, and pour as appropriate) and efficiency in building it into secondary and tertiary packages. Packaging plays a very important role in permitting primary packages to be unitized into secondary packages (e.g., placed inside a corrugated case) and then for these secondary packages to be unitized into a tertiary package (e.g., a stretch-wrapped pallet). As a consequence of this unitizing function, materials handling is optimized, as only a minimal number of discrete packages or loads need to be handled.

1.1.1.4 Communication

There is an old saying that "a package must protect what it sells and sell what it protects"; that is, the package functions as a "silent salesman." The modern methods of consumer marketing would fail were it not for the messages communicated by the package through distinctive branding and labeling, enabling supermarkets to function on a self-service basis. Consumers make purchasing decisions using the numerous clues provided by the graphics and the distinctive shapes of the packaging. Other communication functions of the package include a Universal Product Code (UPC) that can be read accurately and rapidly using modern scanning equipment at retail checkouts, nutritional and ingredient information (including E-numbers for additives), and country of origin.

1.1.1.5 Attributes

There are also several attributes of packaging that are important (Krochta, 2007). One (related to the convenience function) is that it should be efficient from a production or commercial viewpoint, that is, in filling, closing, handling, transportation, and storage. Another is that the package should have, throughout its life cycle from raw material extraction to final disposal after use, minimal adverse environmental impacts. A third attribute is that the package should not impart to the food any undesirable contaminants. Although this last attribute may seem self-evident, there has been a long history of so-called food-contact substances migrating from the packaging material into the food (Grob et al., 2006). Not surprisingly, food packaging materials are highly regulated in many countries to ensure consumer safety.

1.1.2 Package Environments

The packaging has to perform its functions in three different environments (Lockhart, 1997). Failure to consider all three environments during package development will result in poorly designed packages, increased costs, consumer complaints, and even avoidance or rejection of the product by the consumer.

1.1.2.1 Physical Environment

This is the environment in which physical damage can be caused to the product, including shocks from drops, falls, and bumps; damage from vibrations arising from transportation modes, including road, rail, sea, and air; and compression and crushing damage arising from stacking during transportation or storage in warehouses, retail outlets, and the home environment.

1.1.2.2 Ambient Environment

This is the environment that surrounds the package. Damage to the product can be caused as a result of exposure to gases (particularly O_2), water and water vapor, light (particularly UV radiation), and the effects of heat and cold, as well as microorganisms (bacteria, fungi, molds, yeasts, and viruses) and macroorganisms (rodents, insects, mites, and birds), which are ubiquitous in many warehouses and retail outlets. Contaminants in the ambient environment such as exhaust fumes from automobiles and dust and dirt can also find their way into the product unless the package acts as an effective barrier.

1.1.2.3 Human Environment

This is the environment in which the package is handled by people, and designing packages for this environment requires knowledge of the strengths and frailties of human vision, human strength and weakness, dexterity, memory, cognitive behavior, and so on (Yoxall et al., 2007). It also includes results of human activity such as liability, litigation, legislation, and regulation. Since one of the functions of the package is to communicate, it is important that the messages are received clearly by consumers. In addition, the package must contain information required by law, such as nutritional content and net weight. To maximize its convenience or utility functions, the package should be simple to hold, open, use, and (if appropriate) reclose by the consumer.

1.2 FOOD PACKAGING MATERIALS

The materials used to manufacture food packaging comprise a heterogeneous group, including glass, metals, plastics, and paper, with a corresponding range of performance characteristics. The properties of these various materials will not be described here in detail, as they have been well documented elsewhere (Lee et al., 2008; Piringer and Baner, 2008; Robertson, 2006; Yam, 2009). However, some general points will be made.

In the selection of suitable packaging materials for a particular food, the focus is typically on the barrier properties of the packaging material. Foods can be classified according to the degree of protection required, such as the maximum moisture gain or O_2 uptake. Calculations can then be made to determine whether a particular packaging material would provide the necessary barrier required to give the desired product shelf life. Metal cans and glass containers can be regarded as essentially impermeable to the passage of gases, odors, and water vapor, provided that a metal end has been correctly seamed on in the case of cans or a satisfactory closure applied in the case of glass containers. Aluminum foil has excellent barrier properties, provided it is at least 25 μm thick; below this thickness the likelihood of pinholes increases. It is common to laminate plastic polymers to aluminum foil to provide mechanical support and heat sealability. Paper-based packaging materials can be regarded as permeable and for this reason are normally coated with a plastic polymer to ensure adequate barrier properties for the packaging of foods. This then leaves plastics-based packaging materials, which provide varying degrees of protection, depending largely on the nature of the polymers used in their manufacture.

1.2.1 Polymer Permeability

In contrast to packaging materials made from glass or metal, packages made from thermoplastic polymers are permeable to varying degrees to small molecules such as gases, water vapor, organic vapors, and other low molecular weight compounds.

The following expression can be derived from Fick's first law (Robertson, 2006):

$$Q = \frac{DS(p_1 - p_2)At}{X} \tag{1.1}$$

Here Q is the quantity of gas or vapor permeating a polymer of thickness X and surface area A in time t under a pressure gradient of p_1 on one side and p_2 on the other, where $p_1 > p_2$. D is the diffusion coefficient and S the solubility coefficient of the permeant; the product DS is referred to as the permeability coefficient (or constant) or permeation coefficient, or simply the permeability, and is represented by the symbol *P*. Thus:

$$P = \frac{QX}{At(p_1 - p_2)} \tag{1.2}$$

or

$$\frac{Q}{t} = \frac{P}{X} A(\Delta p) \tag{1.3}$$

The term *P*/*X* is called the permeance. A plastic polymer that is a good barrier to gases and water vapor has a low permeability coefficient.

Four assumptions are made in this simple treatment of permeation: diffusion is at steady state; the concentration–distance relationship through the polymer is linear; diffusion takes place in one direction only (i.e., through the film with no net diffusion along or across it); and both *D* and *S* are independent of concentration. However, as with all simplifying assumptions, there are many instances when the assumptions are not valid, and in such cases the predictions made are not subsequently borne out in practice. Although steady state is usually attained in a few hours for small molecules such as O_2, larger molecules in barrier polymers (especially glassy polymers) can take a long time to reach steady state, this time possibly exceeding the anticipated shelf life. Although *D* and *S* are independent of concentration for many gases, such as O_2, N_2, and, to a certain extent, CO_2, this is not the case where considerable interaction between polymer and permeant takes place (e.g., water and hydrophilic films such as polyamides [PA], or many solvent vapors diffusing through polymer films). Typical values for the permeability coefficient of commercial food packaging polymers are presented in Table 1.1.

TABLE 1.1
Permeability to Oxygen, Carbon Dioxide, and Water Vapor of Some Plastic Films

Permeability to	$O_2 \times 10^{11}$ mL cm cm^{-2} s^{-1} (cm Hg)$^{-1}$ at 23°C, 0% RH	$CO_2 \times 10^{11}$ mL cm cm^{-2} s^{-1} (cm Hg)$^{-1}$ at 23°C, 0% RH	$H_2O \times 10^{11}$ g cm cm^{-2} s^{-1} at 23°C, 100% RH
LDPE	15–30	60–160	5–10
HDPE	5–17	150	1.8–3.5
EVA (15%VA)	30–40	—	21–25
Ethylene acid copolymer (ionomer)	20–35	—	5–11
PP	9–16	30–50	4–10
PET	0.14	1.2	4–6[a]
PS	18–25	60–90	9–46
PVC plasticized	1.7–100[b]	6–180	—
PVC rigid	0.3–1.2[b]	1.2–3	14
PA6	0.09–0.11	0.6–0.8	46[a]
PA66	0.2	—	86
PVdC	0.006[b]	—	0.7[a]
EVOH (32% C_2H_4)	0.0015	0.018	17.5[a]

Source: Adapted from Massey L. 2003. *Permeability Properties of Plastics and Elastomers*, 2nd edn. New York: Plastics Design Library.

[a] 40°C, 90% RH.

[b] 23°C, 50% RH.

The permeability coefficient defined in the preceding text is independent of thickness, as the thickness is already accounted for in the calculation of *P*. However, the total amount of protection afforded per unit area of a barrier material approaches zero only asymptotically. Consequently, as polymer thickness *X* is increased beyond a certain value, it becomes uneconomical to increase it further to obtain lower permeability. For example, to equal the O_2 barrier of a 25-μm film of a high-barrier material such as poly(vinylidene chloride) copolymer (PVdC) would require 62,500 μm of polypropylene (PP), 1250 μm of poly(ethylene terephthalate) (PET), 1250 μm of poly(vinyl chloride) (PVC), or 250 μm of nylon 6.

In general, permeability of a penetrant through a polymer depends on many factors, including the nature of the polymer, thickness of the film, size and shape of the penetrant, pressure, and temperature. The structural attributes that can influence the permeability of polymers include polarity, unsaturation, symmetry, lateral chains, steric hindrance, degree of cross-linking, hydrogen bonding, intermolecular forces, comonomers present, crystallinity, glass transition temperature, and orientation.

Literature data for gas transport coefficients (permeability, diffusion, and solubility coefficients) vary generally with parameters that are intrinsic to the polymer such as degree of crystallinity, nature of the polymer, and the thermal and mechanical histories of samples such as orientation. Sorption and diffusion phenomena take place exclusively in the amorphous phase of a semicrystalline polymer and not in its crystalline zones.

The effect of crystallinity on the permeability coefficient of high density polyethylene (HDPE) to O_2 was illustrated by Pauly (1999), who showed that $P \times 10^{11}$ decreased from 54.9 mL (STP) cm cm^{-2} s^{-1} $(cm\ Hg)^{-1}$ at 60% crystallinity to 20.9 at 69% and 10.6 at 81% crystallinity.

The effect of orientation on the O_2 permeability coefficient was also illustrated by Pauly (1999), who showed that $P \times 10^{11}$ decreased for polystyrene from 25.0 to 17.9 mL (STP) cm cm^{-2} s^{-1} $(cm\ Hg)^{-1}$ when oriented 300%; comparable figures for PP were 9.0 to 4.8; for PET, *P* decreased from 0.60 to 0.22 when oriented 500%.

The relative effect of diffusive flow through holes on the atmosphere inside the package can be appreciated by comparing the permeability of gases in air with their permeability in polymers, as shown in Table 1.2. Air is much more permeable than polymeric films, so even a very small hole in a polymeric package can affect the package atmosphere very significantly. This phenomenon is used to advantage with microporous or perforated films. The effect of thin layers and droplets of water on the inside surface of films can also be appreciated by reference to Table 1.2, which shows that the permeability of gases is much higher in water than in polymers. As a result, thin layers and droplets of water (condensation) forming inside polymeric packages do not significantly affect the gas atmosphere in the package (Kader et al., 1998). The permeability ratio β (also referred to as the

TABLE 1.2
Permeability Data of Some Polymeric Films, Air, and Water

	$P \times 10^{11}$ mL (STP) cm cm^{-2} s^{-1} $(cm\ Hg)^{-1}$		Permeability Ratio (β)
	O_2	CO_2	CO_2:O_2
Polyethylene (density 0.914)	30.0	131.6	4.39
Polypropylene	17.4	75.5	4.34
Poly(vinyl chloride)	0.47	1.64	3.49
Poly(vinylidene chloride)	0.055	0.31	5.64
Air	2.5×10^8	1.9×10^8	0.76
Water	9.0×10^2	2.1×10^4	23.33

Source: Adapted with permission from Kader A.A., Singh R.P., Mannapperuma J.D. 1998. Technologies to extend the refrigerated shelf life of fresh fruits. In: *Food Storage Stability*. Taub I.A. and Singh R.P. (Eds). Boca Raton, Florida: CRC Press, Chap 16. Copyright CRC Press, Boca Raton, Florida.

permselectivity) is the ratio of P for CO_2 to that for O_2 and is particularly useful information when designing modified atmosphere packages.

In the published literature it is rare to find many details about a particular plastic packaging material apart from its name, sometimes the name of the resin supplier, and perhaps whether it has been oriented. This makes it virtually impossible to replicate the experimental conditions described in the literature, as the range of polymers available is vast. For example, the web site www.ides.com contains data sheets on more than 77,000 commercial polymers from 694 resin manufacturers. Of course, many of these polymers are not approved or suitable for use in food packaging.

Consider PP, a polymer used increasingly in food packaging. The properties of PP have improved considerably over the past few decades as a result of a wide range of technical advances ranging from new metallocene catalysts to co-monomers. PP and its copolymers can be classified into three categories (Begley et al., 2008): monophasic homopolymer (h-PP), monophasic random copolymer (r-PP), and heterophasic copolymer (heco-PP). The h-PP can be either isotactic, syndiotactic, or atactic, but the isotactic h-PP is particularly useful due to its stereoregularity and the resulting high crystallinity. Therefore, commercially produced h-PP is up to 95% isotactic. Ides lists data sheets for 19 h-PP food-grade polymers.

The linear polymer chains of r-PP contain copolymers such as ethylene and butene in a random manner, which reduces crystallinity and thus improves the optical clarity, the main commercial advantage of r-PP over h-PP. The heco-PP is a block copolymer made up of h-PP phases and, usually, ethylene-propylene rubber (EPR) phases. This combination leads to superior impact strength at low temperatures. Owing to the variety of PP formulations mentioned here, along with their variety of applications in food packaging, a wide range of diffusion behavior is observed in PP; for example, diffusion coefficients for r-PP are at least one order of magnitude higher than those of h-PP and comparable with those for heco-PP (Begley et al., 2008).

The permeability coefficient of a specific polymer–permeant system may increase or decrease with increases in temperature, depending on the relative effect of temperature on the solubility and diffusion coefficients. Generally, the solubility coefficient increases with increasing temperature for gases and decreases for vapors, and the diffusion coefficient increases with temperature for both gases and vapors. For these reasons, permeability coefficients of different polymers determined at one temperature may not be in the same relative order at other temperatures.

1.2.2 Transmission Rate

The aforementioned treatment of steady-state diffusion assumes that both D and S are independent of concentration, but, in practice, deviations do occur. Equation 1.3 does not hold when there is interaction such as that which occurs between hydrophilic materials [e.g., EVOH (ethylene vinyl alcohol copolymers) and some of the PAs] and water vapor, or for heterogeneous materials such as coated or laminated films. The property is then defined as the transmission rate (TR) of the material, where:

$$TR = \frac{Q}{At} \tag{1.4}$$

Here Q is the amount of permeant passing through the polymer, A is the surface area, and t is the time. Permeabilities of polymers to water and organic compounds are often presented in this way, and in the case of water and O_2, the terms water vapor transmission rate (WVTR) and gas transmission rate (GTR), or more specifically oxygen transmission rate (OTR or O_2TR), are in common usage. It is critical that the thickness of the film or laminate, the temperature, and the partial pressure difference of the gas or water vapor be specified for a particular TR.

The specialized instruments commonly used to determine the OTR of plastic packaging materials, such as those manufactured by MoCon, use pure O_2 on one side and measure how much permeates into a carrier gas on the other side (the O_2 gradient is therefore 1 atm). In real life, where there is typically air on one side (O_2 is present at 21% in air) and essentially no O_2 inside the package, the O_2 gradient is 0.21 atm or 16 cm Hg. Thus, to convert OTR values expressed in units of mL m^{-2} day^{-1} atm^{-1} to "true" OTR units of mL m^{-2} day^{-1}, it is necessary to multiply by 0.21; this has been done for the OTR values quoted in this book. An exception to this convention is used in the case of CO_2, where, because of its very low concentration in air (0.03%), CO_2TR units are often given in mL m^{-2} day^{-1} atm^{-1}. In modified atmosphere packaging (MAP), concentrations of CO_2 inside the package are typically 40–60%.

Often the units for WVTR include a thickness term, in which case the WVTR should, strictly speaking, be referred to as the thickness normalized flux, or TNF (Robertson, 2009). To convert a measured WVTR or OTR to *P*, it is necessary to multiply by the thickness of the film and divide by the partial pressure difference used to make the measurement.

Example: Calculate the permeability coefficient of an amorphous polyethylene terephthalate (PET) film to O_2 at 23°C given that the OTR through a 2.54×10^{-3}-cm-thick film with air on one side and inert gas on the other is 8.8×10^{-9} mL cm^{-2} s^{-1}.

O_2 partial pressure difference across the film is 0.21 atm = 16 cm Hg.

$$P = \frac{OTR}{\Delta p} \times \text{thickness}$$

$$= \frac{8.8 \times 10^{-9}\,\text{mL cm}^{-2}\text{s}^{-1}}{16(\text{cm Hg})} \times 2.54 \times 10^{-3}\,\text{cm}$$

$$= 1.4 \times 10^{-12}\left[\text{mL(STP)cm cm}^{-2}\text{s}^{-1}(\text{cm Hg})^{-1}\right]$$

$$= 0.14 \times 10^{-11}\left[\text{mL(STP)cm cm}^{-2}\text{s}^{-1}(\text{cm Hg})^{-1}\right]$$

Therefore

$$P \times 10^{11} = 0.14\left[\text{mL(STP)cm cm}^{-2}\text{s}^{-1}(\text{cm Hg})^{-1}\right]$$

which is the value given in Table 1.1.

The OTR of packaging materials used for MAP of chilled products varies extensively with temperature, relative humidity (RH), and material thickness after the thermoforming of packages. Gnanaraj et al. (2005) reported OTRs, together with the Arrhenius pre-exponential factor k_0 and activation energy E_a, for a range of films at 10°C, 15°C, 23°C, 30°C, and 35°C and 0% and 50% RH. The OTRs at 10°C were typically half those at 23°C. Jakobsen et al. (2005) studied two different polymer combinations: amorphous PET/low density polyethylene (APET/LDPE) (tray) and PA/LDPE (lid). A temperature reduction of 8°C (in the interval 7–23°C) caused an OTR reduction of 26–48%, depending on material type, degree of thermoforming, and RH. An increased OTR was observed as a result of material thinning; however, the increase was not always directly proportional to the degree of material thinning. The changes in OTR observed emphasize the necessity of evaluating the performance of packaging materials under realistic storage conditions to estimate the real O_2 content of a chosen package solution.

1.2.3 Surface Area:Volume Ratio

The dimensions of the package for a given weight of food can have a significant influence on shelf life. Although a spherical shape will minimize the surface area of the package (and thus the quantity of moisture or O_2 that will permeate the package wall), it is not a practical shape for commercial use, and, in practice, most packages tend to be rectangular or cylindrical. Table 1.3 gives the surface areas for a range of different shapes with the same volume (~450 mL). Compared with the surface area of a sphere,

TABLE 1.3
Surface Areas of Different Package Shapes, All with a Volume of ~450 mL

Shape	Dimensions cm	Surface Area		Increase %	Surface Area: Volume Ratio
		cm^2	m^2		
Sphere	Diameter 9.52	285	0.0285	0	0.63
Cylinder	Diameter 7.3 Height 10.8	331	0.0331	16	0.73
Cube	Sides 7.67	353	0.0353	24	0.78
Tetrahedron	Sides 15.65	424	0.0424	49	0.94
Rectangular pack	Height 3 Length 15 Width 10	450	0.0450	58	1.0
Thin rectangular pack	Height 1 Length 20 Width 22.5	985	0.0985	246	2.18

the surface area of a cylinder is 16% greater, a cube 24% greater, a tetrahedron 49% greater, a rectangular shape 58% greater, and a thin rectangular shape 246% greater. Extremely thin packages have a much greater surface area:volume ratio and thus require a plastic with better barrier properties to get the same shelf life than if the same quantity of product were packaged in a thicker format. For different quantities of the same product packaged in different-sized packages using the same plastic material, the smallest package will have the shortest shelf life as it inevitably has a greater surface area per unit volume. Many food companies still seem unaware of this fact as they continue to launch smaller packages without changing the packaging material and then wonder why the shelf life is shorter for the smaller package.

Example: A food powder with a density of 1 is to be packaged in a plastic film that has a WVTR of 2.1 g m^{-2} day^{-1} at 25°C and 75% RH. The initial moisture content of the powder is 3%, and the critical moisture content is 7%. Assuming that each pack will contain 450 g of powder and will be exposed to an external environment at 25°C and 75% RH, calculate the shelf life if the shapes of the packs are the same as those listed in Table 1.3. For simplicity, assume that the driving force for water vapor transmission (WVT) remains constant and that there are no moisture gradients in the powder.

$$\text{Weight of dry solids} = 97\% \text{ of } 450 = 436.5 \text{ g}$$

$$\text{Initial weight of water in powder} = 3\% \text{ of } 450 = 13.5 \text{ g}$$

$$\begin{aligned}\text{Final weight of water in powder} &= 436.5/0.93 - 436.5\\ &= 469.35 - 436.5\\ &= 32.85 \text{ g}\end{aligned}$$

Therefore, weight of water permeating into powder is

$$32.85 - 13.5 = 19.35 \text{ g}$$

For a spherical-shaped package:

Quantity of water permeating into package per day is

$$0.0285 \times 2.1 = 0.05985 \text{ g day}^{-1}$$

$$\text{Therefore shelf life } \theta_s = \frac{19.35}{0.05985} = 323 \text{ days}$$

For the other package shapes

$$\text{Cylinder: } \theta_s = 278 \text{ days}$$

$$\text{Cube: } \theta_s = 261 \text{ days}$$

$$\text{Tetrahedron: } \theta_s = 217 \text{ days}$$

$$\text{Rectangle 1: } \theta_s = 204 \text{ days}$$

$$\text{Rectangle 2: } \theta_s = 93.5 \text{ days}$$

1.3 SHELF LIFE

The quality of most foods and beverages decreases with storage or holding time. Exceptions include distilled spirits (particularly whiskeys and brandies) that develop desirable flavor components during storage in wooden barrels, some wines that undergo increases in flavor complexity during storage in glass bottles, and many cheese varieties in which enzymic degradation of proteins and carbohydrates, together with hydrolysis of fat and secondary chemical reactions, lead to desirable flavors and textures in aged cheeses.

For the majority of foods and beverages in which quality decreases with time, it follows that there will be a finite length of time before the product becomes unacceptable. This time from production to unacceptability is referred to as *shelf life*. Although the *Wizard of Id* thought that shelf life related to the time until the shelf displaying the food rotted out (see Figure 1.1), shelf life refers to the time on the retailer's shelf as well as the consumer's shelf. Although the shelf lives of foods vary, they are routinely determined for each particular product by the manufacturer or processor. Manufacturers generally attempt to provide the longest practicable shelf life consistent with costs and the pattern of handling and use by distributors, retailers, and consumers. Supermarkets will generally not accept the product into their distribution centers unless at least 75% of the shelf life remains. Inadequate shelf life will often lead to consumer dissatisfaction and complaints. At best, such dissatisfaction will eventually affect the acceptance and sales of brand name products, while, at worst, it can lead to malnutrition or even illness. Therefore, food processors give considerable attention to determining the shelf lives of their products.

1.3.1 Definitions

Despite its importance, there is no simple, generally accepted definition of shelf life in the food technology literature. The Institute of Food Technologists (IFT) in the United States has defined shelf life as "the period between the manufacture and the retail purchase of a food product, during which time the product is in a state of satisfactory quality in terms of nutritional value, taste, texture and appearance" (Anon., 1974). This definition overlooks the fact that the consumer may store the product at home for some time before consuming it yet will still want the product to be of acceptable quality.

The Institute of Food Science and Technology (IFST) in the United Kingdom has defined shelf life as "the period of time during which the food product will remain safe; be certain to retain desired sensory, chemical, physical, microbiological and functional characteristics; and comply with any label declaration of nutritional data when stored under the recommended conditions" (Anon., 1993).

Another definition is that "shelf life is the duration of that period between the packing of a product and the end of consumer quality as determined by the percentage of consumers who are displeased by the product" (Labuza and Schmidl, 1988). This definition accounts for the variation in consumer

Wizard of Id By Brant Parker & Johnny Hart

FIGURE 1.1 Shelf life according to the *Wizard of Id*. (Used with permission of John L. Hart FLP and Creators Syndicate, Inc.)

perception of quality (i.e., not all consumers will find a product unacceptable at the same time) and has an economic element, in that, because it is not possible to please all consumers all of the time, a baseline of consumer dissatisfaction must be established. In the branch of statistics known as survival analysis, consumer dissatisfaction can be related to the survival function, defined as "the probability of a consumer accepting a product beyond a certain storage time." Models permitting the application of survival analysis to the sensory shelf life of foods have been published and are discussed further in Chapter 3.

Simply put, shelf life is the time during which all of the primary characteristics of the food remain acceptable for consumption. Thus, shelf life refers to the time for which a food can remain on the retailer's and then the consumer's shelf before it becomes unacceptable.

Until recently, the EU had no definition of shelf life or legislation on how shelf life should be determined; the consolidated EU Directive on food labeling (2000/13/EC) required prepackaged foods to bear a date of "minimum durability" or, in the case of foods that, from a microbiological point of view, are highly perishable, the "use by" date. The date of minimum durability was defined as the "date until which a foodstuff retains its specific properties when properly stored," and any special storage conditions (e.g., temperature not to exceed 7°C) must be specified. This concept allows the processor to set the quality standard of the food, as the product will still be acceptable to many consumers after the "best before" date has passed. More recently, shelf life was defined for the first time in EU legislation, in Commission Regulation (EC) No. 2073/2005 thus: "shelf life means either the period corresponding to the period preceding the 'use by' or the minimum durability date, as defined respectively in Articles 9 and 10 of Directive 2000/13/EC."

According to Cheftel (2005), the date of minimum durability is defined as the date until which the food retains its specific properties when properly stored. It must be indicated by the words "Best before" followed by the date (or a reference to where the date is given on the labeling). Depending on how long the food can keep, the date can be expressed by the day and the month, the month and the year, or the year alone. A list of foods and beverages exempted from date-marking is given in article 9(5) of Directive 2000/13/EC. Foods that are highly perishable microbiologically (and therefore likely to be dangerous for health after a short period) must be labeled with the words "Use by" followed by the date (day and month) or a reference to where the date is given on the labeling. Any distribution after this date is forbidden. The "use by" date must be followed by a description of the storage conditions that should be observed.

In many countries a "best before" date is required on the label. However, if the food is highly perishable from a microbiological point of view and therefore likely, after a short period, to constitute an immediate danger to human health, then the "best before" date must be replaced by a "use by" date. It is illegal to sell food after the "use by" date; food consumed after the "best before" date will still be edible, but its quality will have deteriorated to a level below what the manufacturer considers desirable. Recently, use of the hybrid term "best by" has become popular. A major US brewer now labels bottles of beer with the "born on" date, that is, the date of manufacture, leaving consumers to decide when the beer is no longer acceptable.

1.3.2 Factors Controlling Shelf Life

The shelf life of a food is controlled by three factors:

1. Product characteristics, including formulation and processing parameters (intrinsic factors)
2. Environment to which the product is exposed during distribution and storage (extrinsic factors)
3. Properties of the package

Intrinsic factors are discussed in Chapter 2 and include pH, water activity, enzymes, microorganisms, and concentration of reactive compounds. Many of these factors can be controlled through the selection of raw materials and ingredients, as well as the choice of processing parameters.

Extrinsic factors include temperature, RH, light, total and partial pressures of different gases, and mechanical stresses, including consumer handling. Many of these factors can affect the rates of deteriorative reactions that occur during the shelf life of a product.

The properties of the package can have a significant effect on many of the extrinsic factors and thus indirectly on the rates of the deteriorative reactions. Thus, the shelf life of a food can be altered by changing its composition and formulation, processing parameters, packaging system, or the environment to which it is exposed.

1.3.2.1 Product Characteristics

On the basis of the nature of the changes that can occur during storage, foods can be divided into three categories—perishable, semiperishable, and nonperishable or shelf stable—which translate into very short shelf life products, short to medium shelf life products, and medium to long shelf life products (Robertson, 2006).

Perishable foods are those that must be held at chill or freezer temperatures (i.e., 0°C to 7°C or −12°C to −18°C respectively) if they are to be kept for more than a short period. Examples of such foods are milk; fresh flesh foods such as meat, poultry, and fish; minimally processed foods; and many fresh fruits and vegetables.

Semiperishable foods are those that contain natural inhibitors (e.g., some cheeses, root vegetables and eggs) and those that have undergone some type of mild preservation treatment (e.g., pasteurization of milk, smoking of hams, and pickling of vegetables) that produces greater tolerance to environmental conditions and abuse during distribution and handling.

Shelf stable foods are considered "nonperishable" at room temperatures. Many unprocessed foods fall into this category, and are unaffected by microorganisms because of their low moisture content (e.g., cereal grains, nuts, and some confectionery products). Processed food products can be shelf stable if they are preserved by heat sterilization (e.g., canned foods), contain preservatives (e.g., soft drinks), are formulated as dry mixes (e.g., cake mixes), or are processed to reduce their water content (e.g., raisins or crackers). However, shelf stable foods retain this status only if the integrity of the package that contains them remains intact. Even then, their shelf life is finite due to deteriorative chemical reactions that proceed at room temperature independently of the nature of the package, and the permeation of gases, odors, and water vapor through the package.

1.3.2.2 Distribution and Storage Environment

The deterioration in product quality of packaged foods is often closely related to the transfer of mass and heat through the package. Packaged foods may lose or gain moisture; they will also reflect the temperature of their environment, because very few food packages are good insulators. Thus, the climatic conditions (i.e., temperature and humidity) of the distribution and storage environment have an important influence on the rate of deterioration of packaged foods.

1.3.2.3 Package Properties

Foods can be classified according to the degree of protection required from the package, such as maximum moisture gain or O_2 uptake. This enables calculations to be made to determine whether a particular packaging material would provide the barrier required to give the desired product shelf life. Metal cans and glass containers can be regarded as essentially impermeable to the passage of gases, odors, and water vapor, but paper-based packaging materials can be regarded as permeable. Plastics-based packaging materials provide varying degrees of protection, depending largely on the nature of the polymers used in their manufacture.

In Section 1.2.1 the permeability of thermoplastic polymers was discussed. A discussion of how this information can be used to select the most appropriate polymer for a particular product can be found elsewhere (e.g., see Robertson, 2006).

For a product where the end of shelf life can be directly related to a gain in moisture (e.g., loss of crispness in a snack food), the end of product shelf life is reached when the moisture content m (initially m_i) reaches the critical moisture content m_c, and the following equation applies:

$$\theta_s = \frac{X}{P}\frac{W_s}{A}\frac{b}{p_o}\ln\frac{m_e - m_i}{m_e - m_c} \tag{1.5}$$

where m_e is the equilibrium moisture content of the food if exposed to the RH outside the package; W_s is the weight of dry solids enclosed; p_o is the vapor pressure of pure water at the storage temperature (*not* the actual vapor pressure outside the package); and b is the slope of the moisture sorption isotherm when treated as a linear function.

Equation 1.5 and the corresponding equation for moisture loss have been extensively tested for foods and found to give excellent predictions of actual weight gain or loss. These equations are also useful when calculating the effect of changes in the external conditions (e.g., temperature and humidity), the surface area:volume ratio of the package, and variations in the initial moisture content of the product.

The gas of major importance in packaged foods is O_2, as it plays a crucial role in many reactions that affect the shelf life of foods, for example, microbial growth, color changes in fresh and cured meats, oxidation of lipids and consequent rancidity, and senescence of fruits and vegetables.

The transfer of gases and odors through packaging materials can be analyzed in an analogous manner to that described for water vapor transfer, provided that values are known for the permeance of the packaging material to the appropriate gas and the partial pressure of the gas inside and outside the package. However, the simplifying assumptions made in the derivation of Equation 1.4 can lead to significant errors in the calculated shelf life compared to the actual shelf life. For example, in the case of CO_2 loss from carbonated beverages in PET bottles, the assumption that the gas partitioning between the gas phase and the polymer is described by Henry's law and that mass transfer inside the bottle wall is governed by Fick's law gives rise to the underestimation of the barrier properties of the materials, and, consequently, the predicted shelf life of the carbonated beverage is much shorter than the true one (Masi and Paul, 1982).

Del Nobile et al. (1997) showed the importance of another aspect that is often neglected in predicting the shelf life of carbonated beverages bottled in glassy polymer containers: the influence of the thermal history of the bottle during the period between filling and consumption. In their first example, the shelf life of the beverage was estimated assuming that the storage temperature was constant and equal to room temperature for the entire storage period; the calculated shelf life was 352 days. In the second example, it was assumed that the temperature of the bottle varied during the storage period, but for the sake of simplicity in performing the calculation, the temperature was kept constant and equal to the average temperature of storage; the calculated shelf life was 206 days. In the second example, the actual temperature of the bottle of carbonated beverage under conditions comparable to those occurring during distribution led to an estimated shelf life of less than 2 months, significantly less than that predicted by neglecting the temperature rise due to outdoor storage and sunlight exposure. By averaging the temperature and using the corresponding parameters in the calculations, it is implicitly assumed that the diffusion and sorption parameters change linearly with temperature, and this is far from true.

Packaging can control two variables with respect to O_2, and these can have different effects on the rates of oxidation reactions in foods. One variable is the total amount of O_2 present. This influences the extent of the reaction, and in impermeable packages (e.g., hermetically sealed metal and glass containers), where the total amount of O_2 available to react with the food is finite, the extent of the reaction cannot exceed the amount corresponding to the complete exhaustion of the O_2 present inside the package at the time of sealing. This may or may not be sufficient to result in an unacceptable product quality after a certain period, depending on the rate of the oxidation reaction. This rate will, of course, be temperature dependent. With permeable packages (e.g., plastic packages) into which ingress of O_2 will occur during storage, two factors are important: there may be sufficient O_2 inside the package to cause product unacceptability when it has all reacted with the food, or there may be sufficient transfer of O_2 through the package over time to result in product unacceptability through oxidation. The other variable is the concentration of O_2 in the food. In many cases, relationships between the O_2 partial pressure in the space surrounding the food and the rates of oxidation reactions can be established. If the food itself is very resistant to diffusion of O_2 (e.g., very dense products such as butter), then it will probably be very difficult to establish a relationship between the O_2 partial pressure in the space surrounding the food and the concentration of O_2 in the food.

With certain products packaged with certain materials, the end of shelf life comes when an unacceptable degree of interaction between the package and the product has occurred. Two examples will be given to illustrate the nature of the problem.

The first example is that of a tomato product processed under typical conditions and packaged in a three-piece can with a plain tinplate body and enameled electrolytically chromium-coated steel (ECCS) ends. Over a storage period of 24 months at ambient temperature, several degradative reactions occur. The concentration of tin ions in the product increases rapidly during the first 3 months, from approximately 20 to 160 ppm, reaching 280 ppm after 24 months. Iron also dissolves, increasing slowly from 8 ppm initially to 10 ppm after 18 months, to reach 14 ppm after 24 months. The flavor score declines as a result of the increasing quantities of dissolved tin and iron; the color value shows a decrease owing to an increase in brown pigments, but remains acceptable. The limiting factor for this product is the deterioration in flavor resulting from the dissolution of tin and iron from the package into the product, giving an acceptable shelf life of 24–30 months. If a longer shelf life were required, it would be necessary to use a full enamel-lined can. Alternatively, the product could be stored at chill temperatures to reduce the rate of the degradative reactions.

A second example involves the migration of plasticizers from packaging materials into food such that the legal limit for the migrant in the food is exceeded. For example, gaskets in the lids for glass jars can release epoxidized soy bean oil (ESBO) into meat-containing infant food, and plasticized PVC cling-films have released di-(2-ethylhexyl) adipate (DEHA) into cheese (Grob et al., 2006).

1.3.3 Shelf Life Determination

Methods to determine the shelf life of packaged foods have been published elsewhere (Robertson, 2006) and will not be repeated here. One challenge with shelf life testing is to develop experimental designs that minimize the number of samples required (thus minimizing the cost of the testing) while still providing reliable and statistically valid answers; this is discussed further in Chapter 3. Accelerated shelf life testing (ASLT) applies the principles of chemical kinetics to quantify the effects that extrinsic factors such as temperature, humidity, gas atmosphere, and light have on the rate of deteriorative reactions. By subjecting the food to controlled environments in which one or more of the extrinsic factors is maintained at a higher-than-normal level, the rates of deterioration are speeded up or accelerated, resulting in a shorter-than-normal time to product failure. Because the effects of extrinsic factors on deterioration can be quantified, the magnitude of the acceleration can be calculated and also the "true" shelf life of the product under normal conditions.

The reason behind the need for ASLT of shelf stable food products is simple: as these foods typically have shelf lives of at least one year, evaluating the effect on shelf life of a change in the product (e.g., a new antioxidant or thickener), the process (e.g., a different time/temperature sterilization regime), or the packaging (e.g., a new polymeric film) would require shelf life trials lasting at least as long as the required shelf life of the product. Companies cannot afford to wait for such long periods to know whether the new product, process, or packaging will give an adequate shelf life, and therefore ASLT is used. However, the use of ASLT in the food industry is not as widespread as it might be, due in part to the lack of basic data on the effect of extrinsic factors on the rates of deteriorative reactions, in part to ignorance of the methodology required, and in part to a skepticism about the advantages to be gained from using ASLT procedures. Of course, in tropical countries, the ambient temperatures and humidities experienced during distribution and in warehouses and homes are in the upper range used for ASLT in temperate climates (45°C and 95% RH); therefore, ASLT is not applicable in such situations, as temperature cannot be accelerated beyond 45°C without the risk of introducing deteriorative reactions that are unrepresentative of what may occur under real circumstances. Although high O_2 pressures can be used to accelerate reactions involving oxidation, this method is not used very often, as oxidation reactions typically become independent of the O_2 concentration above a certain level, which varies with temperature and other conditions. However, Cardelli and Labuza (2001) reported that increasing O_2 concentrations from 0.5 to 21.3 kPa accelerated deterioration of roast and ground coffee 20-fold. If both temperature and O_2 concentration accelerated, then the decreased solubility of O_2 at higher temperatures must be factored into any calculations of shelf life.

In shelf life testing there can be one or more criteria that constitute sample failure. One criterion is an increase or decrease by a specified amount in the mean sensory panel score. Another criterion is microbial deterioration of the sample to an extent that it is rendered unsuitable or unsafe for human consumption. Finally, changes in odor, color, texture, flavor, and so on that render the sample unacceptable to either the panel or the consumer are criteria for product failure. Thus, sample failure can be defined as the condition in which the product exhibits either physical, chemical, microbiological, or sensory characteristics that are unacceptable to the consumer, and the time required for the product to exhibit such conditions is the shelf life of the product.

However, a fundamental requirement in the analysis of data is knowledge of the statistical distribution of the observations, so that the mean time to failure and its standard deviation can be accurately estimated, and the probability of future failures predicted. The shelf life for food products is usually obtained from simple averages of time to failure, on the assumption that the failure distribution is symmetrical. If the distribution is skewed, estimates of the mean time to failure and its standard deviation will be biased. Furthermore, when the experiment is terminated before all the samples have failed, the mean time to failure based on simple averages will be biased because of the inclusion of unfailed data. To improve the methodology for estimating shelf life, knowledge of the statistical distribution of shelf life failures is required, together with an appropriate model for data analysis. This important aspect is discussed further in Chapter 3.

Microbial spoilage of foods is an economically significant problem for food manufacturers, retailers, and consumers. Depending on the product, process, and storage conditions, the microbiological end of shelf life can be determined by either the growth of spoilage or pathogenic microorganisms. Over recent years the development and commercialization of predictive models have become relatively widespread. Predictive models have been used to determine the likely shelf life of perishable foods such as meat, fish, and milk. Despite their increasing sophistication and widespread availability, models should not be relied on completely but should rather be used as a tool to assist decision making. Models do not completely negate the need for microbial testing, nor do they replace the judgment of trained and experienced food microbiologists. The use of such models can reduce the need for shelf life trials, challenge tests, product reformulations, and process modifications, thus saving both time and money. The ultimate test for predictive models is whether they can be used to predict outcomes reliably in real situations. For a detailed discussion the reader is referred to Chapter 4.

REFERENCES

Anonymous. 1974. Shelf Life of Foods. Report by the Institute of Food Technologists' Expert Panel on Food Safety and Nutrition and the Committee on Public Information, Institute of Food Technologists, Chicago, Illinois. *Journal of Food Science* 39: 861–865.

Anonymous. 1993. *Shelf Life of Foods: Guidelines for Its Determination and Prediction*. London, England: Institute of Food Science and Technology, Inc.

Begley T.H., Brandsch J., Limma W., Siebert H., Piringer O. 2008. Diffusion behaviour of additives in polypropylene in correlation with polymer properties. *Food Additives Contaminants* 11: 1409–1415.

Cardelli C., Labuza T.P. 2001. Application of Weibull Hazard Analysis to the determination of the shelf life of roasted and ground coffee. *LWT—Food Science & Technology* 34: 273–278.

Cheftel J.C. 2005. Food and nutrition labelling in the European Union. *Food Chemistry* 93: 531–550.

Del Nobile M.A., Mensitieri G., Nicolais L., Masi P. 1997. The influence of the thermal history on the shelf life of carbonated beverages bottled in plastic containers. *Journal of Food Engineering* 34: 1–13.

Gnanaraj J., Welt, B.A., Otwell W.S., Kristinsson H.G. 2005. Influence of oxygen transmission rate of packaging film on outgrowth of anaerobic bacterial spores. *Journal of Aquatic Food Product Technology* 14(4): 51–69.

Grob K., Biedermann M., Scherbaum E., Roth M., Rieger K. 2006. Food contamination with organic materials in perspective: packaging materials as the largest and least controlled source? A view focusing on the European situation. *Critical Reviews in Food Science and Nutrition* 46: 529–535.

Jakobsen M., Jespersen L., Juncher D., Miquel Becker E., Risbo J. 2005. Oxygen and light barrier properties of packaging materials used for modified atmosphere packaging. Evaluation of performance under realistic storage conditions. *Packaging Technology and Science* 18: 265–272.

Krochta J.M. 2007. Food packaging. In: *Handbook of Food Engineering*, 2nd edn. Heldman D.R. & Lund D.B. (Eds). Boca Raton, Florida: CRC Press, pp. 847–927.

Labuza T. P., Schmidl M.K. 1988. Use of sensory data in the shelf life testing of foods: principles and graphical methods for evaluation. *Cereal Foods World* 33: 193–205.

Lee D.S., Yam K.L., Piergiovanni L. 2008. *Food Packaging Science and Technology*. Boca Raton, Florida: CRC Press.

Lockhart H.E. 1997. A paradigm for packaging. *Packaging Technology and Science* 10: 237–252.

Masi P., Paul D.R. 1982. Modelling gas transport in packaging applications. *Journal of Membrane Science* 12: 137–151.

Massey L. 2003. *Permeability Properties of Plastics and Elastomers*, 2nd edn. New York: Plastics Design Library.

Pauly A.S. 1999. Permeability and diffusion data. In: *Polymer Handbook*, 4th edn. Brandrup J., Immergut E.H. & Grulke E.A. (Eds). New York: Wiley, Section VI/543.

Piringer O.-G., Baner A.L. (Eds). 2008. *Plastic Packaging Interactions with Food and Pharmaceuticals*, 2nd edn. Weinheim, Germany: Wiley-VCH.

Robertson G.L. 2006. *Food Packaging Principles and Practice*, 2nd edn. Boca Raton, Florida: CRC Press.

Robertson G.L. 2009. Food packaging. In: *Textbook of Food Science and Technology*, Campbell-Platt G. (Ed). Oxford, England: Wiley-Blackwell, pp. 279–298.

Yam K.L. (Ed). 2009. *The Wiley Encyclopedia of Packaging Technology*, 3rd edn. New York: John Wiley & Sons Inc.

Yoxall A., Luxmoore J., Rowson J., Langley J., Janson R. 2007. Size does matter: further studies in hand-pack interaction using computer simulation. *Packaging Technology and Science* 21: 61–72.

2 Food Quality and Indices of Failure

Gordon L. Robertson
University of Queensland and
Food•Packaging•Environment
Brisbane, Australia

CONTENTS

2.1 FOOD QUALITY AND SAFETY

The term "food quality" has a variety of meanings to professionals in the food industry, but the ultimate arbiters of food quality must be the consumers. This notion is embodied in the frequently cited definition of food quality as "the combination of attributes or characteristics of a product that have significance in determining the degree of acceptability of the product to a user." Another definition of food quality is "the acceptance of the perceived characteristics of a product by consumers who are regular users of the product category or those who comprise the market segment." The phrase "perceived characteristics" includes the perception of the food's safety, convenience, cost, value, and so on, and not just its sensory attributes (Cardello, 1998).

For the majority of foods and beverages, quality decreases over time. Therefore it follows that there will be a finite length of time before the product becomes unacceptable. This time from production to unacceptability is referred to as shelf life and was discussed in Chapter 1. Quality loss during storage may be regarded as the result of a form of processing at relatively low temperatures that goes on for rather a long time.

Knowledge of the kinds of changes that influence food quality is the first step in developing food packaging that will minimize undesirable changes in quality and maximize the development and maintenance of desirable properties. Once the nature of the reactions is understood, knowledge of the factors that control the rates of these reactions is necessary in order to minimize the changes occurring in foods during storage, that is, while packaged.

The nature of the deteriorative reactions in foods and the factors that control the rates of these reactions will be briefly outlined. Deteriorative reactions can be enzymic, chemical, physical (typically as a result of moisture gain or loss), and biological (both microbiological and macrobiological, that is, due to insect pests and rodents). Biochemical, chemical, physical, and biological changes occur in foods during processing and storage, and these combine to affect food quality. The most important quality-related changes are (van Boekel, 2008) as follows:

- Chemical reactions, mainly due to either oxidation or nonenzymic browning reactions.
- Microbial reactions: microorganisms can grow in foods. In the case of fermentation this is desired; otherwise, microbial growth will lead to spoilage and, in the case of pathogens, to unsafe food.
- Biochemical reactions: many foods contain endogenous enzymes that can potentially catalyze reactions leading to quality loss (enzymic browning, lipolysis, proteolysis, and more). In the case of fermentation, enzymes can be exploited to improve quality.
- Physical reactions: many foods are heterogeneous and contain particles. These particles are unstable, and phenomena such as coalescence, aggregation, and sedimentation usually lead to quality loss. Also, changes in texture can be considered physical reactions, although the underlying mechanism may be of a chemical nature.

The principal aim of this chapter is to provide a brief overview of the major chemical, biochemical, biological, and physical changes that occur in foods during processing and storage and to show how these combine to affect food quality. Reactions in foods affecting food quality are summarized in Table 2.1. Knowledge of such changes is essential before a sensible choice of packaging materials can be made, as the rate and magnitude of such changes can often be minimized by selection of the correct packaging materials. At the end of the chapter, the concept of indices of failure (IoFs) of food is introduced. IoFs are the quality attributes that will indicate that the food is no longer acceptable to the consumer.

The deterioration of packaged foods (and this includes virtually all foods, because today very few foods are sold without some form of packaging) depends largely on transfers that can occur between the external environment, which is exposed to the hazards of storage and distribution, and the internal environment of the package. For example, there may be transfer of moisture vapor from a humid atmosphere into a dried product, or transfer of an undesirable odor from the external atmosphere into a high-fat product, or development of oxidative rancidity if the package is not an effective oxygen (O_2) barrier. Also, flavor compounds can be absorbed by some types of plastic packaging materials (a phenomenon referred to as scalping), and chemical contaminants can migrate from the packaging material into the food (e.g., plasticizers from plastic film). In addition to the ability of packaging materials to protect and preserve foods by minimizing or preventing the transfers referred to, packaging materials must also protect the product from mechanical damage and prevent or minimize misuse by consumers (including tampering).

Although certain types of deterioration will occur even if there is no transfer of mass (or heat, as some packaging materials can act as efficient insulators against fluctuations in ambient temperatures)

TABLE 2.1
Overview of Reactions in Foods Affecting Quality

Example	Type	Consequences
Nonenzymic browning	Chemical reaction (Maillard reaction)	Color, taste and aroma, nutritive value, formation of toxicologically suspect compounds (acrylamide)
Fat oxidation	Chemical reaction	Loss of essential fatty acids, rancid flavor, formation of toxicologically suspect compounds
Fat oxidation	Biochemical reaction (lipoxygenase)	Off-flavors, mainly due to formation of aldehydes and ketones
Hydrolysis	Chemical reaction	Changes in flavor, vitamin content
Lipolysis	Biochemical reaction (lipase)	Formation of free fatty acids and peptides, bitter taste
Proteolysis	Biochemical reaction (proteases)	Formation of amino acids and peptides, bitter taste, flavor compounds, changes in texture
Enzymic browning	Biochemical reaction of polyphenols	Browning
Separation	Physical reaction	Sedimentation, creaming
Gelation	Combination of chemical and physical reaction	Gel formation, texture changes

Source: Adapted from van Boekel M.A.J.S. 2008. Kinetic modeling of food quality: a critical review. *Comprehensive Reviews in Food Science and Food Safety* 7: 144–158.

between the package and its environment, it is possible in many instances to prolong the shelf life of the food through the use of packaging (Baner and Piringer, 2008).

It is important that food packaging not be considered in isolation from food processing and preservation, or indeed from food marketing and distribution: all interact in a complex way, and concentrating on only one aspect to the detriment of the others is a sure-fire recipe for commercial failure.

The development of an analytical approach to food packaging is strongly recommended, and to achieve this successfully, a good understanding of food safety and quality is required. The more important of these is, without question, food safety, which is the freedom from harmful chemical and microbial contaminants at the time of consumption. Packaging is directly related to food safety in two ways.

First, if the packaging material does not provide a suitable barrier around the food, microorganisms can contaminate the food and make it unsafe. However, microbial contamination can also arise if the packaging material permits the transfer of, for example, moisture or O_2 from the atmosphere into the package. In this situation, microorganisms present in the food but posing no risk because of the initial absence of moisture or O_2 may subsequently be able to grow and present a risk to the consumer.

Second, the migration of potentially toxic compounds from some packaging materials to the food is a possibility in certain situations and gives rise to food safety concerns. In addition, migration of other components from packaging materials, although not harmful to human health, may adversely affect the quality of the product.

The major quality attributes of foods are texture, flavor, color, appearance, and nutritive value, and these attributes can all undergo undesirable changes during processing and storage. With the exception of nutritive value, the changes that can occur in these attributes are readily apparent to the consumer, either before or during consumption. Packaging can affect the rate and magnitude of many of these quality changes. For example, the development of oxidative rancidity can often be minimized if the package is an effective O_2 barrier; flavor compounds can be absorbed by some plastic polymers but not by others; the particle size of many food powders can increase (i.e., particles can clump) if the package is a poor moisture barrier.

2.2 DETERIORATIVE REACTIONS IN FOODS

Knowledge of the kinds of deteriorative reactions that influence food quality is the first step in developing food packaging that will minimize undesirable changes in quality and maximize the development and maintenance of desirable properties. Once the nature of the reactions is understood, knowledge of the factors that control the rates of these reactions is necessary in order to minimize the changes occurring in foods during storage, that is, while packaged (Robertson, 2006). The nature of the deteriorative reactions in foods is reviewed in this section, and the factors that control the rates of these reactions are discussed in the following section.

Preservation is a means of protecting a product, usually against microbiological deterioration. It is important to understand the differences between biotic deterioration, which refers to changes in a food product brought about either by a biological function (e.g., ripening of fruit, respiration of vegetables) or attack by microorganisms (e.g., molds, bacteria, and yeasts) and abiotic deterioration, which is brought about by physical or chemical agents (e.g., atmospheric O_2, moisture, light, odors, and temperature). Both biotic and abiotic deterioration can lead to food spoilage, albeit by different methods. Packaging can be used to provide a barrier to those agents that lead to deterioration.

Deteriorative reactions in foods are influenced by two factors: the nature of the food and its surroundings. These factors are referred to as intrinsic and extrinsic parameters.

2.2.1 Intrinsic Parameters

Intrinsic parameters are an inherent part of the food and include water activity (a_w), pH, oxidation-reduction potential (E_h), O_2 content, and product formulation, including the presence of any preservatives or antioxidants.

2.2.1.1 Water Activity

The parameter a_w is defined as the ratio of the water vapor pressure of a food to the vapor pressure of pure water at the same temperature. Mathematically:

$$a_w = p/p_o \tag{2.1}$$

where p is the vapor pressure of water exerted by the food and p_o is the saturated vapor pressure of pure water at the same temperature. This concept is related to equilibrium relative humidity (ERH) in that ERH = 100 × a_w. However, whereas a_w is an intrinsic property of the food, ERH is a property of the atmosphere in equilibrium with the food. The a_w of most fresh foods is above 0.99. Every microorganism has a limiting a_w value below which it will not grow, form spores, or produce toxic metabolites.

Water can influence chemical reactivity in different ways. It may act as a reactant (e.g., in the case of sucrose hydrolysis), or as a solvent, where it may exert a dilution effect on the substrates, thus decreasing the reaction rate. Water may also change the mobility of the reactants by affecting the viscosity of the food systems and form hydrogen bonds or complexes with the reacting species. Thus, a very important practical aspect of a_w is controlling undesirable chemical and enzymic reactions that reduce the shelf life of foods. It is a well-known generality that rates of changes in food properties can be minimized or accelerated over widely different values of a_w, as shown in Figure 2.1. Small changes in a_w can result in large changes in reaction rates.

When a food is placed in an environment at a constant temperature and relative humidity (RH), it will eventually come to equilibrium with that environment. The corresponding moisture content at steady state is referred to as the equilibrium moisture content. When this moisture content (expressed as mass of water per unit mass of dry matter) is plotted against the corresponding RH or a_w at constant temperature, a moisture sorption isotherm results (see Figure 2.2). Such plots are very useful in assessing the stability of foods and selecting effective packaging. As a_w is temperature dependent, it follows that moisture sorption isotherms must also exhibit temperature dependence. Thus, at constant moisture content (which is the situation existing in a food packaged in an impermeable package), a_w increases with increasing temperature. As rates of deteriorative

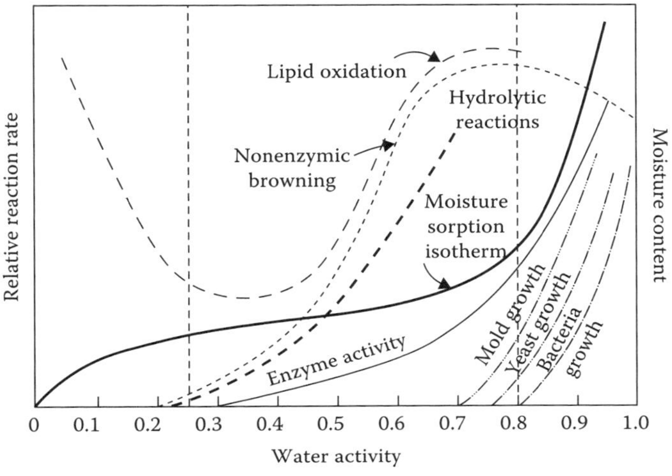

FIGURE 2.1 Rates of reactions as a function of water activity. (Redrawn with permission from Rockland L.B., Beuchat L.R. (Eds). 1987. In: *Water Activity: Theory and Applications to Food.* New York: Marcel Dekker, p. vii. Copyright CRC Press, Boca Raton, Florida.)

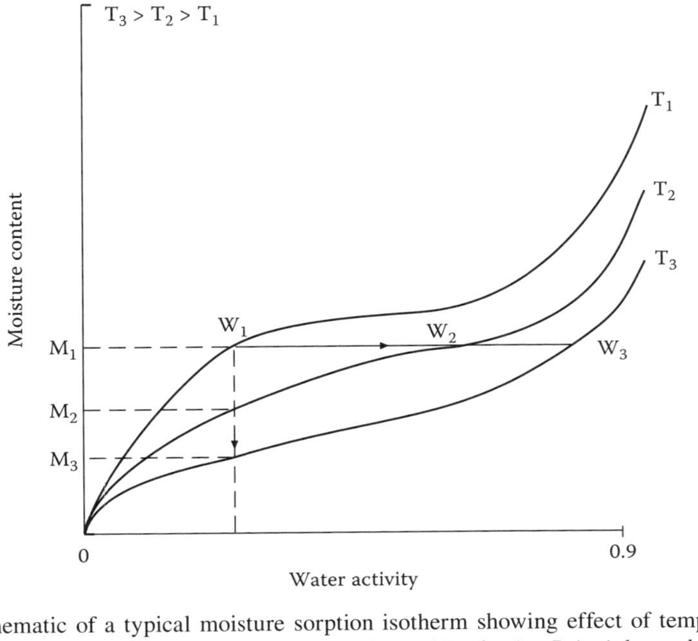

FIGURE 2.2 Schematic of a typical moisture sorption isotherm showing effect of temperature on water activity and moisture content. (From Robertson G.L. 2006. *Food Packaging Principles and Practice*, 2nd edn. Boca Raton, Florida: CRC Press, with permission. Copyright CRC Press, Boca Raton, Florida.)

reactions depend on both a_w and temperature, the increase in rate in such situations will typically be greater than that due solely to an increase in temperature. This has important implications for shelf life.

2.2.1.2 Oxidation-Reduction Potential

The oxidation-reduction potential (also referred to as the redox potential and abbreviated E_h or ORP) is a physicochemical parameter that determines the oxidizing or reducing properties of the medium, and it depends on the composition of the food, pH, temperature, and, to a large extent, the concentration of dissolved O_2 (DO). E_h plays an important role in the cellular physiology of microorganisms, such as growth capacity, enzyme expression, and thermal resistance. Alwazeer et al. (2003) demonstrated that reducing the E_h of orange juice using gas (N_2 and H_2) immediately after heat treatment maximized microbial destruction during pasteurization, prevented the development of microorganisms, and stabilized color and ascorbic acid during storage at 15°C.

The relationship between ORP values and DO levels in milk is not well understood. Several modifications that occur in milk during its processing and storage are driven by different oxidation-reduction reactions. Electrolysis treatments have been applied to milk to produce milk powder with better flavor quality. ORP and DO levels in enriched milk are mainly responsible for the oxidation of unsaturated fatty acids and the loss of viability of probiotic strains such as bifidobacteria. Decreasing the E_h in milk could allow an improvement in the quality of these products. Recent studies on electroreduction of milk by membrane electrolysis have shown that this electrochemical process decreased the E_h of milk without changing the organoleptic and nutritive values (Schreyer et al., 2008).

2.2.2 Extrinsic Parameters

Extrinsic factors that control the rates of deteriorative reactions include temperature, RH, gas atmosphere, and light; packaging can, to varying degrees, influence the impact of these factors on the rates of deteriorative reactions, depending on the specific packaging material.

2.2.2.1 Temperature

Temperature is a key factor in determining the rates of deteriorative reactions, and in certain situations the packaging material can affect the temperature of the food. This is particularly so with packaging materials that have insulating properties, and these types of packages are typically used for chilled and frozen foods. For packages that are stored in refrigerated display cabinets, most of the cooling takes place by conduction and convection. Simultaneously, there is a heat input by radiation from the fluorescent lamps used for lighting. Under these conditions, aluminum foil offers real advantages because of its high reflectivity and high conductivity.

Several models have been developed to represent the effect of temperature on the rates of deteriorative reactions.

2.2.2.1.1 Linear Model

This simple expression relating the rate of reactions and temperature has been used for many years:

$$k = k_o e^{b(T-T_o)} \tag{2.2}$$

where

k_o = rate at temperature T_o (°C)
k = rate at temperature T (°C)
b = a constant characteristic of the reaction
e = 2.7183.

2.2.2.1.2 *Arrhenius Relationship*

The most common and generally valid relationship for the effect of temperature on the rates of deteriorative reactions is that of Arrhenius. The relationship in the integrated form is

$$k = k_o e^{-E_a/RT} \tag{2.3}$$

where

k = rate constant for deteriorative reaction
k_o = constant, independent of temperature (also known as the Arrhenius, pre-exponential, collision, or frequency factor)
E_a = activation energy (J mol^{-1})
R = ideal gas constant (8.314 J K^{-1} mol^{-1})
T = absolute temperature (K)

The integrated relationship contains the inherent assumption that the activation energy and the pre-exponential factor do not change with temperature. Although this assumption is generally true, it is not universally so, and predictions based on this model sometimes fail when applied over a temperature span of greater than ~40°C. Furthermore, when the reaction mechanism changes with temperature, the activation energy may vary substantially. The value of E_a is a measure of the temperature sensitivity of the reaction, that is, how much faster the reaction will proceed if the temperature is raised. The activation energy depends on factors such as a_w, moisture content, solids concentration, and pH.

2.2.2.1.3 *Temperature Quotient*

Another term used to describe the response of biological systems to temperature change is the Q value, a quotient indicating how much more rapidly the reaction proceeds at temperature T_2 than at a lower temperature T_1. If Q reflects the change in rate for a 10°C rise in temperature, it is then called Q_{10}. Mathematically:

$$Q_{10} = \frac{k_{T+10}}{k_T} \tag{2.4}$$

It can be shown that the rate of a deteriorative reaction at two temperatures is related to the shelf life θ at those two temperatures; that is:

$$k_T \theta_{s(T)} = k_{T+10} \theta_{s(T+10)} \tag{2.5}$$

where

$\theta_{s(T)}$ = shelf life at temperature T°C
$\theta_{s(T+10)}$ = shelf life at temperature (T + 10)°C

Therefore,

$$Q_{10} = \frac{\theta_{s(T)}}{\theta_{s(T+10)}} \tag{2.6}$$

If the temperature difference is Δ rather than 10°C, the following equation can be used:

$$(Q_{10})^{\Delta/10} = \frac{\theta_{s(T1)}}{\theta_{s(T2)}} \tag{2.7}$$

For example, if the Q_{10} for the key deteriorative reaction was 3 and the shelf life θ_s at 37°C was 4 months, then the shelf life at 23°C would be

$$\theta_{23} = \theta_{37} \times (Q_{10})^{\Delta/10} = 4 \times (3)^{14/10} = 18.6 \text{ months} \tag{2.8}$$

If, however, the Q_{10} was 2 rather than 3, then

$$\theta_{23} = \theta_{37} \times (Q_{10})^{\Delta/10} = 4 \times (2)^{14/10} = 10.5 \text{ months} \tag{2.9}$$

This example illustrates the importance of having an accurate estimate of Q_{10}.

It can be shown when the Arrhenius model is used that

$$\ln Q_{10} \approx \frac{10E_a}{RT^2} \tag{2.10}$$

Note that Q_{10} is not constant but depends on both the E_a and the temperature; when Q_{10} is reported, the temperature range over which it applies should also be specified.

2.2.2.2 Relative Humidity

The RH of the ambient environment is important and can influence the a_w of the food unless the package provides an excellent barrier to water vapor. Many flexible plastic packaging materials provide good moisture barriers, but none is completely impermeable, thus limiting the shelf life of low a_w foods.

2.2.2.3 Gas Atmosphere

The presence and concentration of gases in the environment surrounding the food have a considerable influence on the growth of microorganisms, and the atmosphere inside the package is often modified. The simplest way of modifying the atmosphere is vacuum packaging, that is, removal of air (and thus O_2) from a package prior to sealing; it can have a beneficial effect by preventing the growth of aerobic microorganisms. Flushing the inside of the package with a gas such as CO_2 or N_2 before sealing is the basis of modified atmosphere packaging (MAP). For example, increased concentrations of gases such as CO_2 are used to retard microbial growth and thus extend the shelf life of foods. MAP is increasing in importance, especially with the packaging of fresh fruits and vegetables, flesh foods, and bakery products.

Atmospheric O_2 generally has a detrimental effect on the nutritive quality of foods, and it is therefore desirable to maintain many types of foods at a low O_2 tension, or at least prevent a continuous supply of O_2 into the package. Lipid oxidation results in the formation of hydroperoxides, peroxides, and epoxides, which will, in turn, oxidize or otherwise react with carotenoids, tocopherols, and ascorbic acid to cause loss of vitamin activity.

With the exception of respiring fruits and vegetables and some flesh foods, changes in the gas atmosphere of packaged foods depend largely on the nature of the package. Adequately sealed metal and glass containers effectively prevent the interchange of gases between the food and the atmosphere. With flexible packaging, however, the diffusion of gases depends not only on the effectiveness of the closure but also on the permeability of the packaging material, which depends primarily on the physicochemical structure of the barrier.

2.2.2.4 Light

Many deteriorative changes in the nutritional quality of foods are initiated or accelerated by light. Light is, essentially, an electromagnetic vibration in the wavelength range between 4000 and 7000 Å; the wavelength of ultraviolet (UV) light ranges between 2000 and 4000 Å. The catalytic effects of light are most pronounced in the lower wavelengths of the visible spectrum and in the UV spectrum. The intensity of light and the length of exposure are significant factors in the production of discoloration and flavor defects in packaged foods.

Modification of plastic materials can be achieved by incorporation of dyes or application of coatings that absorb light at specific wavelengths. Recently nano-sized particles of titanium dioxide have been incorporated into plastic films to absorb UVA and UVB rays. Glass is frequently modified by inclusion of color-producing agents or by application of coatings. In this way a wide range of light transmission characteristics can be achieved in packages made of the same basic material.

There have been many studies demonstrating the effect of packaging materials with different light-screening properties on the rates of deteriorative reactions in foods. Among the most commonly studied foods has been fluid milk, the extent of off-flavor development being related to the exposure interval, strength of light, and amount of milk surface exposed.

2.2.3 Enzymic Reactions

From a food packaging point of view, knowledge of enzyme action is essential to a fuller understanding of the implications of different forms of packaging. The importance of enzymes to the food processor is often determined by the conditions prevailing within and outside the food. Control of these conditions is necessary to control enzymic activity during food processing and storage. The major factors useful in controlling enzyme activity are temperature, a_w, pH, chemicals that can inhibit enzyme action, alteration of substrates, alteration of products, and preprocessing control.

Three of these factors are particularly relevant in a packaging context. The first is temperature: the ability of a package to maintain a low product temperature and thus retard enzyme action will often increase product shelf life. The second important factor is a_w, because the rate of enzyme activity is dependent on the amount of water available; low levels of water can severely restrict enzymic activities and even alter the pattern of activity. Finally, alteration of substrate (in particular, the ingress of O_2 into a package) is important in many O_2-dependent reactions that are catalyzed by enzymes, for example, enzymic browning due to oxidation of phenols in fruits and vegetables.

2.2.4 Chemical Reactions

Many of the chemical reactions that occur in foods can lead to deterioration in food quality (both nutritional and sensory) or the impairment of food safety. Such reaction classes can involve different reactants or substrates, depending on the specific food and the particular conditions for processing or storage. The rates of these chemical reactions are dependent on a variety of factors amenable to control by packaging, including light, O_2 concentration, temperature, and a_w. Therefore, the package can, in certain circumstances, play a major role in controlling these factors, and thus indirectly the rate of the deteriorative chemical reactions.

The two major chemical changes that occur during the processing and storage of foods and lead to a deterioration in sensory quality are lipid oxidation and nonenzymic browning (NEB). Chemical reactions are also responsible for changes in the color and flavor of foods during processing and storage.

2.2.4.1 Lipid Oxidation

Autoxidation is the reaction of molecular O_2 by a free radical mechanism with hydrocarbons and other compounds. The reaction of free radicals with O_2 is extremely rapid, and many mechanisms for initiation of free radical reactions have been described. The crucial role that autoxidation plays in the development of undesirable flavors and aromas in foods is well documented, and autoxidation is a major cause of food deterioration.

Factors that influence the rate and course of oxidation of lipids are well known and include light, local O_2 concentration, high temperature, the presence of catalysts (generally transition metals such as iron and copper, but also heme pigments in muscle foods), and a_w. Control of these factors can significantly reduce the extent of lipid oxidation in foods.

2.2.4.2 Nonenzymic Browning

Nonenzymic browning (NEB) is one of the major deteriorative chemical reactions that occur during storage of dried and concentrated foods. The NEB, or Maillard, reaction can be divided into three stages: (1) early Maillard reactions involving a simple condensation between an aldehyde (usually a reducing sugar) and an amine (usually a protein or amino acid) without browning; (2) advanced Maillard reactions that lead to the formation of volatile or soluble substances; and (3) final Maillard reactions leading to insoluble brown polymers.

2.2.4.3 Color Changes

Acceptability of color in a given food is influenced by many factors, including cultural, geographical, and sociological aspects of the population. However, regardless of these many factors, certain food groups are acceptable only if they fall within a certain color range. The color of many foods is due to the presence of natural pigments such as chlorophylls, anthocyanins, carotenoids, flavonoids, and myoglobin.

2.2.4.4 Flavor Changes

In fruits and vegetables, enzymically generated compounds derived from long-chain fatty acids play an extremely important role in the formation of characteristic flavors. In addition, these types of reactions can lead to important off-flavors. Enzyme-induced oxidative breakdown of unsaturated fatty acids occurs extensively in plant tissues, and this yields characteristic aromas associated with some ripening fruits and disrupted tissues (Lindsay, 2008).

Fats and oils are notorious for their role in the development of off-flavors through autoxidation. Aldehydes and ketones are the main volatiles from autoxidation, and these compounds can cause painty, fatty, metallic, papery, and candlelike flavors in foods when their concentrations are sufficiently high. However, many of the desirable flavors of cooked and processed foods derive from modest concentrations of these compounds. The permeability of packaging materials is of importance in retaining desirable volatile components within packages and in preventing undesirable components entering the package from the ambient atmosphere.

2.2.4.5 Nutritional Changes

In addition to the chemical changes described earlier, which may have a deleterious effect on the sensory properties of foods, there are other chemical changes that can affect the nutritive value of foods. The four major factors that influence nutrient degradation and can be controlled to varying extents by packaging are light, O_2 concentration, temperature, and a_w. However, because of the diverse nature of the various nutrients as well as the chemical heterogeneity within each class of compounds and the complex interactions of these variables, generalizations about nutrient degradation in foods are unhelpful.

2.2.5 Physical Changes

The physical properties of foods can be defined as those properties that lend themselves to description and quantification by physical rather than chemical means and include geometrical, thermal, optical, mechanical, rheological, electrical, and hydrodynamic properties. Geometrical properties encompass the parameters of size, shape, volume, density, and surface area as related to homogeneous food units, as well as geometrical texture characteristics. Although many of these physical properties are important and must be considered in the design and operation of a successful packaging system, in the present context the focus is on undesirable physical changes in packaged foods.

The major undesirable change in food powders is the sorption of moisture as a consequence of an inadequate barrier provided by the package, resulting in caking. This can occur as a result either of poor selection of packaging material in the first place or of failure of the package integrity during storage. Caking or spontaneous agglomeration of food powders (especially those containing soluble components or fats) occurs when they are exposed to moist atmospheres or elevated storage temperatures. The phenomenon can result in anything from small soft aggregates that break easily to rock-hard lumps of variable size to solidification of the whole powder. For foods containing solid carbohydrates, the greatest effect in physical properties results from sorption of water; such changes can occur in boiled sweets (leading to stickiness or graining) and milk powders (leading to caking and lumpiness).

2.2.6 Microbiological Changes

Microorganisms can make both desirable and undesirable changes to the quality of foods, depending on whether they are introduced as an essential part of the food preservation process (e.g., as inocula in food fermentations) or arise adventitiously and subsequently grow to produce food spoilage. In the latter case, they reach readily observable proportions only when they are present in the food in large numbers. As the initial population or microbial load is usually small, observable levels are reached only after extensive multiplication of the microorganisms in the food.

The two major groups of microorganisms found in foods are bacteria and fungi, the latter consisting of yeasts and molds. Bacteria are generally the fastest growing, so in conditions favorable to both, bacteria will usually outgrow fungi. The phases through which the two groups pass are broadly similar: a period of adjustment or adaptation (known as the lag phase) is followed by accelerating growth until a steady, rapid rate (known as the logarithmic phase, because growth is exponential) is achieved. After a time the growth rate slows until growth and death are balanced and the population remains constant (known as the stationary phase). Eventually, death exceeds growth and the organisms enter the phase of decline.

The species of microorganisms that cause the spoilage of particular foods are influenced by two factors: the nature of the foods and their surroundings. These factors are referred to as intrinsic and extrinsic parameters and were discussed earlier.

Every microorganism has a limiting a_w value below which it will not grow, form spores, or produce toxic metabolites. Water activity can influence each of the four main growth cycle phases by its effect on the germination time, the length of the lag phase and the growth rate phase, the size of the stationary population, and the subsequent death rate. Generally, reducing the a_w of a given food increases the lag period and decreases the growth rate during the logarithmic phase, the maximum of which becomes lower.

Whether a microorganism survives or dies in a low a_w environment is influenced by intrinsic factors that are also responsible for its growth at higher a_w. These factors include water-binding properties, nutritive potential, pH, E_h, and the presence of antimicrobial compounds. Microbial growth and survival are not entirely ascribed to reduced a_w but are also attributable to the nature of the solute. Key extrinsic factors relating to a_w that influence microbial deterioration in foods include temperature, O_2, and chemical treatments. These factors can combine in a complex way to encourage or discourage microbial growth.

Microbiological changes due to the growth of microorganisms are desirable in fermentation but are mostly undesirable in other environments, because microbial growth may lead to spoilage and even health-threatening situations when pathogens come into play. The ability to predict growth of bacteria in foods is very important in predicting shelf life. A frequently used growth model is the modified Gompertz model, which is discussed in Chapter 4.

The temperature of storage is particularly important, and several food preservation techniques (e.g., chilling) rely on reducing the temperature of the food to extend its shelf life. Although there is a very wide range of temperatures over which the growth of microorganisms has been reported (–34°C to 90°C), specific microorganisms have relatively narrow temperature ranges over which growth is possible. Molds are able to grow over a wider range of temperature than bacteria, with many being capable of growth at refrigerator temperatures.

The presence and concentration of gases in the environment has a considerable influence on the growth of microorganisms. Most food pathogens do not grow at refrigerator temperatures, and CO_2 is not highly effective at nonrefrigeration temperatures. Therefore, most MAP food is usually held under refrigeration. Temperature abuse of the product (i.e., holding at nonrefrigerated temperatures) could allow the growth of organisms (including pathogens) that were inhibited by CO_2 during storage at lower temperatures. For these reasons, it is difficult to evaluate MAP safety solely on the growth of certain pathogens at abusive temperatures.

2.3 RATES OF DETERIORATIVE REACTIONS

As discussed in the preceding section, a number of deteriorative chemical, biochemical, physical, and microbiological reactions can occur in foods. The rates of these reactions depend on both intrinsic (compositional) and extrinsic (environmental) factors. As well as understanding the nature of these reactions, it is important to have an appreciation of their rates, so that they can be controlled. Control of deteriorative reactions requires a quantitative analysis based on knowledge of the kinetics of food deterioration. Fortunately, simple chemical kinetics can be applied to such reactions.

Quantitative analysis of the deteriorative reactions that occur in a food during processing and storage requires the existence of a measurable index of deterioration (IoD), that is, a chemical, physical, or sensory measurement or set of measurements that may be used reproducibly to assess the changes occurring. An increase or decrease in the IoD must correlate with changes in food quality. For quantitative analysis of quality changes, the IoD must be expressed as a function of the conditions existing during processing and storage so that the changes can be predicted or simulated. Thus, calculation of quality losses requires a mathematical model that expresses the effect of intrinsic and extrinsic factors on the IoD.

The general equation describing quality loss may be written as

$$-\mathrm{d}D/\mathrm{d}\theta = f(I_i, E_j) \tag{2.11}$$

where

$-\mathrm{d}D/\mathrm{d}\theta$ = rate of change of some index of deterioration D with time θ; a negative sign is used if the concentration of D decreases with time

I_i = intrinsic factors ($i = 1 \ldots m$)

E_j = extrinsic factors ($j = 1 \ldots n$)

As the quality of foods and the rate of quality changes during processing and storage depend on intrinsic factors, it is possible in many cases to correlate quality losses with the loss of a particular component such as a vitamin or pigment. The conversion of a single component or quality factor C to an end-product G (e.g., conversion of chlorophyll to pheophytin, or conversion of ascorbic acid to brown pigments) may be written as:

$$D \rightarrow \text{intermediate products} \rightarrow G \tag{2.12}$$

The absolute concentrations of D or G need not be measured. For example, the production of brown pigments in foods is often measured as the increase in absorbance at 420 nm of an alcoholic extract of the food, and the change in absorbance is used as an indicator of the extent of the reaction. Such quality loss can be represented as being proportional to the power of the concentration of the reactant or product:

$$-\mathrm{d}D/\mathrm{d}\theta = kD^n \tag{2.13}$$

or

$$\mathrm{d}G/\mathrm{d}\theta = kG^n \tag{2.14}$$

where

D and G = concentration of index of deterioration or quality factor
θ = time
k = rate constant (dependent on extrinsic factors)
n = a power factor called the order of the reaction that defines whether the rate is dependent on the concentration of D or G. The value of n can be a fraction or a whole number
$\mathrm{d}D/\mathrm{d}\theta$ and $\mathrm{d}G/\mathrm{d}\theta$ = change in concentration of D or G with time

Equation 2.14 implies that extrinsic parameters such as temperature, a_w, and light intensity are held constant; if they are not, then their influence on the rate constant k must be taken into account in evaluating the equation.

For most deteriorative reactions in foods, the reaction order n has generally been shown to be either 0 or 1, that is, a zero- or first-order reaction. Typical pseudo-zero-order deteriorative reactions include nonenzymic browning (e.g., in dry cereals and powdered dairy products), lipid oxidation (e.g., development of rancidity in snack foods, dry foods, and frozen foods), and enzymic degradation (e.g., in fresh fruits and vegetables, some frozen foods, and some refrigerated doughs). Typical pseudo-first-order deteriorative reactions also include nonenzymic browning (e.g., loss of protein quality in dry foods), lipid oxidation (e.g., development of rancidity in salad oils and dry vegetables), vitamin loss in canned and dry foods, and microbial production of off-flavors and slime in flesh foods.

From a packaging point of view, it is often useful to know the concentration of D or G at which the product is no longer acceptable, for example, when the concentration of a vitamin or pigment has fallen below some level (e.g., 50% reduction in concentration) or the concentration of some undesirable brown color has risen above some level. In these situations, the shelf life of the food (θ_s) is the time for the concentration of D (or G) to reach an undesirable or critical level (D_c or G_c). Examples showing the application of these equations to shelf life calculations can be found in Robertson (2006).

2.4 INDICES OF FAILURE

In designing suitable packaging for foods, it is important first to define the indices of failure (IoFs) of the food, that is, the quality attributes that will indicate that the food is no longer acceptable to the consumer. These may or may not be the same as the IoDs. An IoF could be development of rancid flavors in cereals due to oxidation, loss of red color (bloom) in chilled beef due to depletion of O_2, reduction of carbonation in bottled soda due to permeation of CO_2 through the bottle wall, caking of instant coffee due to moisture ingress, development of microbial taint in chilled poultry, or moisture loss in green vegetables resulting in wilting.

Once the IoFs for a particular food have been defined, the next step is to attempt to quantify the magnitude of the particular degradation, for example, how much moisture or O_2 can react with the food before it becomes unacceptable. The final step is to ascertain which (if any) of the IoFs might be influenced by the packaging material, as packaging cannot prevent all deteriorative reactions in foods. If, for example, the IoF of a snack food was loss of crispness, then the packaging material could influence this by the extent to which it permitted the ingress of moisture. Different plastic films, for example, have different water vapor transmission rates and, thus, the shelf life obtained varies depending on the particular plastic selected. Similar considerations apply to foods for which the IoF is oxidation, as different packaging materials have different O_2 transmission rates (OTRs).

However, it is not just the packaging material itself that can influence shelf life; the method of filling the product into the package is also important. For example, with roasted and ground coffee, vacuum filling into metal cans will remove 95% or more of the O_2 from the can compared with inert gas flush packing in plastic foil laminate pouches, which will remove or displace 80–90% of the O_2 in the package. The residual O_2 in the package at the time of filling will have a major influence on shelf life regardless of the O_2 barrier properties of the package itself.

In the chapters that follow, the IoFs for particular foods are described, and ways in which they can be influenced by packaging are outlined.

REFERENCES

Alwazeer D., Delbeau C., Divies C., Cachon R. 2003. Use of redox potential modification by gas improves microbial quality, color retention and ascorbic acid stability of pasteurized orange juice. *International Journal of Food Microbiology* 89: 21–29.

Baner A.L., Piringer O.-G. 2008. Preservation of quality through packaging. In: *Plastic Packaging Interactions with Food and Pharmaceuticals*, 2nd edn. Piringer O.-G., Baner A.L. (Eds). Weinheim, Germany: Wiley-VCH, pp. 1–13.

Cardello A.V. 1998. Perception of food quality. In: *Food Storage Stability*. Taub I.A., Singh R.P. (Eds). Boca Raton, Florida: CRC Press, pp. 1–32.

Lindsay R.C. 2008. Flavors. In: *Fennema's Food Chemistry*, 4th edn. Damodaran S., Parkin K.L., Fennema O.R. (Eds). Boca Raton, Florida: CRC Press, chapter 10.

Robertson G.L. 2006. *Food Packaging Principles and Practice*, 2nd edn. Boca Raton, Florida: CRC Press.

Rockland L.B., Beuchat L.R. (Eds). 1987. In: *Water Activity: Theory and Applications to Food*. New York: Marcel Dekker, p. vii.

Schreyer A., Britten M., Chapuzet J.-M., Lessard J., Bazinet L. 2008. Electrochemical modification of the redox potential of different milk products and its evolution during storage. *Innovative Food Sciences and Emerging Technology* 9: 255–264.

van Boekel M.A.J.S. 2008. Kinetic modeling of food quality: a critical review. *Comprehensive Reviews in Food Science and Food Safety* 7: 144–158.

3 Shelf Life Testing Methodology and Data Analysis

Michel Guillet and Natalie Rodrigue
Creascience
Montreal, Quebec, Canada

CONTENTS

3.1 INTRODUCTION

Shelf life data possess very specific statistical properties. For this reason, the design and analysis of shelf life studies cannot be handled with classical statistical tools, and special care must be taken. This chapter describes the specific features of shelf life data and the typical methods used to collect and then summarize them efficiently. It is based on course notes that the authors prepared and use in their seminars on shelf life statistics (Guillet and Rodrigue, 2005).

The chapter begins with a general definition of the goal of shelf life studies and discusses the key elements that must be precisely defined in the design phase of any such study. This leads into a presentation of what makes shelf life data different from other experimental data. In the following section, different types of shelf life studies are presented and useful statistical concepts are introduced. All these definitions and concepts are then used, first, to provide an overview of strategies for shelf life studies and issues to address when designing them and, second, to describe the statistical methods suitable for the analysis of shelf life data. Finally, ways to correctly interpret and report study results are suggested.

3.2 DEFINITION AND SPECIFIC FEATURES OF SHELF LIFE DATA

3.2.1 General Definition and Its Implications

3.2.1.1 Definition

The *American Heritage Dictionary of the English Language* (*AHD*, 2000) provides the following definition of product shelf life: "The length of time a product may be stored without becoming unsuitable for use or consumption." In accordance with this definition, the typical goal of shelf life experiments is to determine the time it takes for food product samples to reach a state of unsuitability for consumption.

3.2.1.2 Selecting Characteristics on Which Shelf Life Will Be Assessed

The lack of precision in the *AHD* definition suggests that there are several ways of examining food deterioration. In practice, this applies not only to different products that might become obviously unsuitable for different reasons but even to a single product that might fail for a variety of reasons. This means that the initial step of any shelf life study should be to identify the different potential reasons for product failure and the related measurements that should be followed over time. Typical risk variables are microbial counts, texture measurements using an appropriate instrument, sensory taste panel results, and direct consumer acceptability measures. More generally, any measurement on the product likely to be related to its suitability for consumption should be considered a potential candidate.

However, as will be seen in the data analysis section, using more than one characteristic can make the computations quite complex, so in practice, unless there are several risks of failure that might take place at the same time (which are then called *competing risks*), many shelf life studies rely on a single failure criterion. In this perspective, the selection of the appropriate measurement should primarily be based on the likelihood that it will take place first among all potential risks.

3.2.1.3 Defining Acceptable Values of Risk Variables

Once the primary failure criterion has been selected, a crucial second step is to establish a specific tolerance or cutoff value for the corresponding variable. This cutoff will separate suitable samples from unsuitable ones, and so the statistical goal of a shelf life study actually consists of getting the best possible estimate of the time at which the cutoff value is reached for a product stored under specific conditions.

Examples of failure criteria can be found in the scientific literature. For example, Schmidt and Bouma (1992) defined the failure criterion in their study as when at least 60% of the panelists identified the stored samples as objectionable in two consecutive sessions. The first of the two sessions was considered to be the end of the shelf life. Yamani and Abu-Jaber (1994) used counts of psychrotrophic yeasts $\geq 10^7$ as a failure criterion. Araneda et al. (2008) used 25% and 50% consumer rejection probabilities as indices of failure.

It is important to note that the definition of the cutoff value should not rely on statistical criteria. This is because it is a question of risk management and therefore has its share of subjectivity. A common mistake is to define the shelf life of a product as ending when a specific measurement becomes statistically different from the value at the baseline. The major issue with this practice is the impact of sample size: increasing the number of samples increases the sensitivity of the statistical test, which means that the statistical difference from the baseline will be found earlier than with a smaller number of samples. In other words, with such a criterion, a larger sample size almost systematically leads to a shorter shelf life estimate. Only after a clear failure criterion has been defined can a shelf life experiment be conducted.

3.2.2 Statistical Features of Life Data

3.2.2.1 What Exactly Are Life Data?

The failure times collected and analyzed in shelf life studies can actually be found in a variety of situations. There are many applications in medicine (e.g., to estimate the survival times of patients treated with different drugs), in engineering (e.g., to test the reliability of components under different types of stress), and in economics (e.g., to model and to predict the duration of unemployment). Depending on the context, the data collected are referred to as "time to event data," "failure time data," "survival data," or, more generally, "life data." The analysis of life data requires such specific tools that a complete field of statistics has been devoted to it. The statistical techniques are referred to as "survival analysis methods" in medicine (Kalbfleisch and Prentice, 2002), "reliability methods" in engineering (Meeker and Escobar, 1998), and "duration models" in econometrics (Greene, 2008). The terminology used is slightly different, but the techniques themselves are similar. In all applications, life data consist of a measure of lifetime or length of time until the occurrence of a given event. In the case of shelf life, we are specifically interested in the failure time of a food product.

3.2.2.2 Problem of Censored Observations

One common issue with life data is the impossibility of systematically observing the failure times for all samples. This phenomenon is technically referred to as "censoring" and can happen for various reasons (Meeker and Escobar, 1998). Being aware of its existence and knowing how to identify it is crucial, as it is likely one of the most characteristic features of life data.

To illustrate the different types of censoring, consider the following situation, in which a food sample is followed over time for failure. Figure 3.1 depicts the various problems that might occur when trying to determine the failure time of this sample. To start with, on the top plot, failure time is identified by a star at time 22. This plot corresponds to the ideal situation where it is actually possible to observe the exact failure time (no censoring).

3.2.2.2.1 Right-Censoring

The plot below depicts the first type of censoring, which can occur whenever the duration of the study is fixed. It is called *right-censoring*. In the example, the study ends at time 20. A product that will fail at time 22 has, of course, not yet failed at the end of the study. Therefore, the failure time for this sample is unobservable. The best that can be said about the product is that it has survived until time 20. If the study duration had been longer, the exact time of failure would perhaps have been observable. In practice, the value of 20 will be reported in the result file along with an indication that this is a right-censored value. Twenty is a lower bound of the true failure time for this product sample.

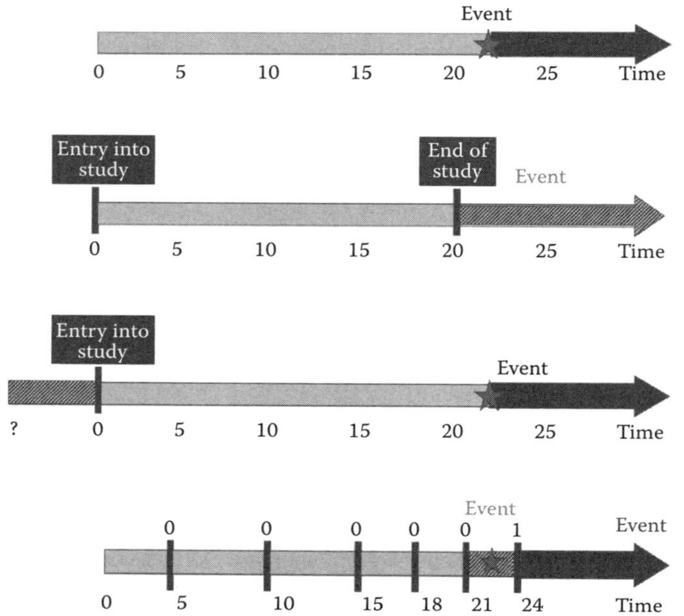

FIGURE 3.1 Different cases of observable and unobservable failure times.

3.2.2.2.2 Left-Censoring

A second situation is termed *left-censoring*. This occurs when only an upper bound of the time to failure can be determined for a given sample. It can happen, for instance, when it is impossible to know when the sample was produced (e.g., testing products from competitors) or when using a destructive testing procedure. This introduces another source of uncertainty into the exact lifetime of the sample.

3.2.2.2.3 Interval-Censoring

A third censoring situation, termed *interval-censoring*, is very common in shelf life testing, because of the noncontinuous monitoring of samples. In this situation, it is virtually impossible to know the exact failure time of each sample. On the last plot, the food sample is tested at times 5, 10, 15, 18, and 21 and has not failed. At the next testing time, 24, it has failed. The exact failure time for this sample is said to be interval-censored between 21 and 24, and these two values should be reported in the data file.

3.2.2.3 Importance of Censored Data

Censored observations are incomplete or partial data, but they do contain relevant information to determine shelf life. Thus, they must not be discarded from the statistical analysis of the data. However, they must also not be treated as if the exact failure time had been observed. Specific statistical methods exist to account for censoring. If censoring is ignored in the data analysis, a biased estimate of shelf life will be obtained (Gacula and Kubala, 1975).

3.2.3 Shelf Life versus Stability Studies

3.2.3.1 What Is a Stability Study?

Many experimenters conduct stability studies but refer to them as shelf life studies. This is not entirely accurate. In a stability study, the evolution or degradation of a product characteristic is measured over time. The evolution of the characteristic can then be modeled using appropriate statistical techniques to determine a mathematical relationship relating time to the values of the characteristic (ICH, 1993, 2003; Simon and Hansen, 2001). As a second step, the failure time of the product can be estimated from stability data by defining a minimum or maximum acceptable value for the characteristic.

3.2.3.2 Difference between Shelf Life and Stability Experiments

The information collected in a stability study is quite different from that collected in a shelf life experiment. It is very important to distinguish between them. Even though they seem to deal with similar topics, their respective goals are not the same, and the tools used to analyze the data have nothing in common.

- In shelf life experiments, the failure time of a food product is of primary interest. Failure time is neither an instrumental nor a sensory measurement. Censored failure times can occur.
- In stability experiments, the evolution or the degradation of a characteristic over time is of primary concern. Failure time is therefore not directly observable. Rather, it is estimated from the data by defining an appropriate cutoff value to indicate failure.

In practice, stability experiments are often conducted as a preliminary step in shelf life studies in order to get estimates of the failure time. These failure times are then gathered in a dataset to proceed with the actual shelf life analysis.

3.3 GOAL OF SHELF LIFE STUDIES: A STATISTICAL PERSPECTIVE

3.3.1 Types of Shelf Life Experiments

Two broad types of shelf life experiments can be defined: simple and comparative experiments.

3.3.1.1 Simple Experiments

In simple experiments, the goal is to estimate the shelf life of a product empirically. Typically, a single product is studied under fixed storage conditions. The single product can be a particular formulation, a new package, and so on. The primary goal of simple experiments is to establish an estimate of the shelf life of the product along with a measure of uncertainty on the estimate.

3.3.1.2 Comparative Experiments

In comparative experiments, more than one condition is tested. Therefore, the primary goal is to compare the conditions or to estimate the effect of a set of factors on the shelf life of a product. Typically, several product formulations or storage conditions are compared. Applications include optimizing a product formulation, selecting the best packaging or closure, and investigating the robustness of the overall product shelf life estimate to variations in the processing, storage, and distribution conditions.

3.3.2 Failure of Classical Methods

For many statistical methods used in research and development applications, such as analysis of variance (ANOVA) and linear regression analysis, assumptions are made about the distribution of the empirical data. For instance, valid interpretation of ANOVA and regression results requires the assumption that the model residuals are normally distributed with constant variability. As technical as such an assumption may appear, it is nevertheless important, because the statistical tests used to interpret the results rely on these assumptions holding to be valid.

In shelf life experiments, the response or outcome variable is the failure time of the product. The normality assumption rarely holds for failure times. In a normal distribution there is a nonzero chance of observing negative values. This clearly does not make sense for failure time data, which will all be positive. Furthermore, the normal distribution is a symmetric distribution. However, the distribution of the failure times is very unlikely to be symmetric. This is a first reason why it does not make sense to use ANOVA or regression to analyze shelf life data.

A second issue with classical statistical techniques arises when censored data are encountered. As mentioned, censored data are not totally informative; for instance, although a lower bound for the failure time is known, its exact value is not. On the other hand, classical techniques assume that each observation in the dataset carries a similar amount of information.

For these reasons, it can be very dangerous to use ANOVA or regression analysis with life data, and even more dangerous to remove censored data to accommodate these methods. Alternative methods are therefore needed.

3.3.3 Useful Statistical Concepts

As shelf life data exhibit specific statistical features, custom statistical tools have been developed to make the most out of these data, and specific statistical concepts need to be introduced to understand fully the underlying mechanism of such tools.

3.3.3.1 Survival Curve

One fundamental idea in shelf life studies is that samples do not all fail at the exact same time. Therefore, to compute an estimate of a product shelf life, the statistical distribution of the failure times needs to be determined. Stated another way, a curve that depicts the probability of the product survival as a function of time needs to be generated. Such a curve is called a *survival curve.* In simple shelf life experiments, estimating the survival curve is the ultimate goal of the statistical analysis of the data. Figure 3.2 shows a typical survival curve.

The curve represents a simple way to visualize the distribution of failures for a product. Once estimated, survival curves can be used for prediction—for example, to determine the percentage of samples that have failed after a given length of time—or for inverse prediction—to determine the time when a given proportion of the samples have failed, say 5% or 10%. As a matter of fact, in most situations, the value retained for a product shelf life is directly derived from the estimated survival curve.

Figure 3.3 illustrates how the previously obtained survival curve can be used for this purpose: if the maximum acceptable failure rate is 10%, then the intersection of the 90% survival probability with the curve suggests a shelf life of approximately 2.5 months.

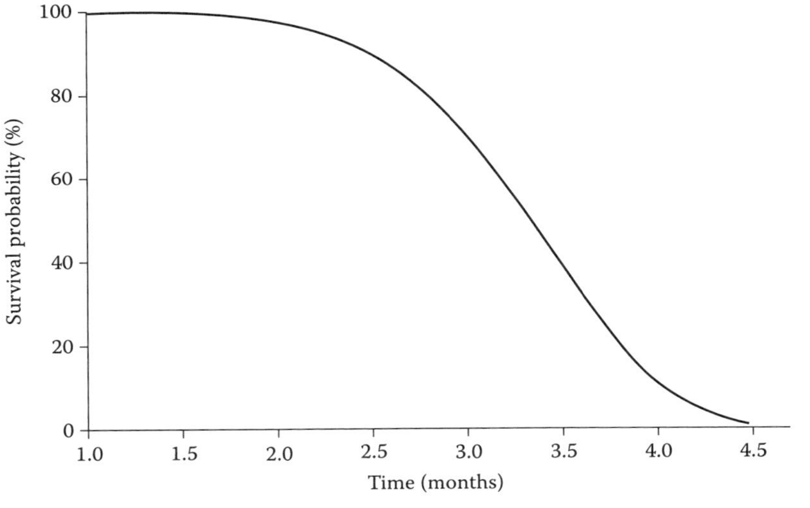

FIGURE 3.2 Example of a survival plot.

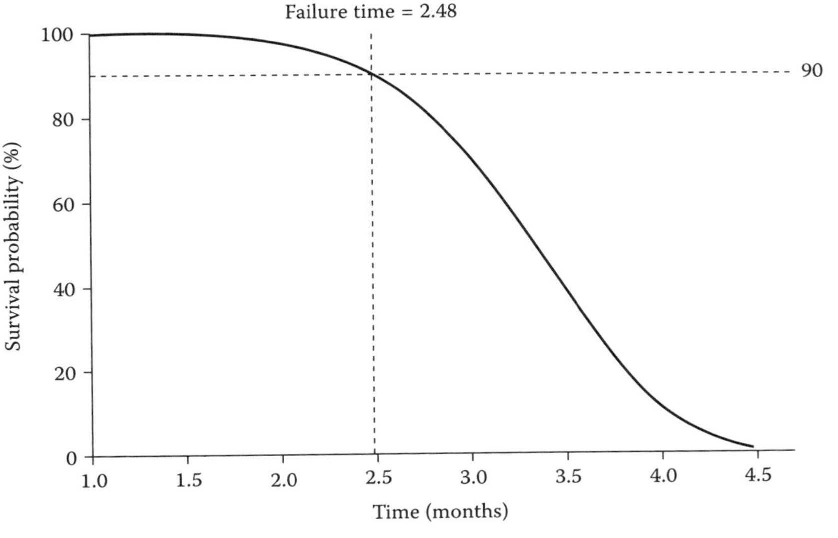

FIGURE 3.3 Using the survival curve to determine failure time for a 10% failure rate or 90% survival rate.

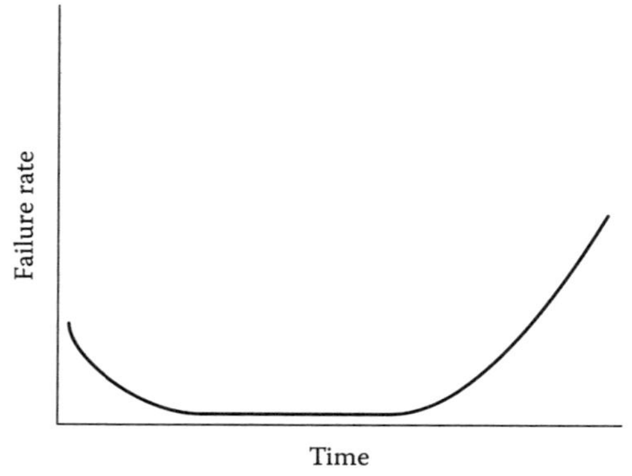

FIGURE 3.4 Bathtub curve illustrating the lifetime of a product.

3.3.3.2 Hazard Function

The hazard function does not have the same importance in the interpretation of results, but it is a key theoretical concept used in life data analysis. Furthermore, it is often part of software output and has some interesting practical interpretations.

The hazard rate at a given time is defined as the risk of failure at that time knowing that the product has survived until that time. The hazard function defines the relationship between time and the hazard rate at that time. In reliability testing the hazard function is referred to as the failure rate. A statistical relationship exists between the hazard function and the survival curve, and so knowing one gives perfect knowledge of the other.

3.3.3.3 Direct Application of the Hazard Function: Bathtub Curve

A classical representation of the risk of failure for manufactured goods and food products is the so-called bathtub curve. Such a curve is shown in Figure 3.4. The bathtub curve actually represents a hazard function that consists of three periods. The first one represents an early failure period corresponding to defective products, for example, faulty package seals. The second is a normal life period, and it concludes with a wear-out or end-of-life period that exhibits an increasing failure rate. The hazard rate is large for small values of time, then decreases to some minimum and stays at that level for some time before increasing again.

3.4 DESIGNING SHELF LIFE STUDIES

Given the specificity of life data, it will come as no surprise that the design of shelf life studies requires slightly different practices than other experimental situations. The questions to address in order to build an efficient design remain the same, namely:

- What is the exact goal of the study?
- Where do the experimental units come from?
- What should be the study duration?
- How many samples are needed to get a precise enough estimate?
- At what time should the measurements be taken?

The following sections discuss these questions along with very specific issues such as the way to deal with destructive testing or to set up experiments over the long run.

3.4.1 Need for Focused Experiments

As is generally the case with experimental designs, it is very important when designing shelf life studies to identify a single primary goal instead of trying to combine different (and often irreconcilable) goals into a single study. For example, a single study should not compare the impact of different packaging materials and at the same time seek reliable estimates of the shelf life of the product stored in each package. Unless the experiment's duration is very long, a much more efficient approach is to design a first experiment to compare the different packaging materials and to select one or two that provide the longest shelf life, and then a second experiment that focuses only on the selected packaging to improve the precision of the shelf life estimate.

One consequence of this is that simple and comparative experiments should not be mixed. Therefore, the following discussion will start with the design of simple experiments, after which issues that are specific to comparative experiments will be emphasized.

3.4.2 Designing Simple Experiments

A mandatory preliminary step in designing any study is to specify the criterion for assessing the failure of the product. Once this has been done, one can really start to address the question of how to design the study, beginning with the definition of the study duration.

3.4.2.1 Study Duration

It goes without saying that the study duration should exceed the expected product shelf life; if no sample has failed at the end of the study, there will not be much to do with the data. However, it is not necessary to wait to terminate the study until all samples have failed. A correct trade-off must be found. The study must not be too short (not precise enough) or too long (too expensive). The way the data are analyzed is also crucial for this decision; as will be seen in the data analysis section, nonparametric methods require only knowledge of past events when computing the survival until a given time. This means such methods can be applied with only a small percentage of failures observed and the duration of the study can be considerably reduced. On the other hand, parametric methods of data analysis require a much larger coverage of all episodes of a product lifetime. Therefore, unless most sample failures happen pretty much at the same time, a longer study duration will be required to estimate with enough precision the survival curve. As soon as the duration is fixed, it should be remembered that censoring may occur; that is, at the end of the study some items will not have failed.

3.4.2.2 Selecting Representative Samples and Fixing Experiment Size

It is crucial that product samples are chosen that are as representative as possible of the production variability. It is not a good idea to select only samples that are as homogeneous as possible. As a matter of fact, the maximum relevant production variability should be integrated into the design, as this will define the scope of the study. Therefore, it is worth investigating what the most important sources of variability in the product are: batches, plants, harvests, producers, and so on. Then, if, for instance, batches have been identified as the primary source of variability, one should ensure that the sampling plan collects samples from several different batches.

A common mistake is to neglect to introduce sample variability into a shelf life experiment, or at least to fail to account for it properly. One needs to keep in mind that the general goal of these experiments is to generalize the results to a larger population, typically all batches produced. For this purpose, the use of several batches is essential to quantify the uncertainty of the measured shelf life. The ICH (2003) suggests that at least three batches should be used for stability models, and this should also be used as a guideline for shelf life modeling. It is also worth noting that when sensory panelists or consumers are used to assess the end of life of a product, the variability among them is in no way an alternative to the product variability and should be handled separately. If it is not, then, as in the papers by Hough et al. (2003, 2004), Araneda et al. (2008), Guerra et al. (2008), and Manzocco and Lagazio (2009), the uncertainty in the shelf life estimate might seem very small. Unfortunately, it is unlikely to be very small, as the size of the experiment has been artificially increased by considering each consumer evaluation as an independent data point. It is actually impossible to measure the error on the estimates, because, apparently, a single batch has been used.

The next logical question to address is how many product samples should be tested. Sample size calculations are necessary to design experiments that are large enough to produce useful information and small enough to be practical. In order to determine the appropriate sample size for a simple experiment, it is necessary to decide what type of analysis will be used to estimate the survival curve (see Section 3.5 on data analysis) and what precision should be achieved for the shelf life estimate. With this information, a growing number of software packages can be used for the computations. More detailed information on the way to determine sample size once the analysis method has been defined can be found in Meeker and Escobar (1998) and NIST/SEMATECH (2008).

It is important to realize that there is often a clear advantage in increasing the sample size in shelf life experiments: given the health risks at stake, shelf life estimates are usually taken as lower bounds of confidence intervals. Therefore, a larger sample size will typically result in a tighter confidence interval and extend the shelf life. If it is not feasible to have a large enough sample size in the short term, the experimenter should keep in mind that it is always possible to "complement" the study in the long term with additional sample testing to validate or improve short-term conclusions (see Section 3.4.4 on dynamic designs).

3.4.2.3 Destructive versus Nondestructive Testing

Some measurements made on a sample during shelf life studies lead to the destruction of the sample; in other cases the sample can be reused to make the same measurement at a later date. Assessing whether the measurement process is destructive or not has important implications for the way the study is conducted. The ideal situation is nondestructive testing, as reusing the same sample allows better control of sample-to-sample variability. It also means that a specific failure time (possibly censored) can be computed for each sample used in the study. With destructive testing, all measures will be censored: if the sample tested has already failed, it is impossible to know exactly when it did (left-censoring as this is an upper bound), and if it has not yet failed, this provides only a lower bound for the failure time (right-censoring). Consideration could be given to including the left- or the right-censored data as is in statistical analysis, but it would make the computations very complex, and so most of the time destructive testing is handled with a two-stage sampling procedure (see FDA, 1987; ICH, 1993, 2003). First, homogeneous samples are selected—for instance, from within the same batch. It is then reasonable to assume that they should fail at roughly the same time. All the samples are tested at several time points using destructive testing. Different techniques are then available to analyze this set of samples (Meeker and Escobar, 1998), but they all share the following property: a single value for the failure time will be obtained out of this series of samples. To gather data to allow an estimate of the failure time distribution, the same procedure is repeated for other batches to capture the variability of the population of products. Obviously, destructive testing requires more samples than nondestructive testing, but otherwise, at the main sampling level, the procedure remains the same as for nondestructive testing.

3.4.2.4 Selecting Sampling Times

When foods are sampled for shelf life determinations, the samples are rarely monitored on a continuous basis. Sampling times need to be specified. In classical experiments, it is common practice to select factor levels that are equally spaced. Most of the time, this is not a good practice for time points in shelf life studies.

If the expected product shelf life is approximately 6 months, it is not very informative to take measurements during the first few months. Conversely, more samples need to be tested during the period in which the product is likely to fail (Gacula and Singh, 1984). As a first consequence, it is far from optimal to select equally spaced measurement times. A second important consequence is that as far as the practical organization of the test allows, it is recommended to adjust the sampling times on the basis of the observed failure rates. Practically speaking, this means that, if feasible, it might prove useful to store a larger number of samples than originally planned to allow for additional testing if needed.

It is therefore difficult to give a general rule for fixing the time points. The two principles outlined here usually provide enough guidance to handle most situations. It can be added that the frequency of sampling times must take into account the precision level required for the shelf life estimate. For instance, if the experimenter wants to estimate the shelf life of a product within ±1 week, it is pointless to test the product every other day.

3.4.3 Designing Comparative Experiments

3.4.3.1 Generalization of Simple Experiments

Most of the principles detailed in the section on simple experiments can easily be generalized to comparative experiments. Classical experimental design strategies may be used in order to make sure that the effect of each factor in the experiment can be assessed and quantified, as this becomes the primary goal for such experiments.

In the same way, sample size has to be determined on the basis of, among other considerations, the magnitude of the difference in shelf life between experimental conditions that the experimenter wants to be able to detect (this parameter is often referred to as the "effect size" in statistical textbooks). Generally speaking, comparative experiments will require a larger overall number of samples to test. However, for each combination of factor levels, the sample size is reduced compared to simple experiments. An additional strategy is to use fractional designs (EMEA, 2002), so, overall, it is feasible to design a comparative experiment with only a slightly larger number of samples (assuming the number of conditions to compare is not too large). The key issue has again to do with staying focused on a single goal rather than trying to mix several goals.

3.4.3.2 Specific Aspects of Comparative Experiments

All factors related to product formulation, packaging, and storage conditions can be examined simultaneously using the principles of factorial designs. However, the time points have to be handled in a distinct way. In classical experimental design, time would be considered as a factor and its levels would be globally defined and applied to all other factor levels. This might not work in many shelf life experiments; for instance, whenever several storage temperatures are tested, more testing should be carried out earlier in the life of the product stored at higher temperatures, as failure is likely to occur more rapidly at higher rather than at lower temperatures. Other factors such as gas atmosphere or light intensity might have similar effects on shelf life (which is why these factors are tested). For this reason, sampling times need not and, often, should not be the same.

3.4.4 Dynamic Designs

Even though shelf life experiments require specific constraints on the way they are conducted, it is worth considering that they also give the experimenter a level of flexibility that is rarely found in other studies. A first remarkable feature of shelf life studies is that the time points can be adjusted if needed. If the first measurements taken suggest that the failure will happen much later than anticipated, it would definitely be better to space out the measurements and start testing more frequently when failures are most likely to take place.

A second very interesting feature of life data is that one can easily improve the precision of the shelf life estimate by getting additional data. As will be discussed more thoroughly in the data analysis section, the addition of failure times for new samples can easily be handled by the analysis procedure and it increases the quality of the estimate of the survival curve. Therefore, provided the experimental procedure remains the same, it is definitely useful and acceptable to improve the precision by collecting additional data in the long run.

3.5 STATISTICAL ANALYSIS OF SHELF LIFE DATA

In this section, the different steps involved in the analysis of shelf life data are covered. First, the presentation of the layout of shelf life data for analysis with statistical software packages is discussed. Section 3.5.2 then covers the analysis of simple experiments, looking at two classical ways of estimating the distribution of failure times of a product. The Kaplan–Meier (KM) methodology, a nonparametric approach, is discussed first, after which parametric methods are covered. A general strategy for using these methods is suggested and their advantages and drawbacks compared with KM methodology are stated. Finally, statistical tests and models that are available to compare different conditions and more generally deal with comparative studies are discussed.

3.5.1 Typical Data Layout for Shelf Life Experiments

Combining the definition of shelf life and the properties of shelf life data naturally leads to a specific presentation for this type of data. Table 3.1 contains the typical layout of the data for simple shelf life experiments with possibly right-censored samples. In such a study, several samples are tested and their failure times are entered into the dataset along with a binary variable to indicate the state of the sample the last time it was examined (0 = failed and 1 = censored). If the sample is right-censored, the last tested time is entered as its largest observed survival time. Unless no censoring at all occurs in the data, the "Final State" column is mandatory.

Interval-censored data use two columns that contain, for each sample, the last time at which it has survived and the first time after failure has occurred respectively. In the case of comparative shelf life experiments, additional columns would be used to identify factors such as storage temperature and package type.

3.5.2 Analysis of Data from Simple Experiments

Given a series of failure times observed for different samples, the goal of the data analysis is to estimate a mathematical function generalizing the distribution of the series to a specific population of products. This distribution replaces the usual normal distribution, and the way the data are used in the computations should allow for censored data. There are two classical ways of estimating such distributions. If no assumption is made about the mathematical form of the distribution of the failure times, the nonparametric KM methodology can be used. The other approach assumes a specific statistical distribution to model the failure times. The latter approach is known as the parametric modeling approach. With both approaches, survival curves can be estimated and predictions of the failure time for different survival rates can be obtained, along with uncertainty measures.

TABLE 3.1
Typical Data Layout for a Shelf Life Experiment

Sample Identification	Failure Time (Days)	Final State (0/1)
1	15	1
2	30	0
3	30	0
4	21	1
5	24	0

3.5.2.1 Nonparametric Approach: KM Methodology

3.5.2.1.1 Principles of KM Estimator

The goal of the KM method is to estimate the proportion surviving (not having failed) at any given time (Kaplan and Meier, 1958), on the basis of what has happened so far, without making any assumption about a specific mathematical formulation. For that purpose, all recorded events (failure and censoring) are first ordered by the time they happened. Then, starting from 100% survival at time 0, every time a failure occurs, the probability of survival is updated as the observed survival rate at that time multiplied by the previous survival rate. As a result, the survival curve is a step function. The drops are randomly located on the horizontal axis, as they depend on the empirical data. Furthermore, due to censoring, the number of samples at risk changes, so the size of the drops also changes.

To illustrate how this method can be applied to shelf life studies, consider a set of 36 samples of cake stored in translucent packages that have been followed for failure over 13 days. Failure was defined as the time at which mold appeared on the surface of the cake. Right-censoring occurred massively at the end of the study as many samples had not yet failed. A couple of other censoring events occurred before the end because the package containing these samples was accidentally opened before they had failed. Table 3.2 shows the time points (in days) at which failures and censored observations occurred.

The KM estimate of the survival curve is shown in Figure 3.5. In several software packages, times at which censoring took place are identified with a circle on the survival curve.

TABLE 3.2
Number of Failed and Censored Samples

Time (Days)	At Risk	Failed	Censored
5	36	1	0
7	35	3	2
8	30	2	0
9	28	7	0
10	21	7	1
11	13	3	0
12	10	3	1
13	6	1	5

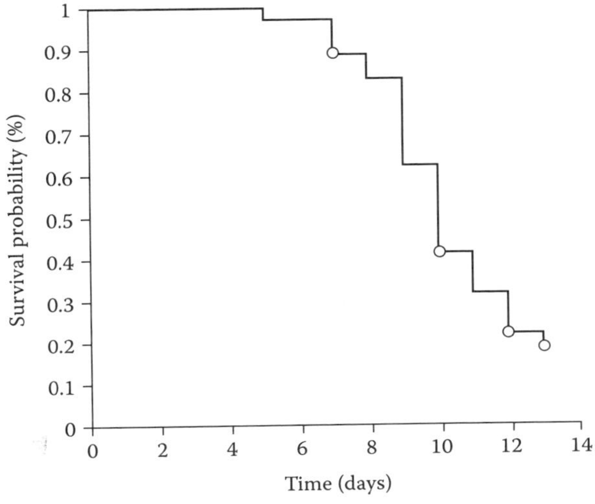

FIGURE 3.5 A nonparametric KM survival curve.

3.5.2.1.2 Assumptions of KM Methodology

As a nonparametric method, the assumptions underlying the KM methodology are not as strong as those for parametric analyses. They do exist, though, and should not be overlooked. First, samples that are censored are assumed to have the same survival chances as those that continue to be followed; otherwise, the estimation of survival probabilities may be biased. Furthermore, survival probabilities are assumed to be the same for all food samples irrespective of whether they enter early or late into the study.

3.5.2.1.3 Estimation of Error

As KM methodology is an inferential method, it is important to quantify the uncertainty on the estimated survival curve. Most software packages that compute KM survival curves also offer as an option the computation of a 95% confidence interval. Figure 3.6 depicts the confidence limits (dotted lines) for the survival curve presented in Figure 3.5.

3.5.2.1.4 Using a Survival Curve to Predict Shelf Life

Once the curve is available, it is easy to use it to determine the failure time for a given survival probability. In medicine, the median survival time is often derived from survival curves based on empirical patient data. The method is to draw a horizontal line at 50% survival and see where it crosses the curve and then look down at the time axis to read off the median survival time. For shelf life applications, the principle remains the same, except that a failure rate of 50% is usually too large, and so most applications use a smaller risk level such as 5% or 10%. Again, this decision is a risk management issue not a statistical one. To account for uncertainty of the prediction, a more conservative estimate obtained by looking at the lower bound of the confidence interval instead of the actual survival curve is recommended.

3.5.2.1.5 Impact of Censored Observations

If censored data are removed from the analysis because their actual failure time is unknown, the resulting survival curve will look like Figure 3.7. When this figure is compared with Figure 3.5, it can easily be observed that the survival probabilities drop more rapidly when the censored observations are removed from the analysis and that the uncertainty on the survival curve is greater due to the smaller number of observations. This curve also suggests that all samples have failed after 13 days, which is clearly not the case.

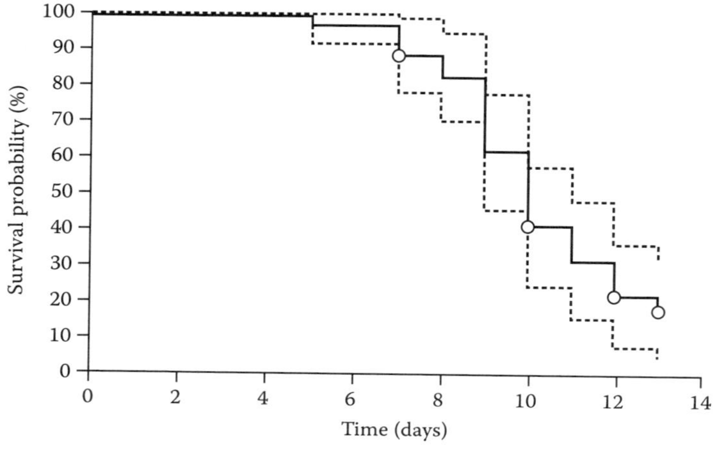

FIGURE 3.6 The nonparametric KM survival curve with a 95% confidence interval.

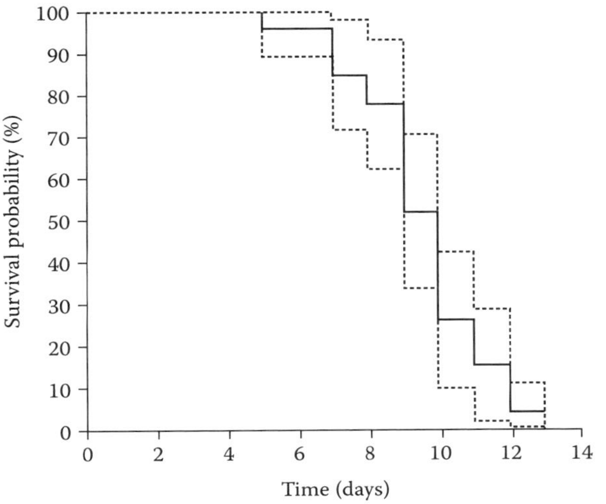

FIGURE 3.7 KM survival curve without censored data.

3.5.2.2 Parametric Approach: Fitting Statistical Distributions

3.5.2.2.1 General Principle

In the parametric analysis of life data, the failure time of the product population is assumed to follow some predefined probability distribution (Hough et al., 2003). Typically, such a distribution is defined by a small number of parameters and a mathematical equation. For instance, the normal distribution is defined by two parameters, the average and the standard deviation. The experimenter needs first to select a distribution and then to use the data collected to estimate the most likely values of the parameters with an appropriate software package. Finally, goodness of fit of the distribution to the data must be assessed using the appropriate tools.

3.5.2.2.2 Some Commonly Used Statistical Distributions

3.5.2.2.2.1 Exponential Distribution The exponential distribution is a commonly used distribution in reliability engineering mainly because of its simplicity (Ross, 1985). It is used to describe units that have a constant failure rate. In simpler terms, this means that an item that has been produced any number of hours, days, weeks, or months ago is as likely to fail as a new item. Although this might make sense for light bulbs and for electronic components more generally, it is clearly inappropriate for food products. This distribution requires the estimation of only one parameter for its application. Figure 3.8 depicts three survival functions based on the exponential distribution for three different values of the parameter.

3.5.2.2.2.2 Weibull Distribution The Weibull distribution is one of the most commonly used distributions in reliability engineering because of the many shapes it can take when its parameters are varied. It can therefore model a great variety of data and life characteristics (Kececioglu, 2003). Gacula and Singh (1984) introduced the Weibull analysis into food shelf life studies. The usual Weibull distribution is defined by two parameters (shape and scale). There is also a three-parameter version where the additional parameter is a threshold parameter. Figure 3.9 presents three Weibull distributions based on different shape parameter values.

The Weibull distribution has been so popular in the past that applying it to a dataset has sometimes been referred to as "Weibull analysis" (Cardelli and Labuza, 2001). In a similar fashion, Calle et al. (2006) justify their choice of the Weibull distribution on the grounds of its flexibility and previous use in food applications. This is an oversimplification, as there are rarely theoretical

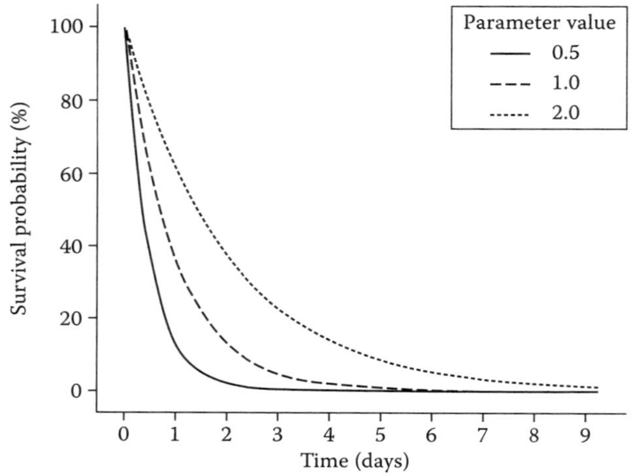

FIGURE 3.8 Three survival curves based on an exponential distribution.

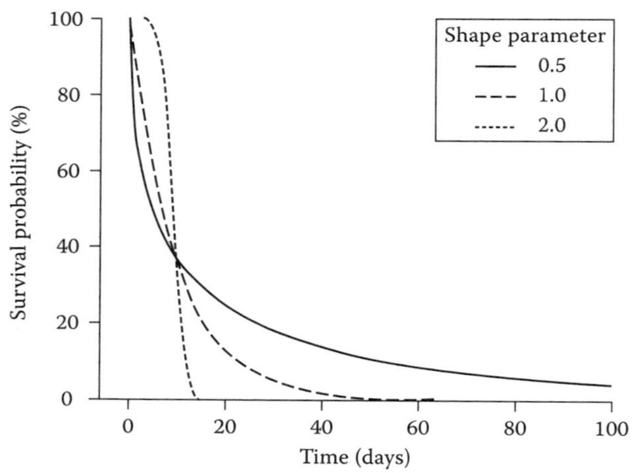

FIGURE 3.9 Three survival curves based on the Weibull distribution for different shape parameter values (scale = 10).

motivations for using the Weibull distribution rather than another one. The main reasons for the Weibull distribution's popularity are its simplicity and the versatility it provides. Another advantage is that it can easily be made linear in order to estimate its parameters from empirical data. Decades ago, without access to fast computers, a graphical analysis was carried out, and it contributed to the almost systematic use of this distribution in many fields of science. However, now that software packages offer a variety of parametric distributions, there is no real reason to fit only a Weibull distribution to data without considering other options.

3.5.2.2.2.3 Lognormal Distribution The lognormal distribution is a common model for failure times. It is in widespread use for the analysis of fracture, fatigue, and material stress (Meeker and Escobar, 1998) but does not seem suitable for many food products (Guerra et al., 2008). It simply assumes that the logarithm of failure times is normally distributed. Therefore, it is characterized by two parameters. Figure 3.10 shows two lognormal survival curves for two different values of the scale parameter.

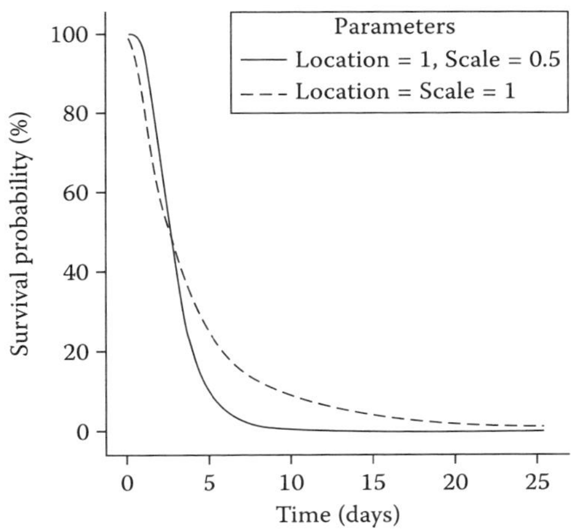

FIGURE 3.10 Survival curves based on the lognormal distribution for different parameter values.

3.5.2.2.2.4 Other Data Distributions The list of distributions discussed thus far is in no way exhaustive, and several other distributions may be considered as well. As emphasized in the next section, unless there is clear evidence or a theoretical background that justifies the use of a specific distribution, a common strategy consists of trying to fit several of the distributions available in the software used (Hough et al., 2004).

3.5.2.2.3 Practical Distribution Fitting Strategy

When parametric distributions are fitted to the data, two questions typically arise: what is the quality of the fit of a given distribution to the data and how does the experimenter select the most adequate distribution?

Two types of tools are usually combined to answer these questions. First, a goodness-of-fit statistic such as the Anderson–Darling statistic (Stephens, 1974) can be used. The Anderson–Darling statistic is a measure of the goodness of fit of the theoretical distribution to the data. It is used to quantify the distance between the KM estimation and the fitted parametric curve. The smaller the value of this test statistic, the better the fit. However, the Anderson–Darling statistic tries to summarize in a single value a complete curve and might actually miss some specific issues, for instance, in the tails of the distribution (and tails are often the most important part of such a distribution in shelf life studies). Therefore, a good complement to this goodness-of-fit measure is a probability plot. On such a plot, the scales are adjusted so that if the fit to the parametric distribution were perfect, all data points would fall on a straight line. Figure 3.11 displays such plots for four different statistical distributions to the dataset presented for the KM analysis: Weibull, lognormal, exponential, and extreme-value. These plots were generated using Minitab software release 14.

The lognormal distribution has the smallest value of the Anderson–Darling statistic, but both the Weibull distribution and the smallest extreme-value distribution are close. The plots help make a final decision as to the best-fitted distribution. Compared to the exponential distribution, which clearly does not fit the data well, both the lognormal distribution and the smallest extreme-value distribution provide a reasonable fit of the data, but not as good as the Weibull for the smallest failure times. Because this portion of the curve is of primary interest in food applications, the Weibull distribution should probably be retained here. However, it is worth mentioning that, overall, the lognormal distribution seems to be closer to the data, so for other applications, as in engineering, where a larger percentage of failures is usually accepted, it would be more appropriate to select this distribution over the Weibull.

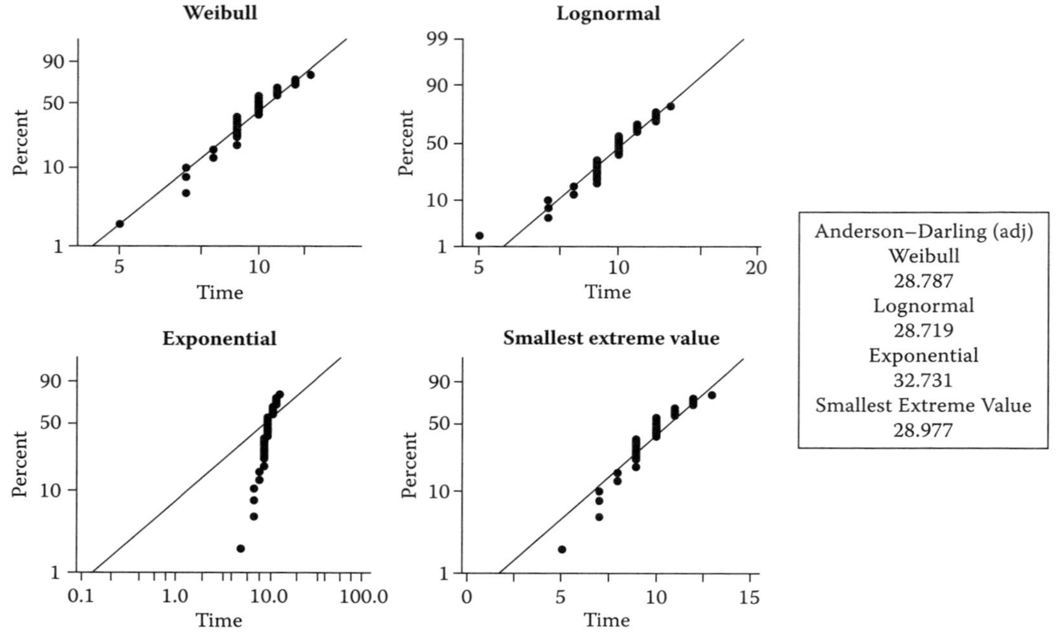

FIGURE 3.11 Probability plots to compare the fit of distributions to empirical failure times.

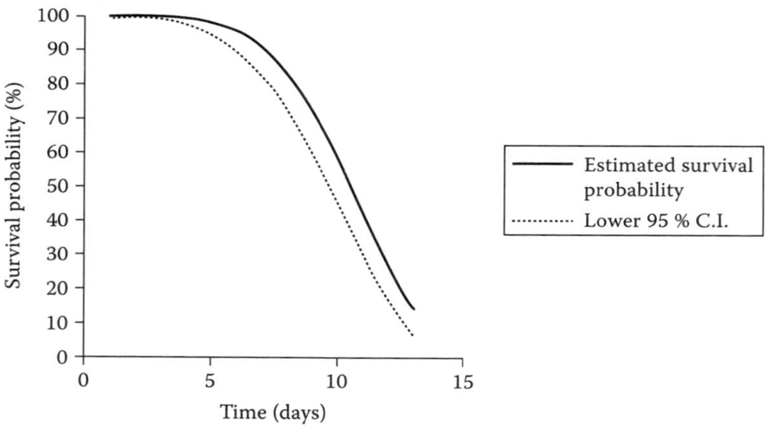

FIGURE 3.12 Parametric survival curve using a Weibull distribution and a one-sided 95% confidence interval.

3.5.2.2.4 Using Survival Curve to Predict Shelf Life

The parametric survival curve based on the Weibull distribution is presented in Figure 3.12. In this figure, a one-sided lower 95% confidence interval has been plotted as well. A one-sided interval is preferred over a two-sided interval because, in shelf life applications, the experimenter is usually more interested in the earliest likely value for shelf life to minimize the risks.

To use these curves to predict shelf life, the same strategy as for the KM estimator can be used: the experimenter needs to define the largest acceptable proportion of defects that can be tolerated. A horizontal line can then be drawn on the plot to determine the corresponding shelf life along with its 95% confidence lower bound. In practice, software packages will provide these results. If a single value has to be given for shelf life, it should be the lower bound of the confidence interval. Finally, it is worth insisting that the relevant variability to compute the 95% confidence interval is the product batch-to-batch variability rather than the variability related to another source such as consumer evaluation.

3.5.2.3 Pros and Cons of KM and Parametric Methodologies

One clear advantage of the KM methodology over a parametric approach is related to its ability to compute estimates based only on past events rather than on the complete product lifetime. This means that it can be used as soon as the first failures have been observed. Conversely, parametric methods need the entire history of failure to provide reliable estimates.

However, the survival curve obtained with parametric methods is smooth and therefore easier to work with than a step function. Also the confidence intervals obtained with KM tend to be wider than those obtained with parametric methods, provided that a satisfactory distribution has been found. Although KM performs well with larger samples, parametric methods can prove a better tool when sample size is limited.

3.5.2.4 Dealing with Competing Risks

When the product under study might fail for several independent reasons (corresponding to different measurements and failure criteria), it is said that competing risks occur. A first way to deal with such a situation is simply to ignore it. If a nonparametric approach is used to estimate the survival function, this will usually not cause any problems. However, survival curves obtained in the presence of competing risks are rarely smooth. Therefore, trying to fit a parametric distribution to failure times for all risks at once does not work well most of the time.

Instead, if the risks can be clearly identified, a more fruitful strategy is to deal with each risk separately. For this purpose, failure times corresponding to a specific risk are isolated and all other failures are included in the dataset as right-censored observations. A parametric distribution can then be fitted to these data to provide a survival curve specific to this risk. This is repeated for each risk, so that in the end the survival curves for each risk can be either kept as such or combined to estimate a global survival function.

3.5.3 Analysis of Comparative Experiments

Until now, survival curves have been estimated for a given condition (simple experiment). In the case of several conditions defined by factor levels, it is desirable to quantify factor effects. There are several model-building tools available to achieve this goal. However, a detailed description of these methods is beyond the scope of this chapter, and the references at the end of the chapter provide further insights into these methods. It is also worth mentioning that all these methods are now readily available in most statistical analysis software packages.

3.5.3.1 Analyzing Comparative Experiments Using Nonparametric Methods

The KM methodology was primarily designed for simple experiments in a case where a single survival curve needs to be estimated. Whenever a comparative experiment involves a relatively small number of conditions (say two, three, or four), this method can be generalized to compare the overall shape of the survival curves across conditions. More specifically, the statistical log-rank test was developed to compare curves (Savage, 1956).

3.5.3.1.1 Illustration of Log-Rank Test to Compare Formulations

In the following example, two groups of 21 cakes corresponding to the current and a new formulation were followed over time to assess their expected shelf life. Failure occurred as soon as mold appeared on a cake. Figure 3.13 shows the survival curve for both product formulations (the solid line is the current formulation; the dotted line is the new formulation).

Table 3.3 shows the test statistic for the log-rank test and the associated p-value. As it is significant at the 5% level ($p = 0.032$), this suggests that the overall shape of the two survival curves is not

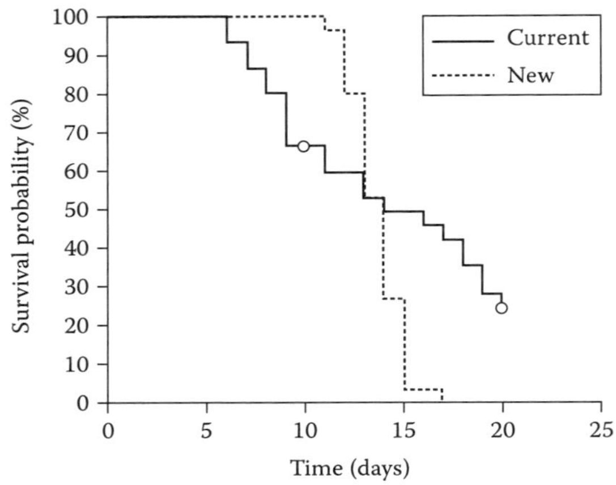

FIGURE 3.13 Survival curves for the two product formulations.

TABLE 3.3
Log-Rank Test and Observed Significance Level

Test of Equality of the Survival Distribution Functions (DF = 1):

Statistic	Observed Value	Critical Value	*p*-Value	α
Log-rank	4.584	3.841	0.032	0.050

the same. The new formulation is stable for a longer period, but when it starts to fail, the failure rate is quite steep. As food shelf life is often concerned with a limited acceptable percentage of failure, in this case the new formulation will likely be preferred over the current one.

Alternatives to the log-rank test can be found in software packages. One of them is the Wilcoxon test. Such tests are also suitable whenever the number of conditions to compare is limited. For more complex experiments including several factors or covariates, the more flexible method of Cox regression is available.

3.5.3.1.2 Semiparametric Approach: Cox Proportional-Hazards Models

Cox proportional-hazards regression allows analysis of the effect of several risk factors on survival (Cox, 1972). It is called a semiparametric method because the time effect is modeled with a nonparametric method, whereas the effects of factors and covariates are modeled in a very similar way to multiple linear regression. As a matter of fact, a risk function estimated using the KM method is used as a baseline, and the risks for a given condition are proportional to the levels of the explanatory variables (factors) entered in the model. The Cox model is by far the most widely used modeling tool in epidemiology for survival data. For shelf life studies, it can prove useful as long as the number of observations is sufficient to get a reliable estimate of the KM part of the model.

Cox regression is also useful whenever competing risks are identifiable and quantifiable. These risks can then be used as explanatory variables in the semiparametric model. This approach permits the simultaneous estimation of the survival functions while accounting for each risk. On the downside, it is rather difficult to compare the shapes of survival curves across different conditions with Cox models. A solution to this problem is to use a parametric model to model the effect of factors on the failure times. Cox models are extensively discussed in most textbooks dealing with survival analysis (see Kalbfleisch and Prentice, 2002; Smith, 2002; Lawless, 2003).

3.5.3.1.3 Parametric Models: Regression with Life Data

The underlying principle in parametric regression for life data is quite simple: a parametric distribution is used to model the failure times, and a regression model is built to explain the failure time as a function of the different factors. It is possible to include censored observations in the analysis, and the interpretation of results is similar to that of multiple-regression model results. When there are few samples in the study, this is a more efficient way of analyzing the data than relying on a nonparametric method. However, all the issues discussed in the section on parametric distribution fitting also apply to this type of model. This implies that the most appropriate parametric distribution must be selected and validated before interpreting any results from the model. Smith (2002) offers in-depth coverage of such models.

3.6 SUMMARY: BEST PRACTICES FOR SUCCESSFUL SHELF LIFE STUDIES

Shelf life data possess properties that make them different from other data collected in research and development. The two most important features are the non-normality of these data and the common occurrence of censored observations. Therefore, they cannot be analyzed using classical statistical tools such as ANOVA or linear regression. For simple experiments in which the shelf life of a product stored in well-defined conditions is tested, one can either use a nonparametric approach (the KM methodology) or fit a parametric distribution to the data. Whenever different conditions need to be compared, a variety of modeling tools generalizing the classical methods are available. These also share the specificity of being based on either nonparametric or parametric models.

Nonparametric models are more flexible as far as assumptions are concerned and can be applied to the data even if the study has not ended. However, they do require more data than parametric models to provide precise estimates. Parametric methods also provide smooth curves instead of step functions.

As far as the design of shelf life studies is concerned, specific attention is also required. First of all, as shelf life studies are focused on the analysis of time-to-event data, an exact event definition is crucial for the success of the study. If failure might occur for several reasons (competing risks), this should be anticipated, recorded in the results, and properly handled in the analysis. In the same way, if censoring is likely to occur during the study (most often right- or interval-censoring), it should not be overlooked and should again be recorded in the results and taken care of in the data analysis. If the retained measurement involves destructive testing, the design must be adjusted according to the sources of variability in the study, and a two-step sampling procedure should be considered.

However, the design of shelf life studies also allows greater flexibility for several aspects than most other experimental situations. First, if the samples are not monitored in real time and time points must be selected to evaluate them, these time points do not need to be equally spaced and should be chosen to be more frequent at times of greater change. A second important feature is that shelf life designs can be easily adjusted or augmented at any moment to improve their performance. It is therefore a good idea to store additional samples and be prepared to make such adjustments.

Several additional aspects should not be overlooked when presenting study results. First, it is crucial to state clearly the scope of the study, especially the experimental conditions and how representative they are of the target real-life situation. Second, the two (often subjective) decisions concerning the exact definition of product failure (to record time-to-event data accurately) and the percentage of acceptable failures (to extract the single shelf life estimate from the survival curve) should be presented, along with a justification of the choices that have been made. Third, when the final shelf life estimate is presented, it should be accompanied by a measure of uncertainty, typically a confidence interval. If a single value has to be given for practical reasons, it should be the lower bound of the confidence interval rather than the estimate itself.

Finally, the presentation of results, especially in industrial applications, should contain suggestions of possible improvements for the estimates. It is rarely possible to have access to unlimited resources for a given study, but the flexibility concerning the experimental design actually extends to the long run. It is acceptable and often worthwhile to set up an ongoing shelf life study by progressively adding more samples (even over several months or years), as long as this is done under similar experimental conditions. It is then possible to re-estimate shelf life by adding new data to the dataset obtained with the original study.

REFERENCES

AHD. 2000. *The American Heritage® Dictionary of the English Language*, 4th edn. Boston, Massachusetts: Houghton Mifflin Company.

Araneda M., Hough G., De Penna E.W. 2008. Current-status survival analysis methodology applied to estimating sensory shelf life of ready-to-eat lettuce (*Lactuca sativa*). *Journal of Sensory Studies* 23: 162–170.

Calle M.L., Hough G., Curia A., Gómez G. 2006. Bayesian survival analysis modeling applied to sensory shelf life of foods. *Food Quality and Preference* 17: 307–312.

Cardelli C., Labuza T.P. 2001. Application of Weibull Hazard Analysis to the determination of the shelf life of roasted and ground coffee. *LWT—Food Science and Technology* 34: 273–278.

Cox D.R. 1972. Regression Models and Life-Tables (with Discussions). *Journal of the Royal Statistical Society Series B* 34: 187–220.

EMEA (European Medicines Agency). 2002. ICH Topic Q 1 D. *Bracketing and Matrixing Designs for Stability Testing of Drug Substances and Drug Products*.

FDA. 1987. *Guideline for Submitting Documentation for the Stability of Human Drugs and Biologics*. Rockville, Maryland: Center for Drugs and Biologics, Office of Drug Research and Review, Food and Drug Administration.

Gacula M.C. 1975. The design of experiments for shelf life study. *Journal of Food Science* 40: 399–403.

Gacula M.C., Kubala J.J. 1975. Statistical models for shelf life failures. *Journal of Food Science* 40: 404–409.

Gacula M.C., Singh J. 1984. *Statistical Methods in Food and Consumer Research*. New York: Academic Press.

Greene W.H. 2008. *Econometric Analysis*, 6th edn. Upper Saddle River, New Jersey: Pearson/Prentice Hall, chapter 20.5.

Guerra S., Lagazio C., Manzocco L., Barnabà M., Cappuccio R. 2008. Risks and pitfalls of sensory data analysis for shelf life prediction: data simulation applied to the case of coffee. *LWT—Food Science and Technology* 41: 2070–2078.

Guillet M., Rodrigue N. 2005. *Efficient Design and Analysis of Shelf Life and Stability Studies*. Montreal, Canada: Course Notes from Creascience Inc.

Hough G., Garitta L., Sanchez R. 2004. Determination of consumer acceptance limits to sensory defects using survival analysis. *Food Quality and Preference* 15: 729–734.

Hough G., Langohr K., Gomez G., Curia A. 2003. Survival analysis applied to sensory shelf life of foods. *Journal of Food Science* 68: 359–362.

ICH. 1993. Stability Testing of New Drug Substances and Products. *Federal Register* 59: 48754–48759 (ICH Q1A).

ICH. 2003. International Conference on Harmonization: Evaluation of Stability Data. *Federal Register* 69(110): 32010–32011.

Kalbfleisch J.D., Prentice R.L. 2002. *The Statistical Analysis of Failure Time Data*, 2nd edn. Hoboken, New Jersey: John Wiley Series in Probability and Statistics.

Kaplan E.L., Meier P. 1958. Nonparametric estimation for incomplete observations. *Journal of the American Statistical Association* 53: 457–481.

Kececioglu D. 2003. *Reliability Engineering Handbook*, Vol 1. Englewood Cliffs, New Jersey: PTR Prentice Hall.

Lawless J.F. 2003. *Statistical Models and Methods for Lifetime Data*, 2nd edn. New York: Wiley-Interscience.

Manzocco L., Lagazio C. 2009. Coffee brew shelf life modelling by integration of acceptability and quality data. *Food Quality and Preference* 20: 24–29.

Meeker W.Q., Escobar L.A. 1998. *Statistical Methods for Reliability Data*. New York: John Wiley & Sons.

NIST/SEMATECH e-Handbook of Statistical Methods. 2008. http://www.itl.nist.gov/div898/handbook/

Ross, S.M. 1985. Statistical estimation of software reliability. *IEEE Transactions on Software Engineering* SE-11: 479–483.

Savage I.R. 1956. Contributions to the theory of rank order statistics—the two sample case. *Annals of Mathematical Statistics* 27: 590–615.

Schmidt K., Bouma J. 1992. Estimating shelf life of cottage cheese using hazard analysis. *Journal of Dairy Science* 75: 2922–2927.

Simon M., Hansen A.P. 2001. Effect of various dairy packaging materials on the shelf life and flavor of ultra-pasteurized milk. *Journal of Dairy Science* 84: 784–791.

Smith P.J. 2002. *Analysis of Failure and Survival Data*. Boca Raton, Florida: CRC Press.

Stephens M.A. 1974. EDF statistics for goodness of fit and some comparisons. *Journal of the American Statistical Association* 69: 730–737.

Yamani M.I., Abu-Jaber M.M. 1994. Yeast flora of labneh produced in bag straining of cow milk set yogurt. *Journal of Dairy Science* 77: 2558–3564.

4 Packaging and the Microbial Shelf Life of Food

Dong Sun Lee
Department of Food Science and Biotechnology
Kyungnam University
Masan, South Korea

CONTENTS

4.1 INTRODUCTION

Packaging has been a key element to preserve the quality of foods in microbiological terms. Thermal preservation became possible with the availability of retortable packaging (initially champagne bottles, then metal containers, and now multilayer plastic pouches). Aseptic packaging relies on isolating the sterilized food inside barrier packaging that has been decontaminated of microorganisms. Dry food products are protected from microbial spoilage by the barrier properties of the package, which prevent moisture transfer into the food. The shelf life of microbiologically perishable foods depends greatly on packaging variables such as gas and water vapor barrier properties, atmosphere modification, and active packaging. These variables affect the microbial flora in the food, the spoilage rate due to organisms of concern, and the time for the food to become microbiologically unacceptable. This chapter discusses the shelf life characteristics of perishable foods in relation to the packaging variables, with an emphasis on the effects of package barrier properties and modified atmosphere packaging (MAP) on microbial shelf life.

Most perishable foods are vulnerable to microbial spoilage even under chilled conditions. Their shelf life is thus, for the most part, terminated when they become unacceptable due to the growth of undesirable microorganisms. Sometimes the growth of certain microbial species may even endanger consumer safety, and therefore the potential proliferation should be avoided or strictly controlled. In addition, certain pathogenic microbes such as *Salmonella* and *Campylobacter* should be totally absent from foods that are microbiologically perishable in nature. Good manufacturing practice is assumed to prevent the contamination of food by these organisms. On the other hand, for some pathogenic bacteria such as *Bacillus cereus* and *Staphylococcus aureus*, the usual practice in food processing and distribution is to reduce the level of contamination, eliminate the potential risk of their growth to harmful levels, and thus avoid subsequent food poisoning. Alongside pathogenic organisms, certain genera of spoilage organisms dominate the microbial flora of foods, and their growth reduces sensory quality, thus limiting shelf life. As the growth of pathogenic organisms should be prevented or controlled below certain critical limits to ensure food safety, so spoilage organisms are also controlled to meet consumers' quality expectations of the food products. In the study of the microbial shelf life of food, it is generally assumed that microbial contamination and growth are controlled to avoid any health hazard. The shelf life of perishable foods is determined by the acceptable growth limit of spoilage organisms or the probability of a safe level of tolerable pathogens.

Microbial spoilage is dependent on the packaging conditions, the effect of which can be either direct or indirect. For example, high concentrations of CO_2 may directly inhibit the growth of certain microbial species, and a package that is highly permeable to water vapor may result in an increase in the moisture content of nonsterile dry food, providing an internal environment favorable for microbial spoilage. Therefore, the impact of packaging variables on the microbial shelf life of food can be understood on the basis of knowledge of the relationship between microbial spoilage and intrinsic and extrinsic factors. This chapter systematically addresses this topic by introducing and analyzing food–package–environment interactions, microbial growth kinetics, and literature data.

4.2 DEFINITION OF PARAMETERS AND TERMS FOR MICROBIAL SHELF LIFE

As with all shelf life studies, the starting point is the selection of the proper quality indices, with the index of most concern being the primary quality index. The microbial shelf life determination of food is also undertaken by identifying a fraction of the total microflora often called the *specific spoilage organisms* (SSOs) (Dalgaard et al., 2002; Koutsoumanis and Nychas, 2000). Among all the microflora, the SSOs are responsible for spoilage under a particular range of environmental conditions. The shelf life is terminated when a certain level of deterioration is reached because of the SSOs, the microbial metabolic product, or both (Mataragas et al., 2006). *Pseudomonas* spp., *Photobacterium phosphoreum*, *Shewanella putrefaciens*, *Brochothrix thermosphacta*, or *Aeromonas* spp. have been recognized as the main SSOs in fish stored under chilled conditions (Dalgaard, 1995; Taoukis et al., 1999). Lactic acid bacteria were reported to be the SSOs for vacuum-packed cooked cured meat products (Mataragas et al., 2006). Molds and yeasts were found to be the predominant microorganisms growing in natural, unpasteurized orange juices (Andres et al., 2001). Yeast growth was suspected to be the main cause of spoilage for cold-filled ready-to-drink beverages (Battey et al., 2002). Microbial spoilage of fruit products is known to be caused mostly by molds such as *Penicillium italicum* and *Penicillium digitatum* (Dantigny et al., 2005). Aerobic bacterial count has been widely used as an index for determining the microbial shelf life of many prepared foods, including meat, fish, vegetables, and cooked dishes (Buys et al., 2000; Corbo et al., 2006; Lee et al., 2008b; Vankerschaver et al., 1996). After the SSO and the range of environmental conditions under which a particular SSO is responsible for spoilage have been identified, the next step in microbial shelf life determination is to decide the population level of the SSO at which spoilage occurs and thus shelf life ends with loss of acceptability (Dalgaard, 1995; Koutsoumanis and Nychas, 2000). This step requires an understanding of the progress of microbial growth as a function of time.

Microbial growth in perishable foods can typically be represented as a function of time by the pattern shown in Figure 4.1. The growth curves are usually divided into lag, exponential, and stationary phases. This kind of segmentation of microbial growth curves is a well-established concept and can be explained by the dynamics of microorganisms in food or culture media (McMeekin et al., 1993). Storage and packaging conditions favorable for microbial spoilage result in shorter lag times and faster growth rates during the exponential phase. The cell density of the stationary growth phase may depend on the conditions: it often increases with favorable growth conditions such as increased ambient temperature but sometimes does not change with the storage conditions. Figure 4.1 presents a typical bacterial growth curve, but mold and yeast counts follow a similar

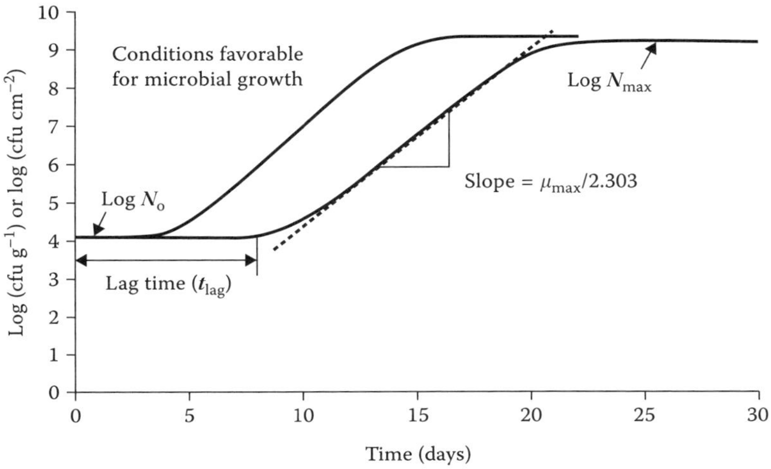

FIGURE 4.1 Typical pattern of bacterial growth on perishable food stored under constant environmental conditions.

pattern. Mold growth in radial diameter or germination percentage increase also follows a shape similar to that of Figure 4.1 (Dantigny et al., 2005).

The acceptable limit of microbial growth that determines the shelf life differs with food type, storage conditions, and defined shelf life. SSO counts of 10^5–10^8 organisms g^{-1} or cm^{-2} are commonly used as a convenient upper limit of quality and are located mostly on the linear exponential phase in Figure 4.1. For pathogenic bacteria such as *Bacillus cereus* and *Staphylococcus aureus*, 10^5 organisms g^{-1} have been used as a limit for risk management of the food supply system for prepared foods (Bahk et al., 2007; Nauta et al., 2003; Rho and Schaffner, 2007). However, the time to reach the limit based on pathogen growth should be understood as the minimum requirement for shelf life control and there should be a safety margin to give a shorter actual shelf life, this time greatly depending on the initial contamination level. Hygienic control of food preparation and processing is required so that shelf life is determined by growth of spoilage organisms rather than pathogens.

Sometimes the shelf life of foods sensitive to microbial proliferation is taken as the lag time. The start of microbial growth is often presumed to be a signal for changes in the hygienic and sensory status of the food, and thus may be taken as a conservative estimate of shelf life. The onset of an increase in bacterial count was used as a criterion for the end of the shelf life of cook-chilled or *sous vide* processed food products (Kim et al., 2002; Simpson et al., 1994). The lag time of mold or yeast growth has been used as the shelf life estimate for a prepared side dish (Lee et al., 2009).

Whether lag time or time to a level of exponential growth of the SSO is used as an estimate of shelf life, a clear and systematic determination of shelf life can be aided by describing the change in microbial population using mathematical functions (often called *primary models*). One of the most frequently used mathematical functions to describe the evolution of microbial density with time is Equation 4.1 (shown graphically in Figure 4.1), proposed by Baranyi and Roberts (1994):

$$\log N = \log N_o + \frac{\mu_{max}}{\ln(10)} \cdot A - \frac{1}{\ln(10)} \cdot \ln\left(1 + \frac{e^{\mu_{max} A} - 1}{10^{(\log N_{max} - \log N_o)}}\right) \tag{4.1}$$

where A is defined as

$$A = t + \frac{1}{\mu_{max}} \cdot \ln\left(\frac{e^{-\mu_{max} t} + 1/(e^{t_{lag}\mu_{max}} - 1)}{1 + 1/(e^{t_{lag}\mu_{max}} - 1)}\right)$$

N is the microbial count [number of organisms, usually measured in colony-forming units (cfu) g^{-1} or cm^{-2}] at time t (day), N_o is the initial density of the microbial cells (cfu g^{-1} or cfu cm^{-2}), μ_{max} is the maximum specific growth rate [inverse of the time required for the cell density to increase e (2.718)-fold, day^{-1}], t_{lag} is lag time (day), and N_{max} is the maximum cell density (cfu g^{-1} or cfu cm^{-2}).

The microbial growth model presented as Equation 4.1 can be rearranged as a differential equation to describe the instantaneous growth rate:

$$\frac{dN}{dt} = \mu_{max}\left(\frac{q}{1+q}\right)\left(1-\frac{N}{N_{max}}\right)N \tag{4.2}$$

where another state variable, q (the physiological state of the cell population), is introduced to represent the normalized concentration of an unknown substance critically needed for cell growth, whose accumulation is exponential with a specific rate of μ_{max} ($dq/dt = \mu_{max}q$).

The four parameters log N_o, t_{lag}, μ_{max}, and log N_{max} describe the progress of microbial growth over time under certain conditions. When the microbial growth pattern in Figure 4.1 is described by Equation 4.1, the parameter log N_o is presumed to be determined by the initial contamination level of the food, which is dictated by raw materials and food manufacturing conditions, whereas log N_{max} represents the maximum cell density attainable under given conditions and is usually beyond the acceptable limit of quality. Lag time (t_{lag}) and maximum specific growth rate (μ_{max}), depending on environmental conditions, directly affect the time taken to reach a certain critical level of microbial density corresponding to acceptable quality. Therefore, in dealing with the effect of packaging conditions on microbial shelf life, these two parameters are most often employed for the analysis and examined for comparative purposes. Even though the growth curve described by Equation 4.1 has curvilinear portions at the beginning and end of the exponential growth phase, the maximum specific growth rate, μ_{max}, can be assumed to represent the main part of the exponential growth. With this simplified treatment, the time (t_s) to reach a critical limit cell density of N_c, located on the exponential growth phase as the shelf life estimate, can be calculated as

$$t_s = t_{lag} + \frac{1}{\mu_{max}}\ln\left(\frac{N_c}{N_o}\right) \tag{4.3}$$

This chapter will frequently use Equation 4.3 to estimate the microbial shelf life from the kinetic parameters of microbial spoilage found in the literature.

There are other widely used primary models, such as the Gompertz and logistic functions, from which lag time and maximum specific growth rate can be similarly obtained and adopted for shelf life analysis (McKellar and Lu, 2004; McMeekin et al., 1993). As this chapter deals with the effect of packaging on the microbial shelf life of food, it will examine quantitatively the microbial growth parameters t_{lag} and μ_{max} as functions of packaging variables, leading to shelf life evaluation and analysis. More intensive treatment using complex mathematical models of the growth curve is sometimes adopted in the discipline of predictive microbiology for accurate description of microbial spoilage phenomena and for handling dynamic environmental conditions; this is beyond the scope of this chapter, and interested readers should consult McMeekin et al. (2002), Van Impe et al. (2005), and Peleg (2006).

4.3 INTRINSIC AND EXTRINSIC FACTORS AFFECTING MICROBIAL GROWTH

The growth of SSOs in packaged foods is affected by intrinsic factors (food properties) and extrinsic factors (environmental conditions inside and outside the package) (Huis is't Veld, 1996). Intrinsic factors include pH, water activity (a_w), structure, initial contamination as a result of processing conditions, and food composition such as the presence of antimicrobials. Extrinsic factors include temperature, gaseous atmosphere, relative humidity (RH), and lighting conditions.

With developments in mathematical modeling of microbial growth, the effects of intrinsic and extrinsic factors on the primary model parameters such as lag time and maximum specific growth rate have been formulated for several spoilage organisms in microbial media or typical foods. Those models are combined and sometimes captured as computer software to predict SSO growth under certain combinations of intrinsic and extrinsic factors. Examples of such software packages are ComBase Predictor® and Seafood Spoilage & Safety Predictor®. Currently, spoilage organisms covered in their growth models include *Br. thermosphacta*, *Pseudomonas* spp., *Ph. phosphoreum*, and *Sh. putrefaciens*.

4.3.1 Intrinsic Factors

The intrinsic properties of foods vary greatly with food type, food formulation, heat treatment, and hygienic status of the processing environment. Water activity and pH are determined mostly by food type and are the main influential domain variables that allow specific microorganisms to grow or proliferate on the food. Generally, microbial growth is reduced with lower a_w, but it has been reported that a high a_w (close to 1.0) sometimes reduces slightly the growth of certain organisms (Braun and Sutherland, 2003; McMeekin et al., 1993). Most foods have a pH in the range of 3–7, and a lower pH in the acidic range usually retards microbial growth. At some lower limits of a_w and pH, microbial growth eventually stops. Figure 4.2 shows the growth rate of a cocktail of fish spoilage bacteria as a function of a_w and pH. Table 4.1 presents the approximate lower limits of a_w and pH for some food poisoning and spoilage microorganisms. Lower pH or a_w favors the growth of yeasts and molds compared with bacteria (Gould, 1996; Huis is't Veld, 1996).

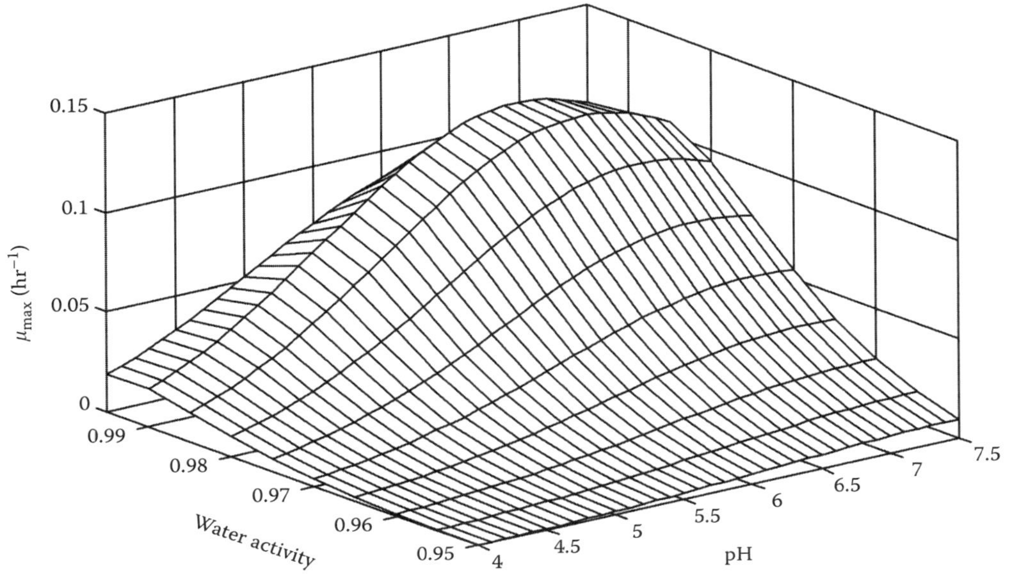

FIGURE 4.2 Maximum specific growth rate (μ_{max}) of a cocktail of *Pseudomonas* spp., *Shewanella putrefaciens*, and *Acinetobacter* spp. as a function of pH and water activity (a_w) at 5°C. (Drawn from a functional relationship reported by Braun P., Sutherland J.P. 2003. Predictive modelling of growth and enzyme production and activity by a cocktail of *Pseudomonas* spp., *Sh. putrefaciens* and *Acinetobacter* sp. *International Journal of Food Microbiology* 86: 271–282.)

TABLE 4.1
Approximate Lowest Limits[a] of Water Activity, pH, and Temperature for Growth of Some Microorganisms

Organism	Lowest Water Activity Limit	Lowest pH Limit	Lowest Temperature Limit (°C)
Bacteria			
Bacillus cereus (mesophilic)	0.93	4.9	10
Bacillus cereus (psychrotrophic)	0.93	4.9	5
Brochothrix thermosphacta	0.94	4.6	0
Campylobacter spp.	0.98	4.9	30
Clostridium botulinum (nonproteolytic)	0.97	5.0	3.3
Clostridium botulinum (proteolytic)	0.94	4.6	10
Clostridium perfringens	0.96	4.5	5
Escherichia coli	0.95	4.4	7
Lactobacillus spp.	0.93	3.0	4
Listeria monocytogenes	0.92	4.3	0
Most lactic acid bacteria	0.95	3.5	5
Pseudomonas spp.	0.97	5.0	–2
Salmonella spp.	0.95	4.0	5
Staphylococcus aureus	0.86	4.0	7
Molds			
Aspergillus flavus	0.78	2.0	3
Most molds	0.80	1.5	<0
Yeasts			
Most yeasts	0.87	1.5	–5
Saccharomyces cerevisiae	0.90	2.3	0

Source: Adapted from Leistner L., Gould G.W. 2002. *Hurdle Technologies*. New York: Kluwer Academic/Plenum Publishers, pp. 1–15; Rahman M.S., Labuza T.P. 1999. Water activity and food preservation. In: *Handbook of Food Preservation*, Rahman M. S. (Ed). New York: Marcel Dekker, pp. 339–382; Rahman M.S. 1999. pH in food preservation. In: *Handbook of Food Preservation*, Rahman M. S. (Ed). New York: Marcel Dekker. pp. 383–396; Shapton D.A., Shapton N.F. 1994. *Principles and Practices for the Safe Processing of Foods*. Oxford, UK: Butterworth-Heinemann, pp. 221–253; and other sources.

[a] Values may vary with food type and microbial strain.

Mild heat processing, such as pasteurization, inactivates vegetative microorganisms and excludes them as the cause of potential spoilage, thus limiting the spoilage microbial flora to spore formers (Gould, 1996; Huis is't Veld, 1996). The presence of preservatives may narrow down the potential list of SSOs in the food; for example, sorbic acid and its salts inhibit the growth of molds and yeasts. Figure 4.3 shows the effect of a preservative (potassium sorbate) on mold growth compared with that of a_w. The spoilage domain of intrinsic factors where SSOs are responsible for the spoilage should be examined specifically for each food item. Combined control of intrinsic factors can be utilized to preserve food safely and ensure good quality; this combination allows the food to have an adequate shelf life by incorporating low levels of additives and applying mild processes of drying and heating. Such a technique is often called *hurdle technology* (Leistner and Gould, 2002).

4.3.2 Extrinsic Factors

Storage temperature is the most influential environmental factor affecting microbial spoilage of foods. All microorganisms have an optimal temperature at which their growth is maximal. Above

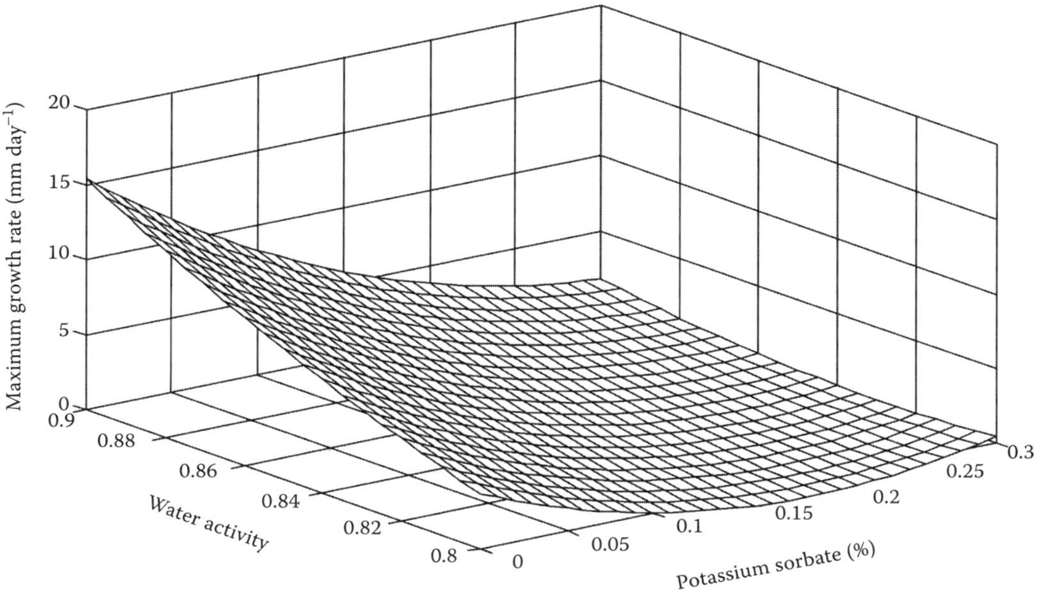

FIGURE 4.3 Maximum growth rate (μ_{max}) of a mold species, *Eurotium rubrum*, as a function of potassium sorbate concentration and water activity (a_w) at pH 5 and 25°C. (Drawn from regression analysis from the data reported by Guynot M.E., Marıin S., Sanchis V., Ramos A.J. 2005. An attempt to optimize potassium sorbate use to preserve low pH (4.5–5.5) intermediate moisture bakery products by modelling *Eurotium* spp., *Aspergillus* spp. and *Penicillium corylophilum* growth. *International Journal of Food Microbiology* 101: 169–177.)

the optimal temperature, microbial enzymes required for their growth start to be denatured or inactivated. Below the optimum, enzyme activity in the microbial system, which is proportional to temperature, is reduced to decrease the microbial growth rate (Figure 4.4). Because packaged foods are usually distributed at or below ambient temperatures, microbial inactivation at high temperatures is not taken into consideration, and thus positive dependence of microbial growth or spoilage on temperature is usually assumed in shelf life determination. Microbial activity or growth is reduced at lower temperatures and stops below certain limit temperatures, which are given for some organisms in Table 4.1.

The Arrhenius equation is widely used for describing the temperature dependence of chemical reactions and is often adopted to represent the effect of temperature on microbial growth, which is given by the inverse of the lag time or the maximum specific growth rate (growth rate):

$$\frac{1}{t_{lag}} \text{ or } \mu_{max} = A\exp\left(\frac{-E_a}{RT}\right) \tag{4.4}$$

where A is a constant (the so-called frequency factor), R is the universal gas constant (8.314 J K^{-1} mol^{-1}), T is the temperature (K), and E_a is the activation energy (J mol^{-1}). E_a values for microbial growth usually range from 50 to 90 kJ mol^{-1}.

The square root (or Bêlehrádek) equation is also widely used to describe the temperature dependence of lag time or growth rate:

$$\sqrt{1/t_{lag}} \text{ or } \sqrt{\mu_{max}} = b(T - T_{min}) \tag{4.5}$$

where b is the slope parameter of temperature dependence ($hr^{-1/2}$ °C^{-1} or $hr^{-1/2}$ K^{-1}) and T_{min} is a hypothetical minimum temperature extrapolated to zero growth (°C or K) (see Figure 4.4).

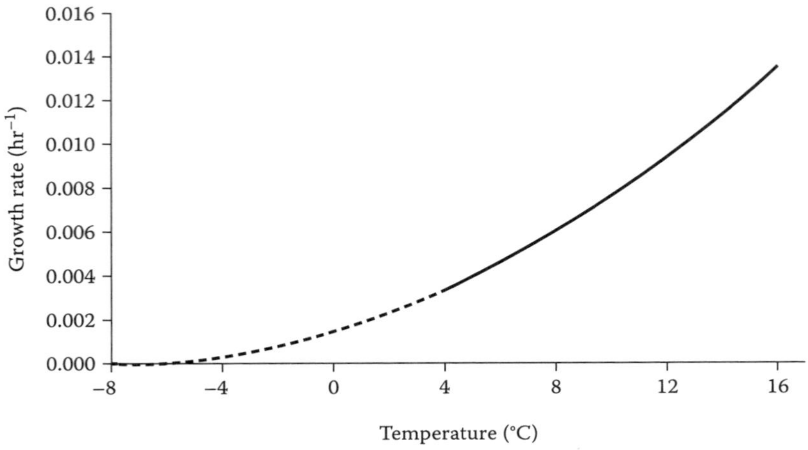

FIGURE 4.4 Temperature dependent growth rate (μ_{max}) of Pseudomonads with b of 0.00484 and T_{min} of –8.0°C (Equation 4.5) with dotted line part of extrapolation. E_a for Equation 4.4 is given as 77.2 kJ mol^{-1}. Based on the analysis of Vankerschaver K., Willocx F., Smout C., Hendricks M., Tobback P. 1996. The influence of temperature and gas mixtures on the growth of the intrinsic microorganisms on cut endive: predictive versus actual growth. *Food Microbiology* 13: 427–440.

Equation 4.5 is assumed to apply to temperature ranges from the minimum temperature to the optimal one for growth, and the T_{min} value is usually 2–3°C lower than the lower limit of growth (McMeekin et al., 1993). The T_{min} value ranges from –10°C to 7°C for most spoilage bacteria, such as Pseudomonads. However, it needs to be emphasized that chill or even freezing temperatures do not destroy or inactivate microorganisms, and storage temperature abuse after refrigeration or freezing can lead to the onset or resumption of microbial spoilage.

Package atmosphere is the second most important factor affecting the microbial growth rate. Carbon dioxide is widely used in MAP because it inhibits a wide range of spoilage and pathogenic microorganisms, particularly gram-negative bacteria and molds. Figure 4.5 shows the effect of CO_2 concentration on the growth of some spoilage bacteria. The suppression of microbial growth by CO_2 has been applied in MAP of perishable foods to extend their shelf life. However, the effectiveness of CO_2 in microbial growth suppression increases under chilled conditions, as the solubility of CO_2 in food is higher at lower temperatures (Lee et al., 2008a).

Other package atmosphere modifications also affect the rate of microbial spoilage: low O_2 concentrations below 0.21 atm slow down the growth of aerobic bacteria, yeasts, and molds. The microbial growth of aerobic microorganisms depends on dissolved O_2 in the media (corresponding to 0–0.21 atm of O_2 partial pressure in the air) and generally follows the empirical Monod equation for microbial growth (Figure 4.6A). The O_2 partial pressure of the package atmosphere, controlling the dissolved O_2 concentration, presumably affects microbial growth in a similar way; however, published studies on the kinetics of food spoilage organism growth as a function of package O_2 concentration are rare, except for simple qualitative reports on microbial spoilage (Hu et al., 2007). High O_2 concentrations (far above the normal atmospheric level) have recently been reported to inhibit some microorganisms (Figure 4.6B), and this approach has begun to be utilized for fresh produce packaging, which normally risks the creation of anoxic conditions due to respiration. The effect of superatmospheric O_2 on extending the lag time and reducing the growth rate for bacteria and yeasts has been reported to be greater in the presence of CO_2, with benefits for fresh produce packaging (Amanatidou et al., 1999; Conesa et al., 2007). It should be noted that microbial growth or activity in a hermetic package may modify the headspace gas composition by consuming O_2 and producing CO_2, which can in turn be utilized for slowing down the growth of microbes such as gram-negative Pseudomonads (Koutsoumanis et al., 2008).

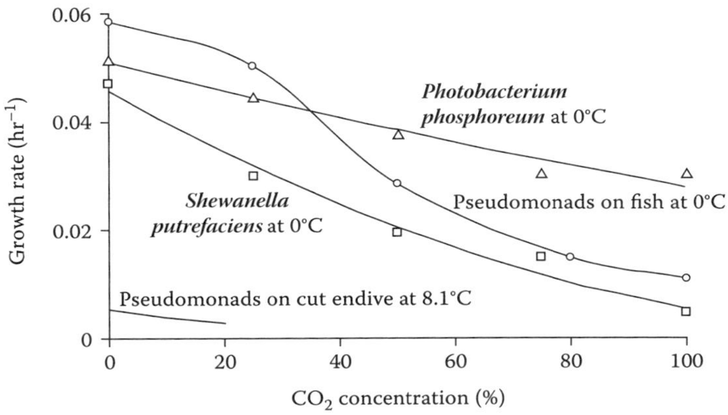

FIGURE 4.5 Effect of package CO_2 concentration on the growth rate (μ_{max}) of some SSOs. Based on experimental data and analyses of Pseudomonads on red mullet fish at 0°C (Koutsoumanis et al., 2000), Pseudomonads on cut endive at 8.1°C (Vankerschaver et al., 1996) and *Sh. putrefaciens* and *Ph. phosphoreum* in cooked fish muscle juice and broth medium, respectively, at 0°C (Dalgaard, 1995).

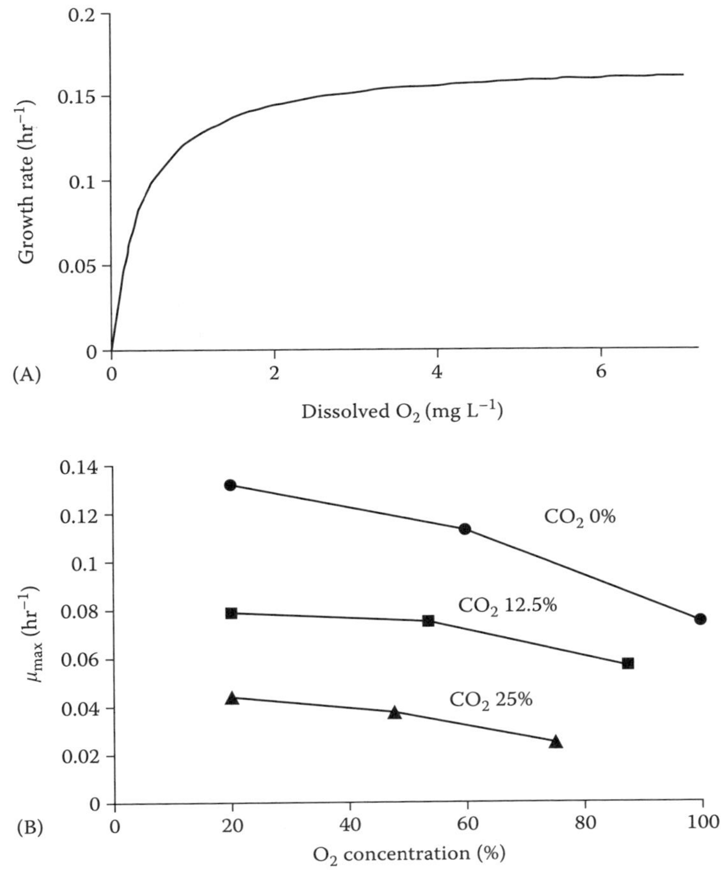

FIGURE 4.6 Effect of oxygen on aerobic bacterial growth. (A) Specific growth rate of a *Pseudomonas* sp. at 30°C as a function of dissolved oxygen concentration (lower range corresponding to 0–0.21 atm of O_2 partial pressure) according to the kinetic information given by Ferreira Jorge and Livingston (1999). (B) Maximum growth rate (μ_{max}) of *Pseudomonas fluorescens* depending on superatmospheric oxygen concentration as reported by Geysen et al. (2005).

Other volatiles produced by fresh produce or deliberately delivered in small amounts into the package headspace by active packaging can inhibit microbial growth and thus extend the shelf life. Ethanol, hexanal, allyl isothiocyanate, and jasmonates are typical volatiles used to reduce fungal decay during the storage of fruits, vegetables, and cakes. Although some of these compounds are produced by senescent metabolic activity of fresh produce or degradation of other foods and may sometimes be undesirable, their selective removal and addition may be helpful for the preservation of perishable foods (Toivonen, 1997). Because active packaging is dealt with in Chapter 20, only a brief mention is made here.

Although the deteriorative effect of light (related to package transparency) on the chemical quality of packaged foods has been studied extensively (Robertson, 2006), its effect on microbial quality is rarely reported in the literature, and package transparency does not appear to have a significant influence on microbial activity. As a specialized application, TiO_2-coated films activated by light have been reported to inhibit microorganisms (Chawengkijwanich and Hayata, 2008). The mechanism is presumed to be oxidation of the polyunsaturated phospholipid component of the microbial cell membranes by hydroxyl radicals and reactive oxygen species generated by illumination of the TiO_2 surface.

4.4 FOOD FACTORS AND MICROBIAL ECOLOGY INFLUENCED BY PACKAGING

Intrinsic food factors are changed by the interaction of the packaged food with its packaging and the external environment. The headspace may work as a buffer between the food and the packaging material. Through all these interactions, moisture content (i.e., water activity), dissolved O_2 and CO_2 contents, and preservative concentration may be changed to affect the microbial flora and growth rate. Figure 4.7 shows a schematic of food–package–environment interactions in relation to microbial growth. In combination with other environmental factors such as temperature, the resulting domain may be located within the growth/no growth boundary or outside it; the possibility for some spoilage and pathogenic organisms to grow in certain foods may be excluded or neglected depending on the domain of intrinsic food factors and temperature. The food factors and temperature will also determine the relative growth rates of microbial species able to spoil the food. In the case of foods exposed to dynamic temperatures, thermal insulation may help to reduce the impact of the temperature variations.

When a packaged food is set up under initial conditions of preparation, microbial contamination, moisture, and package headspace composition, with a specific packaging material and stored at a specific temperature, permeation of water vapor, O_2, and CO_2 determines their respective concentrations inside the package, which in turn determines the moisture, O_2, and CO_2 sorbed or dissolved into the food (Figure 4.7). Usually it can be assumed that the transmission rates of these vapors or gases control the variation of these intrinsic food factors and equilibrium is maintained between the food and the headspace. The high solubility of CO_2 gas in wet or fatty foods may significantly change its initial concentration in the headspace after packaging, particularly when the headspace volume is not large. Whereas the outside gas concentrations of O_2 and CO_2 are taken as constant, humidity in the environment can vary with time. If the packaged food does not make any contribution to O_2 and CO_2 changes in the system, a gas barrier package can be used to maintain constant levels of O_2 and CO_2 in the headspace and thus in the food. High microbial activity, fresh produce respiration, and O_2 absorption by active packaging components may work as other variables affecting microbial growth, as may elaborate migration of preservative or vaporization of volatiles into the headspace.

Together with the intrinsic properties of the food, the storage temperature of the package determines the microbial population outgrowth: ambient storage favors mesophiles, and chilled storage results in the dominance of psychrotrophic organisms. The relative growth rates of microbial species, resulting from the different package conditions, determine the dominant microbial flora on

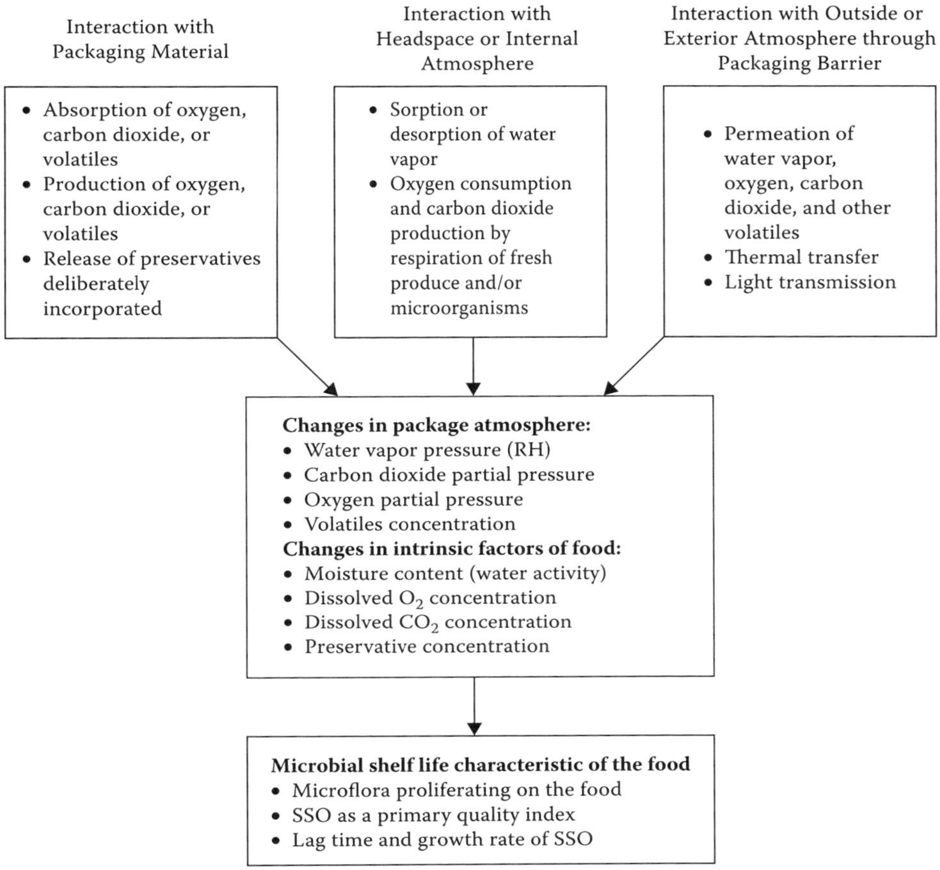

FIGURE 4.7 Food–package–environment interactions affecting the microbial shelf life characteristics of packaged foods.

TABLE 4.2
Influence of Package Atmosphere on the Microbial Group Counts of Pork Meat Stored at 4°C

Package Atmosphere and Storage	Aerobic Bacteria (log cfu cm^{-2})	Lactic Acid Bacteria (log cfu cm^{-2})	Coliforms (log cfu cm^{-2})
Initial	3.3	0.3	0.8
Air for 7 days	8.0	1.4	4.4
N_2 for 7 days	5.9	2.3	3.2
CO_2 for 7 days	3.6	2.4	0

Source: Adapted from Enfores S.-O., Molin G., Ternstrom A. 1979. Effect of packaging under carbon dioxide, nitrogen or air on the microbial flora of pork stored at 4°C. *Journal of Applied Bacteriology* 47: 197–208.

the food. As shown in Tables 4.2 and 4.3, low O_2 or anoxic packages generally favor the growth of anaerobes or microaerophiles on the packaged foods and high CO_2 concentrations favor gram-positive bacteria (such as lactic acid bacteria) over gram-negative ones (such as *Pseudomonas* spp. and coliforms) (Banks et al., 1980; Enfores et al., 1979). Although it strongly inhibits the growth of gram-negative bacteria and molds, the inclusion of CO_2 in the package can also inhibit or delay, to some extent, the growth of certain gram-positive bacteria and yeasts, and thus the microbial flora will differ from that in air (Table 4.3).

Temperature significantly affects the effectiveness of microbial inhibition of modified atmospheres (MAs) (Table 4.4). Compared with the anaerobic conditions of vacuum or N_2, a CO_2 atmosphere is the most effective at low temperatures. As mentioned, intrinsic factors such as pH and a_w also work to confine or decide the microbial spoilage types for perishable foods. Microbial

TABLE 4.3
Influence of Package Atmosphere on the Microbial Flora of Pork Meat Stored at 4°C

Organisms	Percentage of Organisms At			
	Initial State	7 Days in Air	10 Days in N_2	7 Days in CO_2
Acinetobacter calcoaceticus	49			30
Aeromonas hydrophila	3		5	
Bacillus subtilis		3		
Enterobacter liquefaciens	3			
Flavobacterium sp.	8			
Flexibacter sp.	3			
Kurthia zopfii	2		10	5
Lactobacillus plantarum			5	20
Micrococcus varians	2			
Moraxella sp.	5			
Pediococcus pentosaceus	1			
Pseudomonas spp.	24	97	80	45
Staphylococcus xylosus	1			
Aerobic bacterial count (log cfu cm^{-2})	3.3	8.0	7.3	3.6

Source: Adapted from Enfores S.-O., Molin G., Ternstrom A. 1979. Effect of packaging under carbon dioxide, nitrogen or air on the microbial flora of pork stored at 4°C. *Journal of Applied Bacteriology* 47: 197–208.

TABLE 4.4
Effectiveness of Anaerobic Packaging Conditions Compared to Air Packaging at Different Temperatures

Package Atmosphere	Growth of Microorganisms as Percentage of Growth in Air after 23 Days at Temperature (°C):		
	2	6	20
Pseudomonas/Acinetobacter/Moraxella			
Vacuum	95	78	33
N_2	82	69	76
CO_2	0	0	52
Brochothrix thermosphacta			
Vacuum	60	69	63
N_2	45	57	63
CO_2	0	0	46

Source: Adapted from Eklund T., Jarmund T. 1983. Microculture model studies on the effect of various gas atmospheres on microbial growth at different temperatures. *Journal of Applied Bacteriology* 55: 119–125.

ecology resulting from storage temperature, package atmosphere, and intrinsic factors determines the primary microbial index to be used for shelf life determination. For example, counts of lactic acid bacteria have often been used as quality criteria for shelf life determination of chill-stored cooked meats and fresh vegetables packaged in vacuum, low O_2, or high CO_2 MAs, and counts of mesophilic or psychrotrophic aerobic bacteria have been used for aerobic packages of fresh meat, fish, and vegetables. Products with low a_w and pH, such as processed meats and pasta, in air or O_2-permeable packages are spoiled by growth of molds and yeasts. More specifically, *Pseudomonas* spp. were identified as SSOs for minimally processed vegetable products, milk, meat, and fish packaged under O_2-containing atmospheres, with their counts used to determine shelf life; *Ph. phosphoreum* has been used for MA-packed fresh fish; *Lactobacillus sake*, for high-CO_2-packed meat products; *Sh. putrefaciens* or *Br. thermosphacta*, for CO_2-packed fish and meat; and *Leuconostoc* spp., for high O_2 MA-packaged beef steak (Berruga et al., 2005; Devlieghere et al., 1999; Koutsoumanis et al., 2000; McMeekin and Ross, 1996; Sheridan et al., 1997; Vihavainen and Bjorkroth, 2007). Again, it must be emphasized that SSOs and primary microbial quality indices are determined by a combination of intrinsic and extrinsic factors, including packaging conditions.

4.5 EFFECT OF PACKAGE GAS BARRIER ON MICROBIAL SHELF LIFE

For MA packages (vacuum or gas flushed) designed to suppress the growth of aerobic microorganisms and oxidative quality changes by excluding O_2, packaging materials with a poor gas barrier act to promote microbial growth of aerobes and facultative anaerobes (Kotzekidou and Bloukas, 1996; Newton and Rigg, 1979). Even microaerophiles such as *Lactobacillus* spp., which dominate in vacuum and CO_2 packaging of meat products, may have enhanced growth rates with higher O_2 transmission film or packaging (Tsigarida and Nychas, 2006). The effect of gas permeability on microbial spoilage is seen clearly in Figure 4.8, in which a *sous vide* package with a high oxygen transmission rate (OTR) favors the growth of aerobic and anaerobic bacteria. The high microbial load consisting of thermoduric *Bacillus* spp., facultative anaerobes that survived the pasteurization process, was presumed to have been responsible for the microbial spoilage (Gould, 1996; Kim

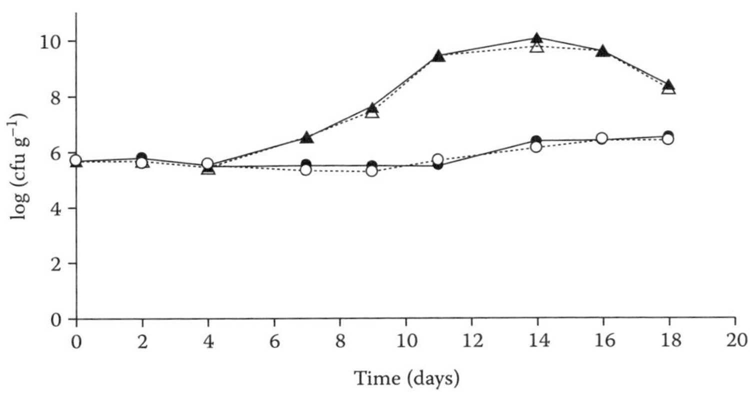

FIGURE 4.8 Effect of gas permeability on evolution of aerobic and anaerobic bacterial counts of *sous vide* packaged seasoned spinach soup (600-g pouch pack) at 10°C containing thermoduric organisms. ▲: aerobic bacteria with high-O_2-permeability film package (OTR 6.3 mL m^{-2} hr^{-1} assumed at 1 atm of O_2 partial pressure differential); Δ: anaerobic bacteria under high-O_2-permeability film package; ●: aerobic bacteria under low-O_2-permeability film package (OTR 2.3 mL m^{-2} hr^{-1}); ○: anaerobic bacteria under low-O_2-permeability film package. (Adapted from Kim G.T., Paik H.D., Lee D.S. 2003. Effect of different oxygen permeability packaging films on the quality of *sous vide* processed seasoned spinach soup. *Food Science and Biotechnology* 12: 312–315.)

et al., 2003). When the microbial lag time was used to estimate shelf life in Figure 4.8, a gas barrier film package with three times less O_2 permeation extended the shelf life to twice that for the more permeable one. In a comparison of an O_2-permeable package in air with a vacuum package with a high gas barrier (two extremes in O_2 transmission), the latter could have an extended shelf life, particularly at lower temperatures, as shown in Figure 4.9.

Figure 4.10 presents shelf life estimates of vacuum-packed meat based on different microbial criteria. Shelf life based on *Pseudomonas* growth (which correlated strongly with sensory odor) was a strong function of the gas permeability of the packaging film, whereas growth of lactic acid

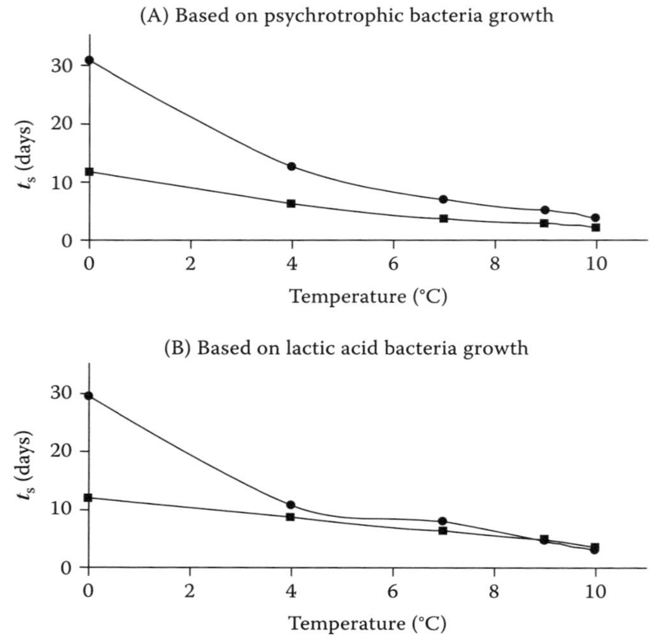

FIGURE 4.9 Microbial shelf life (t_s) of chilled packaged beef estimated from microbial growth model parameters reported by Giannuzzi et al. (1998). The t_s was obtained according to Equation 4.3 as the time for psychrotrophic and lactic acid bacteria to increase by 10^3- and 10^2-fold, respectively. ●: high-barrier poly(vinylidene chloride) (PVdC) vacuum package; ■: gas-permeable low density polyethylene air package.

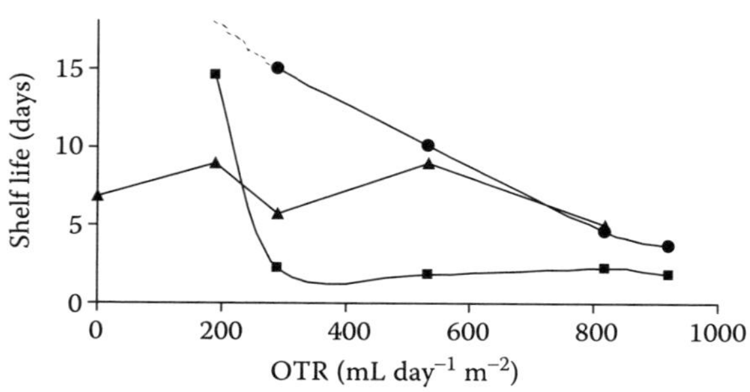

FIGURE 4.10 Different criteria shelf lives of vacuum-packed meat at 0°C as a function of OTR at 25°C and 100% RH. ●: time to reach 5×10^5 organisms g^{-1} of *Pseudomonas*; ▲: time to reach 10^6 organisms g^{-1} of *Lactobacillus*; ■: time to reach 10^5 organisms g^{-1} of *Brochothrix thermosphacta*. Values were read from graphical data of Newton and Rigg (1979), with dotted line being an extrapolated value.

bacteria to 10^6 organisms g^{-1} depended little on the gas permeability. Growth of lactic acid bacteria on vacuum-packed cooked meat was also shown by others to be little affected by the OTR of the packaging film (Cayre et al., 2005; Kotzekidou and Bloukas, 1996). *Br. thermosphacta* growth could be suppressed significantly only with very high gas barrier film (Figure 4.10). The behavior of *Br. thermosphacta* with different gas barrier packages was similar to that reported by Kotzekidou and Bloukas (1996) for vacuum-packed cooked ham, but not to that by Cayre et al. (2005), who reported a dramatic decrease in *Br. thermosphacta* count after earlier maximum with high OTR film for vacuum-packed cooked meat emulsions. A difference in intrinsic factors and microbial interactions may have resulted in different microbial spoilage behaviors even under similar packaging conditions.

In the case of CO_2 packaging of a meat product, high gas barrier packaging was found to be effective in extending the shelf life consistently over the chilled temperature range (Figure 4.11). A comparison between Figures 4.9 and 4.11 shows the temperature dependence of the gas barrier effect to be more pronounced with vacuum packaging than with CO_2 packaging. Some accumulation of CO_2 from microbial activity of the product in vacuum packages may explain the greater effect of barrier packaging at lower temperatures, due to the higher antimicrobial activity of CO_2 at lower temperatures, as explained earlier. Koutsoumanis et al. (2008) also observed a greater effect of high gas barrier film on extending the shelf life of minced pork at lower temperatures that resulted from an MA arising from microbial respiration. Although gas barrier properties are an important variable to reduce the microbial growth rate for vacuum or gas-flushed packages, there seems to be a threshold value of OTR below which microbial growth or shelf life is independent of these. Tsigarida and Nychas (2006) showed that shelf life was not extended further below an OTR of 28 mL day^{-1} m^{-2} based on a 1-atm partial pressure difference (23°C and 75% RH). Generally speaking, the packaging material should provide enough of a barrier to protect against O_2 ingress or CO_2 loss, or both, to have the desired effect from vacuum or gas-flushing.

Respiring fresh produce can retain better microbial quality when the design of the package is such that it can attain the optimal equilibrated atmosphere. This requires a selective range of permeabilities to O_2 and CO_2. As long as the physiological tolerance limits for O_2 and CO_2 concentrations are not violated, higher CO_2 and lower O_2 (below 21%) help ensure lower counts of spoilage organisms such as *Pseudomonas* (Charles et al., 2005; Simon et al., 2005; Vankerschaver et al., 1996).

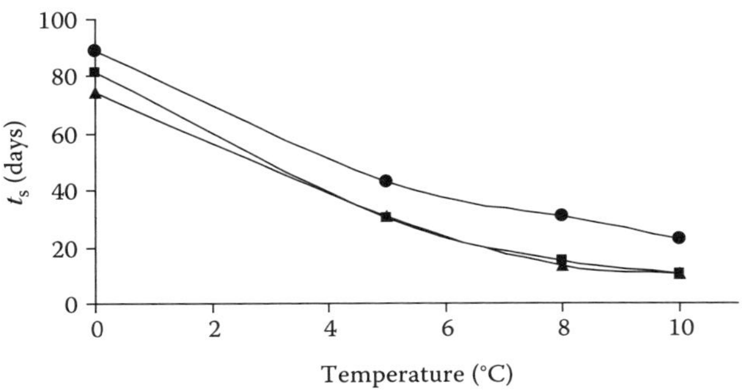

FIGURE 4.11 Shelf life (t_s) of 100% CO_2-packed meat fillets at different temperatures versus OTR of packaging film at 23°C and 75% RH. The t_s was calculated according to Equation 4.3 as the time for *Lactobacillus* spp. to increase by 10^3-fold by using the microbial growth kinetic parameters given by Tsigarida and Nychas (2006). ●: OTR 28 mL day^{-1} m^{-2}; ■: OTR 2600 mL day^{-1} m^{-2}; ▲: air pack.

4.6 MODIFIED ATMOSPHERE PACKAGING TO EXTEND MICROBIAL SHELF LIFE

MAP is an effective technique to preserve perishable chilled foods without resorting to heat processing or chemical preservatives. As discussed earlier, the preserving effect of MAP derives mainly from the use of CO_2 gas, which inhibits or retards microbial growth significantly above a concentration of 20%. Sometimes a lower O_2 concentration is used alone in packaging with a vacuum or in combination with a high CO_2 concentration. Low O_2 and moderate accumulation of CO_2 provide some degree of microbial inhibition in the permeable packaging of fresh produce with a reduction of physiological senescence. Recently, high O_2 concentrations have been applied mainly for fresh-produce packaging to inhibit microbial growth without the creation of anoxic conditions. As many studies have already reported the beneficial effects of MAP, this chapter gives an overview of the effect of MAP conditions on the extension of microbial shelf life by compiling specific data for a variety of foods.

The dependence of microbial shelf life on superatmospheric O_2 and moderate CO_2 concentrations for fresh produce is presented in Figure 4.12, which estimates shelf life according to Equation 4.3 using growth kinetic parameters from the literature (Geysen et al., 2006). Superatmospheric O_2 concentrations are more effective with high CO_2 concentrations in extending microbial shelf life: the maximum extended shelf life at 100% O_2 and 20% CO_2 (hypothetical concentrations corresponding to respective partial pressures of 1.0 and 0.2 atm for O_2 and CO_2) is about six times that in a normal atmosphere (20% O_2 and 0% CO_2). Use of higher levels of CO_2 for fresh fruits and vegetables is limited because of the physiological tolerance limit of the commodity (mostly below 20%). Given the microbial inhibitory effect of superatmospheric O_2, its use eliminates the risk of O_2 depletion inside the fresh produce package because of respiration activity, particularly under temperature-abuse conditions.

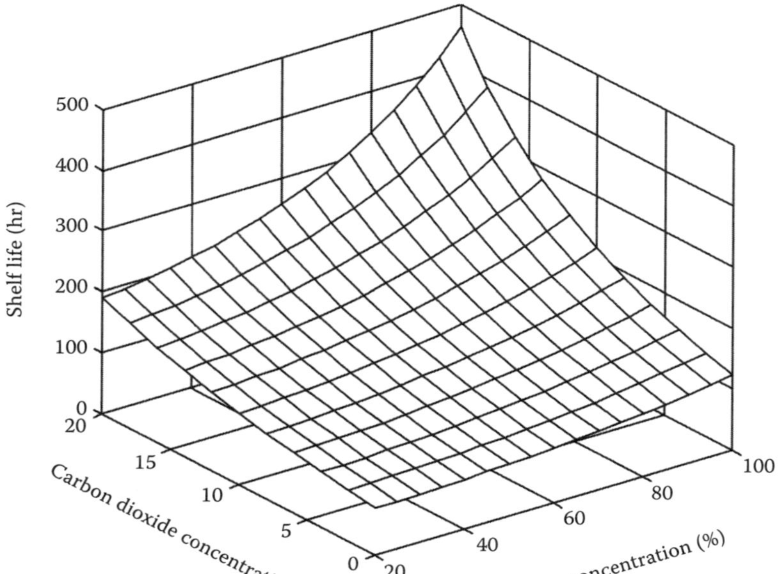

FIGURE 4.12 Estimated shelf life under MA conditions of superatmospheric O_2 and moderate CO_2 concentrations when packaging fresh-cut lettuce at 7°C. Equation 4.3 was used to obtain the time for *Pseudomonas fluorescens* to increase $10^{3.5}$-fold by using the microbial growth model parameters given by Geysen et al. (2006).

Although some level of O_2 may be used to bloom or retain the bright red color of fresh meat, the effectiveness of microbial inhibition for flesh and nonrespiring food products is mostly due to CO_2, the effect of which is more pronounced at lower chill temperatures. Figure 4.13 shows the effect of CO_2 concentration on the extension of microbial shelf life on the basis of the growth of Pseudomonads on fish. The effectiveness of shelf life extension using 100% CO_2 (compared with 0% CO_2) at 0°C was found to be about 480% (1130 vs. 196 hr). This effectiveness of CO_2 on fish Pseudomonads is much higher than the time period estimates for *La. sake* to go from 5×10^2 to 10^7 cells mL^{-1} in modified brain heart infusion medium at 4°C (about a 70% extension with dissolved CO_2 concentration of 2000 ppm when compared to 0 ppm CO_2) (Devlieghere et al., 1998). Gram-negative Pseudomonads would be more sensitive to CO_2 than gram-positive, microaerophilic *La. sake*.

Apart from the theoretical kinetic analysis given earlier, practical MAP applications for extending microbial shelf life use a variety of conditions for preserving the overall quality attributes of commodities. Table 4.5 is a compilation of MAP applications for extending the shelf life of perishable foods. The effectiveness of microbial shelf life extension varies with product and the actual MA applied. However, it is apparent that CO_2 inclusion is an important variable to give a significant shelf life extension for meat, fish, dairy, and prepared products. Bulk packaging or master packs of individual retail gas-permeable packs under a CO_2 atmosphere are often used for extending the total shelf life of flesh products from the wholesaler through retail display to consumers. A residual effect of CO_2 in inhibiting microbial spoilage after moving individual packs to air has sometimes been observed, presumably due to the microbial ecology formulated with MAP (Dixon and Kell, 1989). However, specific data on its effect on shelf life are not available in the published literature.

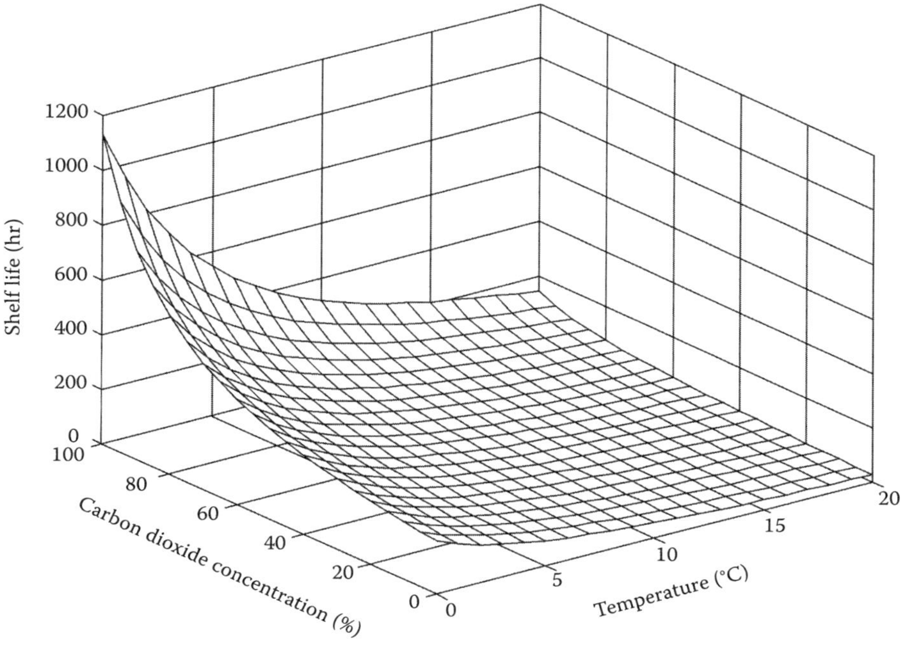

FIGURE 4.13 Shelf life estimates as a function of CO_2 concentration ([CO_2]) and temperature (T), based on the time for Pseudomonads to increase 10^4-fold. The maximum specific growth rate (μ_{max}) given in a Bêlehrádek-type equation $\left(\sqrt{\mu_{max}} = 0.00173(T + 11.4)\sqrt{120.9 - [CO_2]}\right)$ by Koutsoumanis et al. (2000) was used to calculate the shelf life by Equation 4.3 with absence of lag time.

TABLE 4.5
Compilation of Data on Microbial Shelf Life Extension Achieved by Active MAP of Food

Food	Package and Storage Conditions	Criterion for Microbial Quality Limit (cfu g^{-1} or cfu cm^{-2})	Increment of Shelf Life Extension Compared to Control (Air) Package	Reference
Meats				
Lamb meat	80% CO_2/20% N_2 at 2°C	10^7 of aerobic bacteria	140% compared to vacuum pack	Berruga et al. (2005)
Minced chicken meat in combination with irradiation	20% CO_2/80% N_2 at 10°C	10^7 of aerobic bacteria	80%	Grandison and Jennings (1993)
Ostrich steaks	30–80% CO_2/balance N_2 at 2°C	10^7 of aerobic bacteria	50%	Fernandez-Lo´pez et al. (2008)
Pork	70% CO_2/29.5% N_2/0.5% CO at 0–2°C	10^6 of aerobic bacteria	410%	Krause et al., (2003)
Pork sausage	99.6% CO_2/0.4% CO at 4°C	10^7 of aerobic bacteria	100%	Laury and Sebranek (2007)
Smoked turkey breast fillets	30% CO_2/70% N_2 at 4°C	10^7 of aerobic bacteria	30%	Ntzimani et al. (2008)
Fish				
Freshwater crayfish	80% CO_2/20% air at 4°C	10^7 of aerobic bacteria	>100%	Wang and Brown (1983)
Gutted bass	50% CO_2/30% O_2/20% N_2 at 3°C	10^4-fold increase in aerobic bacterial count	>>30%	Torrieri et al. (2006)
Pearlspot fish	60% CO_2/40% N_2 at 0–2°C	10^7 of psychrotrophic bacteria	>120%	Ravi Sankar et al. (2008)
Rock fish fillets	80% CO_2/20% air at 1.7°C	10^7 of aerobic bacteria	>200%	Parkin et al. (1981)
Shrimp	100% CO_2 at 4°C	10^7 of aerobic bacteria	>200%	Lannelongue et al. (1982)
Dairy products				
Cottage cheese	100% CO_2 at 8°C	10^2 of coliform bacteria or 10^3 of yeasts/molds	160%	Mannheim and Soffer (1996)
Sliced Mozzarella cheese	100% CO_2 at 7°C	10^7 of yeasts	>420%	Alves et al. (1996)
Whey cheese	40% CO_2/60% N_2 at 4°C	10^7 of mesophilic bacteria	310% compared to vacuum or air pack	Dermiki et al. (2008)
Fresh fruits and vegetables				
Endive	3% O_2/5% CO_2 at 20°C	10^6 of *Pseudomonas* spp.	>50%	Charles et al. (2005)
Fresh-cut lettuce	100% N_2 at 1, 5, and 10°C	10^7 of total aerobic bacteria	Negligible	Koseki and Itoh (2002)
Fresh-cut pineapple	40% O_2/balance N_2 at 5°C	10^7 of mesophilic bacteria	>30%	Montero-Calderon et al. (2008)
Fresh-cut baby spinach	100% O_2 at 5°C	10^8 of mesophilic bacteria	220%	Allende et al. (2004)
Fresh-cut spinach	0.8% O_2/10% CO_2 at 5°C	10^8 of mesophilic bacteria	80%	Babic and Watada (1996)
Shredded chicory endive	95% O_2/5% N_2 at 4°C	10^5 of yeasts	100% compared to passive MAP	Jacxsens et al. (2001)

(*Continued*)

TABLE 4.5 (*Continued*)

Food	Package and Storage Conditions	Criterion for Microbial Quality Limit (cfu g^{-1} or cfu cm^{-2})	Increment of Shelf Life Extension Compared to Control (Air) Package	Reference
Prepared or miscellaneous foods				
Carrot juice	100% CO_2 at 17°C	10^6 of aerobic bacteria	109%	Alklint et al. (2004)
Fermented, seasoned soused roe of Alaska pollack	30% CO_2/70% N_2 at 10°C	10^7 of yeasts	>500%	Lim et al. (2002)
Fresh wet pasta	22% CO_2/78% N_2 at 8°C	10^6 of aerobic bacteria	>150%	Lee et al. (2001)
Korean braised green peppers with dry anchovies	60% CO_2/40% N_2 at 10°C	10^5 of total aerobic bacteria	130%	Lee et al. (2008b)
Korean braised kidney beans	60% CO_2/40% N_2 at 10°C	Lag time of yeast/mold growth	500%	Lee et al. (2009)

For meat packaging, inclusion of small amounts of carbon monoxide (CO) has been found to give a further extension of microbial shelf life, with the added benefit of red color stabilization (Laury and Sebranek, 2007).

The microbial stability of fresh produce can benefit greatly from a combination of superatmospheric O_2 and self-produced CO_2 (Table 4.5). Although this approach has recently received considerable interest from researchers, it has not yet been adopted by industry. Today, fresh produce packaging depends on low O_2 and slightly increased CO_2 concentrations to prolong freshness by reducing respiration and softening. Some volatile compounds, such as methyl jasmonate, have been reported to decrease the fungal decay of minimally processed fruits in the package, but no specific data on shelf life extension can be found in the literature. Even though nonconventional gases such as Ar, Xe, and N_2O have also been applied for fresh-cut fruits to minimize physiological changes, their effect on microbial quality has not been reported. More work is needed to examine these aspects of microbial inhibition and physiological preservation.

4.7 PACKAGING TOOLS TO MONITOR MICROBIAL SHELF LIFE

Because of its importance in shelf life control and its high dependence on food distribution conditions, real-time monitoring of microbial quality of packaged food in the food supply chain has been desired and tried for a long time. The destructive measurement of microbial counts is very time consuming and requires laborious laboratory testing, and thus is not feasible for practical application in food logistics. Therefore, there have been attempts to use sensor technology to monitor microbial quality, detect spoilage, or predict shelf life under dynamic distribution environments.

A simple but reasonable approach is to predict the microbial quality change on the basis of the temperature history experienced in food supply chain. The quality change in response to temperature fluctuations can be expressed or shown as a color change in a time-temperature indicator (TTI) or the remaining shelf life predicted using a digital device. This approach requires shelf life prediction models for the particular food as a function of environmental conditions. Currently, temperature is the only variable that has been successfully taken into consideration. A TTI can be attached as a label on the package surface to respond to the external temperature, indicating the microbial quality change whose kinetics parallel the indicator color change. Taoukis et al. (1999) developed an algorithm for controlling the stock rotation and shelf life of chilled fish to have better quality delivered to consumers by using a TTI that responded in the same way as microbial deterioration. Smolander et al. (2004) observed a high correlation between microbial quality of chicken cuts and TTI color change, which can be a useful tool to estimate the shelf life in real time. There is a need for a variety of TTIs to represent the microbial quality change due to SSO growth for many different foods. Common TTIs available commercially include Fresh Check™ and Vitsab™. Other temperature-sensing data loggers or devices can also be used for similar purposes. Recently, a radio frequency identification (RFID) tag incorporating a temperature sensor with data communication and calculation functions has been proposed and is being developed to support information management in the food supply chain. For a TTI or RFID tag system to be applied widely for shelf life detection in the food supply chain, more quantitative kinetic data and models for microbial food spoilage need to be accumulated.

Microbial food spoilage accompanies changes in the concentration of metabolic substrates and products, producing discoloration, textural changes, slime formation, and off-flavor development. Attempts to measure or monitor microbial food quality more directly use substrates or products of microbial growth or spoilage as the index. Typical indices for microbial spoilage are glucose, organic acids such as gluconic acid and lactic acid, ethanol, biogenic amines, volatile nitrogen compounds, adenosine triphosphate (ATP) degradation products, several alcohols, and H_2S (Dainty, 1996; Ellis and Goodacre, 2001). Generally, carbohydrates are degraded before amino acids and lactic acid are metabolized by microorganisms to impair the sensory quality of proteinaceous foods. The label or tag attached inside a transparent package surface is equipped with a sensor to measure one of these compounds closely related to food spoilage. The sensor label or tag is designed to cause a color change by reaction with one of these metabolites in the spoiled food. Among the metabolites, measurement of the volatile compounds accumulated in the package headspace is more suited for shelf life control of packaged perishable foods, because this does not require direct contact between the food and the sensor. Some prototype products warning of microbial spoilage are available commercially. Recently an electronic nose sensor has been tried for this purpose.

Regarding volatile compounds as microbial spoilage indicators, CO_2 and 3-methyl butanol have been found to be highly correlated with growth of *Br. thermosphacta* (Sutherland, 2003). Guerzoni et al. (1990) showed a high correlation between CO_2 production and *Saccharomyces cerevisiae* growth, which leads to the spoilage of peach products. According to Haugen et al. (2006), detection of CO_2, acetoin, acetate, or ethanol coincided with the start of the exponential growth of spoilage organisms inoculated into a model milk food. Analysis of volatiles in bakery products using an electronic nose also had the potential to detect and differentiate between spoilage by bacteria, yeasts, or fungi (Needham et al., 2005). Development of relevant sensors with adequate sensitivity to metabolites characteristic of the spoilage organism is required and is expected to be combined with food packaging and logistic tools in the near future, which can be realized by intelligent packaging systems adopting an information transfer function (Yam et al., 2005).

4.8 CONCLUSIONS AND PROSPECTS

Microbial growth or deterioration is considered the most important quality criterion for shelf life determination of most perishable food products because of its high relationship to food spoilage and safety. With the proliferation of chilled foods in the market, there will be more attention to and interest in microbial quality preservation and shelf life control, because there are potential risks and the chance of quality loss due to their intrinsically perishable nature and mishandling during distribution such as temperature abuse. Several packaging technologies, including MAP and intelligent packaging, have been developed to enhance microbial quality stability and safety. These advanced technologies will contribute to the delivery of high-quality food with extended shelf life to consumers. However, there is still a paucity of available quantitative information on shelf life extension conferred by advanced packaging techniques. The effect of new packaging techniques on microbial flora and SSO selection needs to be examined. Accumulation of data describing the dependence of microbial shelf life extension on packaging variables is required for different types of foods with different intrinsic properties. The current practice of trial and error in designing packages to give the desired shelf life would benefit from and be improved by the systematic accumulation and analysis of storage stability data for different packaging and storage conditions. With developments in predictive microbiology, extensive and advanced mathematical models incorporated into computer software that can handle different packaging materials and other variables for a wide range of spoilage organisms will facilitate the estimation of microbial growth and food shelf life.

The traditional approach of shelf life estimation as a fixed time period at a specified temperature, which ignores temperature variations through the food supply chain, is no longer adequate. Online monitoring and display of the remaining shelf life attracts great interest from consumers, retailers, and manufacturers, who are nowadays more concerned about the quality and safety of foods. Predicting or monitoring the growth of spoilage organisms on a real-time basis is required for controlling food shelf life on the basis of microbial food quality. Intelligent packaging devices such as TTIs and other sensors may serve this function effectively, but kinetic models of microbial growth and an understanding of the deterioration mechanisms are prerequisites.

REFERENCES

Alklint C., Wadso L., Sjoholm I. 2004. Effects of modified atmosphere on shelf-life of carrot juice. *Food Control* 15: 131–137.

Allende A., Luo Y., McEvoy J.L., Artes F., Wang C.Y. 2004. Microbial and quality changes in minimally processed baby spinach leaves stored under super atmospheric oxygen and modified atmosphere conditions. *Postharvest Biology and Technology* 33: 51–59.

Alves R.M.V., Sarantopoulos C.I.G.L., Van Dender A.G.F., Faria J.A.F. 1996. Stability of sliced Mozzarella cheese in modified atmosphere packaging. *Journal of Food Protection* 59: 838–844.

Amanatidou A., Smid E.J., Gorris L.G.M. 1999. Effect of elevated oxygen and carbon dioxide on the surface growth of vegetable-associated micro-organisms. *Journal of Applied Microbiology* 86: 429–438.

Andres S.C., Giannuzzi L., Zaritzky N.E. 2001. Mathematical modeling of microbial growth in packaged refrigerated orange juice treated with chemical preservatives. *Journal of Food Science* 66: 724–728.

Babic I., Watada A.E. 1996. Microbial populations of fresh-cut spinach leaves affected by controlled atmospheres. *Postharvest Biology and Technology* 9: 187–193.

Bahk G.-J., Todd E.C.D., Hong C.-H., Oh D.-H., Ha S.-D. 2007. Exposure assessment for *Bacillus cereus* in ready-to-eat Kimbab selling at stores. *Food Control* 18: 682–688.

Banks H., Nickelson II R., Finne G. 1980. Shelf-life studies on carbon dioxide packaged in finfish from the gulf of Mexico. *Journal of Food Science* 45: 157–162.

Baranyi J., Roberts T.A. 1994. A dynamic approach to predicting bacterial growth in food. *International Journal of Food Microbiology* 23: 277–294.

Battey A.S., Duffy S., Schaffner D.W. 2002. Modeling yeast spoilage in cold-filled ready-to-drink beverages with *Saccharomyces cerevisiae*, *Zygosaccharomyces bailii*, and *Candida lipolytica*. *Applied and Environmental Microbiology* 68: 1901–1906.

Berruga M.I., Vergara H., Gallego L. 2005. Influence of packaging conditions on microbial and lipid oxidation in lamb meat. *Small Ruminant Research* 57: 257–264.

Braun P., Sutherland J.P. 2003. Predictive modelling of growth and enzyme production and activity by a cocktail of *Pseudomonas* spp., *Shewanella putrefaciens* and *Acinetobacter* sp. *International Journal of Food Microbiology* 86: 271–282.

Buys E.M., Nortje G.L., Jooste P.J., Von Holy A. 2000. Microbiological shelf life of bulk-packaged *Musculus gluteus medius* supplemented with dietary vitamin E. *Meat Science* 40: 433–441.

Cayre M.E., Garroa O., Vignolo G. 2005. Effect of storage temperature and gas permeability of packaging film on the growth of lactic acid bacteria and *Brochothrix thermosphacta* in cooked meat emulsions. *Food Microbiology* 22: 505–512.

Charles F., Rugani N., Gontard N. 2005. Influence of packaging conditions on natural microbial population growth of endive. *Journal of Food Protection* 68: 1020–1025.

Chawengkijwanich C., Hayata Y. 2008. Development of TiO_2 powder-coated food packaging film and its ability to inactivate *Escherichia coli* in vitro and in actual tests. *International Journal of Food Microbiology* 123: 288–292.

Conesa A., Artes-Hernandez F., Geysen S., Nicolaı B., Artes F. 2007. High oxygen combined with high carbon dioxide improves microbial and sensory quality of fresh-cut peppers. *Postharvest Biology and Technology* 43: 230–237.

Corbo M.R., Del Nobile M.A., Sinigaglia M. 2006. A novel approach for calculating shelf life of minimally processed vegetables. *International Journal of Food Microbiology* 106: 69–73.

Dainty R.H. 1996. Chemical/biochemical detection of spoilage. *International Journal of Food Microbiology* 33: 19–33.

Dalgaard P. 1995. Modelling of microbial activity and prediction of shelf life for packed fresh fish. *International Journal of Food Microbiology* 26: 305–317.

Dalgaard P., Buch P., Silberg S. 2002. Seafood Spoilage Predictor—development and distribution of a product specific application software. *International Journal of Food Microbiology* 73: 343–349.

Dantigny P., Guilmart A., Bensoussan M. 2005. Basis of predictive mycology. *International Journal of Food Microbiology* 100: 187–196.

Dermiki M., Ntzimani A., Badeka A., Savvaidis I.N., Kontominas M.G. 2008. Shelf-life extension and quality attributes of the whey cheese "Myzithra Kalathaki" using modified atmosphere packaging. *LWT—Food Science and Technology* 41: 284–294.

Devlieghere F., Debevere J., Van Impe J. 1998. Effect of dissolved carbon dioxide and temperature on the growth of *Lactobacillus sake* in modified atmospheres. *International Journal of Food Microbiology* 41: 231–238.

Devlieghere F., Van Belle B., Debevere J. 1999. Shelf life of modified atmosphere packed cooked meat products: a predictive model. *International Journal of Food Microbiology* 46: 57–70.

Dixon N.M., Kell D.B. 1989. The inhibition by CO_2 of the growth and metabolism of micro-organisms. *Journal of Applied Bacteriology* 67: 109–136.

Eklund T., Jarmund T. 1983. Microculture model studies on the effect of various gas atmospheres on microbial growth at different temperatures. *Journal of Applied Bacteriology* 55: 119–125.

Ellis D.I., Goodacre R. 2001. Rapid and quantitative detection of the microbial spoilage of muscle foods: current status and future trends. *Trends in Food Science & Technology* 12: 414–424.

Enfores S.-O., Molin G., Ternstrom A. 1979. Effect of packaging under carbon dioxide, nitrogen or air on the microbial flora of pork stored at 4°C. *Journal of Applied Bacteriology* 47: 197–208.

Fernandez-López J., Sayas-Barbera E., Munoz E.S., T., Navarro C., Perez-Alvarez J.A. 2008. Effect of packaging conditions on shelf-life of ostrich steaks. *Meat Science* 78: 143–152.

Ferreira Jorge R.M., Livingston A.G. 1999. A novel method for characterisation of microbial growth kinetics on volatile organic compounds. *Applied Microbiology and Biotechnology* 52: 174–178.

Geysen S., Escalona V.H., Verlinden B.E., Aertsen A., Geeraerd A.H., Michiels C.W., Van Impe J.F., Nicolai B.M. 2006. Validation of predictive growth models describing superatmospheric oxygen effects on *Pseudomonas fluorescens* and *Listeria innocua* on fresh-cut lettuce. *International Journal of Food Microbiology* 111: 48–58.

Geysen S., Geeraerd A.H., Verlinden B.E. 2005. Predictive modelling and validation of *Pseudomonas fluorescens* growth at superatmospheric oxygen and carbon dioxide concentrations. *Food Microbiology* 22: 149–158.

Giannuzzi L., Pinotti A., Zaritzky N. 1998. Mathematical modelling of microbial growth in packaged refrigerated beef stored at different temperatures. *International Journal of Food Microbiology* 39: 101–110.

Gould G.W. 1996. Methods for preservation and extension of shelf life. *International Journal of Food Microbiology* 33: 51–64.

Grandison A.S., Jennings A. 1993. Extension of the shelf life of fresh minced chicken meat by electron beam irradiation combined with modified atmosphere packaging. *Food Control* 4: 83–88.

Guerzoni M.E., Gardini F., Duan J. 1990. Interactions between inhibition factors on microbial stability of fruit-based systems. *International Journal of Food Microbiology* 10: 1–18.

Guynot M.E., Marıın S., Sanchis V., Ramos A.J. 2005. An attempt to optimize potassium sorbate use to preserve low pH (4.5–5.5) intermediate moisture bakery products by modelling *Eurotium* spp., *Aspergillus* spp. and *Penicillium corylophilum* growth. *International Journal of Food Microbiology* 101: 169–177.

Haugen J.E., Rudi K., Langsrud S., Bredholt S. 2006. Application of gas-sensor array technology for detection and monitoring of growth of spoilage bacteria in milk: a model study. *Analytica Chemica Acta* 565: 10–16.

Hu W., Jiang A., Qi H. 2007. Effects of initial low oxygen and perforated film package on quality of fresh-cut cabbages. *Journal of the Science of Food and Agriculture* 87: 2019–2025.

Huis is't Veld J.H.J. 1996. Microbial and biochemical spoilage of foods: an overview. *International Journal of Food Microbiology* 33: 1–18.

Jacxsens L., Devlieghere F., Van der Steen C., Debevere J. 2001. Effect of high oxygen modified atmosphere packaging on microbial growth and sensorial qualities of fresh-cut produce. *International Journal of Food Microbiology* 71: 197–210.

Kim G.-T., Koo K.-M., Paik H.-D., Lyu E.S., Lee D.S. 2002. *Sous vide* processing of seasoned spinach soup. *Food Service Technology* 2: 131–138.

Kim G.T., Paik H.D., Lee D.S. 2003. Effect of different oxygen permeability packaging films on the quality of *sous-vide* processed seasoned spinach soup. *Food Science and Biotechnology* 12: 312–315.

Koseki S., Itoh K. 2002. Effect of nitrogen gas packaging on the quality and microbial growth of fresh-cut vegetables under low temperatures. *Journal of Food Protection* 65: 326–332.

Kotzekidou P., Bloukas J.G. 1996. Effect of protective cultures and packaging film permeability on shelf-life of sliced vacuum-packed cooked ham. *Meat Science* 42: 333–345.

Koutsoumanis K., Nychas G.J.E. 2000. Application of a systematic experimental procedure to develop a microbial model for rapid fish shelf life predictions. *International Journal of Food Microbiology* 60: 171–184.

Koutsoumanis K.P., Stamatiou A.P., Drosinos E.H., Nychas G.-J.E. 2008. Control of spoilage microorganisms in minced pork by a self-developed modified atmosphere induced by the respiratory activity of meat microflora. *Food Microbiology* 25: 915–921.

Koutsoumanis K.P., Taoukis P.S., Drosinos E.H., Nychas G.J. 2000. Applicability of an Arrhenius model for the combined effect of temperature and CO_2 packaging on the spoilage microflora of fish. *Applied and Environmental Microbiology* 66: 3524–3528.

Krause T.R., Sebranek J.G., Rust R.E., Honeyman M.S. 2003. Use of carbon monoxide packaging for improving the shelf life of pork. *Journal of Food Science* 68: 2596–2603.

Lannelongue M., Finne G., Hanna M.O., Nickelson R., Vanderzant G. 1982. Storage characteristics of brown shrimp (*Penaeus aztecus*) stored in retail packages containing CO_2-enriched atmospheres. *Journal of Food Science* 47: 911–913.

Laury A., Sebranek J.G. 2007. Use of carbon monoxide combined with carbon dioxide for modified atmosphere packaging of pre- and postrigor fresh pork sausage to improve shelflife. *Journal of Food Protection* 70: 937–942.

Lee D.S., Paik H.D., Im G.H., Yeo I.H. 2001. Shelf life extension of Korean fresh pasta by modified atmosphere packaging. *Journal of Food Science and Nutrition* 6: 240–243.

Lee D.S., Yam K.L., Piergiovanni L. 2008a. *Food Packaging Science and Technology*. Boca Raton, Florida: CRC Press, pp. 397–424.

Lee K.-E., Kim H.J., An D.S., Lyu E.S., Lee D.S. 2008b. Effectiveness of modified atmosphere packaging in preserving a prepared ready-to-eat food. *Packaging Technology and Science* 21: 417–423.

Lee K.-E., An D.S., Lyu E.S., Chung S.K., Lee D.S. 2009. Identification of hurdles for improving storage stability of braised kidney beans, a Korean seasoned side dish. *Journal of Food Processing and Preservation* 33: 33–46.

Leistner L., Gould G.W. 2002. *Hurdle Technologies*. New York: Kluwer Academic/Plenum Publishers, pp. 1–15.

Lim H.T., Lee W.D., Kim G.N., Lee D.S., Paik H.D. 2002. Extension of shelf-life of the low-salted *myungran joet-gal* (soused roe of Alaska pollack) packaged under modified atmosphere. *Food Science and Biotechnology* 11: 412–416.

Mannheim C.H., Soffer T. 1996. Shelf-life extension of cottage cheese by modified atmosphere packaging. *LWT—Food Science and Technology* 29: 767–771.

Mataragas M., Drosinos E.H., Vaidanis A., Metaxopoulos I. 2006. Development of a predictive model for spoilage of cooked cured meat products and its validation under constant and dynamic temperature storage conditions. *Journal of Food Science* 71: M157–M167.

McKellar R.C., Lu X. 2004. Primary models. In: *Modeling Microbial Responses in Food*, McKellar R.C., Lu X. (Eds). Boca Raton, Florida: CRC Press, pp. 21–62.

McMeekin T.A., Olley J., Ratkowsky D.A., Ross T. 2002. Predictive microbiology: towards the interface and beyond. *International Journal of Food Microbiology* 73: 395–407.

McMeekin T.A., Olley J.N., Ross T., Ratkowsky D.A. 1993. *Predictive Microbiology*. Somerset, UK: Research Studies Press, pp. 11–113, 165–197.

McMeekin T.A., Ross T. 1996. Shelf life prediction: status and future possibilities. *International Journal of Food Microbiology* 33: 65–83.

Montero-Calderon M., Rojas-Grau M.A., Martín-Belloso O. 2008. Effect of packaging conditions on quality and shelf-life of fresh-cut pineapple (*Ananas comosus*). *Postharvest Biology and Technology* 50: 182–189.

Nauta M.J., Litman S., Barker G.C., Carlin F. 2003. A retail and consumer phase model for exposure assessment of *Bacillus cereus*. *International Journal of Food Microbiology* 83: 205–218.

Needham R., Williams J., Beals N., Voysey P., Magan N. 2005. Early detection and differentiation of spoilage of bakery products. *Sensors and Actuators B* 106: 20–23.

Newton K.G., Rigg W.J. 1979. The effect of film permeability on the storage life and microbiology of vacuum-packed meat. *Journal of Applied Bacteriology* 41: 433–441.

Ntzimani A.G., Paleologos E.K., Savvaidis I.N., Kontominas M.G. 2008. Formation of biogenic amines and relation to microbial flora and sensory changes in smoked turkey breast fillets stored under various packaging conditions at 4°C. *Food Microbiology* 25: 509–517.

Parkin K.L., Wells M.J., Brown W.D. 1981. Modified atmosphere storage of rockfish fillets. *Journal of Food Science* 47: 181–184.

Peleg M. 2006. *Advanced Quantitative Microbiology for Foods and Biosystems*. Boca Raton, Florida: CRC Press, pp. 205–276.

Rahman M.S. 1999. pH in food preservation. In: *Handbook of Food Preservation*, Rahman M. S. (Ed). New York: Marcel Dekker, pp. 383–396.

Rahman M.S., Labuza T.P. 1999. Water activity and food preservation. In: *Handbook of Food Preservation*, Rahman M. S. (Ed). New York: Marcel Dekker, pp. 339–382.

Ravi Sankar C.N., Lalitha K.V., Jose L., Manju S., Gopal T.K.S. 2008. Effect of packaging atmosphere on the microbial attributes of pearlspot (*Etroplus suratensis* Bloch) stored at 0–2°C. *Food Microbiology* 25: 518–528.

Rho M.-J., Schaffner D.W. 2007. Microbial risk assessment of staphylococcal food poisoning in Korean *kimbab*. *International Journal of Food Microbiology* 116: 332–338.

Robertson G.L. 2006. *Food Packaging Principles and Practice*. Boca Raton, Florida: CRC Press, pp. 193–224.

Shapton D.A., Shapton N.F. 1994. *Principles and Practices for the Safe Processing of Foods*. Oxford, UK: Butterworth-Heinemann, pp. 221–253.

Sheridan J.J., Doherty A.M., Allen P., McDowell D.A., Blair I.S., Harrington D. 1997. The effect of vacuum and modified atmosphere packaging on the shelf-life of lamb primals, stored at different temperatures. *Meat Science* 45: 101–117.

Simon A., Gonzalez-Fandos E., Tobar V. 2005. The sensory and microbiological quality of fresh sliced mushroom (*Agaricus bisporus* L.) packaged in modified atmospheres. *International Journal of Food Science and Technology* 40: 943–952.

Simpson M.V., Smith J.P., Simpson B.K., Ramaswamy H., Dodds K.L. 1994. Storage studies on a *sous vide* spaghetti and meat sauce product. *Food Microbiology* 11: 5–14.

Smolander M., Alakomi H.-L., Ritvanen T., Vainionpaa J., Ahvenainen R. 2004. Monitoring of the quality of modified atmosphere packaged broiler chicken cuts stored in different temperature conditions: time-temperature indicators as quality-indicating tools. *Food Control* 15: 217–229.

Sutherland J. 2003. Modelling food spoilage. In: *Food Preservation Techniques*, Zeuthen P. and Bogh-Sorensen L. (Eds). Cambridge, England: Woodhead Publishing, pp. 451–474.

Taoukis P.S., Koutsoumanis K., Nychas G.J.E. 1999. Use of time temperature integrators and predictive modelling for shelf life control of chilled fish under dynamic storage conditions. *International Journal of Food Microbiology* 53: 21–31.

Toivonen P.M.A. 1997. Non-ethylene, non-respiratory volatiles in harvested fruits and vegetables: their occurrence, biological activity and control. *Postharvest Biology and Technology* 12: 109–125.

Torrieri E., Cavella S., Villani F., Masi P. 2006. Influence of modified atmosphere packaging on the chilled shelf life of gutted farmed bass (*Dicentrarchus labrax*). *Journal of Food Engineering* 77: 1078–1086.

Tsigarida E., Nychas G.-J.E. 2006. Effect of high-barrier packaging films with different oxygen transmission rates on the growth of *Lactobacillus* sp. on meat fillets. *Journal of Food Protection* 69: 943–947.

Van Impe J.F., Poschet F., Geeraerd A.H., Vereecken K.M. 2005. Towards a novel class of predictive microbial growth models. *International Journal of Food Microbiology* 100: 97–105.

Vankerschaver K., Willocx F., Smout C., Hendricks M., Tobback P. 1996. The influence of temperature and gas mixtures on the growth of the intrinsic microorganisms on cut endive: predictive versus actual growth. *Food Microbiology* 13: 427–440.

Vihavainen E.J., Bjorkroth K.J. 2007. Spoilage of value-added, high-oxygen modified-atmosphere packaged raw beef steaks by *Leuconostoc gasicomitatum* and *Leuconostoc gelidum*. *International Journal of Food Microbiology* 119: 340–345.

Wang M.Y., Brown W.D. 1983. Effects of elevated CO_2 atmosphere on storage of freshwater crayfish (*Pacifastacus leniusculus*). *Journal of Food Science* 48: 158–162.

Yam K.L., Takhistov P.T., Miltz J. 2005. Intelligent packaging: concepts and applications. *Journal of Food Science* 70: R1–10.

5 Packaging and the Shelf Life of Milk

Michael G. Kontominas
Laboratory of Food Chemistry and Technology
Department of Chemistry, University of Ioannina
Ioannina, Greece

CONTENTS

5.1 INTRODUCTION

Milk is a complex mixture of water, proteins, lipids, carbohydrates, enzymes, vitamins, and minerals. Due to its specific composition and a pH close to neutral, it is a highly perishable product with high spoilage potential that can result in rapid deterioration of quality and safety. Quality deterioration may be related to (a) the effect of light and oxygen (O_2) causing light-induced oxidation and autoxidation of milk fat, (b) psychrotrophic bacterial activity/enzymic activity resulting in considerable flavor changes in the product, and (c) pick-up of odorous compounds at any stage of production and processing or interaction with the packaging material resulting in product flavor deterioration. In turn, product safety may be affected either by incomplete destruction of pathogenic microorganisms transferred to milk through the animal or by cross-contamination with a particular pathogen at any stage after collection. Packaging, as an integral part of milk processing operations, can offer effective protection to the product from such hazards (Skibsted, 2000; Papachristou et al., 2006a, 2000b).

Packaging serves a number of different functions, including containment, protection, convenience, and communication, the most important being protection (Robertson, 2006). Packaging protects milk and dairy products against environmental, physical, chemical (i.e., light, O_2, moisture), as well as mechanical hazards. It also protects the product from loss of desirable flavor compounds or pick-up of undesirable odors, and contamination from spoilage or pathogenic microorganisms, insects, or rodents during storage and distribution. In addition to the primary functions of packaging listed here, an effective packaging system should fulfill numerous other requirements, including compatibility with the dairy product it contains, recyclability or reuse, tamper evidence, nontoxicity, aesthetics, machinability, and functionality in terms of shape, size, and disposability (Paine, 1996).

Approximately one-third of the milk produced in the European Union, United States, and Australia is consumed as fluid milk; another third is utilized for the production of butter; 18% is used for cheese production; 10% is used for canned milk, dry whole milk, and ice-cream production; and approximately 6% is fed to livestock. In the United Kingdom, fluid milk sales account for approximately 50% of the total dairy market (Varnam and Sutherland, 1996). Within the fluid milk market, pasteurized milk has a dominant position, followed by ultra-high-temperature (UHT)-treated milk, ultrapasteurized (UP) milk, in-bottle sterilized milk, evaporated canned milk, cultured and flavored (strawberry, cinnamon, coffee, etc.) milk, and microfiltered and bactofuged milk. All of these fluid milk products vary substantially in shelf life as a result of differing composition, thermal processing conditions applied, and packaging materials used. Parameters considered when selecting a particular packaging material for milk include (a) in-depth knowledge of product properties, including deterioration mechanisms, (b) desired shelf life, (c) transportation hazards, and (d) specific properties of available packaging materials and machinery. Contemporary milk packaging materials include glass, metals, plastics, paperboard, fibreboard, and composites. Given the relatively large differences in packaging requirements of specific fluid milk products, each of these products will be dealt with separately.

5.2 PACKAGING OF PASTEURIZED MILK

5.2.1 Definitions and Quality Attributes

Milk is an excellent medium for growth of microorganisms. Its nutrients, including proteins, carbohydrates, and butterfat, as well as its moisture may be used by microorganisms and their enzymes (proteases, lipases) to cause quality deterioration and question the safety of milk. Therefore, some kind of heat treatment is applied to drastically reduce milk's microbial load and inactivate its enzymes, resulting in a more stable product compared to raw milk. The main type of heat treatment applied to milk is pasteurization.

Pasteurization involves heating milk in properly designed and operated equipment for a definite time and to a specific temperature and thereafter cooling it immediately. Pasteurization is applied to milk to destroy all vegetative and pathogenic microorganisms and nearly all other bacteria without significantly altering the flavor or composition of the product. It also kills all the yeasts and molds that might be present in the product. Pasteurization can be achieved by either a batch or continuous process. Under the first, milk is heated to 63°C for 30 min in a double-jacketed vat. The process is called low-temperature long-time (LTLT) pasteurization. This method is usually applied to small quantities of milk (≤100 L) and requires low-cost equipment. Alternatively, high-temperature short-time (HTST) pasteurization refers to heating milk in a continuous flow at a temperature of at least 72°C for at least 15 sec, followed by immediate cooling to 4°C. The entire process is usually automated and is suitable for large-scale operations handling 5000 L hr^{-1} or higher. Under these time–temperature heating conditions, *Mycobacterium tuberculosis* and *Coxiella burnetti*, the most heat-resistant non-spore-forming organisms found in milk, are destroyed (Cerf and Condron, 2006). Thus, these organisms are used as index microorganisms in order to assure complete safety of milk. In recent years concern over possible survival of *Listeria monocytogenes* has led to some processors increasing the pasteurizing temperature to above the legal minimum (Frye and Donnelly, 2005).

Recently, in order to increase the shelf life of pasteurized milk, processes such as bactofugation and microfiltration have been introduced in the dairy industry and are used to complement HTST pasteurization. Bactofugation is a process for separating microorganisms and spores from milk using specially designed centrifuges known as *bactofuges*. Both vegetative microbial cells and bacterial spores have a significantly higher density than milk and are thus easily removed from the product. Bactofugation is carried out at 55–60°C prior to pasteurization. Using bactofugation, the number of microorganisms in milk may be reduced from an initial value of 300,000 colony-forming units (cfu) mL^{-1} to 20,000–30,000 cfu mL^{-1}. Subsequent pasteurization may provide milk with a microbial load of 2000–3000 cfu mL^{-1} (see Table 5.1) (Papachristou et al., 2006a, 2006b).

Bactofuged pasteurized milk has a shelf life in excess of 10 days under refrigeration packaged in either bottles made of poly(ethylene terephthalate) (PET) with a UV blocker or low density polyethylene (LDPE)-coated paperboard cartons stored at 4 ± 2°C (Table 5.2).

Microfiltration is essentially a method of separating suspended particles from dissolved substances in a feed stream. The process is used in the dairy industry to separate bacteria (0.5–5 μm) and viruses (<1000 Å) from milk. The pressure-driven separation process uses porous membranes

TABLE 5.1
Mesophillic Counts (log cfu mL^{-1}) of Whole, Bactofuged, Pasteurized Premium-Quality Milk Packaged in Various Containers during Storage at 4°C under Fluorescent Light

	Days of Storage at 4°C							
Packaging Material	**0**	**1**	**3**	**4**	**6**	**8**	**10**	**13**
Clear PET + UV + transparent label	3.23±0.21[a]	3.21±0.17[a]	3.16±0.16[a]	3.23±0.11[a]	3.26±0.21[a]	4.62±0.41[a]	6.25±0.36	6.70±0.54[a]
Clear PET + UV + white colored label	3.23±0.21[a]	3.21±0.25[a]	3.14±0.21[a]	3.16±0.18[a]	3.15±0.14[a]	4.11±0.29[a]	6.04±0.48[a]	6.95±0.38[a]
Clear PET	3.23±0.21[a]	3.17±0.20[a]	3.22±0.24[a]	3.47±0.22[a]	4.31±0.30[b]	5.55±0.46[b]	6.49±0.33[a]	7.53±0.47[a]
Paperboard carton	3.23±0.21[a]	3.17±0.19[a]	3.21±0.29[a]	3.21±0.25[a]	3.16±0.17[a]	4.46±0.21[a]	6.37±0.28[a]	6.85±0.69[a]

Source: From Papachristou C., Badeka A., Chouliara E., Kondyli E., Kourtis L., Kontominas M.G. 2006b. Evaluation of PET as a packaging material for premium quality whole pasteurized milk in Greece part II. *European Food Research and Technology* 224: 237–247.

[a,b] Values within a column followed by different letters are significantly different ($p < .05$). Values reported are the mean of six determinations ($n = 3 \times 2 = 6$).

TABLE 5.2
Flavor Evaluation of Whole, Bactofuged, Pasteurized Milk Packaged in Various Containers during Storage at 4°C under Fluorescent Light

Packaging Material	Days of Storage at 4°C							
	0	1	3	4	6	8	10	13
Clear PET + UV + transparent label	5^{a}	5^{a}	4.9^{a}	4.9^{a}	4.8^{a}	4.6^{b}	4.3^{b}	3.2^{b}
Clear PET + UV + white colored label	5^{a}	5^{a}	4.9^{a}	4.9^{a}	4.8^{a}	4.5^{b}	4.2^{b}	3.4^{b}/slightly stale
Clear PET bottle	5^{a}	5^{a}	4.9^{a}	4.1^{b}	4.1^{b}	3.9^{a}	2.3^{a}/plastic taste, slightly stale	1^{a}/unacceptable
Coated paperboard carton	5^{a}	5^{a}	4.9^{a}	4.8^{a}	4.4^{ab}	4.4^{b}	3.7^{b}/stale	3.3^{b}/stale, sour

Source: From Papachristou C., Badeka A., Chouliara E., Kondyli E., Kourtis L., Kontominas M.G. 2006b. Evaluation of PET as a packaging material for premium quality whole pasteurized milk in Greece part II. *European Food Research and Technology* 224: 237–247.

[a,b] Values within a column followed by different letters are significantly different ($p < 0.05$). Numerical scale of scoring: very good = 5, good = 4, fair = 3, poor = 2, very poor = 1, unfit for consumption = 0.

with a cut-off pore size in the region of 10^{-6} m. The typical operational pressure is in the range 0.1–0.3 bar. Microfiltered pasteurized milk filled into a nonbarrier package (i.e., pigmented PET bottles at 4 ± 2°C) in a controlled environment has a shelf life in the range 20–30 days under refrigeration (Olympos, 2007).

Quality attributes of milk can be grouped under physicochemical, enzymic, and microbial indices. Physicochemical indices include pH (6.6–6.8), titratable acidity (0.14–0.16) expressed as percentage lactic acid, specific gravity of cow's milk (1.028–1.035 g mL^{-1}) at 15°C, refractive index (1.344–1.349) at 20°C, viscosity (~2 cP) at 25°C, and freezing point (−0.53°C to −0.57°C). Enzymic indices are related to the presence in milk of alkaline phosphatase, which is heat sensitive and is inactivated during pasteurization. The absence of alkaline phosphatase is a criterion for adequate pasteurization. In addition to alkaline phosphatase, lactoperoxidase has been considered an index of milk heat-treated at 78°C for 15 sec for international commerce. Microbial indices refer to national or international standards with regard to the maximum allowable microbial load in pasteurized milk. Such values vary from country to country: for example, in UK freshly pasteurized milk should have a total plate count (30°C for 12 hr) of $\leq 3 \times 10^4$ cfu mL^{-1}; in most of the United States the standard is $\leq 2 \times 10^4$ cfu mL^{-1}, and in Greece it is 10^5 cfu mL^{-1}. The keeping quality of such milk is normally 5 to >15 days under refrigeration (4 ± 2°C) in gable-top LDPE-coated paperboard cartons and high density polyethylene (HDPE) or PET bottles, depending on initial microbial load.

In addition to these quality attributes, milk should possess certain sensory characteristics related to flavor and appearance. The bland mouth-feel of milk is a consequence of the oil-in-water emulsion, whereas the slightly sweet and salty taste results from the balance between lactose and milk minerals. The aroma of milk is a consequence of a component balance involving a large number of compounds, many of which are present in trace levels. Many of these are derived from the fat and the milk fat globule membrane. Classes of such compounds include carbonyls, lactones, esters, alkanals, and sulfur and nitrogen compounds, as well as aliphatic and aromatic hydrocarbons.

The opacity of milk is due to suspended particles of fat, proteins, and certain minerals. Milk color varies from white to yellow according to the carotene content of the fat. Thus, skimmed milk is more transparent, with a slightly bluish tint. Homogenization increases the number and total volume of fat globules, and thus homogenized milk has a whiter color than its unhomogenized counterpart.

5.2.2 Deteriorative Reactions and Indices of Failure

Pasteurized milk quality deterioration is perceived by the consumer through off-flavors that may be caused by physicochemical or microbial changes in the product (Valero et al., 2000; Zygoura et al., 2004). Among these defects, light-induced off-flavors (physicochemical defects) are probably the most common in milk and are attributed to two distinct causes. The first, a burnt sunlight flavor, develops during the first 2–3 days of storage and is caused by degradation of sulfur-containing amino acids (methionine) of the whey proteins (Marsili, 1999). The second is a metallic or cardboardy off-flavor (lack of freshness) that develops 2 days later and does not dissipate. This off-flavor is attributed to light-induced lipid oxidation (Barnard, 1972). Light exposure, especially at wavelengths below 500 nm, also causes destruction of light-sensitive vitamins, mainly riboflavin and vitamin A, as shown in Tables 5.3 and 5.4 (Fanelli et al., 1995; Moyssiadi et al., 2004; Papachristou et al., 2006a, 2000b; Zygoura et al., 2004).

The main chemical defect is lipid peroxidation. Unsaturated fatty acids are attacked by free radicals, which is followed by the addition of O_2 to form peroxides or hydroperoxides (Min and Lee, 1996), resulting in the same sensory changes as light-induced oxidation but through a different mechanism. The mechanism of light-induced oxidation begins with riboflavin, which acts as

TABLE 5.3
Retention of Vitamin A in Whole Pasteurized Milk Packaged in Various Containers during Storage at 4°C

Packaging Material	Vitamin A (μg mL^{-1}) Days of Storage at 4°C				
	0	1	3	5	7
Three-layer pigmented, coextruded HDPE bottle	0.57[a]	0.56[a]	0.55[a]	0.55[a]	0.52[a]
Monolayer pigmented HDPE bottle	0.57[a]	0.57[a]	0.56[a]	0.54[a]	0.51[a]
Clear PET bottle	0.57[a]	0.51[a]	0.43[b]	0.36[b]	0.28[b]
Pigmented PET bottle	0.57[a]	0.54[a]	0.51[a]	0.46[c]	0.40[c]
Coated paperboard carton	0.57[a]	0.57[a]	0.57[a]	0.54[a]	0.49[a]

Source: From Zygoura P., Moyssiadi T., Badeka A., Kondyli E., Savvaidis I., Kontominas M.G. 2004. Shelf life of whole pasteurized milk in Greece: effect of packaging material. *Food Chemistry* 87: 1–9.

[a,b,c] Values within a column followed by different letters are significantly different ($p < 0.05$). Values reported are the mean of six replicates ($n = 6$).

TABLE 5.4
Retention of Riboflavin in Whole Pasteurized Milk Packaged in Various Containers during Storage at 4°C

Packaging Material	Riboflavin (μg mL^{-1}) Days of Storage at 4°C				
	0	1	3	5	7
Three-layer pigmented, coextruded HDPE bottle	1.36[a]	1.30[a]	1.25[a]	1.19[a]	1.11[a]
Monolayer pigmented HDPE bottle	1.36[a]	1.30[a]	1.23[a]	1.15[a]	1.08[a]
Clear PET bottle	1.36[a]	1.14[a]	1.03[b]	0.92[b]	0.72[b]
Pigmented PET bottle	1.36[a]	1.20[a]	1.15[a]	1.04[c]	0.94[c]
Coated paperboard carton	1.36[a]	1.29[a]	1.20[a]	1.15[a]	1.09[a]

Source: From Zygoura P., Moyssiadi T., Badeka A., Kondyli E., Savvaidis I., Kontominas M.G. 2004. Shelf life of whole pasteurized milk in Greece: effect of packaging material. *Food Chemistry* 87: 1–9.

[a,b,c] Values within a column followed by different letters are significantly different ($p < 0.05$). Values reported are the mean of six replicates ($n = 6$).

a photosensitizer, as shown in Figure 5.1 (Skibsted, 2000). Riboflavin absorbs photons to form an excited singlet state (^{1}Rib), which by intersystem crossing forms a triplet state (^{3}Rib). Triplet riboflavin is subsequently deactivated to yield singlet oxygen ($^1O_2$), important in protein oxidation (formation of dimethyldisulfide from methionine) (type II reaction). Alternatively, singlet oxygen acts as an oxidant to initiate free radical processes by electron transfer and formation of substrate radicals, superoxide anions, or both (type I reaction).

Microbiological changes involve growth of psychrotrophic bacteria (gram-negative rods such as Pseudomonads and Alcaligenes) as a result of either inadequate pasteurization or postpasteurization contamination leading to the formation of microbial flavor described as acidic, bitter, fruity, malty, putrid, or unclean. Of the pathogens, *Campylobacter jejuni* has been implicated as the cause of foodborne disease associated with pasteurized milk in UK and the United States. In one incident, underprocessing of milk appeared to be the problem, whereas in another, *C. jejuni* survived batch pasteurization in a privately operated pasteurization plant in a boarding school. *Yersinia*

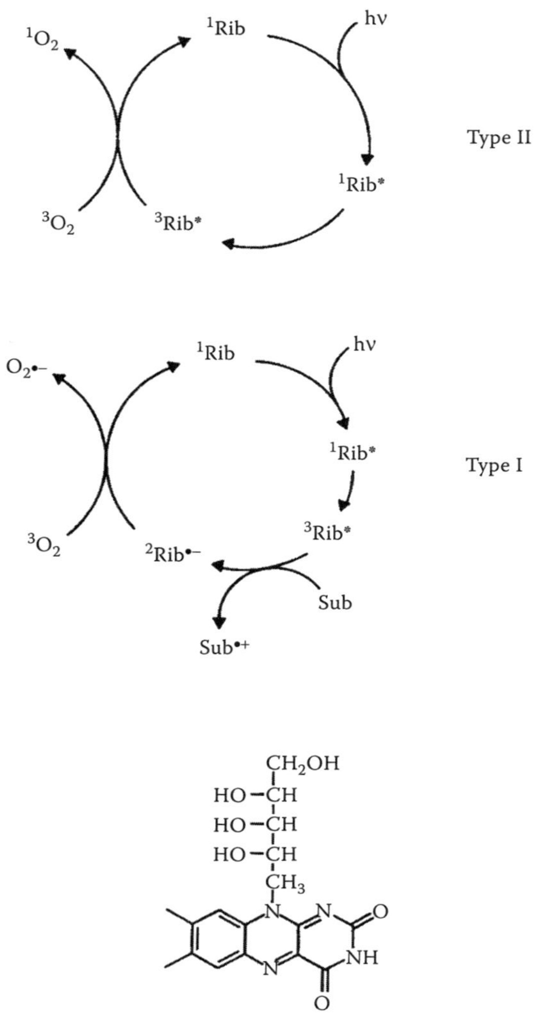

FIGURE 5.1 Role of riboflavin as a photosensitizer in the photo-oxidation of milk. (From Skibsted L.H. 2000. Light induced changes in dairy products. *Bulletin of the International Dairy Federation* No 345, Brussels, with permission.)

enterocolitica has been implicated in three large-scale outbreaks of illness associated with pasteurized chocolate-flavored milk in the United States. It appeared that the pathogen was introduced with the chocolate syrup added to milk after pasteurization. *Salmonella* has also been involved in at least two outbreaks, the first in Chicago, Illinois, and the second in Cambridge, United Kingdom. Both were attributed to contamination of pasteurized milk by raw milk. Finally, *Listeria monocytogenes* was the cause of an outbreak in the United States attributed to incorrect application of HTST pasteurization (Varnam and Sutherland, 1996).

In a study to determine the maximum shelf life of fat-free pasteurized milk, no correlation was found between the microbial count at the end of shelf life and the sensory quality of the milk (Duyvesteyn et al., 2001). It was suggested that microbial counts should not be used to determine the sensory shelf life of milk. The sensory shelf life of the milk stored in paperboard cartons at 2°C, 5°C, 7°C, 12°C, and 14°C was 15.8, 13.7, 12.3, 4.6, and 3.9 days, respectively.

5.2.3. Role of Packaging in Controlling Deteriorative Reactions

Packaging can directly influence the development of the off-flavors described earlier (light-induced oxidation, autoxidation, and microbial flavors) by protecting the product from light, O_2, and microbial cross-contamination (Borle et al., 2001; Sattar and Deman, 1973; Schroeder, 1982; Vassila et al., 2002).

Visible light covers the wavelengths from 380 to 780 nm. Ultraviolet (UV) covers from 280 to 380 nm and is divided into two subregions: UVA (380–320 nm) and UVB (320–280 nm). Both visible and UV light lead to the degradation of foods in general and of milk and dairy products in particular. In order to adequately protect milk against photo-oxidation, industry has turned to containers that are mostly or totally impermeable to light; for example, LDPE-coated paperboard cartons have an average light transmittance of 4% (Zygoura et al., 2004). Opaque HDPE bottles, pigmented (TiO_2, green or blue) PET bottles, and pigmented plastic pouches are among the many commercial packaging materials that are or could be used by the fluid milk industry to protect milk from the effect of light (Cladman et al., 1998; Erickson, 1997; Karatapanis et al., 2006; Mestdagh et al., 2005; Moyssiadi et al., 2004; Papachristou et al., 2006a, 2006b; Van Aardt et al., 2001; Whited et al., 2002; Zygoura et al., 2004), as shown in Figure 5.2. However, the most common HDPE bottles used in Australasia, the United States, and the United Kingdom are not pigmented.

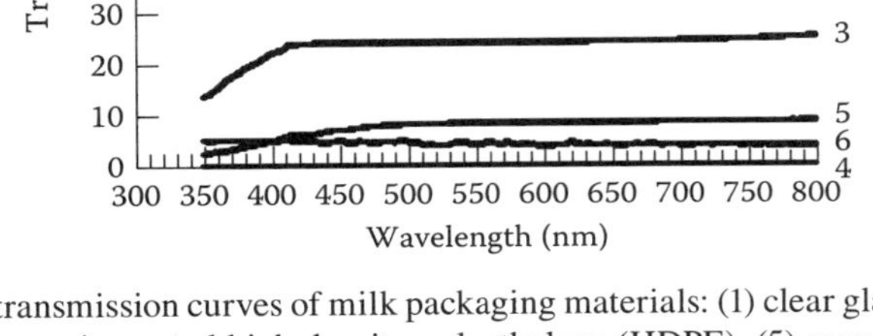

FIGURE 5.2 Spectral transmission curves of milk packaging materials: (1) clear glass, (2) clear PET, (3) pigmented PET, (4) three-layer pigmented high density polyethylene (HDPE), (5) monolayer pigmented HDPE, and (6) coated paperboard carton (T = transmittance, λ = wavelength). (From Karatapanis A.E., Badeka A.V., Riganakos K.A., Savvaidis I.N., Kontominas M.G. 2006. Changes in flavor volatiles of whole pasteurized milk as affected by packaging material and storage time. *International Dairy Journal* 16: 750–761.)

Permeability of O_2 through packaging may also, under specific filling and storage conditions, affect lipid peroxidation in milk (Jung et al., 1998; Mette, 2000; Vassila et al., 2002). However, given the relatively large headspace in milk containers (≥40 mL), O_2 permeability is of minor importance at least in the case of pasteurized milk, which has a short shelf life. The situation is completely different in UP and UHT milk, which have much longer shelf lives. Finally, packaging in combination with refrigerated storage and dispensing protects pasteurized milk from recontamination and provides milk with a satisfactory shelf life by controlling the growth of total and psychrotrophic microorganisms (Erickson, 1997).

5.2.4 Shelf Life of Pasteurized Milk in Different Packages

5.2.4.1 Glass

Once glass was the predominant package for pasteurized milk; only small quantities of pasteurized milk are sold today in glass bottles in several countries, including the United Kingdom, Sweden, and Greece. Typically, a 1-L glass bottle's dimensions for pasteurized milk are 89 mm (base diameter), 35–40 mm (neck diameter), and 267 mm (height), with a weight of 500–510 g. Glass is the most inert of all packaging materials and provides ultimate protection from O_2, moisture, and microorganisms. When colored appropriately (blue, amber, green, and, to a lesser degree, white), glass can protect milk from harmful UV light. Sealing of glass bottles for milk packaging is usually achieved with aluminum foil caps. Most glass bottles are returnable, making on average 30 trips (FAO, 2007).

Major disadvantages of glass are its fragility and weight, although considerable efforts have been made to reduce the weight of glass bottles. According to Dovers et al. (1983), the reusable glass bottle with a trip rate of 25–30 was judged the least environmentally damaging milk packaging material, whereas the single-use glass bottle was judged the worst option environmentally; HDPE bottles and paperboard cartons lay in between. Although returnable glass bottles are seen as an environmentally friendly means of distributing milk at the retail level, it is unlikely that their use will increase. Glass bottles need to be adequately cleaned and sanitized before reuse. Modern bottle washers have five stages, including prerinsing by both immersion and spray-cleaning with a sodium hydroxide solution at approximately 62°C. The bottles are then rinsed with water at approximately 49°C and sanitized with a hypochlorite spray before final rinsing in warm (49°C and 30°C) and cold water.

Sattar et al. (1983) packaged buffalo milk in four different containers: clear glass, green glass, and amber glass bottles, and plastic/alufoil/paperboard brick-shaped cartons. Samples were kept at 5–6°C and 16–24°C for 24 and 16 hr, respectively, under laminate fluorescent light. Analysis revealed that the best protection against ascorbic acid degradation was provided by amber glass, followed by cartons, green glass, and clear glass. The same pattern was observed with respect to the sensory quality of the milk.

5.2.4.2 Plastic Containers

The main plastics used in pasteurized milk packaging are HDPE, PET, polycarbonate (PC), and LDPE.

5.2.4.2.1 High Density Polyethylene Bottles

HDPE bottles of various capacities between 1 and 4 L are widely used for pasteurized milk packaging in several countries, including the United States, Canada, the United Kingdom, and Australia. Unpigmented HDPE bottles transmit 58–79% of the incident light in the wavelength range 350–800 nm. Light transmission can be reduced by pigmenting HDPE with TiO_2 at 1–2%, producing an opaque bottle. HDPE jugs are extrusion-blow-molded to provide a thin-walled, lightweight, and tough container. An advantage of this type of packaging, especially in the 2 and 4 L sizes, is the handle on the bottle, which makes it more convenient to hold than, for example, paperboard cartons. Modern dairies blow-mould their own HDPE bottles to avoid shipping costs and storage space in the dairy plant. HDPE bottles are used for pasteurized full-fat, semiskimmed, and skimmed milk.

5.2.4.2.2 Other Plastic Containers

Zygoura et al. (2004) studied the effect of packaging material on shelf life of whole pasteurized milk stored under fluorescent light at 4°C. They tested (a) multilayer pigmented HDPE (HDPE + 2% TiO_2/HDPE + 4% carbon black/HDPE + 2% TiO_2), (b) monolayer pigmented HDPE (HDPE + 2% TiO_2), (c) clear PET, and (d) pigmented PET (PET + 2% TiO_2). Milk quality was monitored using microbial, chemical, and sensory indices of quality over a 7-day period. Milk packaged in LDPE-coated paperboard cartons served as the commercial control sample. Results showed satisfactory protection of milk stored in all packaging materials with regard to microbiological and chemical parameters throughout the entire storage period. Vitamin A losses recorded after 7 days were 8.8%, 10.5%, 29.8%, 50.9%, and 14.0%, respectively, for samples packaged in multilayer HDPE, monolayer HDPE, pigmented PET, and clear PET, and control samples. Losses of riboflavin were 18.4%, 20.6%, 30.9%, 47.1%, and 19.8%, respectively. On the basis primarily of sensory evaluation, the best overall protection to the product was provided by the multilayer and monolayer pigmented HDPE bottles.

In a similar study, Moyssiadi et al. (2004) investigated the effect of packaging material on shelf life of reduced-fat (1.5%) milk stored under fluorescent light at 4°C for a period of 7 days using the same packaging materials as mentioned earlier. After 7 days, vitamin A losses were 11% for the multilayer HDPE, monolayer HDPE, and pigmented PET bottles, 16% for the paperboard cartons, and 31% for clear PET bottles. Respective losses for riboflavin were 28% for the multilayer pigmented HDPE bottles and paperboard cartons, 30% for the monolayer pigmented HDPE bottles, 33% for the pigmented PET bottles, and 40% for the clear PET bottles. The best overall protection for milk was provided by the multilayer HDPE followed by the monolayer TiO_2-pigmented HDPE bottles.

Karatapanis et al. (2006) investigated changes in the volatile profiles of whole pasteurized milk stored under fluorescent light at 4°C packaged in different containers in a study designed to differentiate between light-induced oxidative and purely autoxidative effects related to packaging material. Packaging materials tested included (a) pigmented HDPE (HDPE + 2% TiO_2/HDPE + 4% carbon black pigment/HDPE + 2% TiO_2) multilayer coextruded bottles; (b) monolayer pigmented HDPE (HDPE + 2% TiO_2) bottles; (c) LDPE-coated paperboard cartons; (d) clear PET bottles; (e) pigmented PET (PET + 2% TiO_2) bottles; and (f) clear glass bottles. Two distinct patterns of milk flavor deterioration were observed. In light-exposed samples, a light-induced oxidation mechanism prevailed, whereas in light-protected samples, an autoxidation mechanism was apparent. Sensory data correlated well with selected volatile compounds, pointing to dimethyldisulfide, pentanal, hexanal, and heptanal as potential markers of fresh-milk quality.

Fanelli et al. (1985) investigated the effectiveness of visible and UV light absorbers incorporated into polyethylene dairy resin to protect vitamins in milk from photodegradation. Three pigments and three UV absorbers were chosen for testing on the basis of U.S. Food and Drug Administration (FDA) approval. Good protection of vitamin A and riboflavin was provided by 0.3% w/w of the pigment FD & C Yellow #5. Protection of ascorbic acid was marginal. Two of the UV absorbers (Cyasorb 531 and Tinuvin 326) provided protection of vitamin A but not of riboflavin or ascorbic acid.

5.2.4.2.3 Poly(ethylene Terephthalate) Bottles

PET bottles are stretch-blow-molded from PET preforms in sizes ranging from 500 mL to 2 L. They are superior to HDPE bottles in terms of their mechanical and optical properties, their lower flavor scalping potential, and substantially lower gas permeability values; for example, the oxygen transmission rate (OTR) at 4°C/50% relative humidity (RH) of a commercial 600-mL PET bottle is 19 $\mu L\ day^{-1}$ compared to 390–460 $\mu L\ day^{-1}$ for a commercial 600-mL HDPE bottle (Van Aardt et al., 2001).

Due to the almost complete transparency of PET to light, milk bottles are either labeled or, even better, sleeved using thermoshrinkable polypropylene (PP) labels. Today most PET bottles are wide necked (35–40 mm diameter) and sealed with rigid PP screw caps. Besides full-fat, semiskimmed, and skimmed milk, PET bottles are also used to package flavored milks such as vanilla, chocolate, and strawberry (Dimmick, 2007), cultured milk, and microfiltered milk. Even though PET bottles used by the dairy industry are single use, other industries such as carbonated soft drinks are considering multiuse PET bottles (Demertzis et al., 1997).

Cladman et al. (1998) studied the effect of prolonged light exposure on chemical changes in pasteurized milk stored at 4°C in (a) clear PET bottles, (b) green PET bottles, (c) PET + UV blocker, (d) PET + exterior labels, (e) HDPE jugs, and (f) clear LDPE pouches. Milk stored in green PET bottles experienced less lipid oxidation and vitamin A loss than milk stored in clear PET bottles or LDPE pouches and HDPE bottles. During the first week of storage, vitamin A loss was lower in milk stored in green PET bottles than in milk stored in clear PET bottles and LDPE pouches. The PET bottles with UV absorbers slowed vitamin A degradation but had little effect on lipid oxidation. Blocking visible light with translucent labels helped to inhibit lipid oxidation and vitamin A degradation.

Mariani et al. (2006) monitored sensory changes in pasteurized milk stored under fluorescent light in a supermarket refrigerator shelf for a period of 9 days in relation to different packages: (a) clear PET bottles, (b) cobalt blue PET bottles, and (c) multilayer pigmented gable-top paperboard cartons. Milk packaged in both clear and cobalt blue PET bottles was affected by off-flavor between the first and second day of storage. Milk packaged in paperboard cartons did not develop any off-flavor during the entire storage period.

Wavelengths higher than 340 mm pass through transparent and opaque plastics such as HDPE, PET, and polystyrene (PS) and through uncolored glass. Colored packaging is capable of partially blocking harmful wavelengths up to 500 nm. The protection provided by colors follows the sequence black (highest), brown, green, blue, red, yellow, uncolored (lowest) (Mottar, 1982).

Van Aardt et al. (2001) evaluated the shelf life of whole (3.25% fat) pasteurized milk in glass, HDPE jugs, amber PET, and clear PET + UV absorber after exposure to fluorescent light (1100–1300 lux) for 18 days at 4°C. In light-exposed samples, oxidation off-flavor was significantly lower when the milk was packaged in amber PET than in the other containers. Milk packaged in HDPE containers showed a significantly higher level of oxidation off-flavor than milk packaged in clear PET + UV absorber containers, but not higher than milk packaged in clear PET or glass containers. Milk packaged in either amber PET or clear PET + UV absorber remained sensorily acceptable after 18 days of storage at 4°C.

Papachristou et al. (2006a, 2006b) evaluated PET as a packaging material for bactofuged, pasteurized milk stored either in the dark or under fluorescent light at 4°C for a period of 13 days. Containers tested included (a) clear PET + UV absorber bottles with a transparent label, (b) clear PET + UV absorber bottles with a white label, and (c) clear PET bottles, with (d) LDPE-coated paperboard cartons serving as the commercial control sample. Results showed satisfactory protection of milk packaged in all containers with regard to microbial and chemical parameters assessed over the 13-day storage period. On the basis of sensory analysis the shelf life of bactofuged, pasteurized milk stored in the dark was 10–11 days for samples packaged in clear PET + UV absorber bottles regardless of the type of label used and 9–10 days in clear PET bottles and paperboard cartons. The shelf life of milk stored under fluorescent light was 10–11 days for clear PET + UV absorber bottles and paperboard cartons and 8–9 days for clear PET bottles. Vitamin A losses recorded after 10 days of storage in the dark were 15.9%, 20.6%, and 14.3% for clear PET + UV absorber bottles, clear PET bottles, and paperboard control samples, respectively. Losses for vitamin E were 26.4%, 36.6%, and 35.0% and for riboflavin 32.9%, 38.3%, and 32.5%, respectively. Vitamin E losses recorded after 10 days under fluorescent light were 42.7%, 53.6%, and 43.9% for PET + UV absorber bottles, clear PET bottles, and paperboard cartons, respectively, with losses for riboflavin being 38.7%, 52.5%, and 35.0%. Average losses for vitamin A were 20.6% for all packaging materials.

5.2.4.2.4 *Polycarbonate Bottles*

PC bottles have a high temperature resistance and high impact strength and clarity, and are currently used for multiuse baby bottles, which are sterilized before each use, as well as for packaging of pasteurized milk in several countries. According to Ohst and Goltzmann (1996), PC bottles are lightweight, clear, and shatter resistant; do not impart an after-taste to the contents; and are recyclable. Hoskin and Dimick (1979) evaluated clear PC, tinted PC, HDPE, and glass returnable (1 gallon) containers, as well as LDPE-coated paperboard cartons for their protection against the development of light-induced flavor and degradation of riboflavin. Milk containers were kept at 7°C under fluorescent light (1076 lux) for up to 72 hr. Results showed a lower preference for light-induced flavor in milk held in PC and glass containers compared to paperboard cartons after 12 hr of exposure. The tinted PC container, fabricated using a protection agent that inhibits transmission of light at 380–480 nm, provided milk with the greatest protection against development of off-flavor. Milk exposed to light in paperboard cartons and milk stored in all containers kept in the dark did not develop any off-flavor. Loss of riboflavin after 72 hr of exposure was 27% in glass bottles, 13% in clear PC bottles, 10% in HDPE bottles, 10% in paperboard cartons, and 6% in tinted PC bottles. Milk of acceptable flavor was obtained up until 48 hr of storage under specific experimental conditions.

In a study by Landsberg et al. (1977), glass, HDPE, and PC multiuse containers were treated with 19 common household chemicals to simulate consumer abuse; glass was found to be the most resistant to retention of the contaminants used. Brede et al. (2003) showed that migration of the weak estrogen bisphenol A (BPA) from PC bottles into hot water was in the order of 0.2 g L^{-1} when the bottles were new and increased to 6–8 g L^{-1} after repeated washing. The highest values reached were 16 g L^{-1} (200 mL filling). Recently, Biedermann-Brem et al. (2008) used alkali washing solutions at concentrations typical for dishwashers and found that even rather extreme scenarios do not result in BPA contamination near the level corresponding to the EU tolerable daily intake (TDI).

5.2.4.2.5 *Linear Low Density Polyethylene/Low Density Polyethylene Pouches*

First developed in Canada in the late 1960s, pillow-shaped pouches (also referred to as sachets) for milk are produced by feeding a linear LDPE (LLDPE) film (75–80 μm) into a form–fill–seal machine and creating a tube that is heat-sealed and then filled with milk, after which the top seal is made. The process is continuous, without interrupting the flow of milk. Milk is dispensed from the pouch by placing it in a jug or pitcher and clipping off the top corner with scissors. A disadvantage of the pouch is that it cannot be reclosed, thus exposing the milk to odor absorption in the refrigerator.

LLDPE is the preferred resin for pouches as it possesses high melt strength, excellent seal integrity, and toughness to withstand tears and pinholes. The pouch material should be pigmented to reduce light transmission. For home use, a combination of two pouches is used: an outer one made of either LDPE or LLDPE and an inner one made of LLDPE. The double-ply structure is used to avoid leakage. Alternatively, LLDPE may be coextruded with an ultra low density polyethylene (ULDPE) for improved sealant performance (Falla, 2004). Sizes available range from 500 mL to 2 L for retail packages. Dimensions for a 1-L sachet are 220–240 mm long and 120–140 nm wide. For institutional use such as restaurants and cafeterias, pouch capacity may rise to 20 L. In such cases the pouch is supported by an injection-molded HDPE crate. Pouches have been used for the packaging of pasteurized milk in countries such as India, Mexico, and China.

Deman (1981) studied the vitamin content of pasteurized milk packaged in LDPE pouches before and after exposure to 2200-lux-intensity fluorescent light up to 48 hr at refrigerator temperature. Vitamin A of whole milk dropped to 67.7% of initial content after 30 hr and remained constant for a further 18 hr. In 2% milk, vitamin A dropped to 23.6% and in skimmed milk to 4.2% of its initial content. No sensory evaluation was carried out in this study.

Hotchkiss et al. (1999) inoculated pasteurized milk with a cocktail of spoilage microorganisms packaged in different barrier film pouches (structure not provided) and stored at 6.1°C for up to 28 days. Addition of CO_2 at 8.7 and 22.5 mM increased the time needed to reach 10^6 cfu mL^{-1} from

6.4 days (no CO_2) to 8 and 10.9 days, respectively, in low-barrier pouches. In high-barrier pouches, the time needed to reach 10^6 cfu mL^{-1} was increased to 9.7 and 13.4 days, respectively, at CO_2 concentrations of 8.7 and 21.5 mM. This increase represents an increase in shelf life of approximately 25–100%, with the major variables in milk shelf life being the amount of CO_2 added and the barrier properties of the packaging material.

Mauriello et al. (2005) coated plastic film with nisin and studied the inhibition of *Micrococcus luteus* inoculated into milk stored at 4°C and 25°C for 2 days. A remarkable reduction in *M. luteus* was observed at 25°C, whereas only a slight reduction was observed at 4°C. When pasteurized milk was poured into the nisin-coated pouch, a reduction in the aerobic plate count (APC) equal to 1.3 log cfu mL^{-1} was observed. It was shown that nisin release was favored at low pH and high temperature. Erickson (1997) studied the chemical and microbiological stability of pasteurized milk packaged in 1-L pouches, LDPE-coated paperboard cartons, and 1-gallon HDPE jugs. The paperboard carton gave the greatest protection against light-induced oxidation, and significantly higher microbial populations were found in the larger containers after 1 week of storage. Data suggested that 1 gallon of milk would remain fresher for the consumer when packaged in multiunit packages than when packaged in a single container. Sensory evaluation was not included in the study.

Vassila et al. (2002) studied changes in chemical and microbial quality parameters of whole pasteurized milk stored under fluorescent light at 4°C in pouches made of (a) LDPE (clear and pigmented with TiO_2), (b) coextruded LDPE/polyamide (PA)/LDPE (clear and pigmented with TiO_2), and (c) coextruded (LDPE + 2% TiO_2/LDPE + 2% TiO_2/LDPE + 4% carbon black/LDPE + 2%TiO_2/LDPE + 2% TiO_2), with varying O_2 (see Table 5.5) and light transmittance for a period of 7 days. Results showed good protection of milk packaged in all pouches with regard to microbial and chemical parameters assessed over the 7-day storage period. With regard to vitamin losses, in both clear and TiO_2-pigmented LDPE and LDPE/PA/LDPE pouches, a high degree of vitamin degradation was observed, ranging from 50.9% to 73.6% for vitamin A and from 34.4% to 45.3% for riboflavin. In the coextruded pouches containing an inner layer of carbon black, the respective losses were 15.1% and 18.9% for vitamin A and riboflavin. Sensory evaluation was not included in the study.

5.2.4.3 Paperboard Laminate Cartons

Paperboard laminate cartons are multilayer containers, usually rectangular with a gable top. The material used for pasteurized milk is paperboard extrusion coated with LDPE on both sides. The

TABLE 5.5
Oxygen Transmission Rate of Packaging Materials

Packaging Material	O_2 Transmission Rate (mL $package^{-1}$ d^{-1})
Clear LDPE pouch, 60 µm	38.3
Pigmented (2% TiO_2) LDPE pouch, 60 µm	38.8
Clear LDPE/PA/LDPE pouch, 60 µm	0.9
Pigmented (2% TiO_2) LDPE/PA/LDPE pouch, 60 µm	0.9
LDPE + 2% TiO_2/LDPE + 2% TiO_2/LDPE + 4% carbon black pigment/LDPE + 2% TiO_2/ LDPE + 2% TiO_2 pouch, 60 µm	39.5
LDPE + 2% TiO_2/LDPE + 2% TiO_2/LDPE + 4% carbon black pigment/LDPE + 2% TiO_2/ LDPE + 2% TiO_2 pouch, 110 µm	22.2
Paperboard carton, 450 µm	7.2

Source: From Vassila E., Badeka A., Kondyli E., Savvaidis I., Kontominas M.G. 2002. Chemical and microbiological changes in fluid milk as affected by packaging conditions. *International Dairy Journal* 12: 715–722.

LDPE = low density polyethylene, PA = polyamide.

thickness of the paperboard is usually 420–490 μm; the thickness of the two LDPE layers is 15–20 μm for the outside layer and 20–40 μm for the inside layer. LDPE is used externally to provide protection from moisture and indirectly for mechanical integrity, and internally it prevents interaction of milk with the paperboard and provides effective heat sealing. Rectangular gable top blank containers 0.25–2 L in capacity are precut and precreased ready to be formed into milk containers using the erect–form–fill–seal principle.

Deman (1978) tested, among other milk packaging containers, paperboard cartons with and without inner brown pigmentation and plastic pouches with and without black pigmented overwrap. Milk was exposed to cool white fluorescent light, warm white fluorescent light, and cool white lamps with plastic UV shields at an intensity of 2200 lux and evaluated for sensory defects and losses of ascorbic acid and riboflavin. The use of UV shields or warm white lamps had only a minor influence on light-induced changes. Losses of riboflavin after 48 hr (cool white fluorescent lamp) were 16.6% in the carton, 13.3% in the pigmented carton, 28.4% in the pouch, and 12.9% in the pouch with overwrap. Losses of ascorbic acid were 10.3%, 9.4%, 86.2%, and 13.1%, respectively. No shelf life data were obtained in the study.

Leong et al. (1992) investigated the development of packaging flavor in pasteurized milk (whole, low-fat, and skimmed) packaged in half-pint (236-mL), quart (946-mL), half-gallon (1890-mL), and gallon (3780-mL) LDPE-coated paperboard cartons and stored at 2.2°C for 6 days. Milk packaged in glass containers served as control samples. Results with respect to fat content showed that packaging flavor developed in milk packaged in half-pint cartons after 1 day of storage. No significant increase in off-flavor intensity was noted following 3 days of storage. Milk packaged in half-pint cartons had a more intense packaging flavor than milk in quart and half-gallon cartons after 6 days. Off-flavor was attributed to migration of package components into milk. The intensity of the off-flavor increased with decreasing fat content.

Gruetzmacher and Bradley (1999) investigated shelf life extension of pasteurized milk and concluded that carton-forming mandrels, filling heads, and airborne microorganisms were sources of contamination during the filling process. Eliminating sources of postpasteurization contamination and proper cleaning followed by sanitizing with chlorine significantly increased milk shelf life in paperboard laminate cartons from 9 to 20 days. Changing the sanitizing agent from chlorine to peroxyacetic acid increased milk shelf life to 34 days.

Simon and Hansen (2001a) packaged 2% milk pasteurized at 92.2°C, 84.0°C, and 76.4°C in a variety of paperboard laminate cartons and monitored its microbial load (APC) for a period of 4 weeks. Milk processed at 76.4°C had the lowest bacterial growth rate, and milk processed at 84.0°C had the highest bacterial growth rate. Milk samples stored at 1.7°C maintained a lower APC than those stored at 6.7°C. The shelf life of samples was between 1 and 4 weeks, depending on the temperature of treatment, packaging material, and storage temperature.

Lee et al. (2004) coated paperboard with nisin, chitosan, or both with the aid of a binder in an ethylene vinyl acetate (EVA) copolymer and measured APC and yeasts in pasteurized milk. The coated paperboard significantly improved microbial stability of milk stored at either 3°C or 10°C but not noticeably at 20°C. Of the packages tested, paperboards that included the combination of nisin and chitosan gave the highest microbial inhibition. No shelf life data were obtained in the study.

5.3 PACKAGING OF ULTRAPASTEURIZED AND ULTRA HIGH TEMPERATURE MILK

5.3.1 DEFINITIONS AND QUALITY ATTRIBUTES

Ultrapasteurization enables dairy processors to produce dairy products with an extended shelf life similar to that obtained with UHT processes, but with fewer flavor defects (Simon and Hansen,

2001b). Ultrapasteurization is achieved by heating milk to 115–138°C for 2–4 sec (usually 138°C for 2 sec) and immediately cooling it below 4°C. The objective is to extend the shelf life of milk to 30–35 days. UHT milk typically receives a heat treatment of 139–150°C for 3–6 sec and has a shelf life of up to 9 months at ambient temperature. Both processes require extremely high levels of hygiene to be observed during production. UP and UHT milk are packaged aseptically in presterilized containers and held either under refrigeration (UP milk) to achieve the extended shelf life or at room temperature (UHT milk), as UHT milk is a commercially sterile product. UP milk processing resembles that of UHT milk but results in fewer product flavor defects. Raw milk is continuously heat-processed either directly by steam injection into the milk (Simon and Hansen, 2001b) or by infusion (addition of milk into steam) or indirectly using tubular or plate heat exchangers similar to those used for pasteurization (Varnam and Sutherland, 1996). Usually before mixing with steam (direct heating), milk is preheated to a temperature of 70–80°C. In the case of direct heating by steam injection, the heated milk passes to a holding tube and then to a vacuum vessel. At this stage the temperature of the milk falls rapidly, causing some of the water and other volatiles to vaporize. This process is known as flash-cooling and has the following objectives: (a) very rapid cooling to avoid extensive thermal damage and (b) removal of water to restore the original composition of the milk. The degree of cooling and quantity of water removed are determined by the level of vacuum. Milk is then packaged aseptically in presterilized containers. The sensory attributes of UP milk are similar to those of pasteurized milk, with the exception of flavor, which may be characterized as slightly "hammy" or "cardboardy" as well as slightly "cooked." Such flavors are mostly due to the formation of volatile carbonyl compounds such as alkanals and ketones, products of unsaturated fatty acid oxidation (Simon and Hansen, 2001b).

5.3.2 Deteriorative Reactions and Indices of Failure

Spoilage microflora of UP milk include *Pseudomonas* spp., *Alcaligenes* spp., *Flavobacterium* spp., *Bacillus coagulans*, *B. subtilis*, and *B. licheniformis*. Both proteolytic and lipolytic enzymes are produced by psychrotrophs. Spoilage by proteolytic enzymes results in gelation and bitter flavors, whereas lipolytic enzyme spoilage produces rancid flavors. Of the pathogens, *B. cereus* and *Clostridium* spp. may survive pasteurization.

Massive contaminations of entire commercial lots of UHT and sterilized milk with a then-unknown mesophilic aerobic spore former were first reported in Italy and Austria in 1985 and in 1990 in Germany. This organism was provisionally called a *highly heat-resistant spore former* (termed HHRS or HRS), as the causative organism could be isolated from a bypass directly after the heating section of an indirect UHT-heating device. Contrary to post-heat-treatment contamination, this problem seemed to be caused by survival of the HRS during the UHT process and occurred more frequently in indirect UHT than in direct UHT processing. The problem subsequently spread to other countries in and outside Europe. Affected milk products included whole, skimmed, evaporated, and reconstituted UHT milk, UHT cream and chocolate milk in different kinds of containers, and also milk powders (Scheldman et al., 2006).

The HRS organism may reach a maximum of 10^5 vegetative cells and 10^3 spores mL^{-1} milk after 15 days' incubation at 30°C of unopened packages of consumer milk according to the EC regulation. These levels do not affect the pH of the milk and usually do not alter its stability or sensory quality. However, this contamination level far exceeds the sterility criterion of 10 cfu $(0.1\ mL)^{-1}$, according to the EC regulation. Several HRS strains have been tested and none showed pathogenic potential. Despite its poor growth characteristics in milk, UHT milk can be regarded as a new ecological niche for *B. sporothermodurans* because of the lack of competition from other organisms in this product (Scheldman et al., 2006).

The current hypothesis is that highly heat-resistant spores are adapted by sublethal stress conditions (e.g., hydrogen peroxide, which is used to sterilize packaging material) in the industrial process and selected for by the heating step. As a result, considerable problems may occur through recirculation (reprocessing) of UHT milk that has passed its use-by date, leading to contaminated lots of milk and milk products.

Extensive research has reported the presence and characteristics of heat-resistant enzymes in milk and their effects on UHT products during storage. Proteases and lipases are of greatest concern. Although phosphatase activity is always zero after milk has been sterilized, it may be reactivated after prolonged storage, where the extent of reactivation increases with storage time and temperature (Robertson, 2006).

Age gelation is an irreversible phenomenon that occurs during storage of UHT-processed milk products, ultimately transforming the product into a gel. It is considered the most important index of failure associated with this type of product, because once the product has gelled, it has reached the end of its shelf life. The severity of the heat treatment, both prior to and during the sterilization process, critically affects age gelation in UHT milk products, with gelation being less critical in UHT milk than in UHT concentrated milk. Sterilized milk produced by the direct-heat UHT process is more prone to gelation than that prepared using the indirect method, probably owing to better control over the severity of the heat treatment given in the latter (Rosenberg, 2002).

Researchers are still not sure whether gelation is attributable to enzymic action or chemical and physical processes. For many years, it was considered that coagulation was caused by the slow action of heat-resistant proteases from psychrotrophs such as *Pseudomonas* spp. However, age gelation has occurred where proteolytic activity was not evident and has not occurred on other occasions when proteolytic activity was evident. A mechanism consisting of an enzymic triggering stage followed by a nonenzymic aggregation phase has been suggested. Although proteolysis is involved, nonenzymic mechanisms play a major role in governing the phenomenon of age gelation, especially those affecting interactions between caseins and whey proteins.

The best way of avoiding age gelation is to prevent the development of heat-resistant enzymes in the milk before processing. This can be achieved by preventing contamination by the causal microorganisms, and particularly by keeping the storage time short and the storage temperature low (e.g., <58°C) to prevent the growth of psychrotrophs (Burton, 1988).

Topçu et al. (2006) reported that UHT milks produced in a direct (steam-injection) system from low-quality raw milk with high somatic cell count and psychrotroph counts showed high levels of proteolysis during storage at 25°C for 180 days. This caused bitterness, gelation, and sedimentation, all of which reduced the shelf life of the milk. The proteolysis appeared to be due to both bacterial proteinase and plasmin. Processing the milk at a higher temperature (150°C rather than 145°C) reduced proteolysis, gelation, and bitterness but caused a more intense cooked flavor in the UHT milk, although the overall acceptability was not affected.

Heat treatment of milk results in denaturation of the whey proteins. The extent of denaturation varies according to the severity of heat treatment, from partial during pasteurization to total during in-bottle sterilization. Loss of available lysine is approximately 1–2% in pasteurized milk and 4–5.5% in UHT milk (Varnam and Sutherland, 1996). The denaturation of whey proteins plays a key role in the development of cooked milk flavor. This is insignificant in pasteurized milk but detectable in UP milk and more extensive in UHT-treated milk, in which cooked flavor is a serious quality defect. Nonenzymic browning reactions that result in darkening of milk color are not readily detectable in UP milk but are more intense in UHT milk. Heat also affects ascorbic acid content: in pasteurized milk 10–25% is lost, and in UP and UHT milk more than 25% is lost. Light-induced losses of vitamins are also very important in UP and UHT milk. Given that these types of products will normally have a shelf life of more than 30 days, it is imperative that packaging materials impermeable to light be used. Loss of riboflavin can be extensive, followed by loss of vitamin A.

The sorption of dairy flavor compounds (aldehydes and methyl ketones) by LDPE and PP films has been investigated quantitatively in an attempt to assist aseptic processors select appropriate packaging materials for maximum flavor stability. PP sorbed these compounds to a greater extent than LDPE. Headspace analysis of UHT-processed milk packaged in aseptic paperboard cartons revealed a loss of higher molecular weight flavor compounds after 12 weeks' storage, owing to the interaction between the LDPE packaging material and the milk (Hansen and Arora, 1990).

5.3.3 Role of Packaging in Controlling Deteriorative Reactions

As with pasteurized milk, packaging can directly influence the quality of UP and UHT milk by providing protection from light, O_2, and microbial cross-contamination during its shelf life. Numerous packaging materials are currently being used for UP and UHT milk, including multilayer HDPE bottles, pigmented PET bottles, coextruded pouches composed of polyethylene/polyvinylidene chloride (PE/PVdC) or EVA/PVdC, and plastic/alufoil/paperboard laminate cartons.

5.3.4 Shelf Life of Ultrapasteurized and Ultra High Temperature Milk in Different Packages

5.3.4.1 Paperboard Laminate Cartons

Paperboard laminate cartons are multilayer containers, usually rectangular with a flat top or tetrahedral shape. For UHT milk packaging applications, aluminum foil is added to the conventional LDPE/paperboard/LDPE structure between the paperboard and the internal LDPE layer (LDPE/paperboard/LDPE/alufoil/LDPE/LDPE) to provide the required barrier properties (see Figure 5.3). The innermost LDPE layer is applied at a lower temperature than the adjacent layer to minimize the tendency for LDPE degradation products formed at high temperatures to diffuse into the milk and alter its taste. Commercial sizes for such containers range from 100 mL to 2 L in capacity.

Simon and Hansen (2001b) used (a) standard milk board, (b) standard board including an ethylene vinyl alcohol (EVOH) barrier layer, and (c) standard board including an aluminum foil layer to package 2% UP milk stored at 6.7°C. Quality was assessed over a period of 15 weeks. They found that the flavor of milk packaged in standard board deteriorated at a faster rate than that of milk packaged in barrier and foil boards.

At week 6 of storage, a slightly cardboardy flavor was detected in milk packaged in standard board, and a slightly cooked flavor was detected in milk packaged in barrier and foil boards. The cardboard flavor intensified with storage time, but the cooked flavor had dissipated by week 10 of storage.

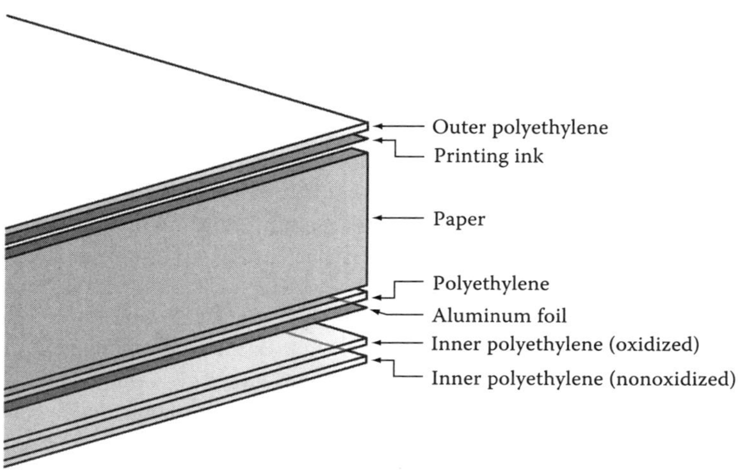

FIGURE 5.3 Structure of paperboard laminate cartons for ultrapasteurized and UHT milk packaging. (From Robertson G.L. 2006. *Food Packaging Principles and Practice*, 2nd edn. Boca Raton, Florida: CRC Press, with permission.)

Rysstad and Kolstad (2006) described the Pure-Lac system developed by Elopak (Spikkestad, Norway) and APV (Silkeborg, Denmark) in the mid-1990s. The Pure-Lac plant is almost identical to the direct-heated steam-infusion UHT plant but is operated at conditions designed to kill heat-resistant psychrotrophic aerobic spores without damaging milk flavor. Holding conditions are 130–145°C for <1 sec with infusion heating and flash-cooling times of <0.2 and <0.3 sec, respectively. Holding conditions and packaging options (clean, ultraclean, or aseptic filling) can be tailored to suit the flavor profile and shelf life required by the processor (Value and Castberg, 1991). A shelf life of >45 days at 10°C can be achieved in plastic/alufoil/paperboard laminate cartons, which provide zero light transmission.

Farrer (1983) compared UHT milk packaged in LDPE-coated paperboard cartons with and without an aluminum foil layer. Results showed that O_2 in the milk packaged in the container with aluminum foil remained almost unchanged at 1 ppm after 44 days, whereas in the milk packaged in the container without aluminum foil, O_2 rose to 8–9 ppm after only a few days. Milk in the carton containing aluminum foil was organoleptically acceptable for 2 months even when stored at 38°C, whereas in the carton without foil, the milk was acceptable only for up to 3 weeks when stored at 15°C.

Rysstad et al. (1998) evaluated the sensory and chemical shelf life of UHT milk stored at room temperature and 6°C in 1-L gable-top cartons with three structures: an aluminum foil barrier; a non-foil, paper-based barrier (X-board); and LDPE-coated paperboard. The OTR of the three structures was 0, 2–4, and >200 mL O_2 m^{-2} day^{-1}, respectively, but, unfortunately, the surface area of the cartons was not given. UHT milk in cartons with an aluminum foil barrier layer stored in the dark had a shelf life of 6 months, whereas milk stored in the X-board and LDPE-coated cartons had a shelf life of 4–5 months. When LDPE-coated and X-board cartons were stored under direct light exposure at 6°C, a light-induced off-flavor was detected after 2 and 8 weeks, respectively. The light-induced off-flavor effect was more pronounced than the effect of autoxidation of unsaturated lipids.

5.3.4.2 Plastics

5.3.4.2.1 Poly(ethylene Terephthalate) Bottles

Mestdagh et al. (2005) compared a number of PET bottles for their ability to prevent photo-oxidation in UHT semiskimmed milk during storage up to 2 months. Three types of PET bottles were tested: (a) a transparent PET bottle with an O_2-scavenging inner layer, (b) a coextruded three-layer (white/black/white) PET bottle with perfect light barrier, and (c) a transparent PET bottle with UV absorber. The results of two different shelf life studies showed that an adequate light barrier was apparently sufficient to prevent light-induced oxidation in milk during extended storage. Oxygen absorbers and UV absorbers did not provide adequate protection against photo-oxidation. If wavelengths detrimental to riboflavin were not completely blocked by the packaging material, incoming light could still give rise to photo-oxidation of milk. Riboflavin and vitamin A were gradually degraded, milk fat was photo-oxidized, O_2 dissolved in the milk was consumed, and the sensory quality substantially decreased.

Saffert et al. (2008) evaluated the effect of package light transmittance on the vitamin content of UHT whole milk. The milk was stored at three different light intensities in PET bottles with a range of light transmittance. Changes in vitamins A, B_2, and D_3 were monitored over a period of 12 weeks at 23°C. Losses of vitamins A and B_2 were more pronounced in transparent PET bottles exposed to the highest light intensity. In these bottles a reduction in light intensity reduced vitamin A loss from 88% to 66%, whereas in the case of vitamin B_2 its complete decomposition was delayed from 4 to 8 weeks. Vitamin D_3 losses in clear PET bottles were almost independent of light intensity. For the pigmented PET bottles, an increase in package light transmittance and light intensity was found to critically affect vitamin B_2 stability. For vitamin D_3 only the increase in light intensity was found to be of relevance, whereas for vitamin A stability no clear effect of light transmittance and light intensity was observed. No sensory evaluation was carried out in the study.

Saffert et al. (2009) stored UHT low-fat milk under light with an intensity of 700 lux in four PET bottle variants representative of those used for milk on the European market in terms of their light transmittance. Changes in the vitamin A, B_2, and D_3 contents over a storage period of 12 weeks at 23°C were monitored. Milk packed in pigmented PET bottles with the lowest light transmittance was stored in the dark under the same experimental conditions and served as the control sample. In clear PET bottles, a reduction of 93% of the initial content was observed for vitamin A and 66% for vitamin D_3, whereas the vitamin B_2 content was completely degraded. In all pigmented PET bottles, the vitamin retention was only slightly higher—the losses ranged between 70 and 90% for vitamin A, between 63% and 95% for vitamin B_2, and between 35% and 65% for vitamin D_3 depending on the pigmentation level. In the dark-stored "control" sample, a 16% loss was observed for vitamin A, but the levels of vitamins B_2 and D_3 remained almost stable.

On the basis of these findings, Saffert et al. (2009) concluded that light barrier properties comparable to the high-pigmented PET bottle seemed to be sufficient to protect the light-sensitive vitamins A, B_2, and D_3 in milk for realistic periods on the retailer's shelf. However, they considered that light-induced sensory changes in UHT milk under commercially relevant storage conditions can only be excluded in light-tight packages.

5.3.4.2.2 Coextruded High Density Polyethylene Bottles

Recently, multilayer HDPE bottles using coextrusion technology have been introduced into the market for UP and UHT milk. In the case of UP milk, a three-layer HDPE bottle is used that consists of an inside and outside white layer (2% TiO_2) and a middle black layer (2% carbon black) (Karatapanis et al., 2006). In the case of sterilized and UHT milk either the barrier of the three-layer container is enhanced by the addition of a PVdC copolymer coating (5–6 μm) or a five-layer HDPE container is used consisting of an outside HDPE white layer (2% TiO_2), adhesive layer, a middle black EVOH (2% carbon black) layer, adhesive layer, and an inside white HDPE layer (2% TiO_2) (INEOS Polyolefins, 2007).

In a study by Mottar (1987) involving packaging of UHT whole milk in these packaging materials and storage at 20°C for 3 months, it was shown that the three-layer HDPE bottle provided adequate protection from light. Although EVOH-coated HDPE provided the most efficient gas barrier, resulting in a higher-quality product, the three-layer HDPE as well as the PVdC copolymer-coated three-layer HDPE provided a shelf life of 3 months at 20°C.

5.3.4.2.3 Plastic Pouches (Sachets)

Perkins et al. (2007) stored UHT milk packed in aseptic pouches with or without O_2-scavenging film at 26°C for 14 weeks. The O_2-scavenging film was shown to reduce dissolved oxygen content by 23–28% during storage. Significant reduction of 23–41% was observed for some stale flavor volatiles. Free fatty acid levels remained far below threshold values, indicating that lipolysis would not alter desirable sensory attributes of the milk. However, the sensory panel failed to detect significant differences in odor between treated and control samples during the entire 14-week storage period.

5.3.4.3 Aluminum Cans

Aluminum cans are two-piece metal containers, one piece making up the can body and bottom end and the second piece making up the top end. The interior of aluminum cans is coated with an enamel or lacquer to protect against corrosion. They are used for the packaging of vitamin-fortified milk for youngsters and flavored milk (e.g., coffee, cinnamon, caramel, nut, or vanilla flavors) in 330-mL containers. The shelf life of such sterilized products is 1 year without refrigeration.

5.4 IN-BOTTLE STERILIZED MILK

5.4.1 Definitions and Quality Attributes

Sterilized milk is available in whole, semiskimmed, and skimmed varieties. It undergoes a more severe form of heat treatment that destroys nearly all bacteria. First, the milk is preheated, sterilized in a tubular heat exchanger at 138–145°C for 2 sec, homogenized, and filled into glass or plastic bottles that are closed with a hermetic (airtight) seal. The bottles are passed through a steam chamber and heated to 113–130°C for approximately 10–12 min. They are then rapidly cooled to avoid further thermal degradation. Typical 1-L bottles used for sterilized milk have narrow necks to ensure an effective seal and the following dimensions: 89 mm (base diameter), 26 mm (neck diameter), and 294 mm (height). Prefabricated crown seals are used to seal these bottles. Such bottles have to withstand the heat sterilization process and subsequent cooling. As plastics technology advances, partial replacement of glass containers by plastics is occurring in various liquid food packaging applications, including milk (INEOS Polyolefins, 2007).

Batch retorts, originally used for milk sterilization, have now been replaced by large-scale continuous retorts of the hydrostatic and rotary types. The process may be used with plastic bottles, but in this case compressed air is used to maintain the total pressure in the sterilizing chamber at 0.3–0.5 bar higher than the pressure of saturated steam to prevent bottles from bursting as they are weakened at high temperature. Unopened bottles of sterilized milk keep for approximately 6 months without the need for refrigeration. Plastic bottles commercially used for in-bottle sterilized milk include three- and six-layer HDPE pigmented containers; the latter contains EVOH as an O_2 barrier (INEOS Polyolefins, 2007).

5.4.2 Deteriorative Reactions and Indices of Failure

The extent of vitamin loss during in-bottle sterilization varies according to the heating process, but in all cases major losses occur. Ascorbic acid is destroyed by 30–100%; thiamine, by 20–50%; vitamin B_6 by 15–30%; vitamin B_{12}, by 20–100%; and folic acid, by 30–50%. Losses are considerably lower when continuous-flow heating is used as part of the process than when the entire heating treatment is in bottle. Lactulose formation increases with the temperature of heat treatment. Typical values for pasteurized milk are approximately 50 mg L^{-1}, for UHT sterilized milk 100–500 mg L^{-1}, and for in-bottle sterilized milk 900–1380 mg L^{-1} (Varnam and Sutherland, 1996). Differences in lactulose content may be used to differentiate in-bottle sterilization from UHT and pasteurized milk.

During heating, lactones and methyl lactones are formed from fat, resulting in a deleterious effect on flavor. Quantities present in pasteurized milk are very small (≥12 nmol g^{-1} fat). Although higher levels of methyl ketones are found in UHT milk (~21 nmol g^{-1}), formation of lactones and methyl ketones is of major significance only in in-bottle sterilized milk (≥100 nmol g^{-1} fat). Nonenzymic browning correlates closely to the severity of the heat treatment. The reaction occurs to a considerable extent during in-bottle sterilization and yields among other end-products melanoidins, which are responsible for brown discoloration, and hydroxymethyl furfural (HMF), which is responsible for acrid flavors.

With respect to microbial spoilage, leakage through caps, and subsequent postprocess contamination, can lead to spoilage of in-bottle sterilized milk. Most problems result from the survival of heat-resistant endospores of mesophilic species of *Bacillus*. Endospores of *B. stearothermophilus* do not normally germinate and outgrow under usual conditions of storage.

5.4.3 Role of Packaging in Controlling Deteriorative Reactions

As with pasteurized and UHT milk, packaging can directly influence the quality of in-bottle sterilized milk by protecting the product from light, O_2, and microbial postprocessing contamination during its shelf life. Packaging materials currently used for in-bottle sterilized milk include narrow-neck glass bottles and three- and six-layer HDPE plastic bottles.

5.4.4 Shelf Life of In-Bottle Sterilized Milk in Different Packages

Gliguem and Birlouez-Aragon (2005) studied the effect of sterilization method, packaging, and storage on the quality of fortified milk samples at ambient temperature in the dark for 4 months.

Two sterilization methods were used: (a) indirect UHT (135°C, 3–4 sec) and (b) in-bottle sterilization (110°C, 20 min). Packaging materials evaluated included three- and six-layer HDPE opaque bottles, the latter containing EVOH as an oxygen barrier. The permeability of HDPE to O_2 was approximately 50,000 mL m^{-2} atm^{-1} at 23°C/0% RH as compared to 4–60 for EVOH-coated HDPE. The use of a three-layer bottle was associated with complete oxidation of ascorbic acid after 1 month of storage, whereas in the six-layer bottle containing an O_2 barrier, the ascorbic acid content slowly decreased to reach 25% of its initial concentration after 4 months of storage. Furosine and the fluorescence of advanced Maillard products and soluble tryptophan (FAST) index, associated with the early and advanced Maillard reaction, respectively, were significantly higher in in-bottle sterilized milk than in UHT samples. The study concluded that only the package comprising an O_2 and light barrier is effective in protecting ascorbic acid in fortified sterilized milk. No sensory evaluation was included in the study.

REFERENCES

Barnard S.E. 1972. Importance of shelf life for consumers of milk. *Journal of Dairy Science* 55: 134–136.

Biedermann-Brem S., Grob K., Fjeldal P. 2008. Release of bisphenol A from polycarbonate baby bottles: mechanisms of formation and investigation of worst case scenarios. *European Food Research and Technology* 227: 1053–1060.

Borle F., Sieber R., Bosset L.O. 2001. Photooxidation and photo-protection of foods with particular reference to dairy products. An update of a review article (1993–2000) *Sciences des Aliments* 21: 576–590.

Brede C., Fjeldal P., Skjevrak I., Herikstad H. 2003. Increased migration levels of bisphenol A from polycarbonate baby bottles after dishwashing, boiling and brushing. *Food Additives and Contaminants* 20: 684–689.

Burton H. 1988. *Ultra-High-Temperature Processing of Milk and Milk Products*. London: Elsevier Applied Science.

Cerf O., Condron R. 2006. Coxiella burnetii and milk pasteurization: an early application of the precautionary principle? *Epidemiology and Infection* 134: 946–951.

Cladman W., Scheffer S., Goodrich N., Griffiths M.W. 1998. Shelf life of milk packaged in plastic containers with and without treatment to reduce light transmittance. *International Dairy Journal* 8: 629–636.

Deman J.M. 1978. Possibilities of prevention of light induced quality loss of milk. *Canadian Institute of Food Science and Technology Journal* 11:152–154.

Deman J.M. 1981. Light induced destruction of vitamin A in milk. *Journal of Dairy Science* 64: 2031–2032.

Demertzis P.G., Johansson F., Lievens C., Franz R. 1997. Studies on the development of a quick inertness test procedure for multi-use PET containers—sorption behavior of bottle wall strips. *Packaging Technology and Science* 10: 45–58.

Dimmick B. 2007. Packing punch with packaging. *Milk Producer*, June: 26–32.

Dovers S., Madden E., Lommon M., Boyden S. 1983. Milk packaging in Australia: a case study in environmental priorities. *Resources, Conservation and Recycling* 9: 61–73.

Duyvesteyn W.S., Shimoni E., Labuza T.P. 2001. Determination of the end of shelf-life for milk using Weibull hazard method. *LWT—Food Science and Technology* 34: 143–148.

Erickson M.C. 1997. Chemical and microbial stability of fluid milk in response to packaging and dispensing. *International Journal of Dairy Technology* 50: 107–111.

Falla D.J. 2004. Using enhanced polyolefin technology in pouches for packaging flowable materials. *Engineering Plastics* 9(65): 384–402.

FAO. 2007. Packaging, storage and distribution of processed milk. Chapter 2. www.fao.org/docrep/003/X6511E/X6511E02.htm#ch2, last accessed 12/05/2009.

Farrer K.T.H. 1983. *Light Damage in Milk*, Farrer Consultants, Blackburn, Victoria 3130, Australia.

Fanelli A.J., Burlew J.V., Gabriel M.K. 1985. Protection of milk packaged in HDPE against photodegradation by fluorescent light. *Journal of Food Protection* 48: 112–117.

Frye C., Donnelly C.W. 2005. Comprehesive survey of pasteurized fluid milk produced in the US reveals a low prevalence of *Listeria monocytogenes*. *Journal of Food Protection* 68: 973–979.

Gliguem H., Birlouez-Aragon I. 2005. Effects of sterilization, packaging and storage on vitamin C degradation, protein denaturation and glycation in fortified milks. *Journal of Dairy Science* 88: 891–899.

Gruetzmacher I.J., Bradley R.L. 1999. Identification and control of processing variables that affect the quality and safety of fluid milk. *Journal of Food Protection* 62: 625–631.

Hansen A.P., Arora D.K. 1990. Loss of flavor compounds from aseptically processed food products packaged in aseptic cartons. In: *Barrier Polymers and Structures*, ACS Symposium Series #423, Koros W.J. (Ed.). Washington, DC: American Chemical Society, chapter 17.

Hoskin J.C., Dimick P.S. 1979. Evaluation of fluorescent light on flavor and riboflavin content of milk held in gallon returnable containers. *Journal of Food Protection* 42: 105–109.

Hotchkiss J.H., Chen J.H., Lawless H.T. 1999. Combined effect of CO_2 and barrier films on microbial and sensory changes in pasteurized milk. *Journal of Dairy Science* 82: 690–695.

INEOS Polyolefins. 2007. Brussels, www.ineospolyolefius.com

Jung M.Y., Yoon S.H., Lee H.O., Min D.B. 1998. Singlet oxygen and ascorbic acid. Effects on dimethyl disulfide and off flavor in skim milk exposed to light. *Journal of Food Science* 63: 408–412.

Karatapanis A.E., Badeka A.V., Riganakos K.A., Savvaidis I.N., Kontominas M.G. 2006. Changes in flavor volatiles of whole pasteurized milk as affected by packaging material and storage time. *International Dairy Journal* 16: 750–761.

Landsberg J.D., Bodyfelt F.W., Morgan M.E. 1977. Retention of chemical contaminants by glass, polyethylene, and polycarbonate multiuse milk containers. *Journal of Food Protection* 40: 772–777.

Lee C.H., Park H.J., Lee B.S. 2004. Influence of antimicrobial packaging on kinetics of spoilage microbial growth in milk and orange juice. *Journal of Food Engineering* 65: 527–531.

Leong C.M.O., Harte B.R., Partridge J.A., Ott D.B., Downes T.W. 1992. Off flavor development in polyethylene-coated paperboard cartons. *Journal of Dairy Science* 75: 2105–2111.

Mariani B., Chiacchierini E., Bucqrelli F.M., Quaglia G.B., Mennesa H.P. 2006. Comparative study of milk packaging materials. Note 2. Sensorial quality change in fresh milk during storage. *Industrie Alimentari* 45(454): 6–10.

Marsili R.T. 1999. Comparison of SPME and dynamic headspace method for the GC/MS analysis of light-induced lipid oxidation products in milk. *Journal of Chromatographic Science* 37: 17–23.

Mauriello G., Deluca E., Lastaria A., Villani F., Ercolins D. 2005. Antimicrobial activity of a nisin-activated plastic film for food packaging. *Letters in Applied Microbiology* 41: 464–469.

Mestdagh F., De Meulenaer B., De Clippeleer J., Devlieghere F., Huyghebaert A. 2005. Protective influence of several packaging materials on light oxidation of milk. *Journal of Dairy Science* 88: 499–510.

Mette M. 2000. Oxygen permeability of polyethylene bottles. *Brauindustrie* 85: 204–208.

Min D.B., Lee H.O. 1996. Chemistry of lipid oxidation. In: *Food Lipids and Health,* McDonald R.E., Min D.B. Eds). New York: Marcel Dekker, pp. 241–268.

Mottar J. 1982. Light transmission: the influence of light on the quality of milk and milk products. In: *Technical Guide for the Packaging of Milk and Milk Products*. Brussels, Belgium: International Dairy Federation Bulletin No.143, chapter 8.

Mottar J. 1987. The usefulness of co-extruded high density polyethylene for packaging UHT milk. *IDF Dairy Packaging Newsletter* No 15. Brussels.

Moyssiadi T., Badeka A., Kondyli E., Vakirtzi T., Savvaidis I., Kontominas M.G. 2004. Effect of light transmittance and oxygen permeability of various packaging materials on keeping quality of low-fat pasteurized milk: chemical and sensorial aspects. *International Dairy Journal* 14: 429–436.

Ohst S., Golzmann G. 1996. Returnable bottles of polycarbonate. *Kunstoffe Plast Europe* 86(5): 27–28.

Olympos SA, 2007. Olympos dairy plant, Larissa, Greece.

Paine F.A. 1996. *The Packaging User's Handbook*. London: Blackie Academic and Professional, chapter 1.

Papachristou C., Badeka A., Chouliara E., Kondyli E., Athanasoulas A., Kontominas M.G. 2006a. Evaluation of PET as a packaging material for premium quality whole pasteurized milk in Greece part I. *European Food Research and Technology* 223: 711–718.

Papachristou C., Badeka A., Chouliara E., Kondyli E., Kourtis L., Kontominas M.G. 2006b. Evaluation of PET as a packaging material for premium quality whole pasteurized milk in Greece part II. *European Food Research and Technology* 224: 237–247.

Perkins M.L., Zerdin K., Rooney M.L., D'Arcy B.R., Deeth H.C. 2007. Active packaging of UHT milk to prevent the development of stale flavor during storage. *Packaging Technology and Science* 20: 137–146.

Robertson G.L. 2006. *Food Packaging Principles and Practice*, 2nd edn. Boca Raton, Florida: CRC Press.

Rosenberg M. 2002. Liquid milk products/sterilized milk. In: *Encyclopedia of Dairy Sciences*, Vol. 2, Roginski H., Fuquay J.W., Fox P.F. (Eds). London, England: Academic Press, pp. 1637–1646.

Rysstad G., Ebbesey A., Eggestad J. 1998. Sensory and chemical quality of UHT milk stored in paperboard cartons with different oxygen and light barriers. *Food Additives and Contaminants* 15: 112–122.

Rystaad G., Kolstad J. 2006. Extended shelf life milk: advances in technology. *International Journal of Dairy Technology* 59: 85–96.

Saffert A., Pieper G., Jetten J. 2008. Effect of package light transmittance on vitamin content of milk part 2. UHT whole milk. *Packaging Technology and Science* 21: 47–55.

Saffert A., Pieper G., Jetten J. 2009. Effect of package light transmittance on vitamin content of milk part 3. Fortified UHT low-fat milk. *Packaging Technology and Science* 22: 31–37.

Sattar A., Deman J.M. 1973. Effect of packaging material on light-induced quality deterioration of milk. *Canadian Institute of Food Science and Technology Journal* 6: 170–174.

Sattar A., Tavanger R., Ahmed S. 1983. Effect of packaging materials on photochemical changes in buffalo milk. *Zeitschrift fur Lebensmittel Untersuchung und Forschung* 177: 121–123.

Scheldman P., Herman L., Foster S., Hendrickx M. 2006. *Bacillus sporothermodurans* and other highly heat-resistant spore formers in milk. *Journal of Applied Microbiology* 101: 542–555.

Schroeder M.J.A. 1982. Effect of oxygen on the keeping quality of milk. *Journal of Dairy Research* 49: 407–424.

Simon M., Hansen A.P. 2001a. Effect of various dairy packaging materials on the shelf life and flavor of pasteurized milk. *Journal of Dairy Science* 84: 767–773.

Simon M., Hansen A.P. 2001b. Effect of various dairy packaging materials on the shelf life and flavor of ultra-pasteurized milk. *Journal of Dairy Science* 84: 784–791.

Skibsted L.H. 2000. Light induced changes in dairy products. *Bulletin of the International Dairy* Federation No 345, Brussels.

Topçu A., Numanoğlu E., Saldamli İ. 2006. Proteolysis and storage stability of UHT milk produced in Turkey. *International Dairy Journal* 16: 633–638.

Valero E., Villamiet M., Sanz J., Martinez-Castro J. 2000. Chemical and sensorial changes in milk quality on the keeping quality of pasteurized milk. *Letters in Applied Microbiology* 20: 164–167.

Value K.B., Castberg H.B. 1991. Processing and packaging aspects of extended shelf life products. *Australian Journal of Dairy Technology* 46: 98–100.

Van Aardt M., Duncan S.E., Marcy J.E., Long T.E., Hackey C.R. 2001. Effectiveness of PET and HDPE in protection of milk flavor. *Journal of Dairy Science* 84: 1341–1347.

Varnam A.H., Sutherland J.P. 1996. *Milk and Milk Products*. London: Chapman and Hall, chapter 2.

Vassila E., Badeka A., Kondyli E., Savvaidis I., Kontominas M.G. 2002. Chemical and microbiological changes in fluid milk as affected by packaging conditions. *International Dairy Journal* 12: 715–722.

Whited L.J., Hammond B.H., Chapman K.W., Boor K.J. 2002. Vitamin A degradation and light-oxidized flavor defects in milk. *Journal of Dairy Science* 85: 351–354.

Zygoura P., Moyssiadi T., Badeka A., Kondyli E., Savvaidis I., Kontominas M.G. 2004. Shelf life of whole pasteurized milk in Greece: effect of packaging material. *Food Chemistry* 87: 1–9.

6 Packaging and the Shelf Life of Cheese

Maria de Fátima Poças and Manuela Pintado
Food Packaging Department, Biotechnology College
Portuguese Catholic University
Porto, Portugal

CONTENTS

6.1 INTRODUCTION

Cheese represents the most diverse group of dairy products. Manufacture and ripening involve a dynamic and synchronized series of biochemical and microbiological processes, leading to a product with typical, refined, and desirable aroma and flavors. Product defects, with off-flavors and odors, textural or surface indices of failure, may result if these events are unbalanced due to internal or external factors. Packaging systems or coating solutions applied either throughout the ripening period or during storage may overcome most of the unbalanced reactions and deficiencies of cheeses.

Packaging is increasingly recognized as an important factor in protecting and controlling the quality and safety of cheese, as well as in addressing consumer issues. Cheese is not an inert product: it will continue to ripen after packaging, and the packaging should ideally not interfere with the natural maturation process, and at the same time it should increase the cheese's shelf life. The balance between these two requirements, and the complex and naturally changing nature of the product, makes cheese packaging design a challenging task.

This chapter presents, first, some of the more common cheese varieties in terms of characteristics, production processes, and the deteriorative reactions that limit the shelf life. Each variety has its own packaging requirements regarding barrier properties to gases, moisture, and light, but they can be broadly divided into two categories: (a) breathable packaging, which plays a critical role in controlling the ripening of the cheese through its moisture and gas permeability characteristics; this packaging is used for cheese varieties with an active surface microflora (e.g., bacterial surface-ripened or mold-ripened cheeses) that have a relative short shelf life; and (b) barrier packaging (e.g., vacuum packages) used for hard varieties of cheese, which generally ripen for long times in an anaerobic environment and have a long shelf life. Detailed considerations of package properties relevant for cheeses in general are considered first and then the packaging and shelf life of hard, soft, fresh, and processed cheeses are analyzed, giving examples of systems used. The chapter concludes with some new solutions for cheese packaging that are at different stages of development.

6.2 CHEESE

6.2.1 Classification, Chemical and Physical Characteristics

Cheese making began about 8000 years ago, and today there are more than 1000 cheese varieties worldwide, each unique with respect to its flavor and form (Beresford et al., 2001). Cheese is a generic name for a group of fermented milk-based food products produced in a wide range of flavors and forms throughout the world. The diversity in the texture, functional properties, flavor, and aroma of the many cheese types is associated mainly with differences in milk composition, key cheese-manufacturing processes (including starter cultures, clotting enzymes, adjuvants, and ripening conditions), and storage temperature and time (Coker et al., 2005).

Classification of cheese is not straightforward, as several criteria can be used: moisture, rheological properties, texture, and various processing factors (Fox, 1999). Three main approaches have been used: (1) texture, which is largely determined by moisture and fat content, (2) ripening indices, and (3) method of coagulation, coupled with other factors (McSweeney, 2007a). Probably the most consensual scheme proposed considers four or five groups based on hardness and moisture content, from very hard to soft. However, from the packaging point of view, it is considered more useful to divide cheeses into hard, soft, and fresh classes. In addition, processed cheese, which is obtained from a blend of natural cheese heated with water and emulsifying salts and packaged while still hot, represents a distinct class. Table 6.1 represents characteristics of some popular varieties of cheeses. Hard cheeses include very hard (e.g., Parmesan from Italy) and hard or semihard cheeses (e.g., São Jorge from Portugal). Soft cheeses include semisoft (e.g., Serra da Estrela from Portugal), soft ripened cheeses (e.g., Brie or Camembert from France), and blue vein mold ripened (e.g., Roquefort from France). Fresh unripened cheese (e.g., Cottage), although soft, is considered separately because of its very high moisture content.

TABLE 6.1
Characteristics of Some Popular Cheeses

Cheese and Origin	Milk	Ripening or Curing Time	[a]Fat in Dry Matter, %	[a]Moisture, %	Texture
Fresh					
Cottage (USA)	Cow, skimmed	Unripened	5–15	80	Soft, curd particles of varying size
Ricotta (Italy)	Cow, whole or partly skimmed; sheep whey	Unripened			Soft, moist, or dry
Soft ripened					
Mozzarella (Italy)	Buffalo; cow	3 weeks	40–50	62–76	Slightly firm, plastic
Brie (France)	Cow	4–8 weeks	45–55	49–58	Soft, smooth when ripened
Camembert (France)	Cow	4–8 weeks	45–55	49–62	Soft, smooth, very soft when fully ripened
Brick (USA)	Cow	2–4 months	50	44	Semisoft to medium firm, elastic, numerous small eyes
Serra da Estrela (Portugal)	Ewe	4–6 weeks	53–58	50–55	Semisoft paste, creamy, white or slightly yellow, uniform (without holes or just a few)
Hard ripened					
Cheddar (UK)	Cow	1–12 months or more	48–55	36–39	Firm, smooth, some openings
Edam (Netherlands)	Cow; partly skimmed	2–3 months	40–50	42–53	Semisoft to firm, smooth, small irregularly shaped or round holes
Gouda (Netherlands)	Cow; whole or partly skimmed	2–6 months	48–55	41–45	Semisoft to firm, smooth, small irregularly shaped or round holes
Provolone (Italy)	Cow	2–12 months or more	45–50	42–49	Firm, smooth
Emmental (Switzerland)	Cow	3–9 months	45–55	37–40	Firm, smooth with large round eyes
Parmesan (Italy)	Cow partly skimmed	14 months–2 years	32–42	36	Very hard, granular
São Jorge (Portugal)	Cow	4–6 months	31–34[b]	37–38[b]	Semihard (8–15 kg and cylindrical in shape), typical "picante" flavor, irregular eyes in the interior cheese
Blue-vein mold ripened					
Blue (France)	Cow	2–6 months	50[b]		Semisoft, pasty, sometimes crumbly
Gorgonzola (Italy)	Cow/goat	3–12 months	53–56[b]	48[b]	Semisoft, pasty, sometimes crumbly
Roquefort (France)	Ewe	2–5 months	52[b]	45[b]	Semisoft, pasty, sometimes crumbly

Source: Adapted from USDA. 1995. How to buy cheese. Home and Garden Bulletin 256. Agricultural Marketing Service. http://www.ams.usda.gov

[a] According to proposed Codex Cheese Standards.

[b] Other sources.

6.2.2 Production

Modern cheese production is the result of a combination of traditional ancient practices and new technological tools and processes. Despite the changes in the scale of cheese production and degree of mechanization, the principles of cheese making have remained the same and are designed to reduce water content, preserve the nutritional properties of the milk, and obtain a safe product with unique sensory characteristics throughout storage and distribution (Coolbear et al., 2008).

The manufacturing process is represented in Figure 6.1. The prepared milk (raw or pasteurized) is brought to temperature and mesophilic or thermophilic starter cultures, or both, and eventually adjunct cultures, are added. The species and strains, selected according to cheese type, are added to transform lactose to lactic acid and to develop characteristic flavors and texture during ripening. In the case of fresh cheeses, which will not undergo ripening, starter cultures are not added. The coagulant rennet is added and the milk is left to coagulate under controlled conditions of type and amount of coagulant (animal, vegetable, or microbial), temperature, and time (Coolbear et al., 2008). Most cheeses are rennet-coagulated, except some fresh cheeses such as Cottage cheese, which may alternatively be acid-coagulated.

Drainage consists of separating the whey and occurs during several operations: cutting of the coagulum, cooking and curd washing, molding, and pressing. After drainage the molded curd is salted by immersion in brine or deposition of salt on the surface. Important variables include the size of the curd particles after cutting, the cooking temperature, whether or not the curd is washed, the final pH of the curd, the temperature of the curd at stretching (for Mozzarella), the method of salting and amount of salt used, and the final pH of the cheese. The definition and control of these variables determine the composition of the cheese, as well as the biochemical events during cheese ripening that define the texture and flavor of the cheese (Broome and Limsowtin, 1998).

Most cheese varieties undergo a period of ripening (curing and maturation), which ranges from about 3 weeks for Mozzarella, for example, to 2 years or more for Parmesan and extra-mature Cheddar or Gouda (Beresford et al., 2001). The duration of ripening is generally inversely proportional to the moisture content of the cheese, although many varieties may be consumed at any of several stages of maturity, depending on the flavor and texture preferences of consumers (Robertson, 2006).

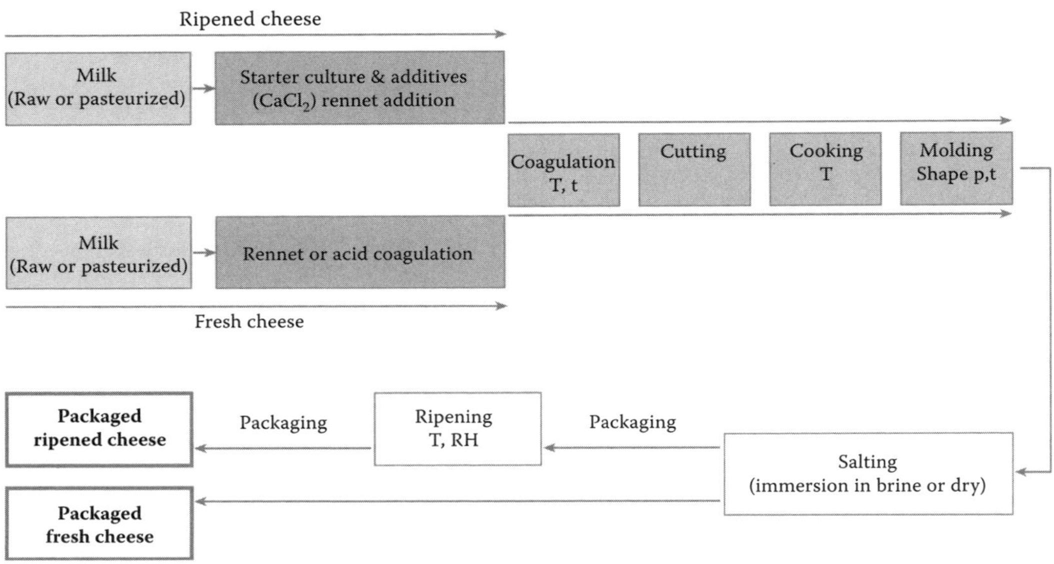

FIGURE 6.1 Schematic representation of fresh and ripened cheese manufacture. T: temperature; t: time, p: pressure, RH: relative humidity.

6.2.3 Microbiology of Ripening

The diversity of microorganisms in the cheese ecosystem establishes interactive associations that contribute to the complexity of cheese flavor. These interactions can result in a balanced flavor but can also contribute unfavorably to the development of off-flavors or other quality defects. Hence, the control of the microorganisms present in cheese is one of the most powerful tools available today to develop unique and characteristic cheese flavors (Pelaéz and Requena, 2005). Examples of cheeses ripened by different microflora are presented in Table 6.2.

The microflora can be divided into two main groups: starter and secondary flora. The starter flora, added at the beginning of manufacture (internal bacteria) or naturally present in the milk, are responsible for acid development during cheese production. Mesophilic and thermophilic starters, with optimal growth temperatures of approximately 30°C and 45°C, respectively, are used. Common mesophilic bacteria are *Lactococcus lactis* subsp. *cremoris*, *Lc. lactis* subsp. *lactis*, *Leuconostoc lactis*, and *Ln. mesenteroides* subsp. *cremoris*; thermophilic lactic acid bacteria are *Streptococcus salivarius* subsp. *thermophilus*, *Lactobacillus helveticus*, and *Lb. delbrueckii* subsp. *Lactis*, which may be used either individually or in combination, depending on the cheese variety. Thermophilic starters are added for the production of semihard and hard cheeses, typical of Italian and Swiss varieties (Marilley and Casey, 2004).

The secondary flora is composed of complex mixtures of bacteria, yeasts, and molds, and contributes significantly to the specific characteristics of a particular cheese variety. This flora may be intentionally added or develop spontaneously through contamination from the surrounding environment (Beresford et al., 2001). Some of these secondary cultures can grow on the surface, as in the case of smear-ripened (Le Gruyère, Limburger, etc.) and surface-mold-ripened (e.g., Camembert and Brie) cheeses; others can grow internally, producing CO_2 (leading to eye formation), propionate, and acetate, as in the case of Swiss varieties (e.g., Emmental).

The surface microflora of smear-ripened cheese have two important functions: (1) production of enzymes (lipases, proteinases, and peptidases) and (2) deacidification of the cheese surface and then the cheese body. The results of various studies confirm that yeasts are the dominant microorganism during the early stages of ripening, followed by bacterial domination in subsequent stages (Corsetti et al., 2001). *Geotrichum candidum* is an important component of the microflora of soft cheeses such as Camembert and semifresh goat's and ewe's milk cheese (Boutrou et al., 2005).

Mold-ripened cheeses are divided into two groups (see Table 6.2): those that are ripened due to the presence of *Penicillium roqueforti*, which grows and forms blue veins within the cheese (e.g., Roquefort, Gorgonzola, Stilton, and Danish Blue), and those that are ripened with *Pe. camemberti*, which grows on the surface of the cheese (e.g., Camembert and Brie).

TABLE 6.2
Examples of Secondary Flora that Support the Ripening of Known Cheeses

	Mold		Surface
Internal Bacteria	**Surface**	**Internal**	**Yeast and Bacteria**
Without eyes			
Parmesan			
Cheddar			
With eyes			Havarti
Emmental	Brie	Roquefort	Limburger
Edam, Gouda	Camembert	Gorgonzola	Münster
Pasta-fillata:		Danish Blue	
Provolone			
Mozzarella			

During the maturation of many cheeses, the starter lactic acid bacteria population declines and the initial small population of adventitious nonstarter lactic acid bacteria (NSLAB) becomes the dominant bacterial population in the maturing cheese. The proteolytic activity of NSLAB appears to complement that of the starter, producing peptides with generally similar molecular weights and free amino acids (Sousa et al., 2001). The NSLAB cohort of Cheddar cheese was shown to be a variable population that changed during ripening but was comprised mainly of strains of *Lactobacillus paracasei* and *Lb. rhamnosus* (Coolbear et al., 2008).

Aroma development in cheese products results from the metabolic activities of cheese bacteria by glycolysis, lipolysis, and proteolysis (Marilley and Casey, 2004). The most complex of these biochemical events (proteolysis) is caused by agents from a number of sources: residual coagulant (usually chymosin), indigenous milk enzymes, starter, NSLAB, and, in many varieties, enzymes from secondary flora. Proteolysis contributes to textural changes of the cheese matrix due to breakdown of the protein network, decrease in a_w through water binding by liberated carboxyl and amino groups, and increase in pH (in particular in surface-mold-ripened varieties), which facilitates the release of sapid compounds during mastication. It contributes directly to flavor and off-flavor (e.g., bitterness) of cheese through the formation of peptides and free amino acids (Sousa et al., 2001).

Lipolysis is also an important biochemical event during cheese ripening, particularly in varieties such as Blue and hard Italian cheeses, where it is very extensive and a major pathway for flavor generation. However, in the case of other cheeses such as Cheddar and Gouda, in which the extent of lipolysis is only moderate, the contribution of lipolytic end products to cheese quality and flavor is less important (Collins et al., 2003).

There are several factors influencing growth of microorganisms during cheese ripening, including water content, salt concentration, pH, and ripening temperature (Beresford et al., 2001). Microbiology of cheese is a complex and dynamic equilibrium; it results in an appreciated product when well balanced, but is responsible for product failure in terms of sensory characteristics and safety issues if unbalanced. The combination of these factors and the extent of their control during cheese manufacture and ripening dictate the success of the final cheese product, which may be a safe and good-quality cheese or an unsafe and defective cheese. Packaging systems can prevent some of these problems by positively controlling the factors that influence microorganisms during ripening or storage of cheese.

6.3 DETERIORATIVE REACTIONS AND INDICES OF FAILURE

6.3.1 Consumer Attributes

Product optimization in terms of consumer preference is the ultimate aim of every food manufacturer, and a company's ability to produce a product that satisfies consumer requirements leads to success and profitability. Consumer acceptance of a food is, however, dependent on many different factors, which may be related to the product itself, the consumer, or the consumer environment. In particular, the sensory appeal of a food product and the visual appearance of its packaging are powerful influences on consumer acceptability (Murray and Delahunty, 2000).

There is a general consumer trend toward natural food products with claimed nutritional and health benefits, and cheese is no exception. Claims for low fat and low salt content, lowering cholesterol and risk of osteoporosis, contributing to dental health, and so on are associated with cheese (Karvonen et al., 2002; NDC, 2006). Convenience is a major decision factor for consumer purchase, both for cheese and for its packaging. Key market trends include demand for elaborate products: sliced, diced, and grated, and assemblies of portions of different cheeses (Sealed Air Co., 2007). Smaller portions have also been developed for many prized cheeses that are too large and expensive for the decreasing size of households. This brings extra requirements in terms of packaging because many products do not sustain a reasonable shelf life when cut. There is a strong emphasis on easy-open and reclosable features, and in addition, in many cases, the package aspect must

TABLE 6.3
Attributes of Cheddar Cheese and Its Packaging

Attributes of Cheddar cheese packaging	
Shape	Unconventional
Aesthetics	Illustrative, graphics work, bold and rich colors
Information on product performance	Sensory information, specialized for diet, maturity level, branding, nutritional information, other
Packaging performance	Convenience (easy open), security (tamper proof)
Presentation in pack	Visibility and color of cheese
Overall packaging features	Hand-made, original, traditional, expensive
Attributes of Cheddar cheese sensory characteristics	
Aroma	Pungent, caramel, sweaty/sour, sweet, creamy, fruity
Flavor	Buttery, rancid, mushroom, moldy, nutty, smoky, soapy, processed, sweet, salty, acidic, bitter, astringent
Overall flavor	Strength, balanced
Appearance	Color, mottling, open, shiny
Texture	Firm, rubbery, crumbly, smooth, moist, grainy, mouth coating

Source: Adapted from Murray J.M., Delahunty C.M. 2000. Mapping consumer preference for the sensory and packaging attributes of Cheddar cheese. *Food Quality and Preference* 11: 419–435.

give a traditional appearance to the product. Hence, cheeses are sometimes overpackaged in poplar wood.

The attributes of Cheddar cheese packaging used to rate consumer preferences and to differentiate between different packaged cheeses are presented in Table 6.3. These attributes were considered by a trained panel and then validated by a consumer panel of 200 individuals. Table 6.3 also includes the attributes used to describe the sensory characteristics of the Cheddar cheese (Murray and Delahunty, 2000).

6.3.2 Flavor and Texture

Flavor is developed during aging as a consequence of microbial and enzymic changes to residual lactose and to lactate and citrate. Parallel to these changes, liberation of fatty acids (lipolysis) and peptides and amino acids (proteolysis) also occurs. The free fatty acids and amino acids undergo subsequent metabolism to volatile flavor compounds. Cheese texture is influenced greatly by the moisture content of the cheese and its calcium, fat, and fat-in-dry-matter levels. However, textural changes during ripening are due to solubilization of calcium phosphate, hydrolysis of the casein matrix, changes to water binding within the curd, and loss of moisture caused by evaporation from the cheese surface (McSweeney, 2007b).

6.3.3 Microbial Deterioration

Owing to ancient cheese practices (and because of the higher sensory quality of unpasteurized milk), most traditional cheeses are produced with unpasteurized milk. To accommodate the population that prefers unpasteurized cheese, the Canadian Food Premises Regulation allows for the sale of unpasteurized cheese provided that it is stored for 60 days above 2°C. This holding period is intended to encourage competitive organisms that will virtually eliminate the pathogens within the cheese (Teng et al., 2004). However, these cheeses are commonly affected by several defects, namely undesirable color and pigmentations associated with the uncontrolled growth of microorganisms such as molds and yeasts on the surface.

Cheeses obtained from pasteurized milk are less prone to loss of quality or safety concerns, because pathogens are destroyed via heat treatment and a specific starter culture is added that leads to a controlled ripening and consequently a more successful final cheese. However, some spoilage and pathogenic microorganisms may be present both in unpasteurized milk cheese after a 60-day ripening period and even in pasteurized milk cheese. Cheese food-poisoning outbreaks in Canada, the United Kingdom, the United States, and Western European countries have resulted from raw milk cheese, from contamination of lactic cultures, and especially from improper pasteurization. Cheeses most commonly responsible for outbreaks in Europe include Brie, Camembert, Vacherin, Farmhouse, Mexican-style, Cheddar, homemade, and soft and unripened cheese varieties (Teng et al., 2004).

The occurrence of cheese defects results, in most cases, from low milk quality or from problems related to technology or to processing, transport, or storage failure. The occurrence of these defects can significantly affect the overall quality and value of cheeses. Unfortunately, some defects associated with poor milk quality or inadequate cheese-making practices do not develop until the aging or distribution stages. Table 6.4 presents some of the factors that cause failures in cheeses that may be due to deficiencies in raw materials, processing practices, environment, and packaging, resulting in microbial growth and changes in texture, flavor, and color. Some defects may not present safety problems to the consumer. However, color and flavor attributes are very important cheese acceptance criteria for consumers. This means that if cheese does not comply with the specifications set for quality or authenticity, it will undoubtedly be rejected by consumers.

Several of these defects may be overcome by suitable packaging, especially visual appearance failures. The main visual appearance failures are related to mold and yeast growth (see Table 6.4) that leads to color spots on the cheese surface or inside. *Debaryomyces hansenii*, *Kluyveromyces marxianus* var. *lactis*, *K. marxianus* var. *marxianus* (the perfect state of *Candida kefyr*), and *Saccharomyces cerevisiae* are all associated with cheeses. Typical defects caused by spoilage yeasts are gas production, yeasty flavor and other off-flavors, discolorations, and changes of texture (Jakobsena and Narvhu, 1996).

The most important spoilage mold species from several countries for hard, semihard, and semisoft cheeses without preservatives added are *Penicillium commune* and *Pe. nalgiovense* (Lund et al., 1995). *Penicillium* (*Pe. brevicompactum*, *Pe. commune*, *Pe. palitans*, *Pe. solitum*, and *Pe. roqueforti* subsp. *roqueforti*) and *Geotrichum candidum* were the most frequently isolated contaminants from semihard cheeses (Kure et al., 2004). *Debaryomyces hansenii* and *Galactomyces geotrichum* prevailed in rennet cheeses, and *Kluyveromyces marxianus* and *Pichia membranaefaciens* were the main species found in acid-cured cheese. The prevalence of *Yarrowia lipolytica* usually indicates an improper yeast population that may lead to poor cheese quality (Valdks-Stauber et al., 1997). There is evidence that this bacterium is responsible for the browning process related to tyrosine transformation in melanoidins (Carreira et al., 1998).

Anaerobic conditions inside packaged cheese must be controlled because *Clostridium* spp. may grow and produce gas due to anaerobic fermentation of lactate (see Table 6.4). Active packaging systems with antimicrobials (AMs), as well as packages with adequate oxygen permeability, may solve some of these frequent microbial failures. Chemical failures associated with excessive water loss or rind decolorization due to excessive exudation, off-flavors, and defatted tastes originate from oxidation (rancidity) or via microbial metabolism (bitterness, rancidity, acidity), and openings and irregular holes formed by uncontrolled microbial fermentation may be prevented via specific and adequate packaging systems as described in the following text.

The selection of the optimum packaging system must consider the fact that cheese is a complex dynamic matrix in which several microbial, physical, and biochemical changes occur during storage. The water activity, fat and salt contents, and cheese microflora regulate the biochemical changes that occur during ripening and determine the flavor, aroma, and texture of the cheese. The shelf life of the cheese will be mostly affected by temperature, light, O_2 and CO_2 concentrations, and relative humidity. Temperature is a factor mainly controlled by the distribution chain, but

TABLE 6.4
Microbial, Textural and Flavor Indices of Failure

Cheese Indices of Failure (IoFs)	Agent(s) Responsible	Possible Cause
	Microbial indices of failure (affecting visual appearance)	
Mold in vacuum-packed Cheddar-type cheeses	*Penicillium* spp. (*Pe. commune* and *Pe. roqueforti*), and black moulds (*Cladosporium cladospoiroides*)	Usually associated with air pockets and free whey drawn from the cheese blocks during vacuum packaging
Mold in other cheese types—within soft or white mold cheeses	*Pe. roqueforti* which is capable of growth in the low O_2 environment *Mucor* spp. and *Rhizopus* spp. (black molds) *Cladosporium herbarum* (dark green spots on cheese surfaces) *Debaryomyces hansenii* (*Candida famata*), *Pe. commune* or *Pe. nalgiovense* (brown spots in Blue cheeses)	Poor hygienic conditions in cheese manufacturing and/or cheese handling and ripening practices
	Yarrowia lipolytica (browning process related to tyrosine)	High relative humidity aerobic conditions during ripening
	Scopulariopsis fusca (brown spots on cheese surfaces)	Colonization on paper and packaging materials stored under unsuitable conditions and spread to cheese surfaces
Slimy and discolored surface	Products of microbial growth (e.g., peptides and polysaccharides); the coryneforms often produce yellow, lavender to light-purple pigments, or orange-colored surface	Surface not dried due to high environment relative humidity
Late gas blowing (hard brine-salted cheeses)	Anaerobic fermentation of lactate by *Clostridium* spp. (*C. tryobutyricum*) to butyrate, H_2, and CO_2	Presence of spores in milk; poor hygiene practices
	Chemical indices of failure (affecting visual appearance)	
Decolorization or disappearance of cheese rind	Exudation of inner moisture to the surface and reduction or elimination of ripened rind	Ripened semisoft cheeses packaged under vacuum or modified atmosphere promote the accumulation of water between the surface and the packaging material
Rind drying and loss of weight	Dehydration of cheese surface caused by excessive water evaporation due to surface exposure	Lack of packaging or package with poor barrier to moisture; too long a storage period
Discoloration of pigments	Photo-oxidation of lipids	Combined effect of O_2 and light during display in retail cabinets; packaging is poor barrier to light and/or O_2
Pink discoloration in annatto-added cheeses	Photo-oxidation of lipids	Combined effect of O_2 and light during display in retail cabinets
Pink discoloration in cheeses without annatto	Products of the Maillard reaction caused by the presence of galactose (which accumulates due to metabolism of *Streptococcus thermophilus*), N-compounds and a critical O_2 concentration	*Streptococcus thermophilus* and O_2 permeability of the packaging material; presence of yeasts or *Enterococcus*
	Flavor indices of failure	
Bitterness	Peptides with molecular mass ca.<6 kDa and a mean hydrophobicity >1400 cal per residue are often bitter	Incorrect patterns of proteolysis

(*Continued*)

TABLE 6.4 (*Continued*)

Cheese Indices of Failure (IoFs)	Agent(s) Responsible	Possible Cause
Rancidity	FFA or derived compounds from the degradation of lipids by the action of lipases on the triacylglyerols of milk fat	Lipolytic activity of enzyme (microbial or from the rennet); ripening at elevated temperatures and/or for prolonged durations
Acidity	Excessive acidity formed during consumption of lactose	Level of starter culture added, temperatures and cheese-making protocol
Off-flavor	Oxidation of other constituents of the cheese, such as riboflavin	Oxygen and light combined effect during retail display cabinets
	Textural indices of failure	
Weak body	Weak casein network structure	High levels of fat and moisture compared with casein levels
Openings and irregular holes	Slits, factures, and irregular holes	Entrapped air in cheese structure or gas production by adventitious microorganisms
Blister under the wax coating	Microbial activity with gas production	Defects on the rind before wax application

Source: Adapted from *Cheese Problems Solved*, McSweeney P.L.H. (Ed). Cambridge: Woodhead Publishing Ltd.

packaging plays an important role in controlling the effect of light, gases, and humidity. Migration of packaging components into the cheese must also be considered for food safety reasons.

6.3.4 Light-Induced Deterioration

Light-induced oxidative processes occur when light from the sun and particularly from illumination on retail shelves passes through the packaging material and reaches the cheese surface, and when simultaneously there is O_2 in the headspace of the package (Mortensen et al., 2004). The sensitivity of dairy products to light depends mainly on the presence of O_2 and the photosensitizing agent riboflavin (vitamin B_2). The latter is capable of absorbing energy and initiating an oxidative chain reaction that can lead to the development of off-flavors, the loss of nutrients such as vitamins and amino acids, and the discoloration of pigments (Alves et al., 2007). Thus, two main factors influence the process: the amount of light and the amount of O_2 present in the package headspace. The former is determined obviously by the light source (not discussed here) and by the barrier the packaging may impose.

Incident light is partially absorbed and partially reflected by the package, and only a portion of it is transmitted to the cheese. The light transmitted through a given packaging material will have characteristics determined by the spectrum of the incident light and the modification determined by the absorption characteristics of the packaging material. Light protection offered by packaging materials depends on the following characteristics (Mortensen et al., 2004).

6.3.4.1 Inherent Absorption Characteristics

Different material categories offer varying degrees of protection: metals offer the best protection, followed by paper and paperboard, various plastics, and finally glass, which transmits about 90%. Differences also exist within each class of material; for example, unbleached paper and paperboard offer a higher barrier than bleached because of the removal of or alterations to the lignin.

6.3.4.2 Thickness

Increasing the thickness of the material leads to lower light transmission; for example, using aluminum foil instead of metalized material, or using a paper with higher grammage, will increase protection.

6.3.4.3 Material Processing Conditions

Materials with a lower degree of molecular orientation or with a higher degree of crystallinity are better barriers to light. Cavitation also promotes reflection of light and pearlized polypropylene (PP) reflects more light than noncavitated films.

6.3.4.4 Coloration, Pigmentation, Printing, and Use of Labels

Incorporation of pigments such as titanium dioxide (white) and carbon black reduces light transmission. Titanium dioxide increases light scattering and reduces transmittance, especially of light at wavelengths shorter than 400 nm. Decreasing degrees of light transmission were reported for different inks: uncolored (lowest), then yellow, green, brown, and black (Mortensen et al., 2002). Attaching labels to the package introduces an extra barrier between the light source and the packaging.

Figure 6.2 presents the light transmission of two packaging materials based on a paper and plastic combination. The material used for Camembert gives an average transmittance of <5%, whereas the material used for Gruyère provides a lower barrier, giving an average transmittance of approximately 12% at wavelengths >400 nm.

6.3.5 Oxygen

Oxygen present in the headspace is available for oxidation reactions, and the amount depends on the initial atmosphere concentration and on the O_2 barrier provided by the packaging material.

6.3.5.1 Initial Amount

Vacuum packaging and flushing with a gas or mixture without O_2 reduce the O_2 content in the headspace. The critical residual levels of O_2 depend on the cheese: levels as low as 1% may result in rapid formation of off-flavors for more sensitive cheeses. Very low levels of residual O_2 can be achieved using O_2 absorbers in addition to modified atmospheres (MAs). The maximum allowable initial amount of O_2 also depends on the headspace-volume-to-cheese ratio, that is, the total amount of O_2

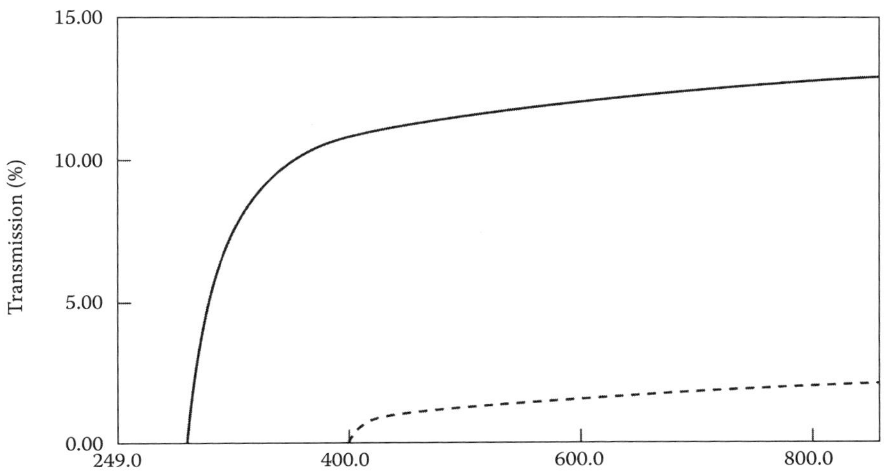

FIGURE 6.2 Light transmission of different packaging materials. (- - -) Camembert, (—) Gruyère.

to the total amount of photo-oxidative substrate in the cheese. This ratio is related to the packaging format and geometry. Increased cheese surface area, in portions, slices, and shredded cheese, leads to a more severe effect due to higher exposure to light and to O_2.

The longer shelf life of cheeses packaged in MAs is due not only to the exclusion of O_2 but also to CO_2 inhibition of the growth of many spoilage microorganisms (Romani et al., 2002). However, very high concentrations of CO_2 can have an adverse effect on cheese taste or aroma, particularly for cheeses with higher fat content. Therefore, the concentration must be optimized, maximizing the (AM) effect and minimizing any adverse sensory effect.

Ineffective packaging processes such as gas flushing and compromised thermoforming or sealing operations and accidental minor damage during handling or transportation may affect a small fraction of packaged products in MAs, which will undergo spoilage if not repacked. Hence, knowledge of the actual gas composition in MA packs, and particularly of the level of O_2, is very important to assess product integrity. An optical O_2 nondestructive analyzer to assess MAs in cheese packages on a small industrial scale was recently developed (O'Mahony et al., 2006).

6.3.5.2 Oxygen Permeability of Package

The permeability of the package depends essentially on the materials, the tightness of the seals or closures, and temperature. When the values of O_2 permeability are presented, it is important to specify the measurement conditions, in particular temperature and relative humidity. The latter is fundamental for hydrophilic materials, such as polyamide (PA) and ethylene-vinyl alcohol (EVOH), whose O_2 barrier is greatly influenced by humidity. The permeability also depends on the temperature. This is important because, although temperature has only a minimal influence on light-induced lipid oxidation, packaging material permeability follows an Arrhenius-type dependence on temperature, allowing higher rates of O_2 ingress at higher temperatures. Thus, at higher temperatures, more O_2 will be available for oxidative processes. In most cases, suppliers present permeability or transmission rate data at standardized conditions (23°C or 38°C) that are not the same as the storage temperatures used for cheese. Very few permeability data are available at the typical temperatures and humidities encountered by packaged cheese.

6.3.6 Humidity and Moisture Loss

At equilibrium, the relative humidity of the surrounding environment inside the package is equal to the cheese water activity, which influences its microbiological and physicochemical evolution over time. Relative humidity has a direct impact on the weight loss and texture of cheese. In unpackaged cheese, water loss depends on the chemical properties of the cheese (particularly salt content) and on the storage conditions, temperature, and relative humidity. In packaged cheese, water loss depends in addition on the permeability to moisture of the packaging material. Figure 6.3 shows the scheme of water interaction in a packaged cheese. The optimal packaging has a permeability to moisture that equilibrates the cheese water loss rate with the flux of moisture permeating out of the package at a water activity that minimizes the surface-mold growth and simultaneously yields good textural and sensory properties (Pantaleão et al., 2007).

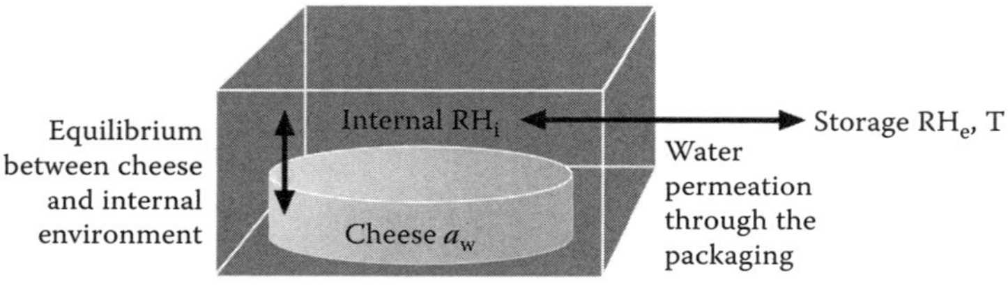

FIGURE 6.3 Water interactions in packaged cheese.

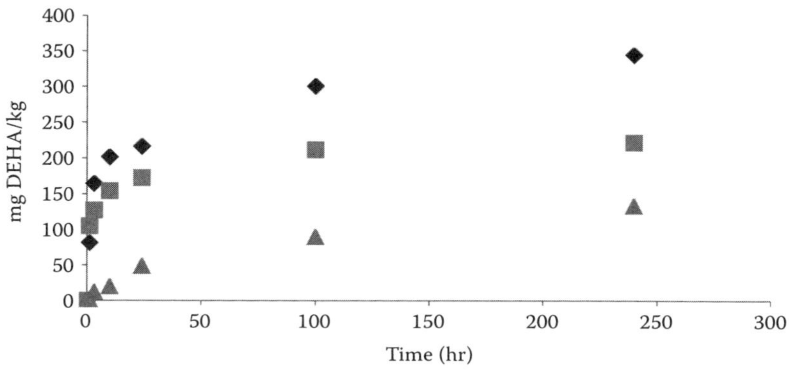

FIGURE 6.4 Migration of DEHA from PVC into Feta (▲), Edam (■), and Kafelotry (◆) cheese. (Adapted from Goulas et al., 2000.)

6.3.7 Migration from Packaging into Cheese

The migration of components from packaging materials must also be considered. Migration of monomers such as styrene from polystyrene (PS) packages (Bendall, 2007) and additives such as plasticizers from films (Goulas et al., 2000; Grob et al., 2007) has been recorded in cheese packaged in plastic polymers. Most of the migration studies are performed in food simulants, because of standardization, analytical determinations, and compliance with legislative issues. For a given migrant, the rate of migration and the amount of migrant transferred from the material into the food depend on the contacting material and on the nature of the food, in this case the cheese. Polyolefins, and particularly polyethylene, are among the plastics showing highest migration rates for most additives. Studies on the migration of diphenylbutadiene, triclosan, and butylated hydroxytoluene (BHT) from low density polyethylene (LDPE) film into cheeses containing different amounts of fat and water showed not only that the fat content influences the migration of lipophilic migrants but also that the ratio of fat to water and the consistency of the cheese play an important role in the whole process (Cruz et al., 2008). Figure 6.4 shows the migration of diethylhexyl adipate (DEHA) from poly(vinyl chloride) (PVC) into three different types of cheeses.

6.4 PACKAGING AND SHELF LIFE OF HARD AND SEMIHARD CHEESES

Hard cheeses such as Parmesan, Mozzarella, Cheddar, Edam, Gouda, and Emmental are ripened by internal bacteria. These cheeses have a slow rate of ripening because of their low moisture and high salt content. They ripen for very long times in an anaerobic environment, usually in packages with very low permeability, such as coatings, vacuum packaging, or modified atmosphere packaging (MAP) with binary mixtures of CO_2 and N_2.

6.4.1 Synthetic Cheese Coatings: Wax and Water-Based Dispersions

Mineral waxes used for coatings on cheeses consist of refined hard paraffin, petroleum jelly, and microcrystalline waxes with various additives. Paraffin wax comprises mostly saturated aliphatic unbranched alkanes with average molecular weights within the range 280–560 Da (carbon chain length C18–C60), with a branched alkane content of 10–40% (molecular weights 450–800 Da). Petroleum jelly is a blend of theses waxes and mineral oil (Castle et al., 1993). Wax can make an important contribution to the image of a product, and a growing range of different colored waxes are used to differentiate cheeses: yellow, orange, black, white, blue, brown, red, purple, and green. Wax protects the cheese from mold growth and weight loss through moisture evaporation and prevents aerobic ripening because of its barrier to O_2. Mineral waxes present a higher barrier to O_2 than acetoglyceride-based waxes, which are also used for cheese. Gouda cheeses are typically coated with yellow or white waxes, whereas Edam cheeses are coated with red wax.

Water-based dispersions are typically copolymers of ethylene and vinyl acetate. These coatings may be uncolored or pigmented with different colors and are used as carriers for antifungal agents such as natamycin (E235), calcium sorbate (E203), and potassium sorbate (E202). These coatings are applied in one or several layers and are also applied before the wax layer.

6.4.2 Vacuum Packaging and Modified Atmosphere Packaging

Vacuum packaging and MAP are used particularly for portioned and sliced hard cheese. Sliced and portioned cheese is more prone to deteriorative changes than the whole cheese because of a larger surface area exposed to light and O_2, and therefore high-barrier packaging must be used. Multilayer films composed of combinations of PA or EVOH as gas barriers and polyolefin-based materials such as linear LDPE (LLDPE), ethylene-vinyl acetate copolymer (EVA), and ionomers as moisture barriers and sealing layers are used.

Parmigiano-Reggiano cheese is traditionally sold unpackaged, but nowadays it is more frequently commercialized in portions of 250–300 g and commonly packaged under vacuum (Romani et al., 2002). Portions of Parmigiano-Reggiano are sold at a recommended temperature of 4–8°C and are given a shelf life of 6 months. Protection against lipid oxidation is obtained when O_2 is excluded by vacuum packaging, even if the cheese is exposed to light (Severini et al., 1998). Gouda cheese cuts are vacuum-packed and have a 10-week shelf life (Kreft, 2008).

Vacuum packaging may have a negative impact on the appearance of some cheeses, particularly those with typical eyes, such as Emmental (Swiss-type) and Edam or Gouda (Dutch-type), which may collapse under reduced pressure. For more resistant cheeses, vacuum skin packaging (VSP) is very common, with packaging structures with varying O_2 barriers.

In MAP, the gas mixture should be optimized for each cheese. Some cheeses withstand mixtures with compositions richer in CO_2; others suffer from sensory problems and package collapse MAs when a higher percentage of CO_2 is used.

Portions of Parmigiano-Reggiano cheese packaged in MAs showed different behavior in the texture according to the CO_2 concentration in the mixture but similar evolution of the flavor profile. Mixtures of 30:70 CO_2:N_2 evolved toward a more cohesive and friable structure than mixtures with a higher CO_2 content (Romani et al., 2002). Grated or diced Parmigiano-Reggiano cheese is sold commercially in a three-layer pouch of PP/EVOH/LLDPE with an MA and is given a shelf life of 3 months. Poly(ethylene terephthalate) (PET) may be used in place of PP for extra mechanical resistance.

In another study, a gas mixture of 30% CO_2 and 70% N_2 best preserved portioned Provolone cheese: the proteolytic and lipolytic phenomena typical of cheese ripening were slowed down. This packaging system extended the shelf life to more than 9 months at 8°C, which represented an increase of 50% with respect to vacuum packaging (Favati et al., 2007). The packaging film was 20-μm PA/80-μm LDPE with an O_2 permeability at 23°C of 77×10^{-11} mL cm^{-2} s^{-1} (cm Hg)$^{-1}$. Although the study indicated that 400-g portions of cheese were packaged, unfortunately the surface area of the packs was not given. In commercial applications, pillow pouches of portions of Provolone are given a 6-month shelf life when refrigerated at 4–8°C.

In order to estimate the shelf life of Provolone cheese using the results obtained in the experiments for the lipolytic process, a mathematical model based on the free fatty acid content was developed. The model considered the kinetics of the evolution over time of the overall concentration of two specific free fatty acids, butyric (C4) and caproic (C6), whose presence has been related to rancidity and perception of a pungent flavor in ripe cheese. The concentration limit for C4 and C6 was set at 1200 ppm (Favati et al., 2007).

Shredded Cheddar cheese stored under high-intensity fluorescent light is more susceptible to light-induced oxidation reactions (which may affect flavor and color) than cheese in blocks due to the dramatic increase in the surface exposed to light. Shreds of cheese packaged in 100% N_2 were highly susceptible to molding, whereas cheese packaged in 100% CO_2 presented negative flavor effects and bixin pigment oxidation (Colchin et al., 2001).

Gas flushing, among other factors, was found to contribute more than vacuum packaging to the development of calcium lactate crystals (CLCs) in Cheddar cheese after 12 weeks of storage at 7°C. CLCs are a quality and appearance defect detrimental to consumer acceptance (Agarwal et al., 2005).

Graviera cheese (Greece) had a shelf life of 9 weeks when stored in the dark at 4°C under atmospheres of 100% N_2 or 50:50 CO_2:N_2, compared to 2–3 weeks for unpackaged cheese. The packaging material consisted of 75-μm LDPE/PA/LDPE with an oxygen transmission rate (OTR) of 11 mL m^{-2} day^{-1} and water vapor transmission rate (WVTR) of 1.29 g m^{-2} day^{-1} (temperature and humidity test conditions not specified). However, samples packaged under 100% CO_2 presented a bitter score after 5 weeks of storage (Trobetas et al., 2008). This negative effect on taste of very high concentrations of CO_2, also mentioned earlier for Cheddar cheese, has been noticed in other cheeses such as sliced Samsø (Juric et al., 2003).

6.5 PACKAGING AND SHELF LIFE OF SOFT CHEESES

Cheeses soft in texture have very variable ripening processes, sensory and physicochemical characteristics, and packaging requirements. Soft cheeses include (a) surface-mold-ripened cheeses such as Camembert and Brie; (b) internal-mold-ripened cheeses, also called blue vein type, such as Roquefort and Gorgonzola; (c) surface-ripened cheeses such as Havarti and Limburger; and (d) internal-bacteria-ripened cheeses such as Mozzarella (McSweeney et al., 2004).

Camembert and Brie are respiring cheeses, and their respiration rate depends strongly on the age of the cheese and hence the development stage of the surface molds, temperature, and the atmosphere inside the package. Therefore, even if no active modification of the atmosphere by flushing the package is performed, the continuous respiration process yields a decrease in O_2 and an increase in CO_2 concentrations. Ultimately, an equilibrium MA is established as a result of the balance between the rate of respiration and the permeability to O_2 and CO_2 of the package. A close match between these processes is required under realistic supply chain conditions. A shelf life can be obtained that ranges from 2 to 3 weeks for cut cheeses to 6 weeks for whole cheeses (Kreft, 2008).

These cheeses are packaged in flexible materials, sometimes using an outer package of wood or paperboard. The permeability of the flexible materials should be tailored to the respiration and moisture transfer needs of the cheese. Perforated films of oriented polypropylene (OPP) and combinations of OPP and paper are used with the number, size, and density of perforations varied according to the barrier required (Figure 6.5). Combinations of regenerated cellulose film (RCF) and paper, wax- or LDPE-coated paper, lacquered aluminum foil, and nonperforated OPP are also used when tighter packs are required. Portions of Brie (200 g) are sold wrapped in laminates of alufoil and paper with a shelf life longer than 1 month.

Roquefort and Danish Blue are slowly respiring cheeses. They are fairly tolerant of conditions of storage, and hence many packaging technologies can be applied to preserve their quality, such as vacuum packaging and MAP. Important criteria for the choice of a package are the softness of the cheese and its respiration rate. Very soft internal-mold-ripened cheeses are best MA-packaged, as vacuum packaging could result in damage to the cheese structure and even in the cheese being liquefied (Kreft, 2008). These cheeses seem to be less dependent on the gas permeability of the

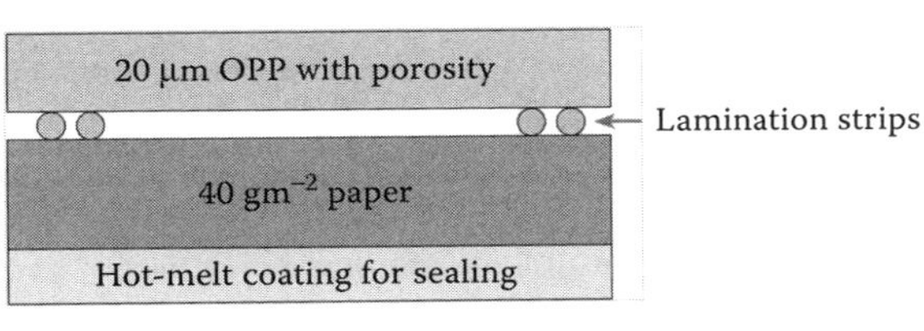

FIGURE 6.5 OPP/paper packaging material combination for respiring cheeses.

packaging than other soft and semisoft cheeses. They can withstand low concentrations of O_2 (as low as 5%) and high CO_2 concentrations (Robertson, 2006). Packages thermoformed with geometry matching the shape of the cheese or cheese portion are made of PS, PP, or PVC, with a sealed lid film. Roquefort in 100-g portions has a shelf life of 5 months when packaged in such packs and stored at 2–6°C. Gorgonzola in the market in such an MA package is given a 2-month shelf life at a recommended storage temperature of 2–8°C.

The quality of smear-coated cheeses (Havarti, Limburger, and Münster) is strongly dependent on the vitality of the surface culture (often *Brevibacterium linens*), which is related to the relative humidity inside the package. A highly controlled humidity inhibits the growth of competing organisms such as molds. Therefore, the packaging barrier to moisture should be well defined in order to allow for equilibrium between the relative humidity surrounding the cheese and the external relative humidity. A shelf life of 6–10 weeks can be obtained (Kreft, 2008).

Light exposure should also be avoided: sliced Havarti cheese stored at 5°C in transparent packages with an atmosphere of 25% CO_2, 75% N_2, and an initial 0.4% O_2 underwent significant changes in sensory attributes and a significant decrease of the riboflavin content compared to cheese stored in the absence of light (Kristensen et al., 2000). The packages had an average percentage transmission above 350 nm of 80% and an OTR of 0.034 mL pkg^{-1} day^{-1} at 23°C and 50% relative humidity. Packaging in black pigmented laminates [oriented polyamide (OPA)/EVOH/LDPE] provided the best protection of Havarti cheese, followed by a white laminate (OPA/LDPE), packaged in 25:75 CO_2:N_2 at 5°C (Mortensen et al., 2002). In cheeses exposed to light and packaged with 0.6% residual O_2, photo-oxidation increased significantly during storage compared to cheeses packaged with only 0.01% residual O_2 (Mortensen et al., 2003).

MAP has also been used for Mozzarella cheese. Mixtures containing 75% CO_2 efficiently stabilized lactic and mesophilic flora while inhibiting staphylococci, molds, yeasts, and psychrotrophs, but to a lower extent (Eliot et al., 1998).

6.6 PACKAGING AND SHELF LIFE OF FRESH CHEESES

Cheeses in the fresh category include Mascarpone, Ricotta, Chevre, Feta, Cream cheese, Quark and Cottage cheese, as well as whey cheeses. Due to their high moisture content, low salt concentration, and high pH, fresh cheeses are susceptible to microbial spoilage and consequently have a limited shelf life. They are also very sensitive to dehydration, and, in fact, most fresh cheeses keep draining slowly. Therefore, they need to be protected from moisture loss by barrier packages. Moreover, light (particularly for cheeses with some fat) and O_2 can also result in quality deterioration. Hence, some of these cheeses are packed under low O_2 conditions in medium-barrier packages with an MA.

Typical packages are plastic cups of high density polyethylene (HDPE) or PP, which provide a good moisture barrier, and PS. Sealed lids for integrity and snap-on outer lids are common. To provide a higher barrier to O_2 and maintain an MA with the correct CO_2 level in the headspace to cause its dissolution into the product, the use of high-barrier materials such as PA/LDPE laminates is essential.

The shelf life of Cottage cheese, without chemical preservatives, stored at 3–4°C is 14–21 days. Flushing the headspace (25%) of commercial packages of Cottage cheese with pure CO_2 extended the shelf life at 8°C by about 150% without altering the sensory properties or causing any other negative effects. Flushing the headspace with pure CO_2 caused dissolution of approximately 8 mmol CO_2 kg^{-1}, which gave additional protection as well as good sanitation against spoilage bacteria, yeasts, and molds (Maniar et al., 1994; Mannheim and Soffer, 1996).

The effects of MAP on physicochemical and sensory characteristics in the Portuguese whey cheese Requeijão (of the Ricotta type) were studied following a response surface methodology using storage time, temperature, and percentage of CO_2 in binary mixtures (CO_2:N_2) of the flushing gas. The cheese was packaged in LDPE/EVA/poly(vinylidene chloride) (PVdC)/EVA bags. Sensory results indicated that temperature was of the utmost importance in reducing lipolysis irrespective of the atmosphere concentration. However, CO_2 gas flushing ensured more constant composition, extending the shelf life for 15 days (Pintado and Malcata, 2000).

Myzithra is one of the most popular whey cheeses in Greece. Based on sensory analysis, MAP containing 40:60 CO_2:N_2 resulted in a shelf life increase of approximately 15 days, whereas MAP containing 60:40 CO_2:N_2 extended the shelf life by 20 days. Furthermore, lipolysis, proteolysis, and lipid oxidation were inhibited due to the presence of CO_2 in the atmosphere (Dermiki et al., 2008).

6.7 PACKAGING AND SHELF LIFE OF PROCESSED CHEESE

Processed cheese is produced by heating a mixture of cheese, water, emulsifying salts (mostly sodium citrates, sodium orthophosphates, or sodium polyphosphates), and further optional ingredients such as butter or spices. Mix constituents and processing conditions are selected to give the desired structure, appearance, color, flavor, and shelf life at an acceptable cost. The mixture is heated in a batch cooker to 70–120°C under a partial vacuum with constant agitation, until a homogeneous mass is obtained, or in a continuous ultra-high-temperature process at 140°C. Generally the hot processed cheese is filled into the desired packages such as pouches or polymer-coated aluminum foils. Thereafter, these packages are sealed and the product is cooled. The structure of the processed cheese depends on the type of cheese used, the fat ratio, the dry matter content, and the ability of the emulsifying salt to sequester the calcium (Schär and Bosset, 2002).

The amount of O_2 initially dissolved in the product depends on the quantity of O_2 introduced into the product due to the manufacturing process and on the filling methods used. This and the packaging barrier to O_2 determine the O_2 available for oxidative reactions. Also, as previously discussed, the barrier to light influences directly the level of energy available for photo-oxidation reactions. During storage, chemical and physical processes occur that impair sensory characteristics and texture: a typical off-flavor (often called *old flavor*) gradually appears, along with structural changes, typically toward a firmer texture. Other indices of failure for processed cheese are polyphosphate hydrolysis, changes in ionic equilibrium, crystal formation, reactions induced by heat-stable enzymes and nonenzymic browning, moisture loss, and interactions with packaging materials, such as migration and corrosion of aluminum foil (Schär and Bosset, 2002).

The typical packaging systems used for spreadable processed cheese are:

- Squeezable nonbarrier tubes made of LDPE, high-barrier tubes made of multilayer materials containing EVOH as a barrier layer, or metal tubes.
- Cups made of PP, PET/LDPE, or PS/EVOH/LDPE heat-sealed with alufoil or plastic laminate.
- Glass cups heat-sealed with an alufoil plastic laminate or with an easy-open tinplate cap.

These packaging systems correspond to different amounts of O_2 available in the package headspace and permeating through the package as a consequence of the different barrier properties of the plastic polymers. A Brazilian spreadable cheese (Requeijão cremoso) was stored in five packaging systems for 60 days at 10°C (see Table 6.5). When all the packages were stored under light, the shelf life in terms of overall sensory quality was only 17 days for the nonbarrier tube. The stability of the product was similar, in terms of thiobarbituric acid (TBA) index, in the coextruded tube, PP, and non-vacuum-sealed glass cups due to the combined effect of the initial O_2 available and the packaging barrier to O_2 and light. Among the types of packages studied, the vacuum–closed glass cup was the type that preserved the initial quality of Requeijão cremoso for the longest period as a result of the minimal amount of available O_2 contained in its headspace. As discussed, it is not enough that a package has good gas barrier properties; it is also necessary to reduce the amount of O_2 available and to use materials that offer a barrier to light (Alves et al., 2007).

TABLE 6.5
Packages for Processed Cheese

Packaging System	OTR, mL $pkg^{-1}day^{-1}$ (23°C, 0% RH)	Initial O_2, mL g^{-1}	Average Light Transmittance, % (λ > 350 nm)	TBA Index,[a] Abs at 532 nm g^{-1}
Tube LDPE + HDPE	0.3	0.006	40	0.100
Coextruded tube LDPE+HDPE/EVOH/LDPE+HDPE	0.008	0.006	40	0.100
PP cups/alufoil	0.07	0.053	70	0.180
Glass cup/alufoil-based laminate	<0.0002	0.020	85	0.160
Glass cup/tinplate cap (vacuum closed)	Very low	0.006	85	0.08

Source: Adapted from Alves R.M.V., Van Dender A., Jaime S., Moreno I., Pereira B. 2007. Effect of light and packages on stability of spreadable processed cheese. *International Dairy Journal* 17: 365–373.

[a] After 60 days of storage exposed to light, except for Tube LDPE + HDPE that was measured after 17 days.

6.8 NOVEL PACKAGING SOLUTIONS FOR CHEESES

6.8.1 Antimicrobial Films and Coatings

In recent years, AM packaging has attracted much attention from the food industry because of the increase in consumer demand for minimally processed, preservative-free products. Reflecting this demand, the preservative agents must be applied to packaging in such a way that only low levels of preservatives come into contact with the food (Cha and Chinnan, 2004). Probably the first application of AM food packaging was in 1954 by Smith and Rollin, who demonstrated that a moisture-proof RCF film coated with sorbic acid was effective in prolonging the shelf life of natural and processed cheeses by retarding surface-mold growth (Ozdemir and Floros, 2004). According to how effectively they influence the food, AM films can be divided into two types: (1) those that contain an AM agent that migrates to the surface of the food and (2) those that are effective against surface growth without the migration of the active agents to the food (Cha and Chinnan, 2004).

Chemical preservatives that can be used in active AM-releasing systems include organic acids and their salts (primarily sorbates, benzoates, and propionates), parabens, sulfites, nitrites, chlorides, phosphates, epoxides, alcohols, ozone, hydrogen peroxide, diethyl pyrocarbonate, antibiotics, and bacteriocins (Ozdemir and Floros, 2004).

New AM packaging materials are continually being developed. The enzyme lysozyme is a single-peptide protein that possesses activity against components of the cell wall of both gram-positive and gram-negative bacteria. Hydrolysis of the cell wall by lysozyme can damage the structural integrity of the cell wall and result in the lysis of bacterial cells (Cha and Chinnan, 2004). Lysozyme and Na_2-ethylenediaminetetraacetic acid (EDTA) were effective in prolonging the shelf life of traditional Mozzarella cheese. These AM compounds were incorporated into the typical conditioning brine of the cheese and significantly inhibited the growth of coliforms and Pseudomonadaceae during the first 7 days of storage at 4°C (Sinigaglia et al., 2008).

The bacteriocin nisin produced by the lactic acid bacterium *Lactococcus lactis* is most effective against lactic acid bacteria and other gram-positive organisms. Nisin is a peptide composed of 34 amino acid residues, with a molecular mass of 3.5 kDa and a highly surface-active molecule capable of binding to different compounds, such as fatty acids of phospholipids. This feature makes it suitable for adsorption to solid surfaces and for killing bacterial cells that subsequently adhere (Sobrino-López and Martín-Belloso, 2008). For example, inserts of greaseproof, wet-strength paper

TABLE 6.6
Restrictions from EFSA for Active Materials Based on Silver Zeolite A

Parameter	Limit
Silver content	2–5%
Maximum content in polymer	10% (w/w) of silver zeolite A containing ≤5% silver
Limit of Ag migration	0.05 mg Ag (kg food)$^{-1}$
Limit of global migration	60 mg (kg food)$^{-1}$

impregnated with nisin were placed between portions of sliced Cheddar cheese packaged under MAP (Scannell et al., 2000).

Natamycin (or pimaricin) is a polyene antifungal antibiotic produced by *Streptomyces natalensis*, commonly used to control fungus growth on the surface of many cheeses. Films with a cellulose polymeric base impregnated with natamycin at concentrations of 2% and 4% inhibited growth of *Penicillium roqueforti* in Gorgonzola cheese (Oliveira et al., 2007).

Antimicrobials based on silver ion zeolites are in use for several applications and have been proposed by AgPolymer as a nonedible cheese coating applied mixed with polyvinyl acetate. Silver has been used throughout history as a means of preventing the transmission of diseases and infections. Silver ions react with negative charges on the bacterial cell wall leading to its destruction. Silver ions stop the cell enzymic respiratory processes and cause cell metabolism to cease. Silver zeolite A (silver zinc sodium ammonium aluminosilicate) is approved for food contact use in the European Union subject to the restrictions presented in Table 6.6. Results presented by the supplier indicate a percentage reduction of viable microorganisms of 99% for *Escherichia coli*, 80% for *Staphylococcus aureus*, and 98% for *Aspergillus niger* after 30 min of application (AgPolymer, 2008).

Many AM systems exploit natural agents. The natural AM compound allyl isothiocyanate (AITC) found in mustard oil is claimed to be effective against cheese-related fungi. It was tested on Danish Danbo cheese, which increases the shelf life from 4 to 13 or 28 weeks, depending on the amount of AM applied and despite an unacceptable mustard flavor that tends to disappear later in the storage period (Winther and Nielsen, 2006). LDPE-based films containing linalool or methyl chavicol (the principal components of basil) have been used to retard microbial growth on Cheddar cheese. Both compounds retained their AM activity against *Escherichia coli* after the film was extrusion-blown. The activity of the films was tested in cheese stored at 4°C, and the changes in mesophilic aerobic bacteria and coliforms, as well as yeast and mold counts, were monitored, with promising results (Suppakul et al., 2008).

6.8.2 Oxygen Absorbers

Active packaging technologies offer the food industry new opportunities for the preservation of foods, and dairy is no exception. For cheese, applications of active packaging currently used (or studied) are O_2 scavengers and moisture absorbers (Ozdemir and Floros, 2004).

Oxygen-absorbing systems provide an alternative to vacuum and gas-flush packaging. Typical systems are based on the oxidation of iron powder by chemical means or scavenging of O_2 through the use of enzymes. In France a matured goat cheese (Le Père Bafien, Poitiers) is distributed in individual units or in six-packs with an O_2 scavenger attached with an adhesive to the clear, peelable lidding material. It reduces the O_2 level in the headspace to zero within 24–36 hr (*Packaging Digest*, 2007).

The control of excess moisture in food packages is important to suppress microbial growth and prevent foggy film formation. If the package has a low permeability to water vapor, water accumulation inside the package is more pronounced. An effective way of controlling excess water accumulation in a food package that has a high barrier to water vapor is to use a moisture scavenger such as silica gel, molecular sieves, natural clays, calcium oxide, calcium chloride, modified starch, or other moisture-absorbing substances. The Humidipak system, typically used to control humidity in cigar cases, was adapted and tested to control the relative humidity inside a package of semihard cheese, with very good results in terms of texture (Pantaleão et al., 2007). This technology is claimed to allow for two-way humidity control by continuously responding to and adjusting the relative humidity by either adding or removing water to maintain a predetermined humidity level. The product consists of a gelled, saturated solution that is filled into a small sachet made of a material with very high permeability, but that does not allow liquid water to pass through or leak into the package (Humidipak, 2008).

6.8.3 Biobased Materials

Biopolymer-based packaging is defined as packaging that contains raw materials originating from agricultural and marine sources. Polylactate (PLA) is a material made from lactic acid produced by fermentation of starch from corn. It has gained increasing interest as a food packaging material for environmental reasons and is particularly studied for use in cheese packaging (Holm et al., 2006a). PLA does not provide a sufficient barrier to O_2 and water vapor to prevent lipid oxidation and moisture loss in many cases. However, if light can be avoided and low storage temperatures used, this material can be a viable option. The performance of PLA compared to amorphous polyethylene terephthalate (APET)/LDPE packaging material on quality preservation of Danbo cheese was assessed during light exposure and storage in the dark. Results showed that moisture loss from cheeses packaged in PLA was approximately 10 times higher than in the reference packages, but dry surface spots were not observed before 56 days of storage in PLA packages. Secondary lipid oxidation products were primarily developed when both O_2 and light were present. During light exposure, lipid oxidation of the cheeses packaged in PLA was limited for the first 56 days of storage at 4°C and 50% relative humidity, and almost negligible when the cheeses were protected from light during 84 days of shelf life (Holm et al., 2006b).

ACKNOWLEDGMENTS

TRUEFOOD—Traditional United Europe Food (Contract number: FOOD-CT-2006-016264)—Integrated Project financed by the European Commission under the 6th Framework Programme for Research and Technological Development.

REFERENCES

Agarwal S., Costello M., Clark S. 2005.Gas-flushed packaging contributes to calcium lactate crystals in Cheddar cheese. *Journal of Dairy Science* 88: 3773–3783.

AgPolymer. 2008. Silver ions based cheese coating. http://www.agpolymer.com/uk/agpolymer_cheese_coating.asp, accessed on 24 July 2008.

Alves R.M.V., Van Dender A., Jaime S., Moreno I., Pereira B. 2007. Effect of light and packages on stability of spreadable processed cheese. *International Dairy Journal* 17: 365–373.

Bendall J.G. 2007. Food contamination with styrene dibromide via packaging migration of leachate from polystyrene cold-storage insulation. *Journal of Food Protection* 70: 1037–1040.

Beresford T.P., Fitzsimons N.A. Brennan N.L., Cogan T.M. 2001. Recent advances in cheese microbiology. *International Dairy Journal* 11: 259–274.

Boutrou R., Guéguen M. 2005. Interests in *Geotrichum candidum* for cheese technology. *International Journal of Food Microbiology* 102: 1–20.

Broome M.C., Limsowtin G.K.Y. 1998. Milk coagulants. *Australian Journal of Dairy Technology* 53: 188–190.

Carreira A., Paloma L., Loureiro V. 1998. Pigment producing yeasts involved in the brown surface discoloration of ewes' cheese. *International Journal of Food Microbiology* 41: 223–230.

Castle L., Kelly M., Gilbert J. 1993. Migration of mineral hydrocarbons into foods. 3. Cheese coatings and temporary casings for skinless sausages. *Food Additives and Contaminants* 10: 175–184.

Cha D.S.C., Chinnan M.S. 2004. Biopolymer-based antimicrobial packaging: a review. *Critical Reviews in Food Science and Nutrition* 44: 223–237.

Coker C.J., Crawford R.A., Johnston K.A., Singh H., Creamer L.K. 2005. Towards the classification of cheese variety and maturity on the basis of statistical analysis of proteolysis data—a review. *International Dairy Journal* 15: 631–643.

Colchin L., Owens S., Lyubachevskaya G., Boyle-Roden E., Russek-Cohen E., Rankin S. 2001. Modified atmosphere packaged Cheddar cheese shreds: influence of fluorescent light exposure and gas type on color and production of volatiles compounds. *Journal of Agricultural and Food Chemistry* 49: 2277–2282.

Collins I.F., McSweeney P.L.H., Wilkinson M.G. 2003. Lipolysis and free fatty acid catabolism in cheese: a review of current knowledge. *International Dairy Journal* 13: 841–866.

Coolbear T., Crow V., Harnett J., Harvey S., Holland R., Martley F. 2008. Developments in cheese microbiology in New Zealand—use of starter and non-starter lactic acid bacteria and their enzymes in determining flavor. *International Dairy Journal* 18: 705–713.

Corsetti A., Rossi J., Gobbetti M. 2001. Interactions between yeasts and bacteria in the smear surface-ripened cheeses. *International Journal of Food Microbiology* 69: 1–10.

Cruz J.M., Sanches Silva A., Sendón García R., Franz R., Paseiro Losada P. 2008. Studies of mass transport of model chemicals from packaging into and within cheeses. *Journal of Food Engineering* 87(1)S: 107–115.

Dermiki M., Ntzimani A., Badeka A., Savvaidis I.N., Kontominas M. 2008. Shelf life extension and quality attributes of the whey cheese Myzythra Kakathaki using modified atmosphere packaging. *LWT—Food Science and Technology* 41: 284–294.

Eliot S., Vuillemard J., Edmond J. 1998. Stability of shredded Mozzarella cheese under modified atmospheres. *Journal of Food Science* 63: 1075–1080.

Favati F., Galgano F., Pace A.M. 2007. Shelf life evaluation of portioned Provolone cheese packaged in protective atmosphere. *LWT—Food Science and Technology* 40: 480–488.

Fox P.F. 1999. Cheese: an overview. In: *Cheese: Chemistry, Physics and Microbiology*, Vol. 1, Fox P.F. (Ed). London: Chapman & Hall, pp. 1–36.

Goulas A.E., Anifantaki K.I., Kolioulis D.G., Kontominas M.G. 2000. Migration of di-(2-ethylhexyl)adipate plasticizer from food-grade polyvinyl chloride film into hard and soft cheeses. *Journal of Dairy Science* 83: 1712–1718.

Grob K., Pfenninger S., Pohl W., Laso M., Imhof D., Rieger K. 2007. European legal limits for migration from food packaging materials: 2. More realistic conversion from concentrations to limits per surface area. PVC cling films in contact with cheese as an example. *Food Control* 18: 201–210.

Holm V.K., Mortensen G., Risbo J. 2006a. Quality changes in semi-hard cheese packaged in a poly(lactic acid) material. *Food Chemistry* 97: 401–410.

Holm V., Mortensen G., Vishart M., Petersen M. 2006b. Impact of poly-lactic acid packaging material on semi-hard cheese. *International Dairy Journal* 16: 931–939.

Humidipak. 2007. http://www.humidipak.com/, accessed on 24 July 2008.

Jakobsena M., Narvhu J. 1996. Yeasts and their possible beneficial and negative effects on the quality of dairy products. *International Dairy Journal* 6: 755–768.

Juric M., Bertelsen G., Mortensen G., Petersen M.A. 2003 Light-induced colour and aroma changes in sliced, modified atmosphere packaged semi-hard cheeses. *International Dairy Journal* 13: 239–249.

Karvonen H.M., Tapola N.S., Uusitupa M.I., Sarkkinen E.S. 2002. The effect of vegetable oil-based cheese on serum total and lipoprotein lipids. *European Journal of Clinical Nutrition* 56: 1094–1101.

Kreft F. 2008. Packaging of Cheese, Agrotechnology & Food Sciences Group of the Wageningen University and Research Centre. http://www.vers.nl/Projects-CheesePackaging.htm, accessed on 24 July 2008.

Kristensen D., Orlien V., Mortensen G., Brockhoff P., Skibsted L.H. 2000. Light-induced oxidation in slice Havarti cheese packaged in modified atmosphere. *International Dairy Journal* 10: 95–103.

Kure C.F., Skaar I., Brendehaug J. 2004. Mold contamination in production of semi-hard cheese. *International Journal of Food Microbiology* 93: 41–49.

Lund F., Filtenborg O., Frisvad J.C. 1995. Associated mycoflora of cheese. *Food Microbiology* 12: 173–180.

Maniar A., Marcy J., Bishop J., Duncan S. 1994. Modified atmosphere packaging to maintain direct-set Cottage cheese quality. *Journal of Food Science* 59: 1305–1308.

Mannheim C.H., Soffer T. 1996. Shelf life extension of Cottage cheese by modified atmosphere packaging. *LWT—Food Science and Technology* 29: 767–771.

Marilley L., Casey M.G. 2004. Flavors of cheese products: metabolic pathways, analytical tools and identification of producing strains. *International Journal of Food Microbiology* 90: 139–159.

McSweeney P.L.H. 2007a. How are cheese varieties classified. In: *Cheese Problems Solved*, McSweeney P.L.H. (Ed). Cambridge: Woodhead Publishing Ltd., pp. 181–183.

McSweeney P.L.H. 2007b. Flavor, texture and flavor defects in hard and semi-hard cheeses. In: *Cheese Problems Solved*, McSweeney P.L.H. (Ed). Cambridge: Woodhead Publishing Ltd., pp. 189–207.

McSweeney P.L.H., Ottogalli G., Fox P.F. 2004. Diversity of cheese varieties: an overview. In: *Cheese: Chemistry, Physics and Microbiology*, Vol. 2, 3rd edn. Fox P.F., McSweeney P.L.H., Cogan T.M., Guinee T.P. (Eds). Amsterdam: Elsevier Academic Press, pp. 1–22.

Mortensen G., Sørensen J., Stapelfeldt H. 2002. Effect of light and oxygen transmission characteristics of packaging materials on photo-oxidative quality changes in semi-hard Havarti cheeses. *Packaging Technology and Science* 15: 121–127.

Mortensen G., Sørensen J., Stapelfeldt H. 2003. Effect of modified atmosphere packaging and storage conditions on photo-oxidation of sliced Havarti cheese. *European Food Research and Technology* 216: 57–62.

Mortensen G., Bertelsen G., Mortensenm B., Stapelfeldt H. 2004. Light-induced changes in packaged cheeses—a review. *International Dairy Journal* 14: 85–102.

Murray J.M., Delahunty C.M. 2000. Mapping consumer preference for the sensory and packaging attributes of Cheddar cheese. *Food Quality and Preference* 11: 419–435.

National Dairy Council (NDC). 2006. *Quick Reference Guide. Nutrition Claims of Dairy Products*. Rosemont, Illinois: Dairy Management Inc.

Oliveira T.M., Soares N.F.F., Pereira R.M., Fraga K.F. 2007. Development and evaluation of antimicrobial natamycin-incorporated film in Gorgonzola cheese conservation. *Packaging Technology and Science* 20: 147–153.

O'Mahony F., O'Riordan T., Papkovskaia N., Kerry J., Papkovsky D. 2006. Non-destructive assessment of O_2 levels in industrial modified atmosphere packaged cheddar cheese. *Food Control* 17: 286–292.

Ozdemir M., Floros J. 2004. Active food packaging technologies. *Critical Reviews in Food Science and Nutrition* 44: 185–193.

Packaging Digest. 2007. Cheese shelf life is lengthened. http://www.packagingdigest.com/article/CA6477330.html, accessed on 24 July 2008.

Pantaleão I., Pintado M.M.E., Poças M.F.F. 2007. Evaluation of two packaging systems for regional cheese. *Food Chemistry* 102: 481–487.

Pelaéz C., Requena T. 2005. Exploiting the potential of bacteria in the cheese ecosystem. *International Dairy Journal* 15: 831–844.

Pintado M.E., Malcata X. 2000. Optimization of modified atmosphere packaging with respect to physicochemical characteristics of Requeijão. *Food Research International* 33: 821–832.

Robertson G.L. 2006. *Food Packaging Principles and Practice*, 2nd edn. Boca Raton, Florida: CRC Press.

Romani S., Sacchetti G., Pittia P., Pinnavaia G.G., Dalla Rosa, M. 2002. Physical, chemical, textural, and sensorial changes of portioned Parmigiano Reggiano cheese packed under different conditions. *Food Science and Technology International* 8: 203–211.

Scannell A.G. M., Hill C., Ross R.P., Marx S., Hartmeier W., Arendt E.K. 2000. Development of bioactive food packaging materials using immobilised bacteriocins Lacticin 3147 and Nisaplin®. *International Journal of Food Microbiology* 60: 241–249.

Schär W., Bosset J.O. 2002. Chemical and physico-chemical changes in processed cheese and ready-made fondue during storage. *LWT—Food Science and Technology* 35: 15–20.

Sealed Air Co. 2007. Cheese event. http://www.sealedair.com/eu/en/products/food/cheese.html, accessed on 24 July 2008.

Severini C., Bressa F., Romani S., Dalla Rosa, M. 1998. Physical and chemical changes in vacumm packaged Parmigiano Reggiano cheese during storage. *Journal of Food Quality* 21: 355–367.

Sinigaglia M., Bevilacqua A., Corbo M.R., Pati S., Del Nobile M.A. 2008. Use of active compounds for prolonging the shelf life of mozzarella cheese. *International Dairy Journal* 18: 624–630.

Smith D.P., Rolin N.J. 1954. Sorbic acid as a fungistatic agent for foods. VII. Effectiveness of sorbic acid in protecting cheese. *Food Research* 19: 50–65.

Sobrino-López A., Martín-Belloso O. 2008. Use of nisin and other bacteriocins for preservation of dairy products. *International Dairy Journal* 18: 329–343.

Sousa M.J., Ardo Y., McSweeney P.L.H. 2001. Advances in the study of proteolysis during cheese ripening. *International Dairy Journal* 11: 327–345.

Suppakul P., Sonneveld K., Bigger S., Miltz J. 2008. Efficacy of polyethylene-based antimicrobial films containing principal constituents of basil. *LWT—Food Science and Technology* 41: 779–788.

Teng D., Wilcock A., Aung M. 2004. Cheese quality at farmers markets: observation of vendor practices and survey of consumer perceptions. *Food Control* 15: 579–587.

Trobetas A., Badeka A., Kontominas M.G. 2008. Light-induced changes in grated Graviera hard cheese packaged under modified atmospheres. *International Dairy Journal* 18: 1133–1139.
USDA. 1995. How to buy cheese. Home and Garden Bulletin 256. Agricultural Marketing Service. http://www.ams.usda.gov/, accessed on 24 July 2008.
Valdks-Stauber N., Scherer S., Seiler H. 1997. Identification of yeasts and coryneform bacteria from the surface microflora of brick cheeses. *Food Microbiology* 34: 115–129.
Winther M., Nielsen P.V. 2006. Active packaging of cheese with allyl isothiocyanate, an alternative to modified atmosphere packaging. *Journal of Food Protection* 69: 2430–2435.

7 Packaging and the Shelf Life of Milk Powders

Elmira Arab Tehrany
Nancy-Université
Laboratoire de Science & Génie Alimentaires
Vandoeuvre lés Nancy, France

Kees Sonneveld
Packaging and Polymer Research Unit
Victoria University
Melbourne, Australia

CONTENTS

7.1 MILK POWDER

7.1.1 Introduction

The use of milk powders for recombination and manufacturing of various food products has increased in recent years, placing greater importance on the definition of the functional properties of powders. The functionality, particularly the reconstitution properties, of milk powders can be highly variable and occasionally unpredictable. Spray-dried dairy powders are common ingredients in many food and dairy products. The nutritional quality of dairy powders depends on the intensity of the various thermal treatments during processing.

Lipid oxidation in whole milk powders (WMPs) is a major cause of deterioration during processing and storage. Light-induced degradation reactions in milk create a serious problem for the dairy industry, through the development of off-flavors, leading to the formation of volatile secondary oxidation products. Lipid oxidation has received much attention because of its undesirable implications for human health and its contribution to a decrease in the nutritional value of foods. Many factors can influence lipid oxidation in milk powder, such as water activity, temperature, O_2, and light.

7.1.2 Manufacture

Drying means that the water in a liquid product—in this case milk—is removed, so that the product acquires a solid form. The water content of milk powder ranges between 2.5% and 5%, and no microbial growth occurs at such a low water content. Drying extends the shelf life of the milk, simultaneously reducing its weight and volume. This reduces the cost of transporting and storing the product (Bylund, 2003). Preheating conditions are used to a large extent to control the functional properties of the powder. A number of changes occur in milk during preheating: whey protein denaturation, association of denatured whey proteins with the casein micelle, transfer of soluble calcium and phosphate to the colloidal phase, destruction of bacteria, and inactivation of enzymes (Singh and Newstead, 1992).

The manufacture of milk powder involves a series of continuous or semicontinuous steps, such as milk standardization, thermal treatment, evaporation, spray drying, and fluidized bed drying, each of which has associated process variables that affect the efficiency of the process and the quality of the product (O'Callaghan and Cunningham, 2005).

The manufacturing process for skim milk powder (SMP) involves heating the skim milk (known as preheating), concentrating the skim milk solids by evaporation to 45–50% total solids, heating the skim milk concentrate, and then spray-drying the milk concentrate to produce a powder (Oldfield et al., 2005). Depending on the intensity of the heat treatment, milk powder is classified into categories related to the temperature–time combinations the skim milk has been exposed to prior to evaporation and drying. High temperature denatures whey proteins, the percentage denatured increasing with the intensity of the heat treatment. The degree of denaturation is normally expressed by the whey protein nitrogen index (WPNI) as milligrams of undenatured whey protein (u.w-p) per gram of powder. SMP is classified into the following three types based on the WPNI, which is correlated to the spray-drying conditions: low-heat powder (70°C/15 sec, WPNI >6.0 mg g^{-1} u.w-p), medium-heat powder (85°C/20 sec, WPNI 5–6.0 mg g^{-1} u.w-p), and high-heat (~135°C/30 sec, WPNI <1.4 0 mg g^{-1} u.w-p) (Bylund, 2003).

In the manufacture of WMP, milk is subjected to a range of processes such as agitation, pumping, heating, concentration, homogenization, and spray drying. These processing treatments cause a number of physical and chemical interactions of the milk components (Ye et al., 2007).

7.1.3 Spray Drying

Spray drying is the most common method of dehydrating milk and milk products. It involves rapid removal of moisture, leading to the formation of amorphous lactose, which forms a continuous matrix in which proteins, fat globules, and air cells disperse (Shrestha et al., 2007). Spray-drying technology in combination with other unit processes plays an important role in responding to market demands for powders with a wide range of functional properties (Kelly et al., 2002). Spray-drying technology involves the transformation of the milk emulsion into a great number of small droplets that are exposed to a fast current of hot air as they fall into the spray chamber. As water is evaporated from the droplets they become powder particles (Birchal et al., 2005).

7.1.4 Properties of Spray-Dried Milk Powders

The quality of food powders is based on a variety of properties, depending on the specific application. In general, the final moisture content, insolubility index, dispersability index, free fat, rheological properties, and bulk density are of primary importance (Straatsma et al., 1999a). These characteristics depend on drying parameters (type of spray dryers, nozzles/wheels, pressure, agglomeration, and thermodynamic conditions of the air: temperature, relative humidity, and velocity) and characteristics of the concentrate before drying (composition/physicochemical characteristics, viscosity, thermosensibility, and availability of water) (Schuck, 2002).

The insolubility index is of primary importance for the quality of instant powders. Insoluble material is formed during spray drying of concentrated milk. The actual amount depends on the temperature and moisture content during the drying period (Straatsma et al., 1999b). An important quality attribute of milk powder is the bulk density. It is obviously of considerable interest from an economic point of view because it influences the cost of storage, packaging, and transport (Robertson, 2006).

7.2 FOOD QUALITY ATTRIBUTES OF MILK POWDERS

7.2.1 Whole Milk Powder

WMP consists mainly of whey proteins (almost 4%), caseins (almost 20%), milk fat (almost 26%), and lactose (almost 38%). The particles of milk powder consist of a continuous mass of amorphous lactose and other low-molar-mass components in which fat globules and proteins are embedded (Walstra et al., 1999). The physical processes, involving mainly milk fat and lactose, together with chemical reactions, have the ability to reduce the shelf life of WMP and other dry products based on milk powder. These other products include infant formula and instant powders for coffee, cocoa, and chocolate-flavored beverages. Long-term storage of milk powder affects the nutritive value, mainly due to loss of lysine, and the sensory qualities of the reconstituted milk, as well as the functional and physical properties that are so important for the use of milk powder as a food ingredient (Thomas et al., 2004). In coffee whitener applications, WMP must contribute adequate whitening ability and stability to the relatively low-pH and high-temperature conditions inherent in coffee solution (Oldfield et al., 2000).

Milk powder quality is influenced by various factors during processing or storage:

1. Manufacturing techniques and parameters
2. Drying techniques and conditions
3. Storage conditions

Three deteriorative reactions determine the shelf life of milk powder in practice: lactose crystallization, lipid oxidation, and Maillard reactions (nonenzymic browning) (Carić, 1994).

The basic properties that determine milk powder quality, and where defects mainly appear, include powder structure, solubility, water content, scorched particles, flowability, oxidative changes, flavor, and color (Carić, 1994; Schuck et al., 2007). Hence, milk powder quality should be quantified by the relationship between the operational process variables that best describe these properties (Birchal et al., 2005).

7.2.2 Low-Fat Milk Powder or Skim Milk Powder

Low-fat milk powder or SMP is produced as a result of the removal of milk fat and water from milk. It contains a maximum of 1.5% fat and a maximum moisture content of 5% (Clark, 1992). The quality of this product is determined by the total heat treatment (i.e., temperature and time of all operations in processing). Most of the properties mentioned (solubility, water content, flavor, color, and others) are strongly affected by the intensity of the applied heat treatment. The storage time and the temperature influence the entire quality of the powder. SMP has low bulk density; an increase in bulk density is accompanied by a corresponding improvement in other quality characteristics such as wetting, sinking, and dispersing abilities (Carić, 1994).

7.3 DETERIORATIVE REACTIONS AND INDICES OF FAILURE

7.3.1 Cohesion/Flowability

Powder deposition on processing equipment is a problem in the dairy industry, particularly in the spray-drying process, and results in economic disadvantages. Cohesion increases with a reduction in particle size; fat also plays an important part in the observed trend toward higher cohesion with increasing temperature. More surface area is available for cohesive forces, in particular, and frictional forces to resist flow (Fitzpatrick et al., 2004). Melting of fat is likely to cause the major increase in cohesion, but there are several possible mechanisms. The liquid fat may have formed bridges between the particles, which increases the bonding strength. Alternatively, fat liquefaction could have softened the powder, resulting in deformation of the powder particles, which would have increased the contact area between the particles, thus enhancing already present attractive forces (Rennie et al., 1999). During processing, the behavior of powders is strongly influenced by particle properties as well as the design and operating conditions of the equipment. The flowability of powders in such equipment is an important issue as it can strongly influence the efficiency and reliable operation of these processes (Moreno-Atanasio et al., 2005). Intuitively, one would expect particle shape to affect flowability, as shape will influence the surface contacts between particles; however, there is not much reported work on the influence of shape on powder flowability (Fitzpatrick et al., 2004).

7.3.2 Caking

Several properties of powders with amorphous lactose can be related to its glass transition temperature T_g. These include surface stickiness and caking, time-dependent lactose crystallization and release of encapsulated lipids, and increasing rates of nonenzymic browning and lipid oxidation. When an amorphous component is given suitable conditions of temperature and water content, powder can mobilize as a high-viscosity flow, which can make it sticky and lead to caking (Fitzpatrick et al., 2007). The changes in mechanical properties and diffusion are responsible for stickiness, caking, and lactose crystallization. Caking is a deleterious phenomenon by which a low-moisture, free-flowing powder is first transformed into lumps, then into an agglomerated solid, and ultimately into a sticky material, resulting in loss of functionality and lowered quality (Aguilera et al., 1995). Amorphous lactose is generally present in high-fat powders and can contribute to flowability problems. However, these problems also arise under conditions [a_w (water activity) and powder temperature] where the amorphous lactose is stable (Foster et al., 2005a). This indicates that milk fat also contributes to caking (McKenna, 1997; Peleg, 1977). The changes in the reaction rates are more complex and are affected by other factors, including pH, heterogeneities in water distribution, and miscibility of proteins and carbohydrates (Roos, 2002).

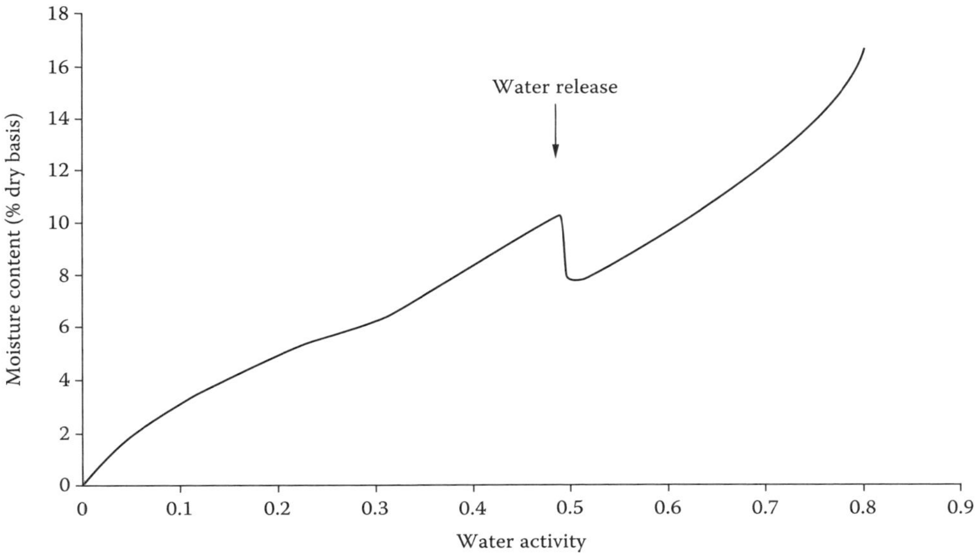

FIGURE 7.1 Generalized moisture sorption isotherm for milk powders showing a break at $a_w = 0.5$ due to lactose crystallization. (Redrawn from Thomas M.E.C., Scher J., Desobry-Banon S., Desobry S. 2004. Milk powders aging: effect on physical and functional properties. *Critical Reviews in Food Science and Nutrition* 44: 297–322, with permission.)

7.3.3 Maillard Reactions

Maillard reactions are an important class of deteriorative reactions in milk products. This type of chemical reaction is initiated by condensation of lactose with the free amino group of lysine in milk proteins (Thomsen et al., 2005). In milk products such Maillard reactions are induced by heating during processing and long-term storage at moderate to high temperatures (O'Brien and Morrisey, 1989). Crystalline forms of lactose depend on the preservation time and many other conditions, such as humidity, storage temperature, and preparation process. The crystalline state is thermodynamically favored as it has a lower free energy due to the structured arrangement of the molecules. During crystallization, the amorphous lactose will initially absorb moisture from the surroundings due to its hygroscopic nature, and subsequently release moisture as it crystallizes, as shown in Figure 7.1. The crystallization kinetics can be determined from the mass change of the powder (Ibach and Kind, 2007). Lactose crystallization modifies the microstructure and chemical composition of the surface of powder particles (Thomas et al., 2004).

7.3.4 Lipid Oxidation

Lipid oxidation in WMPs is also a major cause of deterioration during processing and storage (McCluskey et al., 1997). The reaction of unsaturated lipids with molecular O_2 results in the formation of hydroperoxides, which then break down to off-flavor compounds (Liang, 1999a). Many factors are responsible for the degradation of lipids due to oxidation, and one of the major causes of this defect has been identified as the oxidation of unsaturated lipids (Cadwallader and Howard, 1998). Lipid peroxidation is responsible for changes in the taste and odor of milk powders through the development of off-flavors, which are caused by the formation of secondary reaction products (alkanes, alkenes, aldehydes, and ketones) (Romeu-Nadal et al., 2007). These compounds impart off-flavors and loss of nutrients to milk powders and thus limit their shelf life stability (Fenaille et al., 2003). Oxygen, light exposure, storage temperature, water content, percentage of unsaturated fatty acids, and process parameters are the most important factors that affect oxidation.

7.3.4.1 Water Activity

The shelf life of WMP clearly depends on the preheat treatment of the milk, the a_w of the product, and the storage temperature. One of the factors influencing the rate of autoxidation in milk powder, although less investigated, is a_w (Roos, 2002; Stapelfeldt et al., 1997). Loncin et al. (1968) found that autoxidation in an unspecified milk powder, as measured by peroxide values, was stimulated by an a_w below 0.11 and unaffected by $a_w s$ between this value and 0.75. Stapelfeldt et al. (1997) found that WMP retained its quality best within an a_w range of 0.11–0.23. The preheat treatment of milk prior to the manufacture of milk powder is the major factor controlling the oxidative stability of the product, as heat treatment at high temperatures, apart from increasing the microbial safety, delays the onset of oxidized flavor, which is the limiting factor for the storage of milk powder (Baldwin et al., 1991).

7.3.4.2 Temperature

According to Stapelfeldt et al. (1997), Thomsen et al. (2005), and Augustin et al. (2006), it was expected that long-term stability of milk would be influenced negatively by a low preheat intensity, a high storage temperature, and a high a_w during storage. Although the effect of preheat treatment and storage was in qualitative agreement with earlier findings, the effect of a_w should be noted, especially as these findings were further substantiated by the techniques used to follow different stages of oxidation in the main experiment. There has been increasing interest in the supplementation of milk powder formulas with long-chain polyunsaturated fatty acids (LC-PUFAs) especially with arachidonic acid (C20: 4n-6, AA) and docosahexanoic acid (C22: 6n-3, DHA). High temperatures and the presence of O_2 lead to increased oxidation of PUFAs (Romeu-Nadal et al., 2007).

7.3.4.3 Oxygen

As O_2 is consumed during oxidation, the O_2 content will also influence lipid oxidation. In addition, the O_2 concentration in the headspace and the product is important, as this can influence the oxidation rate. Oxygen concentration could also influence the oxidation pathways and lead to different oxidation products (Grosch et al., 1981). It has been shown by numerous authors that if O_2 in milk powder or infant formula packages is replaced by N_2 and CO_2, the oxidation is not detectable and the peroxide value does not increase (Van Mil and Jans, 1991). Oxidation increases during storage; for example, WMP has a maximum shelf life of 6 months at room temperature (Anon., 1989). However, it was found that WMP could have a shelf life in excess of 12 months if it was packed in cans under vacuum or an inert gas such as N_2 to inhibit the development of off-flavors (Kieseker and Aitken, 1993).

The amount of O_2 needed to cause unacceptable oxidative changes is usually very small (Labuza, 1971). There is little detailed knowledge about what levels are acceptable for specific food products and how the storage stability is related to the amount of O_2 available for oxidation, especially at very low O_2 levels, that is, below 1 mL L^{-1}. Andersson and Lingnert (1997) reported on the influence of O_2 levels down to 0.6 mL L^{-1} on the oxidation of cream powder. An increased temperature also increases the effect of O_2 concentration. At high partial pressures of O_2, the oxidation rate should, theoretically, be independent of O_2 concentration and be directly dependent on substrate concentration. A decrease in O_2 concentration increases the effect of the O_2 partial pressure, which leads to a situation, at low O_2 partial pressures, where the oxidation rate is independent of substrate concentration but directly dependent on O_2 partial pressure (Labuza, 1971). Oxygen levels can be reduced by the traditional method of N_2 flushing or by the more recently developed approach of using O_2 absorbers or scavengers. Nitrogen flushing generally reduces the O_2 to 2–5% (Warmbier and Wolf, 1976), which is not enough to prevent oxidation (Bishov et al., 1971; Labuza, 1971; Lloyd et al., 2004).

7.3.4.4 Light

The rate of lipid oxidation is greatly influenced by light; this has created a serious problem for the dairy industry because of the development of off-flavors, a decrease in nutritional quality, and the severity and speed at which these phenomena develop (Bossett et al., 1994). Most ultraviolet (UV) light damage to lipids occurs at wavelengths less than 200 nm. Although UV light is thermodynamically capable of producing radicals directly in lipids, the process is not a competitive reaction. The principal light-absorbing groups of lipids are double bonds, peroxide O–O bonds, and carbonyls; the last two are most important (Schaich, 2005).

It is well known that exposure of foods and beverages to light may result in oxidation of lipids and other constituents, leading to the formation of off-flavors, discoloration, and loss of vitamins, especially riboflavin and β-carotene. Important factors influencing the deteriorative effect of light are the intensity and spectrum of the light source, the duration of light exposure, and the light transmittance of the packaging material. The effect of light on lipid oxidation and flavor stability of a particular food can be explained by both photolytic auto-oxidation and photosensitized oxidation (Bradley and Min, 1992).

Sattar et al. (1976) investigated the effect of light on the oxidation of milk fat and found that although there was an induction period for light-induced oxidation of milk fat, there was not for light-induced oxidation of vegetable oils. It was suggested that the induction period was due to the presence of α-carotene acting as a built-in light filter. Even though the presence of α-carotene in milk fat slowed down the rate of oxidation at the beginning of the trial, the light-exposed samples still showed a much higher oxidation rate than the samples kept in the dark.

7.4 IMPACT OF PACKAGING ON INDICES OF FAILURE

7.4.1 Moisture Transfer

Absorption or desorption of moisture can significantly affect the shelf life of foods. This is particularly the case for dry, powdery products such as milk powders. The main purpose of packaging is to protect the powder from moisture ingress to preserve the product characteristics. When they gain moisture, powdery products become lumpy or cake. In addition, the moisture may lead to deleterious changes such as structural transformations, enzymic reactions, browning, and oxidation, depending on temperature and the availability of O_2 (Roos, 2001). Moisture or water vapor ingress in combination with light, O_2, and an elevated temperature can result in physical loss of texture and caking due to lactose crystallization, microbial spoilage, nonenzymic reactions (such as Maillard browning), and fat oxidation (Uppu, 2002). The effectiveness of a package can be determined during shelf life testing or by combining information from break-point testing (holding at increasing humidities) and knowledge about the characteristics of the moisture permeability of the packaging material (Brown and Williams, 2003).

Although an $a_w < 0.6$ is considered sufficient to prevent the growth of microorganisms, chemical reactions and enzymic changes may occur at considerably lower levels (Roos, 2001). It is important for the determination of the maximum shelf life for milk powders (especially WMPs) not to exceed a moisture content corresponding to an a_w at which the rate of lipid oxidation is at a minimum (Robertson, 2006). Commonly the a_w of WMP varies from 0.25 (low) to 0.35 (high) (Baechler et al., 2005) and for SMP from 0.32 to 0.43 (Shrestha et al., 2008).

Moisture sorption isotherms (MSIs) for powders describe the equilibrium relationship between the moisture content of the powder and the relative humidity of the surrounding environment at a specific temperature. Such MSIs are major sources of information for optimizing concentration and dehydration processes, microbial growth conditions, and the physical and chemical stability of the product (Hardy et al., 2002). Knowing the MSIs of powdered milk products is essential to be able to predict their stability in association with packaging characteristics (Foster et al., 2005b). Figure 7.2 depicts a stability map for dairy powders containing amorphous lactose.

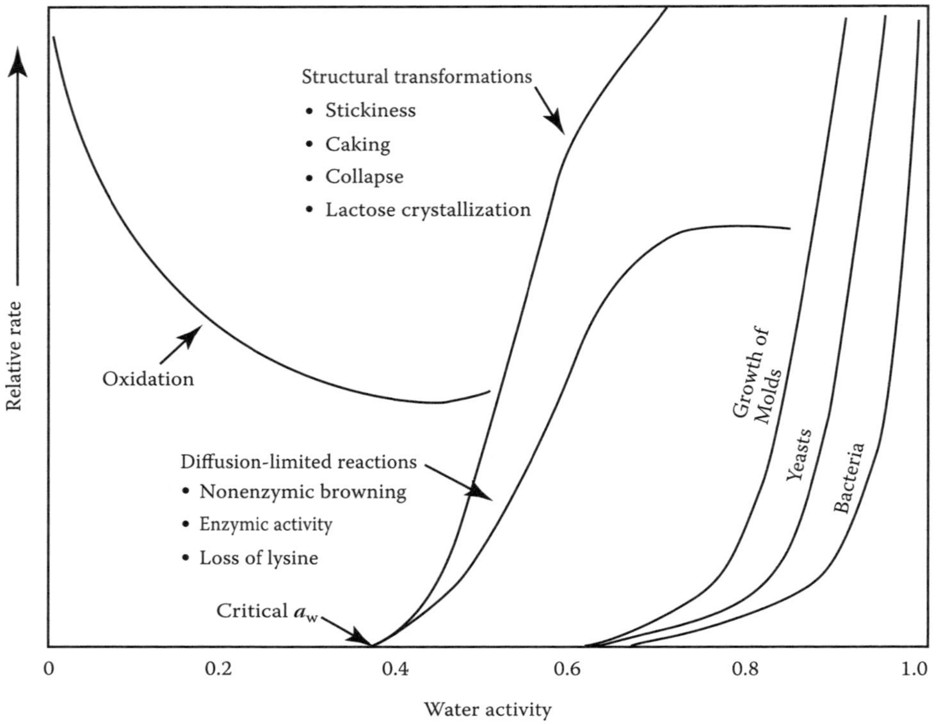

FIGURE 7.2 Stability map for dairy powders containing amorphous lactose. The critical water activity corresponds to the glass transition depression of amorphous lactose to 24°C, which may enhance deteriorative changes and loss of quality. (Redrawn from Roos Y.H. 2002. Importance of glass transition and water activity to spray drying and stability of dairy powders. *Lait* 82: 475–484, with permission.)

Changes in the immediate environment (i.e., temperature, moisture, and gas composition) can cause different types of reactions that may be interrelated and sometimes act synergistically. Therefore, it is very difficult to control a particular reaction (Uppu, 2002).

Moisture content and a_w can often determine the rate of deteriorative reactions as well as microbial growth. As indicated earlier, prevention of microbial growth can be achieved provided $a_w <$ 0.6 (Roos, 2001). However, increased moisture levels due to transmission or condensation of water vapor (due to temperature fluctuations) could result in favorable conditions for microbial growth. Off-flavors, increased acidity, and visual and textural changes may be additional negative effects of microbial growth.

Another significant factor that causes caking in milk powders is lactose. Lactose is highly hygroscopic, but crystallization does not occur if $a_w < 0.43$, the moisture content <8.4%, and storage temperature <20°C (Vernam and Sutherland, 1996). A generalized MSI for milk powders was shown in Figure 7.1 with a break at $a_w = 0.5$, where water is released due to lactose crystallizing (Thomas et al., 2004). With the relatively high lactose content in filled milk powder (FMP) (~35%), the powder may be prone to caking with an increase in free moisture due to lactose crystallization. Difficulties in dispersing the powder in water (i.e., diminishing the instantizing properties) may be the result.

Therefore, in selecting a suitable packaging system for milk powders, three factors must be taken into account: the initial moisture content of the powder, the final acceptable moisture content of the powder, and the required shelf life (Robertson, 2006).

7.4.2 Oxidation

A number of food components react chemically with O_2, affecting the color, flavor, nutritional status, and occasionally the physical characteristics of foods. In some cases, the effects are deleterious and reduce the shelf life of the food; in others they are essential to achieve the desired product characteristics. Many studies have reported the development of off-flavors in milk after various storage times, usually at the end of their shelf life (Chávez-Servín et al., 2008; Contarini and Povolo, 2002; Cormier et al., 1991; Vallejo-Cordoba and Nakai, 1994). Packaging is used to exclude, control, or contain O_2 at the level most suited for a particular product.

Oxidation of powdered milk products is predominantly associated with unsaturated fats present in milk fat. Oxidation of unsaturated fats results in aldehydes and ketones, which are subsequently converted into alcohols (Nursten, 1997). Fat oxidation occurs in the presence of O_2 and moisture and can be catalyzed by light. The O_2 atmosphere inside the package, the presence of antioxidants, the a_w, and the temperature all influence the rate of oxidation (Uppu, 2002).

Powders containing a high percentage of fats, particularly unsaturated fats, are susceptible to sensory effects, collectively called oxidative rancidity, and changes in flavor. Saturated fatty acids oxidize slowly compared with unsaturated fatty acids (Brown and Williams, 2003). The presence of unsaturated bonds in fat will increase oxidation. In general, the higher the level of unsaturation, the greater the chance of fat oxidation.

It is therefore not surprising that to prevent oxidation of milk powder, the packaging should provide a high-level O_2 barrier and be able to retain that barrier during the anticipated shelf life. Gas flushing with a chemically inert gas such as N_2 may be essential to replace O_2 present in the package before closing. This is particularly true for WMP, where the shelf life is governed to a large extent by the rate of oxidation of the unsaturated fats and the consequent development of objectionable flavors (Robertson, 2006). Most of the common spoilage bacteria and fungi require O_2 for growth. Therefore, to increase the shelf life of foods, the internal package atmosphere should contain a minimum concentration of residual O_2.

In addition to fat oxidation, atmospheric O_2 and light are the prime factors influencing the stability of vitamins A and D. These factors, in combination with environmental factors such as temperature and moisture, influence the rate of reduction in the vitamin content (Ottaway, 1993).

7.4.3 Light

Light-induced degradation reactions in milk create a serious problem for the dairy industry because of the development of off-flavors, the decrease in nutritional quality, and the rate at which these phenomena develop (Mestdagh et al., 2005). Like many other foods, milk and dairy products are susceptible to oxidation, as mentioned earlier. Dairy products in particular are very sensitive to light oxidation because of the presence of riboflavin (vitamin B_2). This strong photosensitizer is able to absorb visible and UV light and transfer this energy into highly reactive forms of O_2 such as singlet O_2 (Min and Boff, 2002). Packaging materials that can provide a barrier to light are essential to avoid this particular deteriorative reaction in milk products (Mestdagh et al., 2005).

As mentioned earlier, light in combination with O_2 and moisture affects the quality of milk powder, and therefore light ingress via the package should be avoided. A package with a high barrier to the transmission of visible and invisible wavelengths is important. Therefore, packaging materials that are highly opaque are essential.

In summary, the packaging of powdered milk needs to be considered in terms of its ability to block light, avoid transmission of water and water vapor, and prevent permeation of O_2. The fourth factor influencing the indices of failure of milk powder is the ambient temperature. Although temperature is a prime factor determining the shelf life of milk powders, these products are not usually stored under controlled temperatures. Therefore, storage of milk powders at high ambient temperatures will accelerate deteriorative reactions, particularly if plastic barrier packaging materials are used, as the permeability of O_2 and water vapor increases at higher temperatures.

7.5 SHELF LIFE OF MILK POWDERS IN DIFFERENT PACKAGES

Shelf life is defined as the period between production and the time the food item loses its state of safe and satisfactory quality in terms of nutritional value, microbial status, flavor, texture, and appearance. The packaging plays a fundamental role in maintaining the quality and therefore the shelf life of foods. The package is an integral part of the preservation system and functions as an interface between the food and the external environment; the package should be designed and developed not only to contain the food product but also to protect it and add value to it, as its design may directly affect the purchase decision of the consumer (Da Cruz et al., 2007; Robertson, 2006; Sothornvit and Pitak, 2007). A range of variables influence packaging development. In designing a packaging system, trends, prerequisites, conditions, and developments in the external environment must be taken into consideration (Sonneveld, 2000).

For retailing to consumers, milk powder is packed into either metal cans or multilayer pouches. The type and construction of the package depends on the type of milk powder (e.g., skimmed, whole, filled, vitamin-added), the surface area:volume ratio of the package, the desired shelf life, the ambient storage and transport environment, and the anticipated market environment. WMP, for example, is often packed under N_2 gas to protect the product from fat oxidation, maintain its flavor, and extend shelf life. Packaging performance specifications therefore vary and depend on variations in product characteristics, the ambient distribution environment, and the market environment (Sonneveld, 2000). Essentially, packaging systems for milk powder must protect the powder from exposure to moisture, O_2, and light and anticipate the likely external environmental factors, which include temperature, time, relative humidity, light, and physical hazards.

7.5.1 Metal Cans

Packaging milk powder in metal cans has been highly popular for a long time, particularly for retail packaging. For example, cans are commercially available with capacities of 400, 900, 1800, and 2500 g. The main reason for using metal cans is their excellent physical strength, durability, absolute barrier properties to moisture, O_2, and light, absence of flavor or odor, and rigidity (Robertson, 2006).

Because bare steel is susceptible to corrosion, it is commonly electrolytically coated with a very thin layer of tin; in addition, an organic lacquer is applied to further protect the metal from corrosion and avoid metal–food contact (Robertson, 2006). Among the organic polymeric coatings, epoxy-phenolic lacquers are often used on tinplate, although waterborne polymer coatings have been playing an increasingly important role as well (Manfredi et al., 2005).

A recent concern has been the presence of natural and synthetic chemicals in foods that exhibit estrogenic affects and act as endocrine disrupters. Powdered milk (including infant formulas) may have hormonally active contaminants introduced in the manufacturing process or leached from containers (Casajuana and Lacorte, 2004). Bisphenol A (BPA) has recently been found to be one of the more potent anthropogenic estrogen mimics (Kim et al., 2001). It is a monomer used to produce (among other things) epoxy resins that are widely used to coat the interior of cans, leading to potential human exposure. Kuo and Ding (2004) detected BPA in powdered milk and infant formulas on the Japanese market at concentrations from 45 to 113 ng g^{-1}.

The milk powder steel can is commonly cylindrically shaped and may feature a reclosable (tight fit) lid. In the standard version the can features a cylindrical body with "can ends" on both ends. The can body is welded longitudinally, and the can ends are seamed onto the can body. To obtain appropriate closure (i.e., to maintain the integrity of the pack) an elastomeric compound is included in the end seam. In cans with a reclosable lid it is common to seal the underside of the can end with an aluminum foil laminate to ensure integrity during storage and distribution.

Milk powder has a long shelf life when packed in metal cans due to their excellent barrier properties. The exchange of moisture and O_2 and the influx of light are not possible. Powders with a higher fat content are more susceptible to oxidation, and most powders are susceptible to deteriorative effects such as lumping and caking from moisture ingress. With adequately constructed cans, a shelf life in excess of 5 years is realistic, particularly when FMP products have been gas-flushed with N_2 to minimize the amount of available O_2. However, national food safety authorities often adopt a conservative approach by reducing the nominated shelf life.

Nonfat dry milk (NDM) and powdered whey beverages are available at the retail level in the United States and many other countries packaged in no. 10 cans (157 × 178 mm with a capacity of 3108 mL) in a reduced-O_2 atmosphere to prolong shelf life (up to 54 months). Lloyd et al. (2004) found that in the 10 US brands tested, wide variation existed in headspace O_2, can seam quality, sensory quality, and vitamin A (with 6 of 10 brands entirely lacking the vitamin). The a_w of the brands ranged from 0.14 to 0.28 (a typical range), corresponding to 3–5% moisture content. The brand that scored highest in overall acceptability had an average headspace O_2 of 7% and poor can seams, calling into question the ability of the package to maintain product quality over an extended storage time.

7.5.2 Multilayer Pouches

In recent years, aluminum foil/plastic film laminates have been introduced as a replacement for the tinplate can. The laminates can be formed, filled, gas-flushed, and sealed on a single machine from reel stock (Robertson, 2006). Such flexible pouches or sachets are well positioned to exploit the opportunities for convenience food markets. Flexible packages reduce the volume of traditional packaging such as metal cans, reduce transport costs, reduce the cost of the packaging, and require less material, thus minimizing postconsumer waste (Twede and Goddard, 1998). However, in many developing countries milk powder in metal cans is still the preferred packaging option for larger capacities because of recloseability and the fact that the empty can can be reused as a household utensil.

Milk powder packed in pouches is commercially available in a capacity range of 250–2500 g. In addition, sachets with smaller capacities are also available to provide convenient single-serve portions of up to 35 g. As with metal cans, milk powder packed in multilayer pouches is predominantly destined for retail distribution. The single-serve sachets are mainly distributed in developing countries because of the need to provide an affordable but highly nutritious food product. This type of retail distribution usually entails exposure to high humidity, high temperature, high levels of light, and relatively long storage times (Uppu, 2002). To maintain the quality of the milk powder in such small sachets is a challenge given the very high surface area:volume ratio. A 2-year shelf life for milk powder in portion packs is normally required when distributing in the relatively complex environments of developing countries. In countries with more highly developed economies a maximum shelf life of up to 12 months is common.

Commonly, a laminated multilayer pouch for milk powder must comprise a barrier to water vapor, O_2 (at least for WMP products), and light. Aluminum foil is capable of providing such a barrier provided the foil does not have pin holes in it. Aluminum foil built into a flexible material provides a close-to-absolute barrier. Building into a flexible material is essential because the foil does not have any mechanical strength by itself and therefore needs protection from mechanical damage. A sandwich construction with two plastic layers—one on the inside, such as low density polyethylene (LDPE), so that the pouch can be sealed and one on the outside, such as biaxially oriented polypropylene (BOPP) or poly(ethylene terephthalate) (PET), to provide mechanical protection and also carry information—is common practice (Uppu, 2002).

Alternatively, with pouches for which a shorter shelf life is acceptable, the alufoil layer may be replaced with a high-barrier plastic layer such as a copolymer of ethylene vinyl alcohol (EVOH) or polyvinylidene chloride (PVdC), possibly with the addition of a thin layer of metal or silica oxide

(SiO_x) deposition to enhance its O_2 barrier characteristics (Lange and Wyser, 2003). However, the shelf life will likely be less than that of a pouch containing an alufoil layer. A shelf life of up to 2 years is not feasible with portion pouches in a challenging distribution environment, such as exists in many developing countries, other than with the inclusion of an alufoil layer. Sachets with larger capacity (in excess of 250 g) comprising a high-barrier plastic layer sandwiched between LDPE and BOPP or PET would be able to achieve a similar shelf life to an alufoil-sandwiched portion pack pouch.

As indicated earlier, the shelf life of milk powder is not determined solely by the package construction and the amount of product packed. External factors such as variations in the physical distribution environment, the retail business setting, the demographic, social, and ethnic conditions, the regulatory environment, and, importantly, the costs of the systems affect the required shelf life and the required packaging performance associated with it (Sonneveld, 2000).

REFERENCES

Aguilera J.M., del Valle J.M., Karel M. 1995. Caking phenomena in amorphous food powders. *Trends in Food Science and Technology* 6: 149–155.

Andersson K., Lingnert H. 1997. Influence of oxygen concentration on the storage stability of cream powder. *LWT—Food Science and Technology* 30: 147–154.

Anon. 1989. *Dairy Processing Handbook*. Lund, Sweden: Alfa-Laval.

Augustin M.A., Sanguansri L., Bode O. 2006. Maillard reaction products as encapsulants for fish oil powders. *Journal of Food Science* 71: E25–E32.

Baechler R., Clerc M.F., Ulrich S., Benet S. 2005. Physical changes in heat-treated whole milk powder. *Lait* 85: 305–314.

Baldwin A.J., Cooper H.R., Palmer K.C. 1991. Effect of preheat treatment and storage on the properties of whole milk powder. Changes in sensory properties. *Netherlands Milk and Dairy Journal* 45: 97–116.

Birchal V.S., Passos M.L., Wildhagen G.R.S., Mujumdar A.S. 2005. Effect of spray dryer operating variables on the whole milk powder quality. *Drying Technology* 23: 611–636.

Bishov S., Henick A., Giffee J., Nii I., Prell P., Wolf M. 1971. Quality and stability of some freeze-dried foods in "zero" oxygen headspace. *Journal of Food Science* 36: 532–535.

Bossett J.O., Gallmann P.U., Sieber R. 1994. Influence of light transmittance of packaging materials on the shelf life of milk and dairy products—a review. In: *Food Packaging and Preservation*. Mathlouthi M. (Ed). Glasgow, UK: Blackie Academic & Professional, pp. 222–268.

Bradley D.G., Min D.B. 1992. Singlet oxygen oxidation of foods. *Critical Reviews in Food Science and Nutrition* 31: 211–236.

Brown H., Williams J. 2003. Packaged product quality and shelf life. In: *Food Packaging Technology*. Coles R., McDowell D., Kirwan M.J. (Eds). Oxford, England: Blackwell Publishing, pp. 65–95.

Bylund G. 2003. *Dairy Processing Handbook*. Lund, Sweden: Tetra Pak Processing Systems AB, pp. 361–374.

Cadwallader K.R., Howard C.L. 1998. Analysis of aroma-activated components of light activated milk. In: *Flavor Analysis: Developments in Isolation and Characterization*. Mussinan, C.J., Morello, M.J. (Eds). ACS Symposium Series #705. Washington: American Chemical Society, pp. 343–358.

Carić M. 1994. *Concentrated and Dried Dairy Products*. New York: VCH Publishers, pp. 57–94.

Casajuana N., Lacorte S. 2004. New methodology for the determination of phthalate esters, bisphenol A, bisphenol A diglycidyl ether, and nonylphenol in commercial whole milk samples. *Journal of Agricultural and Food Chemistry* 52: 3702–3707.

Chávez-Servín J.L., Castellote A.I., López-Sabater C. 2008. Volatile compounds and fatty acid profiles in commercial milk-based infant formulae by static headspace gas chromatography: evolution after opening the packet. *Food Chemistry* 107: 558–569.

Clark W.S.J. 1992. Dry milk. In: *Encyclopedia of Food Science and Technology*, Vol. 4. Hui Y.H. (Ed). New York: Wiley & Sons, pp. 656–658.

Contarini G., Povolo M. 2002. Volatile fraction of milk: comparison between purge and trap and solid phase microextraction techniques. *Journal of Agricultural and Food Chemistry* 50: 7350–7355.

Cormier F., Raymond Y., Champagne C.P., Morin A. 1991. Analysis of odor-active volatiles from *Pseudomonas fragi* grown in milk. *Journal of Agricultural and Food Chemistry* 39: 159–161.

Da Cruz A.G., Faria J.A.F., Van Dender A.G.F. 2007. Packaging system and probiotic dairy foods. *Food Research International* 40: 951–956.

Fenaille F., Visani P., Fumeaux R., Milo C., Guy P.A. 2003. Comparison of mass spectrometry-based electronic nose and solid phase microextraction gas chromatography-mass spectrometry technique to assess infant formula oxidation. *Journal of Agricultural and Food Chemistry* 51: 2790–2796.

Fitzpatrick J.J., Iqbal T., Delaney C., Twomey T., Keogh M.K. 2004. Effect of powder properties and storage conditions on the flowability of milk powders with different fat contents. *Journal of Food Engineering* 64: 435–444.

Fitzpatrick J.J., Hodnett M., Twomey M., Cerqueira P.S.M., O'Flynn J., Roos Y.H. 2007. Glass transition and the flowability and caking of powders containing amorphous lactose. *Powder Technology* 178: 119–128.

Foster K.D., Bronlund J.E., Paterson A.H.J. 2005a. The contribution of milk fat towards the caking of dairy powders. *International Dairy Journal* 15: 85–91.

Foster K.D., Bronlund J.E., Paterson A.H.J. 2005b. The prediction of moisture sorption isotherms for dairy powders. *International Dairy Journal* 15: 411–418.

Gacula M.C. 1975. The design of experiments for shelf life study. *Journal of Food Science* 40: 399–403.

Grosch W., Schieberle P., Laskawy G. 1981. Model experiments about the formation of volatile carbonyl compounds from fatty acid hydroperoxides. In: *Flavour '81*. Schreier P. (Ed). New York: Walter de Gruyler, pp. 433–448.

Hardy J., Scher J., Banon S. 2002. Water activity and hydration of dairy powders. *Lait* 82: 441–452.

Ibach A., Kind M. 2007. Crystallization kinetics of amorphous lactose, whey-permeate and whey powders. *Carbohydrate Research* 342: 1357–1365.

Kelly J., Kelly P.M., Harrington D. 2002. Influence of processing variables on the physicochemical properties of spray dried fat-based milk powder. *Lait* 82: 401–412.

Kieseker F.G., Aitken B. 1993. Recombined full-cream milk powder. *Australian Journal of Dairy Technology* 48: 33–37.

Kim J.C., Shin H.C., Cha S.W., Koh W.S., Chung M.K., Han, S.S. 2001. Evaluation of developmental toxicity in rats exposed to the environmental estrogen bisphenol A during pregnancy. *Life Sciences* 69: 2611–2625.

Kuo H.W., Ding W.H. 2004. Trace determination of bisphenol A and phytoestrogens in infant formula powders by gas chromatography-mass spectrometry. *Journal of Chromatography A* 1027: 67–74.

Labuza T.P. 1971. Kinetics of lipid oxidation in foods. *CRC Critical Reviews in Food Science and Technology* 2: 355–405.

Lange J., Wyser Y. 2003. Recent innovations in barrier technologies for plastic packaging—a review. *Packaging Technology and Science* 16: 149–153.

Liang J.H. 1999a. Fluorescence due to interaction of oxidizing soybean oil and soy proteins. *Food Chemistry* 66: 103–108.

Lloyd M.A., Zou J., Farnsworth H., Ogden L.V., Pike O.A. 2004. Quality at time of purchase of dried milk products commercially packaged in reduced oxygen atmosphere. *Journal of Dairy Science* 87: 2337–2343.

Loncin M., Bimbenet J.J., Lenges L. 1968. Influence of the activity of water on the spoilage of foodstuffs. *Journal of Food Technology* 3: 131–142.

Manfredi L.B., Ginés M.J.L., Benítez G.J., Egli W.A., Rissone H., Vázquez A. 2005. Use of epoxy-phenolic lacquers in food can coatings: characterization of lacquers and cured films. *Journal of Applied Polymer Science* 95: 1448–1458.

McCluskey S., Connolly J.F., Devery R., O'Brien B., Kelly J., Harrington D., Stanton C. 1997. Lipid and cholesterol oxidation in whole milk powder during processing and storage. *Journal of Food Science* 62: 331–337.

McKenna A.B. 1997. Examination of whole milk powder by confocal laser scanning microscopy. *Journal of Dairy Research* 64: 423–432.

Mestdagh F., De Meulenaer B., De Clippeleer J. 2005. Protective influence of several packaging materials on light oxidation of milk. *Journal of Dairy Science* 88: 499–510.

Min D.B., Boff J.M. 2002. Chemistry and reaction of singlet oxygen in foods. *Comprehensive Reviews in Food Science and Safety* 1: 58–72.

Moreno-Atanasio R., Antony S.J., Ghadiri M. 2005. Analysis of flowability of cohesive powders using distinct element method. *Powder Technology* 158: 51–57.

Nursten H.E. 1997. The flavour of milk and dairy products: 1. Milk of different kinds, milk powder, butter and cream. *International Journal of Dairy Technology* 50: 48–56.

O'Brien J.M., Morrisey P.A. 1989. The Maillard reaction in milk products. *International Dairy Federation* 238: 53–61.

O'Callaghan D., Cunningham P. 2005. Modern process control techniques in the production of dried milk products—a review. *Lait* 85: 335–342.

Oldfield D.J., Teehan C.M., Kelly P.M. 2000. The effect of preheat treatment and other process parameters on the coffee stability of instant whole milk powder. *International Dairy Journal* 10: 659–667.

Oldfield D.J., Taylor M.W., Singh H. 2005. Effect of preheating and other process parameters on whey protein reactions during skim milk powder manufacture. *International Dairy Journal* 15: 501–511.

Ottaway P.B. 1993. Stability of vitamins. In: *The Technology of Vitamins in Foods*. Ottaway P.B. (Ed). New York: Blackie Academic & Professional, pp. 91–113.

Peleg, M. 1977. Flowability of food powders and methods for its evaluation—a review. *Journal of Food Process Engineering* 1: 303–328.

Rennie P.R., Chen X.D., Hargreaves C., Mackereth A.R. 1999. A study of the cohesion of dairy powders. *Journal of Food Engineering* 39: 277–284.

Robertson G.L. 2006. *Food Packaging Principles and Practice*, 2nd edn. Boca Raton, Florida: CRC Press.

Romeu-Nadal M., Chávez-Servín J.L., Castellote A.I., Rivero M., López-Sabater M.C. 2007. Oxidation stability of the lipid fraction in milk powder formulas. *Food Chemistry* 100: 756–763.

Roos Y.H. 2001. Water activity and plasticization. In: *Food Shelf Stability: Chemical, Biochemical and Microbiological Changes*. Eskin M.N.A., Robinson D.S. (Eds). Boca Raton, Florida: CRC Press, pp. 4–36.

Roos Y.H. 2002. Importance of glass transition and water activity to spray drying and stability of dairy powders. *Lait* 82: 475–484.

Sattar A., Deman J.M., Alexander J.C. 1976. Effect of wavelength on light-induced quality deterioration of edible oils and fats. *Journal of the Canadian Institute for Food Science and Technology* 9: 108–113.

Schaich K.M. 2005. Lipid oxidation: theoretical aspects. In: *Bailey's Industrial Oil and Fat Products*. Shahidi F. (Ed). New Jersey: John Wiley, pp. 269–355.

Schuck P. 2002. Spray drying of dairy products: state-of-the-art review. *Lait* 82: 375–382.

Schuck P., Mejean S., Dolivet A., Jeantet R., Bhandari B. 2007. Keeping quality of dairy ingredients. *Lait* 87: 481–488.

Shrestha A.K., Howes T., Adhikari B.P., Bhandari B.R. 2007. Water sorption and glass transition properties of spray dried lactose hydrolysed skim milk powder. *LWT—Food Science and Technology* 40: 1593–1600.

Shrestha A.K., Howes T., Adhikari B.P., Bhandari B.R. 2008. Spray drying of skim milk mixed with milk permeate: effect on drying behavior, physicochemical properties, and storage stability of powder. *Drying Technology* 26: 239–247.

Singh H., Newstead D.F. 1992. Aspects of proteins in milk powder manufacture. In: *Advanced Dairy Chemistry*, Vol. 1. Fox P.F. (Ed). England: Elsevier Science Publishers, pp. 735–765.

Sonneveld K. 2000. What drives (food) packaging innovation? *Packaging Technology and Science* 13: 29–35.

Sothornvit R., Pitak N. 2007. Oxygen permeability and mechanical properties of banana films. *Food Research International* 40: 365–370.

Stapelfeldt H., Nielsen B.R., Skibsted L.H. 1997. Effect of heat treatment, water activity and storage temperature on the oxidative stability of whole milk powder. *International Dairy Journal* 7: 331–339.

Straatsma J., Van Houwelingen G., Steenbergen A.E., De Jong P. 1999a. Spray drying of food products: 1. Simulation model. *Journal of Food Engineering* 42: 67–72.

Straatsma J., Van Houwelingen G., Steenbergen A.E., De Jong P. 1999b. Spray drying of food products: 2. Prediction of insolubility index. *Journal of Food Engineering* 42: 73–77.

Thomas M.E.C., Scher J., Desobry-Banon S., Desobry S. 2004. Milk powders ageing: effect on physical and functional properties. *Critical Reviews in Food Science and Nutrition* 44: 297–322.

Thomsen M.K., Lauridsen L., Skibsted L.H., Risbo J. 2005. Two types of radicals in whole milk powder. Effect of lactose crystallization, lipid oxidation, and browning reactions. *Journal of Agricultural and Food Chemistry* 53: 1805–1811.

Twede D., Goddard R. 1998. *Packaging Materials*, 2nd ed. Surrey, England: Pira International.

Uppu P. 2002. *Determination of Packaging Material Requirements for Optimum Shelf Life of Packaged Filled Milk Powder.* Masters Dissertation, Melbourne, Australia: Victoria University of Technology.

Vernam A.H., Sutherland J.P. 1996. Concentrated and Dried Milk Products. In: *Milk and Milk Products Technology, Chemistry and Microbiology*. London: Chapman & Hall, pp. 136–148.

Vallejo-Cordoba B., Nakai S. 1994. Keeping-quality assessment of pasteurized milk by multivariable-analysis of dynamic headspace gas chromatographic data. 1. Shelf-life prediction by principal component regression. *Journal of Agricultural and Food Chemistry* 42: 989–993.

Van Mil P.J.J.M., Jans J.A. 1991. Storage stability of whole milk powder: effects of process and storage conditions on product properties. *Netherland Milk and Dairy Journal* 45: 145–167.

Walstra P., Geurts T.J., Noomen A., Jellema A., Van Boekel M.A.J.S. 1999. Milk powder. In: *Dairy Technology. Principles of Milk Properties and Processes*. New York: Marcel Dekker, pp. 445–470.
Warmbier H.C., Wolf M.J. 1976. Pouch for oxygen-sensitive products. *Modern Packaging* 49: 38–41.
Ye A., Anema S.G., Singh H. 2007. Behaviour of homogenized fat globules during the spray drying of whole milk. *International Dairy Journal* 17: 374–382.

8 Packaging and the Shelf Life of Yogurt

Roger D. MacBean
Parmalat Australia Ltd.
South Brisbane, Australia

CONTENTS

8.1 ORIGINS AND DEVELOPMENT

Yogurt is a modern food with ancient origins. It originated perhaps 10,000–15,000 years ago, probably in the Middle East, from adventitious contamination by lactic acid bacteria of milk in traditional containers such as earthenware pots or containers made from animal skins. Over time, the preservation of milk by making fermented milk products spread around the world, wherever milking animals were domesticated, and numerous types of traditional fermented milk products emerged, depending on the source of milk and the microbial cultures and manufacturing methods used. Tamime and Robinson (1999a), Chandan (2006), and Vasiljevic and Shah (2008) have provided detailed accounts of the origins and development of yogurt and fermented milk products.

During the twentieth century, yogurt became popular in Western countries, first in a fairly traditional form, as a "set" product, fermented in the containers used for its sale; then, to add variety, with a fruit preparation (similar to a jam) added to the bottom of the containers, with the yogurt above the fruit—the so-called fruit-on-the-bottom yogurts; then, as "stirred" or "Swiss-style" yogurts, with the yogurt coagulum made in fermentation tanks, cooled, and filled into the final containers, often with fruit preparation mixed through the yogurt. The latter half of the twentieth century saw a great increase in the variety of products in terms of fat content, texture, acidity level, and culture. During this period, packaging technology also developed significantly, and the forms of packaging for yogurt products became diverse as well. Product differentiation and marketing drove consumer interest and sales, so that by the end of the twentieth century global food companies were making and marketing spoonable and drinkable yogurt products around the world under globally recognized brand names.

Although yogurt is still made by traditional means in some parts of the world, industrially yogurt products in all their various forms are made and packaged using sophisticated methods, making use of state-of-the-art technology for culture preparation, processing, packaging, and distribution. From a product that in ancient times allowed milk to be preserved for a few days, there now exist products containing live starter and probiotic cultures that have refrigerated shelf lives of 6 weeks or more. There are also pasteurized yogurt products in some markets in which the cultures have been killed by heat treatment after the product has been made, and these may have refrigerated or ambient shelf lives measured in months, depending on the manufacturing method and packaging material.

This chapter addresses modern spoonable and drinking-yogurt products and technology and the influence packaging materials and systems have on their shelf life.

8.2 YOGURT MANUFACTURING PROCESSES

The common types of yogurt are set, stirred, and drinking yogurt, and they are found in most developed markets around the world. The main production steps are shown in Figure 8.1.

For all types of yogurt, the first step is the preparation of a yogurt mix and the heat treatment of this mix. The milk mix is formulated to have the required amount of milk fat, milk protein and other nonfat milk solids, and added sugars, stabilizers, flavors, and colors, if these are used. The solids level, particularly the protein level, affects the texture and firmness of the yogurt coagulum, and so it is common to increase the nonfat solids of the mix for set and stirred yogurts. This can be done by adding skim milk powder or skim milk concentrate during the preparation of the milk mix or by using membrane separation techniques such as reverse osmosis or ultrafiltration to concentrate skim milk or the protein in skim milk. Where regulations permit their addition, stabilizers such as

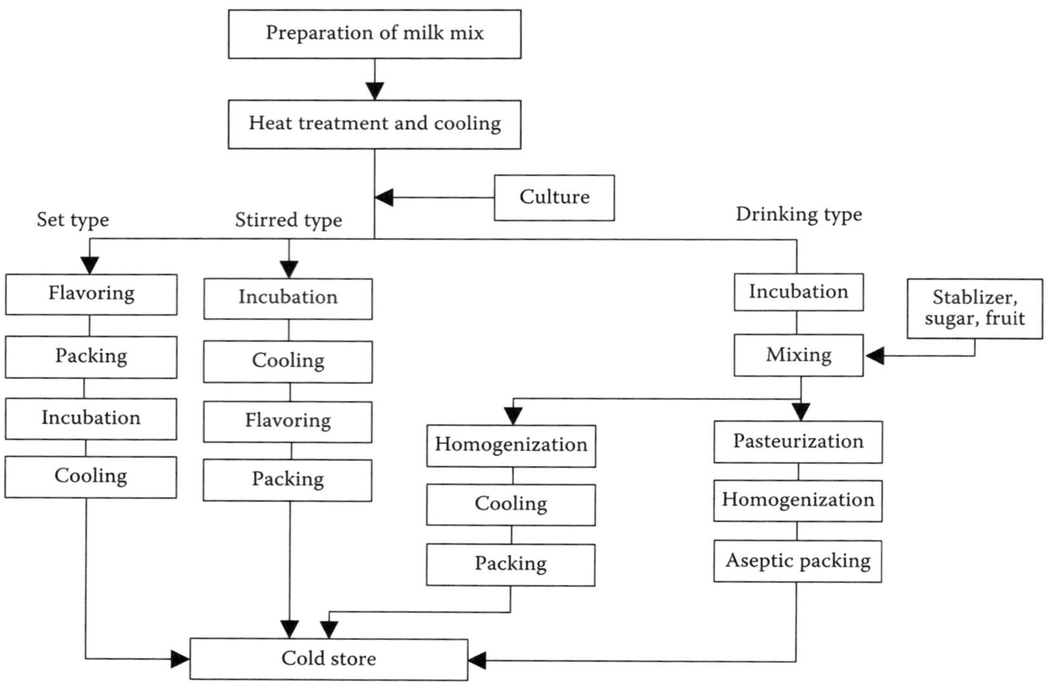

FIGURE 8.1 Production steps for set, stirred, and drinking yogurts. (Redrawn from Bylund G. 1995. Cultured milk products. In: *Dairy Processing Handbook*. Lund, Sweden: Tetra Pak Processing Systems AB, p. 248.)

gelatin, modified starches, pectin, and gums can also be used to make the coagulum firmer and to improve its water-binding properties, reducing the tendency for serum separation from the coagulum (so-called syneresis).

The yogurt mix is heat-treated for two reasons: first, to kill pathogens and spoilage microorganisms and, second, to denature the milk proteins and bring about interactions between them that increase their water-binding capacity and hence the firmness of the yogurt coagulum (Chandan and O'Rell, 2006; Tamime and Robinson, 1999b). It is common to heat-treat the milk mix for yogurt making at around 90–95°C with a holding time of about 5 min (Bylund, 1995). This heat treatment is normally applied in a continuous flow system that includes a homogenization step to ensure the milk fat remains in a homogeneous emulsion throughout the milk mix during fermentation.

Stirred yogurt is the most widely made and sold type of yogurt. Following heat treatment, the milk mix is cooled to the fermentation temperature of around 42–43°C (or lower if probiotic organisms, such as *Bifidobacterium* subsp., with lower optimum growth temperatures are added) and pumped into a stainless steel fermentation tank. Yogurt starter culture, consisting usually of a mixed culture of *Streptococcus thermophilus* and *Lactobacillus delbrueckii* subsp. *bulgaricus*, is inoculated into the yogurt mix and fermentation takes place. In a few hours (the time depending on the rate of addition of starter culture, its activity, and the temperature), a coagulum will form due to acid produced by the starter organisms. When the pH of the coagulum has reached 4.2–4.6, depending on the exact production system used, it is "broken" by pumping it from the fermentation tank (sometimes with slow stirring) and then cooled in a heat exchanger and pumped to a storage tank, where it is ready for mixing with a fruit or flavor preparation and packing. In many production systems the temperature to which the coagulum is cooled is between 15°C and 22°C (i.e., a precooling), with final cooling to the distribution temperature of less than 5°C achieved in the final package. However, the exact cooling temperatures used for precooling vary among production systems.

Pasteurized or long shelf life yogurt may either be heated in the container after filling, at temperatures above 50°C for times of around 15–30 min, or heated before filling and hot-filled at temperatures of up to 70°C. The shelf life will vary according to the exact conditions used. For pasteurized yogurt, an appropriate stabilizer system must be used to ensure a satisfactory texture following the heating and cooling steps.

The process for drinking yogurt shares the steps of yogurt mix preparation and heat treatment, and of inoculation and fermentation, with stirred yogurt. Following these steps the coagulum is mixed with a flavoring mix, often incorporating some fruit juice, and the resultant mixture is homogenized. There is usually a stabilizer present in the formulation that enables a smooth beverage product to be produced, free of any graininess in texture.

Frozen yogurt products (either hard-frozen or soft-serve) are also produced and sold in significant quantities in some markets. The process for making these products relies essentially on ice-cream manufacturing technology, except that yogurt is used as a main component in making the ice-cream mix before churning and freezing. The packaging materials, containers, and filling equipment are the same as for hard-frozen or soft-serve ice-cream.

Yogurt can also be dried into a powder for use in manufacturing other formulated foods. Packaging materials, containers, and filling equipment are the same as for powdered products such as whole milk powder.

8.3 YOGURT CONSUMPTION

In some countries, including the Netherlands, Finland, Sweden, Denmark, France, and Germany, the consumption of fermented milk products has historically been very high. Chandan (2006) reported that in all these countries consumption of fermented milk in 1998 was in excess of 25 kg/person/year. Moreover, in the period 1998–2004, the per-capita consumption of yogurt in Western Europe increased from approximately 13 to about 15 kg/person/year (USDA, 2006). In recent years there has been a strong growth in yogurt consumption in many other countries, including the United

States, up from 0.9 kg/person/year in 1975 to 3.7 kg/person/year in 2002 (USDA, 2004); Canada, up from 3.04 L/person/year in 1987 to 7.13 L/person/year in 2007 (Canadian Dairy Information Centre, 2008); and Australia, up from 5.6 kg/person/year in 2001/2002 to 6.9 kg/person/year in 2007/2008 (Dairy Australia, 2009).

Yogurt is well recognized as a healthy product, and it may be expected that consumption will continue to grow in many markets in line with trends toward healthy eating, including in Asian markets, where the per-capita consumption of milk and dairy products is currently low. In established dairy countries, the market for yogurt products is also characterized by innovations in formulations (e.g., the addition of probiotic cultures) and packaging (e.g., multipacks; pack size and shape differentiation using, for example, open-mold, form–fill–seal technology; tubes for products for children; reclosable pouches; and label styles, including shrink-wraps and paperboard tear-off labels), which helps to bring new consumers into the market and to drive consumption.

8.4 QUALITY ATTRIBUTES OF YOGURT

During the production of yogurt, the starter cultures grow and produce lactic acid. This reduces the pH to below the isoelectric point of the milk proteins and causes coagulation, producing the characteristic gel structure and contributing to the clean, acid taste of the product. The starter organisms also produce a range of other flavor compounds comprising volatile and nonvolatile organic acids, and carbonyl compounds such as acetaldehyde, acetone, acetoin, and diacetyl. Many starter cultures also produce polysaccharide materials, which increase the thickness and stability of the yogurt gel (Shah, 2003).

Tamime and Robinson (1999c), Karagül-Yüceer and Drake (2006), and Lewis and Dale (2000) have given good descriptions of the desirable sensory properties of yogurt. Set yogurt should have a smooth, junket-like appearance and a firm, cuttable texture. Stirred yogurt should have a smooth, glossy, thick texture. There should not be excessive whey at the surface of either product. Both set and stirred yogurt should have a fresh, clean, slightly acidic taste, with the characteristic note of acetaldehyde, sometimes described as "green." There should not be excessive postacidification in the product caused by continuing acid development by the cultures in the package following final cooling to refrigeration temperatures.

Drinking yogurt products vary widely in viscosity and mouthfeel, from products that are quite thick to those that have a thinner, more refreshing mouthfeel. Generally, the solids levels of drinking yogurts are lower than those of spoonable products.

The composition of yogurt products varies according to the tastes of the market and the specific segment in the market. The milk fat content may vary from virtually zero for skim milk yogurt products to more than 4% for "indulgence" products and even up to 8% or 9% for Greek-style products. Protein levels may be the same as those of the milk from which the yogurt is made (between 3% and 3.5%) or significantly greater in order to increase the firmness of the coagulum. Yogurt contains the nutrients of the milk fat, milk protein, lactose, and the minerals and vitamins of its milk components, as well as other nutrients, such as those from added fruit preparations, and is widely recognized as a nutritious product.

Probiotic microorganisms have been defined by Fuller (1989) as live microbial food supplements that benefit the health of consumers by maintaining or improving their intestinal microbial balance and, when added to yogurt, can add to its health-promoting properties. Metchnikoff (1908), in his famous book *The Prolongation of Life*, proposed a theory that the regular consumption of milk soured with lactic acid bacteria could prolong life by combating toxic bacteria in the gut. Metchnikoff's work provided the original inspiration for the search for and discovery of probiotic microorganisms. Many such organisms have been discovered and found to have positive health effects, and in recent years it has become increasingly common to add probiotic cultures to both spoonable and drinking-yogurt products (Shah, 2006). Talwalkar and Kailasapathy (2004) and Vasiljevic and Shah (2008) have reviewed the use of probiotics in commercial yogurt products and concluded that *Lactobacillus* spp. and *Bifidobacterium* spp. are the most common probiotic bacteria added to yogurt. Probiotics are an example of "functional" ingredients or supplements added to foods, and for them to be effective, it is generally agreed (Shah, 2000; Talwalkar and Kailasapathy, 2004) that there needs to be a concentration of at least 10^6 live organisms per gram in the yogurt at the time of consumption; higher levels are advocated by some (Vasiljevic and Shah, 2008). As some probiotic organisms are sensitive to O_2 and their survival in yogurt may be related to O_2 levels present (Dave and Shah, 1997; Talwalkar and Kailasapathy, 2004), the packaging material may be important in maintaining the level of probiotic organisms in the product (Miller, 2003; Miller et al. 2002; Talwalkar, 2003; Talwalkar and Kailasapathy, 2004), although this is still open to some conjecture.

Yogurt is also a very suitable product for the addition of other functional ingredients and supplements, such as vitamins and omega-3 long-chain fatty acids. These may be sensitive to O_2 levels in the product, as some (e.g., omega-3 fatty acids) are susceptible to oxidation.

The shelf life of yogurt products is determined by the time the product remains safe to eat, the time its functional claims remain true to label or to regulatory requirements, and the time its sensory properties remain acceptable to consumers. Fresh yogurt is at its best in the first few weeks of shelf life, after which there is a discernible reduction in sensory characteristics. For example, a Spanish study by Salvador and Fiszman (2004) on whole milk and skim milk flavored, set yogurts showed a gradual deterioration in sensory properties over a 91-day period of storage such that the probability of consumer acceptance was found to be around 40% for the whole milk yogurt and only 15% for the skim milk yogurt after 91 days' storage at 10°C.

8.5 INDICES OF FAILURE FOR YOGURT

8.5.1 Microbial Spoilage

Yeasts and molds are the principal agents of microbial spoilage of yogurt (Craven et al., 2001). In fresh yogurt products, yeasts and molds may be present due to contamination in the processing operations, including from added fruit preparations, from the packaging materials, or, the filling operations. Also, if the package itself lacks integrity—for example, as a result of faulty seals—then spoilage organisms can enter the product. Filling equipment for yogurt can range from relatively open fillers where no special precautions are taken to prevent atmospheric contamination to ultraclean fillers where the product is filled under an atmosphere of high-efficiency particulate air (HEPA)-filtered, laminar-flow air (usually class 10 to 100) and the packaging materials are subjected to a decontamination treatment, such as UVC light, infrared light, or even H_2O_2 vapor or steam sterilization. In order to achieve excellent quality throughout shelf lives of up to 6 or 7 weeks, which are required in some markets, most large, modern yogurt production operations use ultraclean fillers, whether of the form–fill–seal or preformed cup type, and great care is taken to prevent contamination from yeasts and molds. For example, monitoring of the factory air and of fruit preparations for yeasts and molds may be conducted.

As mentioned, fruit preparations can be a source of yeasts and molds. The technology for production and supply of fruit preparations and for dosing them into yogurt either in-line or at the filler ranges from relatively simple to highly sophisticated systems. Fruit preparations may be made so as to be "commercially sterile," meaning free from viable yeasts and molds, and supplied in bulk stainless steel containers or plastic bags (which may be metalized to provide a light barrier to protect against fading of light-sensitive fruit colors) contained in specially designed pallet-sized containers or in smaller, flexible, bag-in-box systems. When these "aseptic" fruit preparations are used, there must be strict attention to proper processing and contamination-free filling by the fruit preparation supplier and strict attention to the prevention of contamination in coupling the fruit preparation container to the fruit dosing system at the yogurt manufacturing facility. Contamination by yeasts and molds almost inevitably leads to growth, with the potential for spoilage of the yogurt product to which the fruit is added. For this reason, for bulk rigid containers of fruit preparation, N_2 may be used as the headspace gas and changes in headspace CO_2 concentration may be monitored as an indicator of yeast growth. Where procedures cannot be relied on to ensure contamination-free operation, preservatives such as sorbic acid may be added to the fruit preparation where regulations permit.

With good control during manufacturing, the shelf life of yogurt products should not be limited by yeast and mold contamination from the processing and filling operations. Where contamination does take place, the results can be catastrophic, including expensive product retrievals or recalls from the market. For this reason, there has been some work on high pressure processing (HPP) techniques aimed at the postpackaging destruction of yeasts and molds without affecting the viability of the yogurt starter and probiotic organisms. In particular, Carroll (2006) has outlined work done by Fonterra Co-operative Group Limited. However, the technology is expensive. Also, packages, including closures, must be selected for suitability for HPP, as the packages are submerged in water during processing. For simpler factory operations, such as those using open fillers with no decontamination of packaging materials, the reliable shelf life of products is likely to be less than 6 weeks, and if the factory environment contains significant levels of yeasts and molds, the shelf life may be considerably less than this.

Some deterioration of the product during its shelf life due to bacterial action is inevitable: first, due to the continuing action of the yogurt culture bacteria and, second, due to spore-forming bacteria that survive the heat treatment, as the yogurt mix is not usually sterilized. The normal result is the slow development of a cheesy flavor. For highly flavored products, this cheesy flavor may not be apparent for some weeks after production, but for plain products, it may start to become apparent within a couple of weeks. Modified atmospheres in the package headspace have been used to retard microbial growth and hence improve shelf life (Lewis and Dale, 2000). Nitrogen has been used to retard the growth of any yeasts or molds present, and CO_2 has been used to retard the growth of spoilage bacteria. However, industrially the use of modified atmospheres in package headspaces seems to have found limited use, with manufacturers concentrating instead on cleanliness and sanitation to minimize contamination, including the use of ultraclean filling machines.

For pasteurized yogurt products, microbial spoilage due to yeasts and molds should not be an issue if the processing and packaging operations are well controlled. Depending on the pasteurization conditions, spoilage due to bacterial action is also delayed.

Postacidification will reduce the product pH and also the sensory appeal if the yogurt cultures continue to produce significant amounts of acid after packaging and final cooling. In recent years there has been a trend toward milder, less acidic yogurt products, and culture suppliers have developed yogurt cultures with low postacidification characteristics. Maintaining a good cold chain is also necessary to reduce postacidification.

Yogurt is a delicate product with a pleasant texture. However, if the product is not well made, or if the package does not provide sufficient protection during transport and distribution, syneresis can occur, with a resulting separation of whey or serum. Syneresis may also occur if there is continuing acid production by the yogurt starter cultures, and so it is important to select starter cultures with low postacidification properties. It is also important to minimize the stresses the yogurt coagulum is subjected to during physical distribution. Serum separation is unsightly, and in extreme cases the yogurt coagulum may visibly contract in the container. Polystyrene (PS) is the most common material used for yogurt packaging. However, PS is brittle and, so, usually contains a rubberizing compound such as butadiene to give the package flexibility; the resulting material is known as high-impact polystyrene (HIPS). The flexibility of HIPS both provides some degree of cushion to the product and protects against splitting of the container during transport, which can occur, especially if the design or forming of the package creates thin or weak spots.

8.5.2 Viable Organisms

An important index of failure for fresh yogurt is the level of viable organisms in the product. This is true for both the basic yogurt starter organisms and for added probiotic cultures. According to regulations in various parts of the world, yogurt must contain either a certain level of viable organisms or viable starter culture organisms at an unspecified level. For example, the Codex Alimentarius (2003) provides a guide that there should be a minimum of 10^7 colony-forming units (cfu) g^{-1} of the microorganisms constituting the starter culture and a minimum of 10^6 cfu g^{-1} of any "labeled" microorganisms if a claim is made for their presence. The regulations in France (French Decree, 1988) and Japan (Japanese Ordinance, 1951) also specify 10^7 cfu g^{-1}, whereas those in China (China National Standard GB 2746, 1999) and Australia (ANZFSC, 2008) specify 10^6 cfu g^{-1}. It is generally agreed that probiotic organisms should be present at 10^6 cfu g^{-1} or more in order to be effective in gut health (Shah, 2000; Talwalkar and Kailasapathy, 2004; Vasiljevic and Shah, 2008).

Thus, there is a need to ensure that throughout the shelf life the starter organisms and any added probiotic organisms retain sufficient viability to meet national regulations and label claims made by the manufacturer. The viable counts of most cultures used in making yogurt or added as probiotics to yogurt decline slowly throughout the shelf life whatever package is used. However, some strains of probiotics are known to be O_2-sensitive in broth culture, although the situation in yogurt is less clear (Talwalkar, 2003). Thus, considerable work has been done to investigate the effect of O_2 on various cultures, the effect of packaging materials on survival of cultures in yogurt, and the use of techniques such as encapsulation and O_2 scavengers to protect probiotic organisms from exposure to O_2. This work is discussed in the following text.

8.5.3 Flavor Changes

Most yogurt products are fresh rather than pasteurized and contain live cultures. Unless they are highly flavored, the flavor is delicate and subtle. Accordingly, there is a natural staling of the product throughout the shelf life due to bacterial metabolism, even in an impermeable package. However, this staling process may be increased if the package is permeable to gases, allowing product volatiles such as acetaldehyde to escape. If light is able to affect the product, light-induced oxidation may also occur, and the colors of flavored or fruited yogurt may fade or change during shelf life, reducing the eye appeal of the product. Anthocyanin pigments in red fruits and β-carotene in yellow products are examples of colors whose fading is accelerated in the presence of light. Both O_2 and light can affect fading, so choice of packaging material can influence the rate of deterioration. In practice, selection of the package type and design depends on a number of criteria, including cost, as the margins obtainable from the market are generally modest. So HIPS, although relatively permeable to O_2 (Robertson, 2006a), is the most common packaging material due to its other advantages, such as ease of thermoforming, mechanical properties, and competitive cost.

8.5.4 Oxidation

Vitamins such as riboflavin and ascorbic acid, and functional ingredients such as long-chain omega-3 fatty acids, that are susceptible to oxidation may be components of functional yogurt products. Protection from O_2, light, or both may be needed in order to ensure that the potency of these active ingredients is maintained and also that off-flavors due to degradation products are avoided. Tamime and Robinson (1999d) discussed changes in vitamins during the production process and subsequent storage and summarized the situations as follows: dissolved O_2 can reduce significantly the content of vitamins C, B_6, B_{12} and folic acid; depending on the particular strains of yogurt starter cultures, vitamins such as thiamin and biotin may be utilized and hence decreased, although some starter cultures produce folic acid, niacin, biotin, and riboflavin during metabolism, causing an increase. Deschênes (2006), in a recent review, stated that data for the effect of packaging on the shelf life of functional foods and their bioactive components are scarce, fragmented, and dispersed within the literature and suggested that a review of the current literature would be useful. She could, though, report that conjugated linoleic acid levels in yogurt did not vary over the normal shelf life. Sanguansri and Augustin (2006) reported that omega-3 fatty acids, because of their high degree of unsaturation, are very susceptible to oxidation and the development of objectionable off-flavors and odors. Sharma et al. (2003) showed that in a drinking yogurt, omega-3 fatty acids from tuna oil in an encapsulated form gave higher acceptability scores than those in the free form.

8.6 PACKAGING MATERIALS FOR YOGURT

A wide range of packaging materials is used for yogurt products (Brody, 2006; Nilsen et al., 2002; Tamime and Robinson, 1999e).

The most popular material by far in current use for spoonable yogurt (either set or stirred) is thermoformed HIPS in the form of small cups or larger tubs, with either an aluminum foil/plastic laminate or a paper/plastic laminate heat-seal lid or closure. These containers may be produced in form–fill–seal packaging machines or be delivered preformed from packaging materials suppliers. It is normal to add pigments such as TiO_2 to the HIPS in order to improve the appearance of the package and to provide some barrier to light. This also helps in heating and softening the HIPS sheet for thermoforming when radiant heating is used (Robertson, 2006b). White is most often used, but other colors are also common. Also, often for form–fill–seal containers and sometimes for preformed cups, labels are applied that provide a further barrier to light. A HIPS sheet for forming into small containers in form–fill–seal fillers is usually around 1.0–1.4 mm thick and the containers produced have wall thicknesses of around 0.2 mm. Preformed HIPS containers may have wall thicknesses around 0.25–0.5 mm. Injection-molded polypropylene (PP) containers can be around 0.5–1 mm in wall thickness [Pritchard W.J. (www.itechnik.com.au) 2008, personal communication].

Rectangular paperboard cartons or cups (with or without an aluminum foil layer), glass containers, PP, and blow-molded high density polyethylene (HDPE) containers are all in common use; poly(ethylene terephthalate) (PET), polyvinyl chloride (PVC), polyvinylidene chloride copolymer (PVdC), and polylactate (PLA) have also been used or proposed, and for some specialty products in some markets, ceramic containers have been used (Frederiksen et al., 2003; Tamime and Robinson, 1999e).

For pasteurized, spoonable yogurt products, laminated materials are desirable if a long shelf life is needed, with some having shelf lives of 4–6 months at ambient temperatures. For these products, a low water vapor transmission rate (WVTR) is required to stop the product losing water during shelf life. A good O_2 barrier will help to protect the product from oxidation, and a good light barrier will help to delay fading of light-sensitive colors and to avoid light-induced oxidation.

Drinking-yogurt products are becoming increasingly popular. For these products, the most popular containers are HDPE bottles sealed with either aluminum foil laminate heat-seal closures or with low density polyethylene (LDPE) snap or screw caps. Bottles made from other plastics (e.g., PET) may also be used. For bottles it is common for shrink-sleeves to be used for labeling and decoration. For long-life, heat-treated drinking-yogurt products, plastic/alufoil/paperboard laminate cartons with good water vapor, O_2, and light barrier properties are also often used.

8.7 SHELF LIFE OF YOGURT IN DIFFERENT PACKAGES

As with virtually all packaged food products, the package in which a yogurt product is provided to the consumer is of major importance. It must provide a safe, convenient, attractive, functional, and cost-effective means for protecting the product throughout distribution and merchandising, for presenting it to the consumer, and for enabling easy consumption. Factors affecting the environmental impact of packaging must also be taken into consideration, and these are likely to become even more important in the future. Thus, selection of the packaging materials and of the package design must take into consideration physical product protection, protection of sensory properties, food safety, and aesthetic, functional, environmental, and cost issues. The market for yogurt products is a competitive one in which innovation plays a significant role. In common with many segments of the food industry, there are several subsegments, ranging from those that offer value propositions to consumers and in which cost considerations are paramount (remembering that the package cost is likely to be a large part of the ex-factory cost) to those in which novel offerings may require either special properties for product protection or design characteristics to ensure the product stands out from the competition on retail shelves.

In recent years, the addition of probiotic cultures to yogurt has become increasingly popular. There has been much research on the health-giving properties of various probiotic cultures, and it is now clear that certain probiotic cultures, consumed regularly and in sufficient quantities, can assist health and well-being (Mattila-Sandholm et al., 2002; Vasiljevic and Shah, 2008). Probiotic cultures vary in their sensitivity to O_2 (Dave and Shah, 1997; Talwalkar, 2003; Talwalkar and Kailasapathy, 2004), at least in broth cultures, and it has been shown that numbers of some species of *Bifidobacterium* and *Lactobacillus* reduce significantly during the shelf life of probiotic yogurts. Although this can be due to a number of factors, including the acidic pH, it has stimulated research aimed at elucidating the effect of O_2 on survival of probiotics during shelf life and on approaches to reducing the O_2 content of yogurt, including approaches involving packaging.

Several approaches have been considered: the use of packaging materials that are less permeable to O_2 (Miller, 2003), the use of O_2-scavenging packaging materials (Miller, 2003), the addition of an O_2 scavenger to the yogurt (Dave and Shah, 1997), the encapsulation of O_2-sensitive probiotic cultures to protect them from O_2 in the yogurt (Kailasapathy, 2006), variations in production methods to reduce product O_2 levels (Miller, 2003), the addition of prebiotic compounds (Donkor et al., 2007a; Juhkam et al., 2007), and the addition of whey protein concentrate (Antunes et al., 2005). However, it should be noted that the most widely used approach in commercial practice is to select probiotic strains with robust technological properties, meaning the ability to maintain good viability throughout shelf life under normal commercial manufacturing and storage conditions (Mattila-Sandholm et al., 2002). For example, Miller (2003) showed that the survival of both *Lactobacillus acidophilus* strain 2409 and *Bifodobacterium infantis* strain 1912 was not significantly reduced in either normal set yogurt or in O_2-reduced yogurt, with levels being maintained well in excess of 10^7 cfu g^{-1} over 42 days of refrigerated storage.

The application of a shrink-sleeve label may provide a significantly increased barrier to O_2 transmission (and to light transmission too, if it is opaque). Shrink-sleeves of PS and PVC are in common use to label and decorate small bottles of probiotic drinking yogurt. These bottles are often closed with an aluminum foil lid, and it is possible that the combination of shrink-sleeve and foil lid provides a measure of protection against O_2 permeation. Miller (2003) showed that the O_2 level in yogurt depended on a number of factors, including aspects of the production method as well as permeation through the packaging material. For HIPS containers, the thickness of the material at various points of the package appeared to influence the O_2 content in the near vicinity, and the O_2 level increased throughout the package during storage to levels of around 12.5 ppm by day 35 of a total of 42 days of storage. This O_2 level was similar to the saturation point of O_2 in water at the storage temperature of 4°C. These results were not surprising given the relative permeability of HIPS to O_2. This study also showed that using a HIPS/tie/EVOH/tie-layer/LDPE (where EVOH is ethylene vinyl alcohol copolymer) laminate material (NuPak) resulted in reduced O_2 content in the yogurt compared with straight HIPS over 42 days of storage, as shown in Figure 8.2.

Miller (2003) also showed that adding an O_2 scavenger (ZerO_2) to the packaging material further reduced the O_2 content, particularly for set yogurt over the first few weeks of a 6-week storage trial.

Adding ascorbic acid to yogurt has been shown to protect a *La. acidophilus* strain (strain not specified) to some extent, but the counts of bifidobacteria remained unaffected (Dave and Shah, 1997). A similar effect was reported with l-cysteine and some bifidobacteria (Collins and Hall,

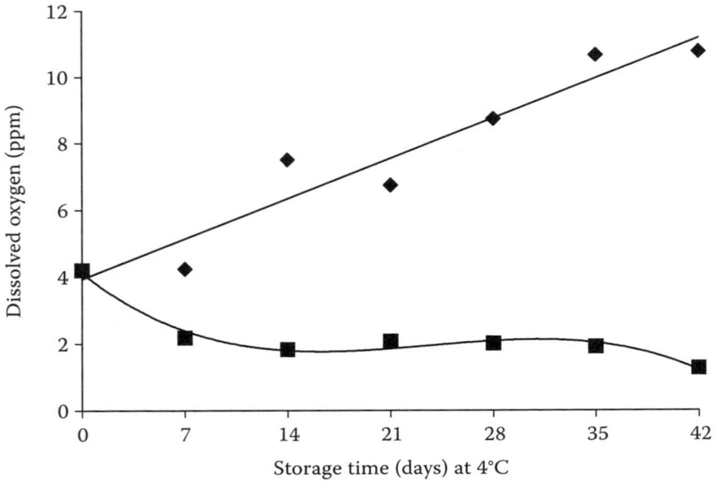

FIGURE 8.2 Oxygen levels in yogurt stored for 42 days at 4°C in HIPS (♦) and NuPack (■) packaging material. (From Miller C.W. 2003. *A Study of Packaging Methods to Reduce the Dissolved Oxygen Content in Probiotic Yoghurt*. Sydney, Australia: Centre for Advanced Food Research, University of Western Sydney, Ph.D. Thesis, with permission.)

1984). Encapsulation in an alginate polymer assisted the survival of *La. acidophilus* strain DD910 and *Bifidobacterium lactis* strain DD920 (Kailasapathy, 2006). However, Donkor et al. (2007b) found, in a study on a commercial yogurt containing the probiotics *Lactobacillus rhamnosis* GG and *Bifidobacterium animalis* Bb-12, that there were stable counts of these cultures throughout 4 weeks of storage at levels exceeding the accepted required minimum of 10^6 live cells mL^{-1} of product. Furthermore, there was very little change in viable populations of the yogurt starter cultures *Streptococcus thermophilus* and *La. delbrueckii* subsp. *bulgaricus*. This commercial yogurt (brand name Vaalia) was packaged in HIPS containers. This Vaalia yogurt did, however, contain the prebiotic inulin, which may have provided a protective effect. There have been studies showing positive effects on the survival of probiotics due to the presence of prebiotics such as inulin, oligofructose, and Hi-Maize high-amylose corn starch (Donkor et al., 2007a; Juhkam et al., 2007), and also for whey protein concentrate (Antunes et al., 2005).

A Polish study of *La. acidophilus*, *Bifidobacterium bifidum*, and *Streptococcus thermophilus* counts in goat and cow milk bioyogurt packaged in glass, HDPE, PS, and PP over 21 days found that for *La. acidophilus* the package type had no effect; for *B. bifidum* HDPE gave the best results, and for *St. thermophilus* HDPE was best for cow milk bioyogurt and PS was best for goat milk bioyogurt (Kudelka, 2006).

A recent review on the subject of packaging systems and probiotic dairy foods discusses some of the principles and also differing results found in various studies (da Cruz et al., 2007). It appears that there are significant strain differences in the sensitivity to O_2 of various probiotic cultures; also, differences in formulations and processing systems can influence the survival of probiotic cultures. Adding to the difficulties of interpreting some of the data is the fact remarked upon by some authors that enumeration methods for probiotic cultures can give misleading results unless properly verified. All these factors must be borne in mind when selecting probiotic cultures for yogurt, designing the formulations, and selecting suitable materials for packaging the product. The results discussed above for the effect of O_2 levels on probiotic cultures may also be useful when choosing packaging materials for yogurts containing O_2-sensitive functional ingredients such as long-chain omega-3 fatty acids. From a practical point of view, there can be no substitute for careful trials with the actual product in the proposed packaging material and format, ensuring that analysis methods are proven and able to give clear results.

A recent comprehensive study and discussion of the effect of some packaging materials on the flavor of strawberry-flavored, stirred yogurts of 0% and 4% fat content during their shelf life showed interesting differences between glass packaging as a reference, PP, and 50/50 PS/HIPS (Saint-Eve et al., 2008). In common with a previous study by Salvador and Fiszman (2004), mentioned earlier, 0% fat yogurt was found to deteriorate faster than 4% fat yogurt regardless of the packaging type. It was also concluded that PS/HIPS seemed to be preferable to PP for avoiding the loss of fruity notes and for hindering the development of odor and aroma defects, particularly for 4% fat yogurts. Earlier studies on the influence of light transmittance and gas permeability on the quality of whole natural yogurt during storage have been reviewed and summarized by Tamime and Robinson (1999e). Their summary showed that the greatest protection from light and O_2 combined was provided by colored glass, with decreasing degrees of protection as the packaging material was either more permeable to O_2 or more transparent to light. They also concluded that packaging materials of low O_2 permeability should be used for long-life or pasteurized yogurt. However, although packaging materials can have some effect on the shelf life of yogurt, the hygiene of the production and filling operations and the storage temperature have major impacts. Tamime and Robinson (1999e) stated that in the absence of aseptic production and filling, and at a storage temperature of 8°C, shelf life should not exceed 16–18 days. Such a shelf life would not be commercially viable in most developed markets today.

Light can induce oxidation of dairy products, including yogurt. Riboflavin, naturally present in yogurt, can absorb visual light and react as a photosensitizer. Oxygen is also necessary, and the end results of this light-induced oxidation are protein and lipid degradation with resultant off-flavors. Opaque or semi-opaque packaging materials, most usually containing a white pigment such as TiO_2, are normally used for yogurt. However, for clear glass or plastic packages and for thin-walled HIPS packages, significant light may penetrate into the product. This can be a consideration for plain yogurt products with delicate flavors. Secondary packaging such as overwraps will help to reduce light penetration. PLA as a biobased packaging material has been used for some yogurt products with an environmental or organic market positioning, and a study of the effect of light-induced changes in plain yogurt gave similar or better results than PS (Frederiksen et al., 2003).

There are yogurt products available that comprise two parts: one part containing yogurt and the other containing an auxiliary product, for example, a cereal product such as muesli that can be mixed with the yogurt, or fruit or a flavored syrup or similar foods. Some packages present the two components side by side within the same package; others present the two components as separate packages joined together in "piggyback" formats, for example. Selection of the packaging material must have regard to the particular properties of these components. For example, WVTRs are not of critical importance in a yogurt product with a shelf life of a few weeks but are of critical importance for a cereal product, which must retain its "crunch" in order to appeal to consumers. For an auxiliary product susceptible to oxidation, oxygen transmission rates (OTRs) may be critical, and protection from light may be important for a fruit product containing light-sensitive color compounds such as anthocyanin pigments, which give the red–purple–blue color of some fruits. It is important to assess the needs of the total product for protection before deciding on the packaging materials (both the container material and the closure material) for a side-by-side package or for a two-part package. With good-quality raw materials and appropriate selection of packaging materials, it should be possible to ensure that the shelf life of the auxiliary product is no less than that of the yogurt. It is worth stating again that the standard and hygiene of processing and filling operations is what normally determines the shelf life of yogurt products. The shelf life of "fresh" yogurt may be only a couple of weeks for unprotected operations and up to 6 weeks or more for well-operated, ultraclean operations.

8.8 CONCLUSIONS

Many packaging materials mentioned in this chapter are successfully used as primary packages for spoonable or drinking-yogurt products. The choice of a material and a package form will depend on several factors, including physical product protection; the required shelf life and the indices of failure for the particular product; legislation and guidelines regarding packaging materials; marketing considerations regarding appearance, product presentation, and product usage; special requirements of the retail trade, such as desired package dimensions for optimal shelf space utilization or shelf-ready packaging; economic considerations relating to capital costs for filling equipment and packaging line equipment, operating costs for packaging materials, including waste and filling machine operating costs, and commercial margins available in the relevant market; and, increasingly, environmental considerations.

Packaging technologists must work with other members of development teams to find the optimal solutions for each project, whether for developing new products or for value management reviews of existing products. Collaboration across the supply chain between suppliers of packaging equipment, packaging materials suppliers, food manufacturing and marketing companies, and retailers is becoming increasingly common and may even be regarded as a necessity in order to achieve the best results in package design and implementation. First principles should be applied whenever possible, but there is no substitute for adequate premarket trials to ensure the proposed packaging system is performing as intended.

REFERENCES

Antunes A.E.C., Cazetto T.F., Bolini H.M.A. 2005. Viability of probiotic micro-organisms during storage, post-acidification and sensory analysis of fat-free yogurts with added whey protein concentrate. *International Journal of Dairy Technology* 58: 169–173.

ANZFSC. 2008. Standard 2.5.3 Fermented Milk Products. Australia New Zealand Food Standards Code. Food Standards Australia New Zealand. www.foodstandards.gov.au/_srcfiles/Standard_2_5_3_Fermented_milk_v103.pdf, accessed May 2009.

Brody A.L. 2006. Fermented dairy packaging materials. In: *Manufacturing Yogurt and Fermented Milks*. Chandan R.C., Kilara A., Hui Y.H. (Eds). Oxford, England: Blackwell Publishing, chapter 8.

Bylund G. 1995. Cultured Milk Products. In: *Dairy Processing Handbook*. Bylund G. (Ed). Lund, Sweden: Tetra Pak Processing Systems AB, pp. 241–262.

Canadian Dairy Information Centre. 2008. http://www.dairyinfo.gc.ca/pdf/dpconsumption.pdf, accessed December 2008.

Carroll T. 2006. Exploiting high pressure in dairy applications. http://www.hi-events.com/content/uploads/Fonterra-HiE2006.pdf, accessed December 2008.

Chandan R.C. 2006. History and consumption trends. In: *Manufacturing Yogurt and Fermented Milks*. Chandan R.C., Kilara A., Hui Y.H. (Eds). Oxford, England: Blackwell Publishing, pp. 3–15.

Chandan R.C., O'Rell K.R. 2006. Principles of yogurt processing. In: *Manufacturing Yogurt and Fermented Milks*. Chandan R.C., Kilara A., Hui Y.H. (Eds). Oxford, England: Blackwell Publishing, pp. 195–197.

China National Standard GB 2746. 1999. *Standardization Administration of China*.

Codex Alimentarius. 2003. Codex Alimentarius Standard for Fermented Milks (Codex Standard 243-2003). www.codexalimentarius.net/download/standards/400/CXS_243e.pdf, accessed December 2008.

Collins E.B., Hall B.J. 1984. Growth of bifidobacteria in milk and preparation of *Bifidobacterium infantis* for a dietary adjunct. *Journal of Dairy Science* 67: 1376–1380.

Craven H.M., Broome M.C., Chandler R.E., Jensen N. 2001. Dairy products. In: *Spoilage of Processed Foods*. Moir C.J., Andrew-Kabilafkas C., Arnold G., Cox B.M., Hocking A.D., Jenson I. (Eds). Sydney, Australia: Australian Institute of Food Science and Technology Inc., pp. 160–161.

da Cruz A.G., Faria J. de A.F., Van Dender A.G.F. 2007. Packaging system and probiotic dairy foods. *Food Research International* 40: 951–956.

Dairy Australia. 2009. www.dairyaustralia.com.au/Our-Dairy-Industry/Industry-Statistics/Consumption-Summary.aspx, accessed May 2009.

Dave R.I., Shah N.P. 1997. Effectiveness of ascorbic acid as an oxygen scavenger in improving viability of probiotic bacteria in yoghurts made with commercial starter cultures. *International Dairy Journal* 7: 435–443.

Deschênes L. 2006. Packaging technologies of functional foods. In: *Functional Food Ingredients and Nutraceuticals: Processing Technologies*. Shi J., King J.W. (Eds). Boca Raton, Florida: CRC Press, pp. 329–337.

Donkor O.N., Nilmini S.L.I., Stolic P., Vasiljevic T., Shah N.P. 2007a. Survival and activity of selected probiotic organisms in set-type yoghurt during cold storage. *International Dairy Journal* 17: 657–665.

Donkor O.N., Tsangalis D., Shah N.P. 2007b. Viability of probiotic bacteria and concentrations of organic acids in commercial yoghurts during refrigerated storage. *Food Australia* 59: 121–126.

Frederiksen C.S., Haugaard V.K., Poll L., Becker E.M. 2003. Light-induced quality changes in plain yoghurt packed in polylactate and polystyrene. *European Food Research and Technology* 217: 61–69.

French Decree. 1988. Fermented Milks and Yaourt or Yoghurt; French Decree No. 88-1203 of 30th December 1988. *Official Journal of the French Republic*: December 31, 1988.

Fuller R. 1989. Probiotics in man and animals. *Journal of Applied Bacteriology* 66: 365–378.

Japanese Ordinance. 1951. Ministerial Ordinance on Milk and Milk Products Concerning Compositional Standards, etc. *Ministry of Health and Welfare Ordinance No. 52:* December 27, 1951. http://www.mhlw.go.jp/english/topics/foodsafety/dl/t-1.pdf, accessed December 2008.

Juhkam K., Elias P., Roasto M., Tamme T. 2007. Viability of *Lactobacillus acidophilus* in yoghurt containing inulin or oligofructose during refrigerated storage. *Milchwissenschaft* 62: 52–54.

Kailasapathy K. 2006. Survival of free and encapsulated probiotic bacteria and their effect on the sensory properties of yoghurt. *LWT—Food Science and Technology* 39: 1221–1227.

Karagül-Yüceer Y., Drake, M. 2006. Sensory analysis of yogurt. In: *Manufacturing Yogurt and Fermented Milks*. Chandan R.C., Kilara A., Hui Y.H. (Eds). Oxford, England: Blackwell Publishing, pp. 265–276.

Kudelka W. 2006. Types of natural bioyoghurt raw products and of their packaging and their impact on the survival rate of probiotic bacteria in bioyoghurt. *Milchwissenschaft* 61: 419–421.

Lewis M., Dale R.H. 2000. Chilled yogurt and other dairy desserts. In: *Shelf-Life Evaluation of Foods*, 2nd edn. Man C.M.D., Jones A.A. (Eds). Gaithersburg, Maryland: Aspen Publishers, pp. 89–109.

Mattila-Sandholm T., Myllärinen P., Crittenden R., Mogensen G., Fondén R., Saarela M. 2002. Technological challenges for future probiotic foods. *International Dairy Journal* 12: 173–182.

Metchnikoff E. 1908. *The Prolongation of Life: Optimistic Studies.* New York: G.P. Putnam's Sons.

Miller C.W. 2003. *A Study of Packaging Methods to Reduce the Dissolved Oxygen Content in Probiotic Yoghurt.* Sydney, Australia: Centre for Advanced Food Research, University of Western Sydney, Ph.D. Thesis.

Miller C.W., Nguyen M.H., Rooney M. 2003. The control of dissolved oxygen content in probiotic yoghurts by alternative packaging materials. *Packaging Technology and Science* 16: 61–67.

Miller C.W., Nguyen M.H., Rooney M., Kailasapathy K. 2002. The influence of packaging materials on the dissolved oxygen content of probiotic yoghurt. *Packaging Technology and Science* 15: 133–138.

Nilsen K.O., Evavold S., Solgaard R.T. 2002. Influence of packaging materials on the keeping quality of fermented milk. In: *Fermented Milk—Proceedings of the IDF Seminar on Aroma and Texture of Fermented Milk*, International Dairy Federation Special Issue 0301, pp. 215–224.

Robertson G.L. 2006a. Permeability of thermoplastic polymers. In: *Food Packaging Principles and Practice*, 2nd edn. Boca Raton, Florida: CRC Press, pp. 55–78.

Robertson G.L. 2006b. Structure and related properties of plastic polymers. In: *Food Packaging Principles and Practice*, 2nd edn. Boca Raton, Florida: CRC Press, pp. 9–42.

Saint-Eve A., Lévy C., Le Moigne M.L., Ducruet V., Souchon I. 2008. Quality changes in yogurt during storage in different packaging materials. *Food Chemistry* 110: 285–292.

Salvador A., Fiszman S.M. 2004. Textural and sensory characteristics of whole and skimmed flavored set-type yogurt during long storage. *Journal of Dairy Science* 87: 4033–4041.

Sanguansri L., Augustin M.A. 2006. Microencapsulation and delivery of omega-3 fatty acids. In: *Functional Food Ingredients and Nutraceuticals: Processing Technologies.* Shi J., King J.W. (Eds). Boca Raton, Florida: CRC Press, pp. 297–327.

Shah N.P. 2000. Probiotic bacteria: selective enumeration and survival in dairy foods. *Journal of Dairy Science* 83: 894–907.

Shah N.P. 2003. The exopolysaccharide production by starter cultures and their influence on textural characteristics of fermented milk. In: *Fermented Milk—Proceedings of the IDF Seminar on Aroma and Texture of Fermented Milk, International Dairy Federation* Special Issue 0301, pp. 101–115.

Shah N.P. 2006. Probiotics and fermented milks. In: *Manufacturing Yogurt and Fermented Milks.* Chandan R.C., Kilara A., Hui Y.H. (Eds). Oxford, England: Blackwell Publishing Ltd., pp. 341–354.

Sharma R., Sanguansri P., Marsh R., Sanguansri L., Augustin M.A. 2003. Applications of microencapsulated omega-3 fatty acids in dairy products. *Australian Journal of Dairy Technology* 58: 211.

Talwalkar, A. 2003. *Studies on the Oxygen Toxicity of Probiotic Bacteria with Reference to* Lactobacillus acidophilus *and* Bifidobacterium *spp.* Sydney, Australia: Centre for Advanced Food Research, University of Western Sydney, Ph.D. Thesis.

Talwalkar A., Kailasapathy K. 2004. A review of oxygen toxicity in probiotic yogurts: influence on the survival of probiotic bacteria and protective techniques. *Comprehensive Reviews in Food Science and Food Safety* 3: 117–124.

Tamine A.Y., Robinson R.K. 1999a. Historical background. In: *Yoghurt Science and Technology*, 2nd edn. Tamine A.K., Robinson R.K. (Eds). Cambridge, England: Woodhead Publishing, pp. 1–10.

Tamine A.Y., Robinson R.K. 1999b. Background to manufacturing practice. In: *Yoghurt Science and Technology*, 2nd edn. Tamine A.K., Robinson R.K. (Eds). Cambridge, England: Woodhead Publishing, pp. 61–71.

Tamine A.Y., Robinson R.K. 1999c. Assessment of organoleptic characteristics. In: *Yoghurt Science and Technology*, 2nd edn. Tamine A.K., Robinson R.K. (Eds). Cambridge, England: Woodhead Publishing, pp. 572–578.

Tamine A.Y., Robinson R.K. 1999d. Biochemistry of fermentation. In: *Yoghurt Science and Technology*, 2nd edn. Tamine A.K., Robinson R.K. (Eds). Cambridge, England: Woodhead Publishing, pp. 467–473.

Tamine A.Y., Robinson R.K. 1999e. Packaging. In: *Yoghurt Science and Technology*, 2nd edn. Tamine A.K., Robinson R.K. (Eds). Cambridge, England: Woodhead Publishing, pp. 90–103.

USDA, 2004. Marketing Bulletin, July 2004. http://www.fmmacentral.com/, accessed December 2008.

USDA, 2006. http://www.ers.usda.gov/publications/err28/err28c.pdf, accessed December 2008.

Vasiljevic T., Shah N.P. 2008. Cultured milk and yogurt. In: *Dairy Processing and Quality Assurance.* Chandan R.C, Kilara A., Shah N.P. (Eds). Ames, Iowa: Wiley-Blackwell, pp. 219–251.

9 Packaging and the Shelf Life of Water and Carbonated Drinks

Philip R. Ashurst
Dr. P.R. Ashurst & Associates
Ludlow
Shropshire, United Kingdom

CONTENTS

9.1 INTRODUCTION

Bottled water is now widely available for sale, and its consumption has risen dramatically over the past 15 years. The growth in bottled water is influenced by three principal public concerns: declining quality from often overworked municipal water supplies, possible toxic contamination of groundwater sources, and a generally increased interest in personal health. Bottled water has also become a "must have" fashion accessory for many consumers. There is a public perception that bottled water is safe, natural, and free from additives such as fluoride and chlorine (Robertson, 2006). Wilk (2006) discussed the ways that the rich cultural meanings of water are used in marketing and branding, and the forms of consumer resistance that oppose bottled water as a commodity. In his view, the contrast between tap water and bottled water can be seen as a reflection of a contest for authority and public trust between governments and corporations, in a context of heightened anxieties about risk and health. He concluded that bottled water is a case where sound cultural logic leads to environmentally destructive behavior.

Natural Mineral Waters are sold with the understanding (and in Europe the legal requirement) that they have not been subjected to any treatment that would remove natural, indigenous bacteria, which are believed by some to have medicinal and therapeutic qualities. It has never been proven that the ingested levels of indigenous microorganisms in bottled water have an adverse effect on health. Despite this, much controversy surrounds the question of the potential pathogenicity of indigenous microorganisms in mineral waters (Warburton, 2002).

Bottled water is defined by the US Food and Drug Administration (FDA) as "water that is intended for human consumption and that is sealed in bottles or other containers with no added ingredients except that it may contain safe and suitable antimicrobial agents" (Warburton and Austin, 2000). However, antimicrobial agents are not permitted in many other countries. The water may be subjected to a number of treatments, including distillation, carbonation, ozonation, chlorination, and filtration, depending on the quality of the source water, the type of bottled water being manufactured, and where it is being manufactured (Senior, 2004).

Traditionally, soft drinks were prepared by dissolving granulated sugar in specially treated water, or alternatively by diluting liquid sugar with this water. A variety of ingredients, including preservatives, flavoring and coloring agents, and acidulants (invariably either citric or phosphoric acid), were then added. Other constituents such as fruit juice or comminuted fruit, artificial sweeteners, antioxidants, and ingredients to deliver clouding and foam were added, depending on the particular product being made. Recently, "diet" soft drinks in which the sugar has been replaced with an artificial sweetener (typically aspartame or sucralose) have become very popular (Robertson, 2006).

Soft drinks are now prepared almost exclusively using the premix system, whereby the blended syrup, after flash pasteurization if necessary, is mixed in appropriate proportions with carbonated, treated water prior to delivery to the filler. Although traditionally the product has been cooled to 1–3°C before arrival at the filler in order to minimize loss of carbonation and facilitate filling, fillers and ancillary equipment capable of handling the product at ambient temperatures have recently been introduced (Ashurst, 2005).

Most food and drink products are supplied to consumers in some form of primary packaging and, in many cases, secondary packaging as well. Beverages, however, are totally dependent on primary packaging as a means of containing the product. This is true for all types of beverages, but for carbonated drinks the packaging plays a greater role by retaining not only the liquid but also the CO_2, which gives the product one of its principal characteristics. This chapter provides an insight into the nature of the packaging materials used for beverages and the shelf life of the product that is contained.

Development of the beverage industry for soft drinks and water is paralled by the development of suitable packaging that provides physical support for the product in addition to providing a reasonable shelf life for the contents. All early carbonated products were packaged in glass, which even today provides the performance benchmark for product protection despite its principal disadvantages of weight and brittleness. Today, a significant proportion of all beverages are packed in either some form of plastic container, plastic-laminated paperboard, or other flexible packing, most of which have made a significant contribution to the markets only since the last quarter of the twentieth century. Metal cans still provide an important alternative to the other types of packaging. The packaging of carbonated products is, with few exceptions, limited to glass, metal cans, and plastics.

It is self-evident that the primary function of any beverage packaging, which must include the closure, is to provide the physical retention of the contents, and this is therefore the first index of failure that must be considered. The consumer expects the package to retain the amount of product that he or she purchased until the time of its consumption. Container leakage may also result in damage to other property. The primary evaluation of beverage packaging is thus concerned with the retention of liquid content. This performance characteristic is determined not only by the container itself but also by the effectiveness of the seal between container and closure. It is rare for containers themselves to be produced with a defect in the body that permits leakage. For packages that are produced on line, such as in form–fill–seal operations, there is a significantly increased risk of the failure of seals, and quality checks need to be strengthened accordingly.

Assuming that the contents of a beverage container are retained satisfactorily, there are other quality attributes that determine the suitability or otherwise of the beverage–packaging combination.

In most countries there is now a statutory requirement to put some form of product durability marking on the label, and this is usually in the form of a "best before" date, although with some very short shelf life products, such as freshly squeezed juices, a "use-by" date is more appropriate (see Section 9.4). The performance of any package is thus measured by its ability to keep the contents in a condition that is as close to the taste, appearance, and nutritional or other standards required by the manufacturer within the period between the dates of manufacture and expiry.

This chapter will, therefore, review the way in which water and carbonated beverages deteriorate and examine the indices of failure and how such failures can be detected and minimized. It will then examine the performance of different packaging types and the impact that they may be expected to have on those indices. The final section of the chapter will look at the shelf life of packaged waters and carbonated beverages and the performance that may be expected in different types of packaging.

9.2 INDICES OF FAILURE AND DETERIORATION OF WATER AND CARBONATED BEVERAGES

9.2.1 Indices of Failure

Indices of failure are summarized in Table 9.1. The major deteriorative reaction in bottled water is microbial growth. To avoid this, the water is usually treated prior to bottling with chlorine or ozone; the latter is preferred as it is much faster acting than the former. Such treatment is not usually permitted for water designated "Natural Mineral Water." The two major deteriorative reactions in carbonated beverages are loss of carbonation and oxidation or acid hydrolysis of the essential flavor oils. The first is largely a function of the effectiveness of the package in providing a barrier to gas permeation, and the latter can be prevented to a large extent by the use of high-quality flavorings and antioxidants and by de-aerating the mix prior to carbonation (Robertson, 2006).

9.2.2 Physical Deterioration

The primary consideration for any container for carbonated beverages is retention of the liquid product and then retention of the CO_2 content at an acceptable level over the shelf life of the product.

TABLE 9.1
Summary of Indices of Failure

Failure Index	Characteristic	Determined By
Physical	Loss of contents	Net weight
	Loss of carbonation	Carbon dioxide level
Physicochemical	Taste deterioration	Organoleptic assessments
	Change of appearance	Unacceptable visual appearance
	Presence of contaminants	Analytical techniques
Microbiological	Presence of unwanted microorganisms	Microbial count or gross effects
Packaging	Damage or deformation	Visual inspection

This is a function not only of the container itself but also of the nature and quality of the closure and its effective application to the container body. Although leakage from container bodies is not unknown, modern automated manufacture and testing operations coupled with rigorous standards of inspection now make this unusual.

The degree of carbonation of soft drinks is typically expressed in volumes or g L^{-1} of CO_2. One volume equals approximately 2 g L^{-1}, and at room temperature each volume produces about 1 atm (101 kPa) of internal pressure. Temperature has a significant effect on internal pressure, a 4-volume beverage such as a cola rising to 7 atm pressure at 38°C and to 10 atm at maximum storage or pasteurization temperatures (Anon., 1997). The typical carbonation level of beverages ranges from 1.5 volumes for citrus and other fruit-based soft drinks to 4 volumes for common cola drinks and 5 volumes for club soda, ginger ale, and other mixer drinks (Anon., 1997).

A review of the performance of different container types is included in Section 9.3.

9.2.3 Physicochemical Deterioration

For most beverages the physical containment of the liquid contents and CO_2 only becomes an issue if the container is damaged or at the end of its shelf life. Of greater significance for carbonated soft drinks are the deteriorative effects of O_2, heat, and light. These influences are usually much less for carbonated water.

9.2.3.1 Oxygen

For most carbonated beverages, O_2 is likely to be very damaging, and every effort should be taken to exclude the gas during product mixing and packaging. In addition to the oxidative damage that is likely to occur to product ingredients such as flavorings and colors, O_2 affects the level of carbonation that may be achieved in the product.

9.2.3.1.1 Effect on Carbonation

Air is the usual source of O_2 in products and it contains principally 79% N_2 and 21% O_2, ignoring inert gases. Because of the differential solubility of O_2 and N_2, any air dissolved in the product will contain 35% O_2 and 65% N_2. In practice, the presence of dissolved O_2 will have the effect of giving a false reading of the CO_2 level, although that is usually within the level of tolerance of most analytical methods in use. More significantly, it is likely to give rise to the phenomenon of "fobbing," as nucleation sites are created as a direct consequence of air in the product. This causes the product to gush uncontrollably out of the container when pressure is released.

9.2.3.1.2 Effect on Components

9.2.3.1.2.1 Carbonated Water Unless the source of water used in bottling contains significant levels of organic material or some unusual inorganic compounds, the effect of dissolved O_2 on bottled and carbonated water is minimal. Some slight changes of perceived taste may be noted. If the water source is a normal drinking water supply, this is likely to be the situation, whereas some sources that are certified as "Natural Mineral Water" may contain specific components that are well in excess of normal drinking water levels and may cause greater changes of taste if significant levels of O_2 are present. In some noncarbonated bottled waters a green appearance may become apparent during storage as a result of high light levels and headspace O_2, creating the conditions for the development of algal spores. The presence of CO_2 will usually obviate this problem.

9.2.3.1.2.2 Carbonated Soft Drinks Some components of carbonated soft drinks are especially vulnerable to the effects of O_2. These include, in particular, some of the flavor components that are widely used in citrus-flavored products. Essential oils of orange, lemon, and, to a lesser extent, other citrus fruits are widely used as starting materials for related flavorings, and many of their components are very vulnerable to oxidation. The terpene hydrocarbon fractions are most vulnerable to oxidation, and, although in most citrus flavors for beverage use the majority of less-soluble terpenes have been removed, sufficient remains to initiate the process of auto-oxidation. Oxidized citrus oil components have undesirable flavors and can rapidly render the end product unacceptable. These oxidative effects are complex and related to the flavor degradation that also occurs in the presence of light (Ashurst and Hargitt, 2009).

Another very important effect of the presence of O_2 in products is the degradation of colorings that may occur. Colorings of both natural and artificial origin may be affected, and the usual result is a bleaching effect. Many manufacturers add ascorbic acid to act as an O_2 scavenger. Although the problem of oxidation is much more serious in noncarbonated soft drinks, every effort should still be made to minimize the presence of O_2 in carbonated beverages.

9.2.3.2 Light

The effects of light are very damaging to most beverages, with both color and flavors being affected. As far as possible, the exposure of product to all forms of light should be minimized, but any product exposed to direct sunlight is likely to be affected in a very short time. Canned beverages are at an obvious advantage, as no light exposure ever occurs once the product is in the container. For bottled beverages there are many courses of action that can be taken to minimize light exposure or at least to exclude the most damaging part of the light radiation. These include the use of colored bottles, covering as much of the bottle as possible with label(s), and the use of secondary packaging such as boxes or shrink-wrap (which can be colored).

9.2.3.2.1 Bottled Water

Depending on the water source and the level of dissolved constituents, the effect of light on bottled waters is usually minimal. However, increasingly most water sources incorporate very small traces of complex organic molecules, which themselves may not have any discernible effect on taste. The exposure of such molecules to light energy may cause a breakdown to much lower molecular weight moieties, which may have extremely low flavor thresholds and cause the generation of an off-taste in bottled water. As already indicated, the presence of light may also produce a green appearance as algal spores develop, although this effect is very unusual in carbonated water as the presence of O_2 is required.

On rare occasions, a secondary effect of light has been known to generate a flocculent mass in bottled water when CO_2 is present. This phenomenon has been found to occur when water from a surface reservoir is used and the reservoir has experienced significant algal growth. Complex polysaccharides can be released into the water, and traces may survive subsequent water treatment to appear in the bottled water. Addition of CO_2 alters the pH to a point where the polysaccharides then flocculate, producing the visual mass. This phenomenon may also occur in noncarbonated beverages with a low pH.

9.2.3.2.2 Carbonated Soft Drinks

9.2.3.2.2.1 Effect on Color Light, and particularly direct sunlight, usually causes rapid degradation of a range of components in soft drinks. For the most part manufacturers go to considerable lengths to minimize the exposure of their product to light, but there will come a point when the product needs to be shown to the consumer. Once the product is in the hands of the consumer, it will often be continually exposed to light.

The most obvious effect of exposing soft drinks to light is often the loss of color. This may be a complete bleaching of the color or simply a fading. If the color is faded, it may not be immediately apparent to consumers, unless they have an unfaded example of the product available for comparison.

Color loss is sometimes the effect of color alone, but it may also be aggravated by the presence of O_2 (see previous section). If the preservative sulfur dioxide (SO_2) is present in the product, it too may become involved in the complex chemistry of color fading, but SO_2 is not now in widespread use as a beverage preservative. Color loss can usually be minimized by the use of ascorbic acid, although this is normally present as an O_2 scavenger and the interaction between the presence of O_2, the presence of antioxidants, and color fading is complex and will vary from product to product depending on its pH, packaging, and other components. The most vulnerable colors are those in the yellow/orange/red range.

9.2.3.2.2.2 Effect on Flavor and Taste Although the effect of light on the color of a product is usually self-evident, the impact on flavor is unlikely to be obvious until the product is opened and consumed. The fact that a product has suffered color fade or loss may also indicate a flavor deterioration.

Flavorings added to soft drinks may be natural or artificial (i.e., synthetic), although the term "nature identical" is used to describe flavor components that are synthetic but otherwise identical to those occurring in nature. The term "artificial" is, within the flavor industry, reserved for those substances that are not found (thus far) in nature. The flavorings that are added to beverages normally consist of a complex mixture of aroma chemicals in a permitted food solvent such as ethanol. The stability of the flavor in the product is closely related to the origin of the aroma chemicals, as the greater the purity of individual aroma chemicals, the more stability they are likely to possess. Aroma chemicals that are chemically unsaturated are more vulnerable to breakdown than those that are fully saturated, but once the process of breakdown has started it may create a chain reaction as a result of highly reactive chemical species being released by the process and affecting other molecular stability.

Thus, where the flavor is derived from a natural source such as the essential oil of lemons, there is likely to be a much more complex spectrum of aroma chemicals present than would be there if the flavor is created from a small number of aroma chemicals of known purity. In practice, flavors of natural origin tend to show greater instability to light than their synthetic counterparts. Natural citrus flavors are particularly vulnerable to the effects of light (and O_2) because they contain many terpenic compounds that readily decompose.

9.2.3.2.3 Contaminants

One consideration in the selection of the type of packaging is the risk of contaminants entering the product either directly from the container (e.g., metal pickup from cans) or through product interaction with the container or its components. This aspect will be discussed in Section 9.3, where individual container types are reviewed.

9.2.3.3 Heat and Aging

The effects of heat and aging on beverages appear to be closely linked and largely occur irrespective of the type of package used. The effects on products are mainly associated with the development of characteristic "aged" flavors, but where a product has been exposed to significant heat, a noticeable "cooked" character will often be present. These flavors are thought to be products of Maillard reactions and are often more noticeable when carbohydrates and more complex components are used in the product formulation.

The flavor effects that occur as a result of heat and aging will be more noticeable on the shelf life of a product that is to be sold in tropical markets. In these cases, the type of processing employed should be selected with more consideration for the overall amount of heat exposure that a product receives than would be the case in more temperate markets.

Typically, most carbonated beverage products will have a shelf life of at least 6 months in a temperate market, but in tropical markets it is likely to be 3 or 4 months.

9.2.3.3.1 Processing

Depending on the type of product and its ingredients, pasteurization may be necessary to ensure microbiological stability (see Section 9.2.4.). The choice of using either flash or in-pack pasteurization or hot-filling will be based on the ingredients used, the type of packaging selected, and the available plant. A factor that should be taken into consideration is the thermal effect of the different processes on the product's shelf life and aging characteristics. For example, and leaving aside other considerations, although the pasteurizing effect of 90°C for 2 or 3 sec may be of a similar nature to an exposure at 70°C for 20 or more minutes, the longer time period will have more effect on reducing overall shelf life because of greater exposure of the product to heat.

If in-pack pasteurization or hot-filling is employed, great care should be taken to ensure that products are adequately cooled. Most modern plant is likely to have an integral cooling section that will reduce the temperature to below 25°C, at which temperature there is little risk of significant heat damage. Products that are not adequately cooled before being stacked may cool sufficiently on the outside of the stack or pallet but may retain heat in the middle for sufficient time to develop a noticeable cooked flavor and darkened color. Product affected in this way is said to have suffered "stack burn."

9.2.4 Microbiological Deterioration

9.2.4.1 Beverages

All beverages are at risk of suffering deterioration as a result of microbial action, and this presents probably the highest threat to product stability. As almost all beverages are acidic in nature, the principal risk is of microbial spoilage rather than contamination by a pathogenic organism. The spoilage organisms of particular concern are yeasts and molds, although some bacteria such as *Lactobacillus* and *Alicyclobacillus* species can cause significant problems, albeit mainly in noncarbonated products. The main effects of microbial contamination are likely to be the development of off-flavors and the change of physical appearance. For organisms such as some types of resistant yeasts, there is the possibility of CO_2 production to a level that, added to the introduced carbonation, will cause bottle bursting and the significant possibility of damage and injury.

The initial processing of a product to reduce or eliminate any contamination is vital, but once the product is within its container, the packaging plays the essential role of both retaining CO_2 and protecting the contents from subsequent contamination. In carbonated drinks the presence of CO_2 substantially reduces the risk of spoilage for the following reasons:

- Carbon dioxide is a metabolite of many types of yeast, and its presence at pressure will, in many instances, reduce the activity of these organisms to a point where they cease to grow.
- The presence of CO_2 produces a blanket of inert gas in the headspace of the product in which there is little or no O_2 present. This blanket both minimizes the risk of infection and suppresses the growth of organisms requiring O_2 (e.g., molds).

- Carbonated water is likely to contain less O_2, which again will minimize the growth of any organism that requires its presence.
- Carbon dioxide produces a weak acid when dissolved in water, and its presence will reduce the pH of the product. This lowering of pH will increase the effectiveness of any chemical preservatives that have been added to the product.

Carbonated beverages can be manufactured in one of three ways:

1. The ingredients of the end product (except carbonated water) are made up as a syrup that is typically five or six times concentrated. This syrup will often be flash-pasteurized and then mixed in a proportioning system with the required amount of water, which has been carbonated in a separate operation. The carbonated end product is then filled into the required containers. This approach is known as the *premix method.*
2. The concentrated syrup is dosed into each container on the filler and the container topped up with carbonated water. After leaving the filler, containers will be mechanically inverted to ensure adequate mixing of syrup and carbonated water. This is the *postmix method.*
3. A less frequently used method is to make up the product at drinking strength, inject CO_2 in a suitable system, and then fill the containers.

Each of these methods carries a different level of risk of microbial contamination, and the choice of method to be used will normally be made on the basis of the sensitivity of the product to microbial contamination, the processing required, the available plant, and the type of container into which the product is to be packed.

The highest-risk product—for example, a carbonated fruit juice free from any chemical preservatives—would require full in-pack pasteurization and would thus have to be made in a plant where a tunnel pasteurizer was available. This, in turn, would demand a container that when heated to the pasteurizing temperature (typically 70°C) would not deform as a result of the increased internal pressure. The choice is normally limited to a glass bottle or metal can.

Slightly lower-risk products, such as carbonated drinks containing a lower proportion of fruit juice as well as chemical preservatives, will normally be successfully packaged in any appropriate container by either pre- or postmixing, provided that the syrup used is subjected to flash pasteurization.

The lowest-risk carbonated drinks are those made with flavors that do not contain fruit juice. Such products can be successfully made without any use of pasteurization, and there is, in many countries, a long history of small bottling operations with minimal plant hygiene producing stable carbonated drinks that have rarely if ever suffered from microbial contamination. Chemical preservatives would always be employed in such products.

9.2.4.2 Bottled Water

Although bottled waters are not generally at risk from the effects of the yeasts and moulds that affect other beverages, they do carry a significant risk of microbial contamination. Because the pH of bottled water can vary widely, depending on its composition and level of carbonation, it carries, in general, the risk of contamination by pathogens as well as by nonpathogenic organisms from environmental sources. The level of risk will depend on the initial level of contamination within a water source as well as that from subsequent bottling operations.

9.2.4.2.1 Natural Mineral Water

In many countries and in those countries within the European Union (EU) in particular, the term "Natural Mineral Water" is a description reserved in statute. It carries a requirement that the water source must meet certain stringent criteria but must, in particular, be free from pathogenic organisms such as total coliforms, fecal coliforms, fecal streptococci, and staphylococcal organisms. It is also a requirement that the water source is free from parasites such as *Giardia* and *Cryptosporidium*. The specific requirements for Natural Mineral Water in the EU are defined by Council Directive 80/777/EC.

In other countries the definition of Natural Mineral Water may have different requirements in terms of water quality. However, there is a Codex Alimentarius definition for Natural Mineral Water that is based on 80/777/EC, so, for countries that are Codex members, there is a requirement for designated Natural Mineral Water to be free from pathogenic bacteria and parasites and thus safe to drink. There are stringent testing requirements over a considerable period of time that must be fulfilled before water can be designated a Natural Mineral Water.

From this preamble, it will be immediately apparent that the standards of plant hygiene and operation in a factory packing Natural Mineral Water must be of the highest order, as no further water treatment (with some limited exceptions) is permitted before the product is packaged. As with carbonated beverages, the presence of CO_2 will create an effective barrier to the subsequent introduction and growth of pathogens (and indeed many other microorganisms), but this should not be a substitute for good manufacturing practice.

Assuming that Natural Mineral Water is free from pathogens, the presence of other microorganisms that arise from the environment from which the water is taken is accepted as a normal part of its composition. These environmental organisms are those characterized by growth in media at 20–22°C, although many bacterial species occurring naturally in groundwater will grow over a much wider temperature range. The presence of such flora has raised a wide range of questions and debates about effects on health, primarily because the flora is not particularly well characterized. However, as no disinfection that modifies or eliminates the biological constituents of water is permitted, the bacteria must be regarded as natural components in exactly the same way as their chemical counterparts.

The chemical composition of a groundwater plays a major role in determining the range and numbers of organisms, as most microbial processes are consistent with oxidation–reduction reactions that can be viewed as the microbial food chain. Probably the most limiting substrate in groundwater systems is organic carbon. This fraction can arise from polyphenolic substances that, in turn, derive from humic materials that reach the groundwater through hydrological recharge and groundwater flow.

In noncarbonated water the numbers of bacteria are known to increase rapidly after bottling. A typical example will show the level of colony counts of water emerging from the spring as about 1–4 colony-forming units (cfu) mL^{-1}. Immediately after, bottling numbers are only slightly higher. During storage at 20°C, bacterial populations increase in numbers to reach a peak of more than 10^5 cfu mL^{-1} by the end of 1 week. During the next 4 weeks bacterial populations slowly decline or remain fairly constant. At the end of a 2-year storage period, colony counts are still likely to be in the order of 10^3 cfu mL^{-1}. This level of bacteria can produce changes to the flavor of bottled water, although in many instances this is not regarded as a defect but rather as a characteristic of the particular water.

In carbonated Natural Mineral Water the levels of bacteria are much lower, and this is probably due to the reduction in pH. As a consequence, the risk of flavor defects arising is much lower in carbonated than in uncarbonated Natural Mineral Water.

9.2.4.2.2 Other Bottled Waters

In countries around the world there are a number of alternative definitions for various water types, including now various "flavored waters." Bottled water is defined by the US FDA as "water that is intended for human consumption and that is sealed in bottles or other containers with no added ingredients except that it may contain safe and suitable antimicrobial agents" (Warburton and Austin, 2000). The discussion below is devoted to bottled drinking water that meets the European Drinking Water Directive (80/778/EC) and similar regulations in other countries. The EU Drinking Water Directive has much stricter overall limits on bacterial counts, as shown in Table 9.2.

There is an absolute requirement that water for human consumption should not contain pathogenic organisms, and additional testing should include examination for salmonella, pathogenic

TABLE 9.2
EU Specifications for Microbial Contamination of Drinking Water

Parameter	Sample Volume (mL)	Guide Level	Maximum Admissible Concentration by Membrane Filtration	Maximum Admissible Concentration by Multiple Tube Method
Total coliforms	100	—	0	MPN < 1
Fecal coliforms	100	—	0	MPN < 1
Fecal streptococci	100	—	0	MPN < 1
Sulfite-reducing *Clostridia*	20	—	—	MPN ≤ 1
Total bacterial count for water for human consumption (37°C)	1	10	—	
Total bacterial count for water supplied for human consumption (27°C)	1	100	—	
Total bacterial count for water in closed containers (37°C)	1	5	20	
Total bacterial count for water in closed containers (27°C)	1	20	100	

staphylococci, fecal bacteriophages, and enteroviruses. Nor should the water contain parasites, algae, or other organisms such as animacules.

The attainment of these low levels of microbial components will normally be achieved by one or more disinfection processes, including treatment with chlorine and exposure to UV light irradiation. As a consequence of such treatments, there is likely to be a very low level of microorganisms present in carbonated water that has originated from public water supply sources, and the risk of any microbial deterioration of bottled water of this type is extremely small.

The most obvious manifestation of spoilage in bottled water is the appearance of floating pieces of mold mycelium, with *Penicillium* species, along with *Cladosporium* and *Phaeoramularia*, which are the most commonly isolated fungi (Hocking and Jensen, 2001). *Pseudomonas aeruginosa*, an organism associated with soil contaminated with human and animal feces, is not generally associated with spoilage of bottled water, but its presence can affect water color, clarity, and taste. It is capable of growth to high numbers in minimal-nutrient environments such as deionized and demineralized water and has been implicated in food and waterborne disease. Most *Ps. aeruginosa* strains are resistant to commonly used antibiotics and sanitizers or disinfectants but should be inactivated by pasteurization. The organism has been isolated from bottled waters from Brazil, Canada, France, Germany, Indonesia, Spain, and the United States (Hocking et al., 2001).

9.3 IMPACT OF PACKAGING

From the previous sections of this chapter it will be apparent that packaging plays a critical role in protecting carbonated beverages and water from most aspects of deterioration. As indicated at the outset, carbonated beverages cannot exist as a product unless they are packaged in a container that retains CO_2, so gas retention is an essential protective function. No container yet developed can, however, protect the product from the effects of heat exposure and aging, although the complex nature of the aging process of beverages may be accelerated by O_2 ingress and the effects of light.

9.3.1 Metal Containers

Although the market for carbonated beverages in metal cans has declined to some extent in recent years, they remain popular containers for single-serve beverages mainly because of their convenience, robustness, and marketing image. Most beverage cans are now manufactured from aluminum as two-piece cans (can body and one end) using the "drawn and wall ironed" (DWI) process. In this process, a coil of aluminum is initially formed into shallow cups, which are placed into a punch in which a single stroke carries the cup through a redraw ring followed by four ironing rings.

The final wall thickness of 0.120 mm is achieved at this stage. The bottom domed profile is formed at the completion of the stroke. The newly formed can will have a ragged edge, which is then trimmed to present a perfect edge and height for subsequent processes. Following formation, cans must be cleaned, washed, and dried to remove all traces of lubricants used in previous processes and to prepare the inner surface for coating (if required) and the outer surface for decoration.

Earlier types were produced as three-piece cans. These consist of a cylindrical body and two ends. In this process, blank rectangular can bodies were produced and formed into cylinders that could be welded or soldered, with the fixed end attached before filling. Can blanks were usually decorated before formation into bodies. The equipment for producing three-piece cans is less capital intensive and more suitable for short production runs than that for two-piece cans.

Because two-piece cans lack the side seam that is present in three-piece containers, they are inherently less prone to leakage of liquid and CO_2. Provided that the single end is applied and sealed correctly, the risk of leakage is very small, and cans arguably provide the best retention of CO_2 of all container types. The possible exception is the risk of leakage as a result of can corrosion and eventual pinholing. Corrosion may occur either from within or without, and the selection of a can enamel or lacquer that is compatible with the beverage is critical. The most likely source of corrosion from the exterior may occur if the can is subjected to a tunnel pasteurization process and is not completely dried before being packed in secondary packaging. There is a similar vulnerability if cans are inappropriately stored. Can closure is also critical, and the application of the end in such a way as to ensure no leakage of gas or liquid requires seaming equipment adjusted to very fine tolerances, with regular quality checks to provide the necessary assurance.

For beverage cans made from tinplate, control of migration of iron into the beverage is critical, because extremely small levels of iron (0.5 ppm) can compromise the flavor of the beverage. One study of the influence of two different compound systems applied during the end seaming operation (can closing) on the iron pickup from tinplate cans after storage with a cola soft drink for 180 days at 37°C found levels between 0.15 and 1.0 ppm depending on the compound/lacquer combination (Bernardo et al., 2005).

Metal cans for beverages have an easy-open end consisting of a scored portion in the end panel and levering tab (formed separately) that is riveted into a bubble-like structure fabricated during pressing. The aluminum alloy used to manufacture easy-open ends for beverage cans is specially developed to give the required mechanical properties but is subject to environmental stress cracking (ESC) corrosion due to reaction with moisture. The score area is particularly susceptible because of the tensile stress to which this part of the end is subjected (Page, 2006).

Cans have some disadvantages. Cans for carbonated beverages do not normally exceed 500 mL, and for soft drinks they mostly do not exceed 330 mL, so for any market requiring larger unit volumes they are impractical. Cans are usually decorated and printed at the time of manufacture, and, because modern filling plants often run at speeds in excess of 1000 cans per minute, which yields around 300 million units per annum, the level of business needed to justify the investment in stock and the logistical consequences are substantial. Provided used cans are collected, they have excellent value and potential for recycling.

Close attention must be paid to the details of product formulations and a good liaison maintained between the product formulator and can maker. The details of any internal compound or enamel (lacquer) to be applied to the internal surface of the can must be agreed, and the formulation must be compatible with the can. For example, SO_2 must be rigorously excluded from the formulation because, if present, it may be chemically reduced at the interface with the metal of the can to generate hydrogen sulfide, which will likely make the product unacceptable.

Metal uptake may also become a problem and can cause metallic off-flavors as well as put unacceptable levels of metal into the product. When cans were mostly produced in three pieces, solder was often used for the side seam, and there have in the past been examples of significant levels of heavy metals such as lead entering the product from the solder. This is no longer a potential problem in two-piece cans as no solder is used.

Adequate shelf life testing is an essential element in determining the suitability of formulation and can specification.

9.3.2 Glass Containers

9.3.2.1 Container

Glass was for many years the only practical means of packing carbonated beverages. The first known glass bottles for carbonated products were made and patented by Hiram Codd in 1870. The bottle had a glass ring inside the neck and an internal glass ball that, when the bottle was filled with carbonated product, was held against the neck ring by the internal pressure of the CO_2. By today's standards, the retention of both liquid and gas was probably very limited, but when first introduced, the bottle made packaging of carbonated beverages a viable operation.

Glass containers became (and still are) the benchmark standard by which all other containers for carbonated beverages are evaluated. Despite its many qualities, glass has significant disadvantages that have provided one of the driving forces for the development of other packaging materials for beverages. Glass is heavy and liable to break in a way that can potentially cause damage and injury. These factors effectively limit the maximum size of container when glass is used to contain carbonated beverages to about 1.0 to 1.5 L.

For carbonated products that require in-pack pasteurization, glass is frequently still the preferred container as it can be designed to withstand the excess pressures that develop without deformation or compromising its strength or clarity. However, for such products, there is usually a practical volumetric limit of about 0.3 to 0.5 L.

Glass containers are tested during manufacture, and any defective glass that is missed by such testing is likely to fail during the process of filling carbonated beverages. Thus, the most likely cause of failure of glass containers is leakage at the interface between cap and bottle body. The rigid nature of glass means that any molding defect of the neck rim that presents an uneven surface to the closure becomes a potential source of leakage.

Glass is attractive as it allows the consumer to see the product but offers little protection against the adverse effects of visible light on the product. Some protection of the product can be achieved by using colored glass or wrap-round labels or by the application of a film to all or part of the outside of the bottle.

The principal advantages of glass include its quality image; low-cost production tooling; brand differentiation through shape, design, and texture; product compatibility; impermeability; odor resistance; good transparency; tamper resistance; resaleability; recyclability; reuse opportunity; sleeving and decorative opportunities; protection against UV light; suitability for in-pack pasteurization; and good top-load strength and rigidity.

The principal disadvantages to the use of glass are its weight and fragility.

9.3.2.2 Selection of Closure

Most glass bottles containing carbonated beverages are likely to have one of four closure types (Theobald, 2006):

- Crown cork
- Aluminum roll-on pilfer-proof (ROPP) cap
- Molded plastic cap
- Ring pull

Crown cork closures originally had cork liners, but these have been largely replaced with typically board liners coated with various surface coatings or laminates. The board used is capable of deformation under compression and will thus form a good seal over an irregular or uneven glass neck finish. ROPP caps have either a flowed-in liner or a wad; the latter is typically made from expanded polyethylene or expanded polystyrene coated or laminated with a suitable material to give the desired moisture and gas barrier properties (Theobald, 2006).

The selection of an appropriate closure is essential to the correct performance of a glass container and the retention of its contents. Too large a closure can create leakage due to the force generated by the internal pressure of CO_2 and particularly the conditions that may arise if the product is to be subjected to in-pack pasteurization. Too small a closure can give rise to problems at the time of filling.

The closure must therefore be capable of pressure retention, and the alternatives are metal or plastic with a composite liner to make the seal between closure and bottle body. Metal closures are either preformed (crown or twist-off crown) or rolled on to the thread of the glass to create the seal, tamper evidence, and pilfer proofing. These latter closures are known as ROPP caps. Plastic closures are preformed and screwed into position with or without a tamper-evident ring (Larbey, 2006).

For glass (or plastic) bottles tail-end blow-off (TEBO) has become an important issue. This is also known as "missiling" of caps and can occur when a consumer is removing the cap from a carbonated beverage product. The release of pressure can cause the cap to suddenly part company with the bottle and become a missile that can cause damage or injury. The problem has been solved by a redesign of the thread of the bottle body where two or three vertical slots are introduced. This allows excess pressure to vent to atmosphere at the instant the seal between bottle and cap is broken but before the cap loses contact with the thread.

9.3.3 Plastic Bottles

Over many years, plastics of various kinds have been assessed for suitability as containers for beverages. When, early in the 1960s, the use of plastic bottles for soft drinks was first considered, it soon became apparent that only the polyester and nitrile families of plastics had the necessary physical and chemical characteristics required. Poly(ethylene terephthalate) (PET) was the preferred polyester, but acrylonitrile–methylmethacrylate copolymer, methacrylonitrile–styrene copolymer, and rubber-modified acrylonitrile–styrene copolymer were also suitable (Turtle, 1984). Because nitrile plastics could be made into bottles using existing blow-molding equipment and PET could not be, due to its inclination to crystallize and go hazy at higher temperatures, early market development work in the 1970s was carried out with nitrile bottles. Coca-Cola successfully launched a 950-mL nitrile bottle in 1975, but the release in 1977 of toxicological data showing that acrylonitrile (AN) monomer could be carcinogenic at high dosage led to the removal of the nitrile bottle from the market (Turtle, 1984).

Meanwhile, attempts to successfully manufacture PET bottles using a stretch–blow molding process were continuing. In the spring of 1977, the plastic PET bottle for soft drinks was launched by Pepsi-Cola, followed soon after by Coca-Cola and other beverage producers. It has been described as probably the biggest single development in the soft drinks industry since the introduction of the ring-pull can a decade earlier (Turtle, 1984). Early PET containers were made with hemispherical bases that required the addition of a base cup typically made from high density polyethylene (HDPE) to allow the container to be positioned vertically during filling and storage. Later developments produced the "petaloid" base that enabled the base cup to be dispensed with (Giles, 2005).

Today PET bottles are made by a two-step procedure: an amorphous preform is made first by injection molding and is subsequently stretch–blow-molded to form a biaxially oriented, semicrystalline but transparent bottle. Modern PET bottle resins are usually copolyesters, with the most common comonomer being isophthalic acid (Bashir et al., 2002).

All plastics are permeable to gases, and most manufacturers of carbonated soft drinks require a specified maximum loss of CO_2 of 15% over 26 weeks for a 1.5 or 2.0 L bottle. For smaller bottles, which have a less favorable surface area:volume ratio, the loss of 15% of CO_2 is likely to occur over a period of about 10–12 weeks. Attempts to improve the barrier properties of PET bottles have focused on four areas:

1. Make a plastic article and coat it.
2. Blend high-barrier plastics into medium-barrier ones.
3. Use multilayer structures containing barrier layers.
4. Upgrade plastics with barrier-enhancing additives.

Thin, glass-like coatings are produced by physical vapor deposition (PVD) or plasma-enhanced chemical vapor deposition (PECVD) using oxides of silicon (SiO_x) with a typical thickness of 150–300 nm. However, the intrinsic brittleness of SiO_x films requires rather complex conversion processes in order to withstand mechanical stress and stretching on a PET bottle surface. Recently, high-gas-barrier PET bottles coated with a diamond-like coating (DLC), more precisely defined as hydrogenated amorphous carbon films, have become available (Shirakura et al., 2005). For example, amorphous carbon treatment on internal surface (ACTIS) from Sidel (www.sidel.com) with a thickness of 100 nm gave a 20-fold reduction in the permeation rate of O_2. In a comparative test for a 17.5% CO_2 loss, a coated PET container and a noncoated PET container were stored at 21°C. The uncoated container had 10 weeks' shelf life, and the coated container, 44–45 weeks' shelf life; this performance could be improved by increasing the thickness of the coating (Boutroy et al., 2006). Although SiO_x coatings have the advantage of transparency, the process window for PECVD is so small that it makes it impossible to guarantee a reliable process; in contrast, the DLC coatings have a very large process window.

Two commercial approaches using barrier-enhancing additives are O_2 scavengers and nanoclays. In the case of O_2 scavengers, the upgraded plastic may be the only component. In the case of nanoclays, it is usually, but not always, used as a barrier layer of a multilayer structure. Commercial O_2 scavengers are incorporated into PET or polyamides (PAs), and commercial nanoclays are incorporated primarily into PAs. For example, ValOR ActivBloc100 (www.valspar.com) combines a high-performance carbonation barrier with an O_2 scavenger and can be used in any PET multilayer bottle application requiring high carbonation retention and low O_2 content. A similar approach is offered by Oxbar (www.constar.net).

Since the arrival of the one-piece bottle, and continuing today, there has been the problem of ESC in PET beverage bottles. ESC is a problem that afflicts carbonated soft drinks packaged in PET bottles. It does not happen to a substantial extent for still beverages such as juices or bottled water. There have, typically, been two approaches to minimizing the risk and incidence of ESC in beverage bottles: first, to make bottles that are less susceptible to failure and, second, to minimize contact of the filled bottles with chemical environments such as cleaners, sanitizers, spilled beverages, and lubricants that promote deterioration.

Stress cracking of PET bottles filled with carbonated soft drinks is a complex process that is influenced by a large number of variables, including climate, chemistry, polymer quality, and bottle engineering design. Because of the multiple dependencies, the mechanics of stress cracking are still not well understood. There is a growing awareness that "alkalinity," including naturally occurring water alkalinity, is connected to stress crack failure of PET bottles. Morrison et al. (2008) observed that bottle failure has a strong positive correlation to both water alkalinity and relative humidity, indicating that hydrolysis of ester linkages is a predominant mechanism in the failure of PET bottles.

Overall, PET provides a good combination of physical strength, particularly when containers are full of carbonated liquid, and the retention of both liquid and CO_2 gas contents. If large PET containers are used in tropical climates for highly carbonated beverages, serious package deformation can occur. As with any container, the interface between the closure and body of the container is critical for ensuring retention of contents. The factors influencing taste and odor of carbonated beverages packaged in plastic containers are shown in Figure 9.1.

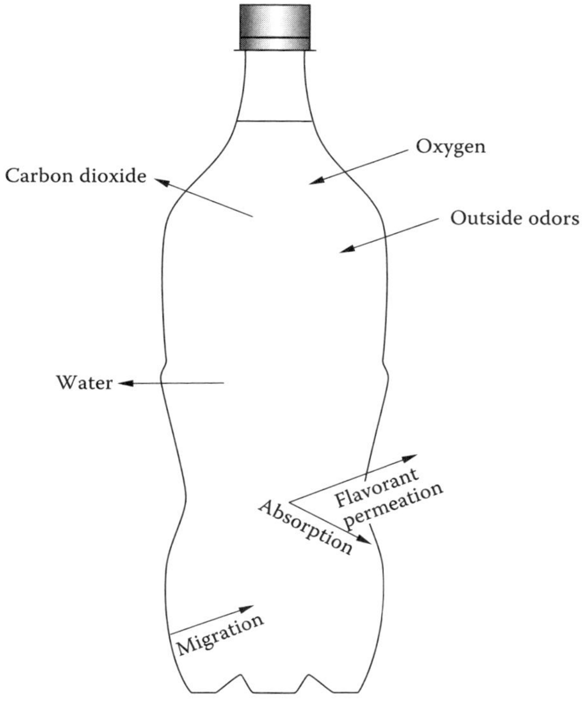

FIGURE 9.1 Factors influencing taste and odor of carbonated beverages packaged in plastic containers. (From Robertson G.L. 2006. *Food Packaging Principles and Practice*, 2nd edn. Boca Raton, Florida: CRC Press, with permission.)

9.3.3.1 Poly(ethylene Terephthalate) Performance

All plastics are permeable to some extent to gases, and most bottlers will require a performance that is tolerable with respect to the required shelf life. For carbonated beverages the specified loss rate is typically 15% of the CO_2 over a 26-week shelf life for a 1.5- or 2.0-L container. Smaller containers have a less favorable surface area:volume ratio and thus show a similar loss in about 10 weeks (see Figure 9.2). PET shows one of the highest CO_2 gas barriers for all plastics used for packaging and is an order of magnitude better than polyolefins or polycarbonates. Acrylates, naphthalates, and amide-based plastics generally show better barrier performance than PET. The barrier develops during the stretch–blow orientation of the material. PET shows less favorable retention of moisture than polyolefins and poorer resistance to heat than polycarbonates but overall has the most favorable balance of performance for carbonated beverages (Gunning, 1999).

Problems of bottle deformation may occur when PET containers of carbonated beverages are stored at higher temperatures. This effect is not usually observed until temperatures consistently over 35°C are encountered. Such effects can be minimized by increasing the weight of plastic in the bottle.

The O_2 barrier performance of PET is low, but, with high levels of carbonation and the shelf life required for most carbonated beverages, it is regarded as acceptable. For products that are particularly sensitive to the effects of O_2, such as beer, isotonic sports drinks, and fruit juices, poly(ethylene naphthalate) (PEN) provides a much more effective O_2 barrier but at a significantly increased cost, which may be as much as four or five times that of PET.

Product packed in PET is normally regarded as being free from taints. The major significant volatile compound in PET is acetaldehyde (AA), and this was of concern in odor quality; AA is also the major cause of the color change in PET during aging (Kim-Kang, 1990). AA is present in PET

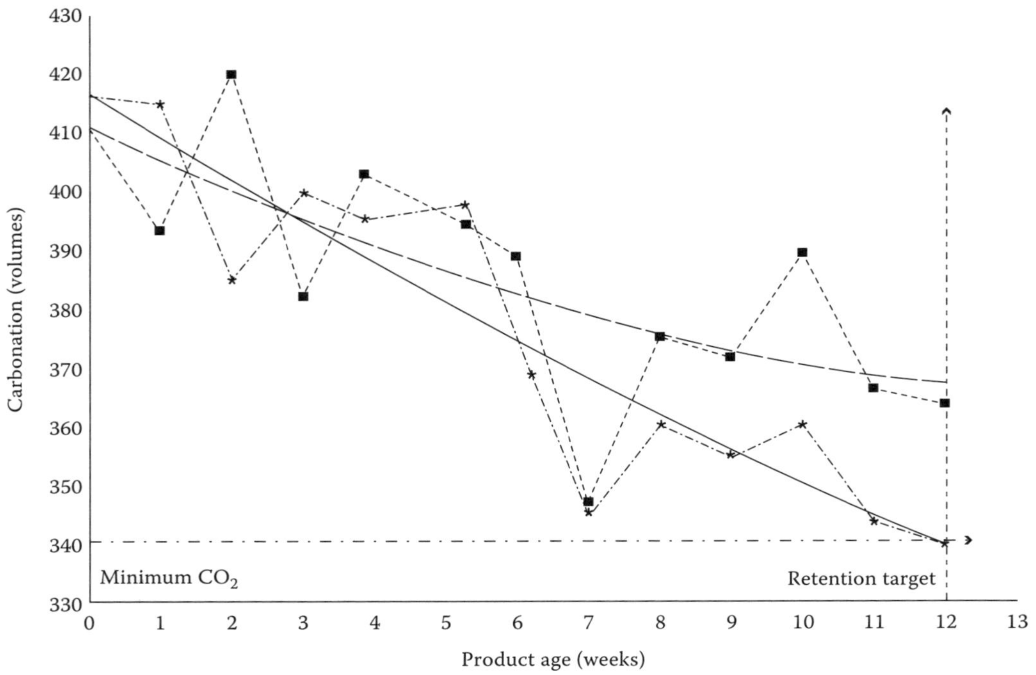

FIGURE 9.2 Carbonation loss from PET bottles. Bottles held under ambient storage conditions typical for soft drinks (----▪----). Bottles held at 20°C then chilled to 5°C for 24 hr prior to carbonation measurement (-·-·-*-·-·-). Trend line for bottles held under ambient storage conditions (— — —). Trend line for bottles held at 20°C then chilled to 5°C for 24 hr prior to carbonation measurement (———). (Redrawn from Syrett D. 2006. Bottle design and manufacture and related packaging. In: *Carbonated Soft Drinks Formulation and Manufacture*. Steen D.P., Ashurst P.R. (Eds). Blackwell Publishing, Oxford, pp. 181–217, with permission.)

as a thermal degradation product formed during the melt condensation reaction and melt processing of PET. AA itself is also unstable. It readily oxidizes and polymerizes when exposed to air. AA possesses a distinct odor and taste, generally described as sweet, plastic-like, and fruity, with a low sensory detection threshold. Using water as a food simulant, Ashby (1988) reported peak values for AA of <50 ppb after 8 days at 55°C and <15 ppb after 10 days at 40°C.

In the mineral water industry it is generally assumed that the odor threshold of AA in water ranges from 20 to 40 pg L^{-1}. The odor detection threshold of AA in soft drinks will be much higher due to the masking effect of other volatiles (if AA is not already present), and therefore an off-flavor due to the release of AA from a PET soft drink bottle is not to be expected (Nijssen et al., 1996). If mineral water is stored in PET bottles, migration of AA into the water will be detected only in carbonated mineral water. Stability experiments on AA in still water showed that within 8 days the AA concentration decreased from 100 to below 1 pg L^{-1}. In boiled (O_2-free) water and water at pH 3.7, a steady-state situation is followed by a rapid decrease in concentration (Nijssen et al., 1996).

Recently, Matsuga et al. (2006) analyzed commercial samples of water bottled in PET from Japan, Europe, and North America and showed that formaldehyde (FA) and AA migrated into the water from the PET bottle. In commercial water without bacteria, the levels of migrated FA and AA remained unchanged, whereas in natural mineral water containing heterotrophic bacteria, the migrated FA and AA was decomposed. Of the carbonated water samples, one sample contained bacteria and showed a reduction in FA and AA; the others had no bacteria and showed no decomposition activity. The authors speculated that the existence of bacteria influenced the concentration of carbonate gas.

Present manufacturing techniques have dramatically reduced residual AA levels in PET packaging. The first generation of resins typically had residual AA levels of 2–3 ppm, but today these are <1 ppm. However, the AA increases when the pellets are remelted and converted to preforms. According to Bashir et al. (2002), today's PET resins can easily meet the requirement for packaging strongly flavored drinks such as colas (<8 ppm of AA in the preforms). However, the most stringent application is the packaging of potable water, for which preforms with less than 3 ppm of AA are required. Bashir et al. observed major differences in the performance of six commercial water-grade PET resins. The intrinsic AA generation potential of the resin controls the AA in the performs, and this is controlled by polymerization conditions. TripleA (www.colormatrix.com) is a scavenger technology that is claimed to reduce AA formation in PET packaging by up to 80%.

Recently Özlem (2008) obtained carbonated beverage samples from four different manufacturers following filling into PET bottles and stored them for 6 months at 5°C, 20°C, and 40°C, representing, respectively, refrigerator temperature, ambient temperature, and high temperature. At zero time the AA content varied from 18.5 to 358.5 ppb. After 6 months, samples stored at 5°C had AA levels from 10 to 10,530 ppb; those stored at 20°C levels varied from 28 to 45,195, and those stored at 40°C varied from 1630 to 130,000 ppb. The EU specific migration limit for AA is 6000 ppb.

PET bottles may contain low levels of residual monomer and low molecular weight oligomers that are formed during the resin polymerization and melting process, as well as additives, reaction by-products, and polymer degradation products. All of these have the potential to migrate into foods. Migration of PET oligomers (consisting mainly of cyclic compounds ranging from dimer to pentamer) has been reported at very low levels from PET bottles into alcoholic and carbonated beverages. Nasser et al. (2005) identified cyclic oligomers in PET bottles used for mineral water and fruit juice in Brazil.

In recent years there has been consumer concern about the likelihood of plasticizers migrating into water when PET bottles are reused in the home. Schmid et al. (2009) reported maximum concentrations of the plasticizers di(2-ethylhexyl)adipate (DEHA) and di(2-ethylhexyl)phthalate (DEHP) of 0.046 and 0.71 mg L^{-1}, respectively, for water in PET bottles exposed to 17 hr of sunlight. These results are in the same range as levels of these plasticizers reported in studies on commercial bottled water.

Antimony trioxide is used as an additive and initiator in the manufacture of 90% of the PET manufactured worldwide, at a maximum level of 0.035% as Sb (antimony). Sb is a potentially toxic trace element with no known physiological function. Shotyk et al. (2006) reported the leaching of Sb from PET bottles into water. For example, 12 brands of bottled natural waters from Canada contained 156 ± 86 ppt, and 3 brands of deionized water contained 162 ± 30 ppt. Comparison of three German brands of water available in both glass bottles and PET containers showed that waters bottled in PET contained up to 30 times more Sb, with a range of 253–546 ppt Sb. One German brand of water in PET bottles yielded 626 ± 15 ppt Sb 6 months after bottling. The median concentration of Sb in 35 brands of water bottled in PET from 11 other European countries was 343 ppt. All of the waters found to contain Sb were at concentrations well below the guidelines commonly recommended for drinking water, which are as follows: World Health Organization, 20 ppb; US Environmental Protection Agency and Health Canada, 6 ppb; German Federal Ministry of Environment, 5 ppb; and Japan, 2 ppb.

9.3.3.2 Polylactic Acid

Polylactic acid (PLA) is a comparatively new biodegradable polyester made from a renewable resource (see Chapter 19). It degrades on contact with moisture due to hydrolytic cleavage of the ester bonds in the PLA polymer, which impairs polymer stability and subsequently barrier and mechanical properties of the package (Holm et al., 2006). Despite the well-known hydrolytic effect of water on PLA (Cairncross et al., 2005), PLA is used commercially on a limited scale for packaging water, the end of shelf life being when the amount of water remaining in the bottle is less than the quantity declared on the label.

9.4 SHELF LIFE OF WATER AND CARBONATED BEVERAGES

For any food or beverage product, the two overriding considerations that determine shelf life are product safety and product quality. The safety of a product usually relates to potential microbial contamination and will normally be determined by either the processing employed or, particularly in the case of Natural Mineral Water, by ensuring the bulk product is free from undesirable contaminants. Product quality is determined by the performance of the product in a particular packaging format over a period of time.

9.4.1 Determination of Shelf Life

The principal reasons for determining the shelf life of a product are driven by both technical and commercial/legal considerations as follows:

Technical:

- To establish the period of time during which a specific combination of product and packaging retains its desired quality and taste
- To ensure that the selected processing technique is adequate
- To check that a substitute ingredient does not impair the desired quality
- To evaluate any changes in packaging components or production facilities
- To facilitate understanding of the changes that occur within a product over time in different storage situations (e.g., tropical conditions)
- To ensure consistency of production and to confirm that quality systems are adequate

Commercial/legal:

- To comply with any legislation requiring an indication of product durability
- To ensure that any nutritional claims are fully met
- To meet customer demands and reduce the risk of product failure and write-off
- To inform the distribution and marketing process
- To facilitate the impact of any product/packaging changes (e.g., for cost reductions)

Some of the more important criteria that affect shelf life are as follows:

- Raw materials
- Product formulation
- Processing
- Hygiene
- Packaging
- Storage and distribution conditions
- Consumer handling
- Display factors

For any product it is, then, necessary to establish the parameters that will enable the manufacturer to measure and record the changes in product quality and to determine the point at which the product is considered to be unacceptable. In the United Kingdom, product shelf life is notified to the consumer either as a "use-by" date or as a "best before" indicator. Use-by dates are employed for products where product safety may be compromised if the product is consumed later than the stated date. They will apply mostly to products where there is a significant microbiological risk factor, and it would be unusual to state a "use-by" date on either bottled waters or carbonated beverages. These products will normally have a "best before" indicator, which does not imply any safety risk but the likelihood that the product will not taste as required or demonstrate its indicated qualities.

9.4.2 Shelf Life Measures

9.4.2.1 Water

As already indicated, water that meets the criteria for the "Natural Mineral Water" standard is deemed to be free from contaminants within the quality factors set, and the only significant risk is any possible contamination that may occur during the packaging operation. It would thus be expected that a water bottler would take samples from each batch of production (however that is determined) and submit them for microbial examination. Samples would also be retained for the duration of the stated shelf life to observe any visible changes and to provide a reference sample in case of complaint. Assuming the taste profile of the water has been established by earlier storage testing, the water would not normally be subjected to regular taste testing unless a problem was suspected.

If the water is carbonated, a testing regime for the level of retained carbonation over the required shelf life would be expected.

For water that does not meet the standard of "Natural Mineral Water," tests would usually be made as required on the incoming bulk water, and it would usually be subjected to disinfection before bottling, after which a similar regime of testing to that indicated earlier for Natural Mineral Water would be adopted.

9.4.2.2 Carbonated Beverages

Microbial quality is, where necessary, checked either by plating tests or, increasingly, by rapid microbiological tests that can detect the presence of very low levels of organisms in a matter of a few hours. Some manufacturers will hold very sensitive products such as lightly carbonated apple juice in a quarantine store until microbiology has been completed.

Other physical characteristics, and in particular the retention of CO_2, will be tested on a regular basis over and beyond the anticipated shelf life, as gas retention is likely to be the limiting factor in the shelf life of a carbonated beverage. Other chemical components will also be checked over the anticipated shelf life. This is particularly important if claims are being made for the presence of ingredients of nutritional value such as vitamins.

The final suite of testing to establish shelf life will relate to the organoleptic qualities of the product. When a product is first formulated and subjected to shelf life testing, a retained sample of freshly made product will often be held in dark refrigerated conditions at around 2–5°C. These conditions are likely to keep the product in a condition that is reasonably representative of the target. However, it may be considered that such freshly made product is not truly representative of the average aged product purchased by the consumer, and some manufacturers may select a reference sample of slightly different age.

Organoleptic shelf testing of product as it ages will be carried out against appropriate reference samples using an appropriate technique. There is scope for the use of both trained and untrained tasting panels, as there is a range of information that must be collected. The information shown in Table 9.3 will usually be required to evaluate a product, as it is important to consider all the sensory clues that a consumer may use to judge acceptability.

Manufacturers may wish to duplicate testing to both exclude and include the influence of color changes in the product. Correct selection of specific types of tests and the number of testers required to obtain statistically significant data also play an important part in the evaluation of shelf life.

In summary, it is essential to carry out a wide range of tests, and particularly organoleptic tests, to ascertain the performance of any beverage product, although for carbonated beverages the retention of CO_2 is probably the overriding factor in determining shelf life.

TABLE 9.3
Typical Organoleptic Tests for Beverages

Information Required	Type of Panel
For how long can a product be stored before a change in sensory perception is noticed?	Discrimination test using either a trained panel or a large number of untrained consumers
How do the sensory attributes of the product change on storage?	Descriptive test using a trained panel
For how long can the product be stored before the changes in sensory attributes render the product unacceptable?	Consumer testing to understand acceptability and descriptive testing from a trained panel

REFERENCES

Anon. 1997. Carbonated beverage packaging. In: *The Wiley Encyclopedia of Packaging Technology*, 2nd edn. Brody A.L., Marsh K.S. (Eds). New York: John Wiley & Sons, pp. 158–161.

Ashby R. 1988. Migration from polyethylene terephthalate under all conditions. *Food Additives & Contaminants* 5: 485–492.

Ashurst P.R. (Ed). 2005. *Chemistry and Technology of Soft Drinks and Fruit Juices*, 2nd edn. Oxford, England: Blackwell Publishing.

Ashurst P.R., Hargitt R. 2009. *Soft Drink and Fruit Juice Problems Solved*. Cambridge, England: Woodhead Publishing.

Bashir Z., Al-Uraini A.-A., Jampoon M., Al-Khalid A., Al-Hafez M., Ali S. 2002. Acetaldehyde generation in poly(ethylene terephthalate) resins for water bottles. *Journal of Macromolecular Science Part A* 39: 1407–1433.

Bernardo P.E.M., dos Santos J.L.C., Costa N.G. 2005. Influence of the lacquer and end lining compound on the shelf life of the steel beverage can. *Progress in Organic Coatings* 54: 34–42.

Boutroy N., Pernel Y., Rius J.M., Auger F., von Bardeleben H.J., Cantin J.L., Abel F., Zeinert A., Casiraghi C., Ferrari A.C., Robertson J. 2006. Hydrogenated amorphous carbon film coating of PET bottles for gas diffusion barriers. *Diamond and Related Materials* 15: 921–927.

Cairncross R.A., Becker J.G., Ramaswamy S., O'Connoer R. 2005. Moisture sorption, transport, and hydrolytic degradation in polylactide. *Applied Biochemistry and Biotechnology* 129–132: 774–785.

Giles G.A. 2005. Packaging materials. In: *Chemistry and Technology of Soft Drinks and Fruit Juices*, 2nd edn. Ashurst P.R. (Ed). Oxford, England: Blackwell Publishing, pp. 200–235.

Gunning P. 1999. Packaging of beverages in polyethylene terephthalate (PET) bottles. In: *Handbook of Beverage Packaging*. Giles G.A. (Ed). Boca Raton, Florida: CRC Press, pp. 71–92.

Hocking A.D., Jensen N. 2001. Soft drinks, cordials, juices, bottled waters and related products. In: *Spoilage of Processed Foods: Causes and Diagnosis*. Moir C.J., Andrew-Kabilafkas C., Arnold G., Cox B.M., Hocking A.D., Jenson I. (Eds). Sydney, Australia: AIFST (NSW Branch) Food Microbiology Group, pp. 84–100.

Holm V.K., Ndoni S., Risbo J. 2006. The stability of poly(lactic acid) packaging films as influenced by humidity and temperature. *Journal of Food Science* 71: E40–E44.

Kim-Kang H. 1990. Volatiles in packaging materials. *CRC Critical Reviews in Food Science and Nutrition* 29: 255–271.

Larbey R. 2006. Closures for plastic bottles and tubs. In: *Packaging Closures and Sealing Systems*. Theobald N., Winder B. (Eds). Oxford, England: Blackwell Publishing Ltd, pp. 158–182.

Matsuga M., Kawamura Y., Sugita-Konishi Y., Hara-Kudo Y., Takatori K., Tanamoto K. 2006. Migration of formaldehyde and acetaldehyde into mineral water in polyethylene terephthalate (PET) bottles. *Food Additives and Contaminants* 23: 212–218.

Morrison E.D., Malvey M.W., Johnson R.D., Anacker J.L., Brown K.A. 2008. Effect of chemical environments on stress cracking of poly(ethylene terephthalate) beverage bottles. *Polymer Testing* 27: 660–666.

Nasser A.L.M., Lopes L.M.X., Eberlin M.N., Monteiro M. 2005. Identification of oligomers in polyethylene terephthalate bottles for mineral water and fruit juice. Development and validation of a high-performance liquid chromatographic method for the determination of first series cyclic trimer. *Journal of Chromatography A* 1097: 130–137.

Nijssen B., Kamperman T., Jetten J. 1996. Acetaldehyde in mineral water stored in polyethylene terephthalate (PET) bottles: odour threshold and quantification. *Packaging Technology and Science* 9: 175–185.

Özlem K.E. 2008. Acetaldehyde migration from polyethylene terephthalate bottles into carbonated beverages in Turkey. *International Journal of Food Science & Technology* 43: 333–338.

Page B. 2006. Closures for metal containers. In: *Packaging Closures and Sealing Systems*. Theobald N., Winder B. (Eds). Oxford, England: Blackwell Publishing Ltd, pp. 68–100.

Robertson G.L. 2006. *Food Packaging Principles and Practice*, 2nd edn. Boca Raton, Florida: CRC Press.

Schmid P., Kohler M., Meierhofer R., Luzi S., Wegelin M. 2009. Does the reuse of PET bottles during solar disinfection pose a health risk due to migration of plasticisers and other chemicals into the water? *Water Research* 42: 5054–5060.

Senior D.A.G. 2004. Bottling water—maintaining safety and integrity through the process. In: *Technology of Bottled Water*, 2nd edn. Senior D.A.G., Dege N.J. (Eds). Oxford, England: Blackwell Publishing, chapter 6.

Shirakura A., Nakaya M., Koga Y., Kodama H., Hasebe T., Suzuki T. 2005. Diamond-like carbon films for PET bottles and medical applications. *Thin Solid Films* 494: 84–91.

Shotyk W., Krachler M., Chen B. 2006. Contamination of Canadian and European bottled waters with antimony from PET containers. *Journal of Environmental Monitoring* 8: 288–292.

Syrett D. 2006. Bottle design and manufacture and related packaging. In: *Carbonated Soft Drinks Formulation and Manufacture*. Steen D.P., Ashurst P.R. (Eds). Blackwell Publishing, Oxford, pp. 181–217.

Theobald N. 2006. Closures for glass containers. In: *Packaging Closures and Sealing Systems*. Theobald N., Winder B. (Eds). Oxford, England: Blackwell Publishing Ltd, pp. 101–117.

Turtle B.I. 1984. The polyester bottle. In: *Developments in Soft Drinks Technology—2*. Houghton H.W. (Ed). Essex, England: Elsevier Applied Science Publishers, chapter 3.

Warburton D.W. 2002. The microbiological safety of bottled waters. In: *Safe Handling of Foods*. Farber J.M. (Ed). New York: Marcel Dekker, chapter 16.

Warburton D.W., Austin J.W. 2000. Bottled water. In: *The Microbiological Safety and Quality of Food*, Vol. 1. Lund B.M., Baird-Parker T.C., Gould G.W. (Eds). Gaithersburg, Maryland: Aspen Publishers, chapter 32.

Wilk R. 2006. Bottled water: the pure commodity in the age of branding. *Journal of Consumer Culture* 6: 303–325.

10 Packaging and the Shelf Life of Orange Juice

Antonio López-Gómez, María Ros-Chumillas, and Yulissa Y. Belisario-Sánchez
Food Engineering and Agricultural Equipment Department
Technical University of Cartagena
Cartagena, Spain

CONTENTS

10.1 INTRODUCTION

Orange juice is the predominant juice manufactured by the juice industry worldwide and is consumed in relatively high quantities in many countries. Fruit juices were originally developed to use up the surplus fresh fruit production, but now in many areas fruit (in particular citrus and apple) is specifically grown for juicing. Today's consumers desire high-quality foods with fresh flavor, texture, and color, and orange juice is the most appreciated and consumed juice because of its pleasant taste and high ascorbic acid content.

The deteriorative reactions for orange juice occur mainly during pasteurization, bulk storage, and packaging. Orange juice suffers a number of significant deteriorative reactions, including ascorbic acid degradation; cloud loss; microbial spoilage; off-flavor development; and changes in color, texture, and appearance, all of which contribute to important loss of quality. Although conventional thermal processing ensures the safety and extends the shelf life of orange juice, it often leads to detrimental changes in the sensory quality of the juice. Reducing the temperature through the use of cold or aseptic packaging rather than hot-filling minimizes undesirable changes in orange juice.

A wide range of packaging materials are used for orange juice, including metal cans, glass bottles, plastic/alufoil/paperboard laminate cartons, plastic bottles and cups, and flexible packages. Their influence on orange juice quality and shelf life is discussed in this chapter.

10.2 ORANGE JUICE MARKETS

Citrus fruits are the largest fruit crop in international trade in terms of value. There are two clearly differentiated markets in the citrus sector: the fresh citrus fruits market, with oranges predominating, and the processed citrus products market, mainly orange juice.

The main citrus-fruit-producing countries are Brazil (which surpassed Florida as the world's number one orange producer in 1983), the Mediterranean countries, the United States (where citrus fruits for consumption as fresh fruit are mainly grown in California, Arizona, and Texas, and most orange juice is produced in Florida), and China. These countries represent more than two-thirds of the global citrus fruit production.

The orange is a favorite fruit in the United States, where it has consistently ranked as the third most consumed fresh fruit, behind bananas and apples. As a juice, it ranks number one. On average, Americans consume 2.5 times more orange juice annually than its nearest competitor, apple juice. Orange juice has been a driving force behind increased orange consumption over the past half-century and is in part the reason behind the decline in consumption of fresh oranges. Consumers substitute orange juice for fresh orange consumption and receive many of the same benefits. Commercial cultivation of oranges intended for large-scale processing into juice began in Florida in the 1920s and accelerated in the late 1940s with the introduction of frozen concentrated orange juice (FCOJ) for home dilution (Anon., 2004). International trade in orange juice was predominantly in the form of FCOJ in order to reduce the volume so that storage and transportation costs were lower. The growing popularity of not-from-concentrate orange juice (NFCOJ) since the mid-1990s has helped maintain a strong demand for orange juice as the popularity of FCOJ has declined (Pollack et al., 2003).

The major feature of the world market for orange juice is the geographical concentration of production. There are only two main players: the state of Florida in the United States and the state of São Paulo in Brazil. The combined production of orange juice from these two players makes up approximately 85% of the world market. The major difference between them is that Brazil exports 99% of its production, whereas 90% of Florida's production is consumed domestically and only

10% is exported (UNCTAD, 2008). Data from the USDA (2008a) indicated that world orange juice production (as 65°Brix concentrate) during 2007/08 in selected major producing countries was estimated at 2.3 million metric tons (MMT), up 56,000 tons from 2006/07. Brazil's production of orange juice during 2008/09 was estimated at 1.32 million tons, down 7% from 2007/08, due to an expected reduced availability of fruit for processing. U.S. orange juice production in 2007/08 was estimated at 789,000 tons, up about 155,000 tons from 2006/07. On average, 95% of Florida's oranges are processed each season. Orange juice production in China was forecast to nearly double, to 20,000 tons, in 2007/08 compared to the previous year. Local oranges for juicing are more readily available as a result of more fruit-bearing trees from plantings in prior years. Although domestic processing companies have built several large juicing facilities, they are not running at full capacity because of a lack of sufficient fresh oranges.

As shown in Figure 10.1, total world imports of orange juice for 2007 were valued at an estimated \$2.9 billion, with FCOJ valued at \$1.4 billion and NFCOJ valued at \$1.5 billion. The EU-27 was the top market, with imports valued at approximately \$1.0 billion in 2007. Over 93% of the EU-27 orange juice imports are NFCOJ. Figure 10.2 indicates that EU-27 consumers appear to be reducing orange juice consumption, even while domestic production remains stable (USDA, 2008b). The United States is the second largest importer of orange juice, with imports valued at \$627 million in 2007. U.S. orange juice imports are nearly 87% FCOJ (USDA, 2008a).

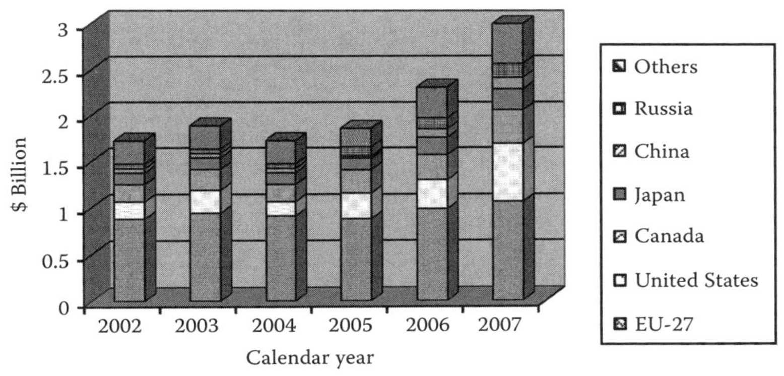

FIGURE 10.1 Top orange juice importers. (From data of USDA. 2008a. *World markets and trade: Orange Juice*. United States Department of Agriculture. Foreign Agricultural Service, Office of Global Analysis, April, 2008.)

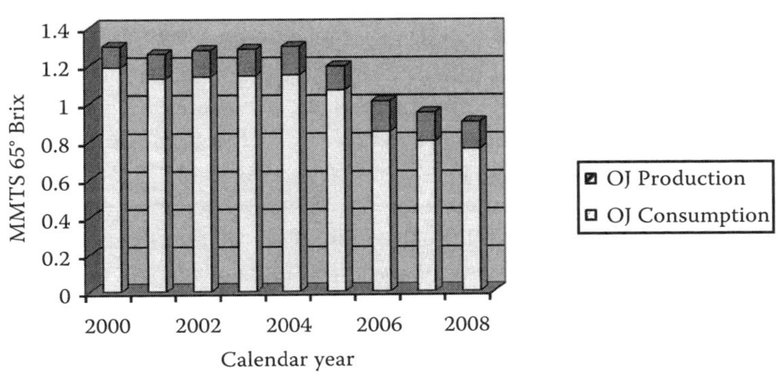

FIGURE 10.2 Orange juice (OJ) production and consumption in the EU-27. (From data of USDA. 2008b. *Citrus market update: European Union—27*. Foreign Agricultural Service, Office of Global Analysis, April, 2008.)

10.3 ORANGE JUICE PROCESSING

A detailed description of juice extraction and subsequent processing operations is outside the scope of this book, and the reader is referred to standard texts on the subject (Anon., 2004; Ashurst, 2005; Barrett et al., 2005; Braddock, 1999; Chen et al., 1993); only a general overview will be presented here.

After extraction, the juice (typically 11°Brix soluble solids with a pulp content of 20–25% by volume) passes through a finisher (a horizontally mounted screen drum), where the pulp content is reduced to approximately 10–12% by the removal of coarse particles such as cell walls, rag, and other fibrous materials. The juice is then deaerated, pasteurized, and subsequently stored in refrigerated bulk tanks, filled into containers as single-strength juice or NFCOJ, or evaporated in order to obtain FCOJ. Approximately 10–12 tons of fruit are necessary to produce 1 ton of concentrated (65°Brix) orange juice (Braddock, 1999; Schöttler et al., 2002).

10.3.1 Deaeration

A key step in the processing of orange juice is deaeration, which is generally applied immediately prior to pasteurization to remove air from the juice. Deaeration is important both to minimize oxidative reactions in the juice (e.g., oxidation of ascorbic acid and flavor compounds) and to reduce corrosion if the juice is subsequently packaged in a metal container (Castberg et al., 1995; Ebbesen, 1998).

Jordán et al. (2003) showed that during the industrial processing of orange juice the biggest losses in the concentration of volatile components occurred during deaeration. By the addition of aromatic fractions recovered during deaeration it is possible to obtain processed orange juice with an aromatic profile closer to that of fresh juice (Nisperos-Carriedo and Shaw, 1990). The pasteurization process did not change the analytical composition of deaerated orange juice in a significant way for any of the 42 compounds measured (Jordán et al., 2003).

Soares and Hotchkiss (1999) showed that both deaeration and package barrier properties are major factors in maintaining ascorbic acid in refrigerated orange juice. The rate of ascorbic acid degradation is inversely correlated with the permeation rate for both deaerated and nondeaerated juices, regardless of the initial dissolved oxygen (DO) content. Juice in high-O_2-permeability containers showed a faster decrease in ascorbic acid content, independent of initial DO content.

10.3.2 Pasteurization

Pasteurization involves heating the juice in tubular or plate heat exchangers to temperatures in the region of 90–100°C for 12–45 sec (Chen et al., 1993), although some authors give other conditions, such as 85°C for 15 sec to 95°C for 2 sec (Lewis and Hepell, 2000). Pasteurization was originally used as a means of controlling microflora, but it is also important for stabilizing the cloud of orange juices, as consumers regard orange juices without a stable cloud as inferior and unacceptable. Although pasteurization ensures the safety and extends the shelf life of orange juice, it often leads to detrimental changes in the sensory qualities of the product.

The major enzyme responsible for destabilizing the cloud is pectinmethylesterase (PME), which must be inactivated as soon as possible after extraction of the juice. This is generally done by pasteurizing the juice at 90–95°C for 15–30 sec; the precise time depends on the pulp content. During pasteurization, enzymes responsible for the oxidation of ascorbic acid in natural orange juice, such as cytochrome oxidase, ascorbic acid oxidase, and peroxidase (POD), are also destroyed.

POD is recognized as one of the most heat-stable enzymes in higher plants and is involved in reactions that are mainly associated with loss of flavor quality in orange juices (Bruemmer et al., 1976). Therefore, long heat treatment times to ensure POD inactivation are recommended (Tomás-Barberán and Espín, 2001). Unfortunately, intensive thermal methods lead to important undesirable effects such as color changes, cooked flavors, and loss of vitamins and nutrients. Moreover, consumers demand safer and better quality products with minimal processing and "fresh-like" characteristics.

Pasteurization can trigger a series of undesirable reactions such as the destruction of vitamins and provitamins, the acceleration of the reaction between sugars and amino acids (Maillard or nonenzymic browning reaction, giving rise to products with a dark color and bitter taste), the destruction of pigments (carotenoids), the denaturing of proteins, the acceleration of the oxidation of fats, and the formation of toxic products (Braddock, 1999). Pasteurization also has adverse effects on the aromatic fraction of orange juice (Moshonas and Shaw, 1997).

Several effective new process technologies are available to accomplish a microbial reduction in juices without the use of heat (Sizer and Balasubramaniam, 1999). Pulsed electric fields, ultraviolet light, minimal thermal processes, and batch and continuous high pressure processing systems have been offered commercially. The applicability of each technology to a specific juice depends on the characteristics of the product and the pathogens of interest that may be resistant to the process. Each of the minimal processes is intended to reduce pathogens and does not accomplish a kill adequate for commercial sterility. As such, products must be maintained under refrigerated storage and distribution to slow spoilage. Combining emergent technologies such as pulsed electric fields, high pressures, or flash pasteurization with other techniques such as aseptic storage and aseptic packaging is becoming increasingly common (Polydera et al., 2003; Torregrosa et al., 2006).

10.3.3 Hot-Filling

Hot-filling is a well-proven and recognized method to ensure the shelf stability of orange juice at ambient temperatures for more than 180 days. This method is used extensively in the citrus industry for filling hot (>84°C) pasteurized juice into glass and some plastic containers [e.g., heat-set poly(ethylene terephthalate) (PET)] as well as metal cans. The hot-filling sterilizes the inner surface of the container. The necessary filling temperature and holding time in the package prior to cooling depend on the type and size of container and its degree of initial microbial contamination (Anon., 2004). After a specific holding time, the containers are cooled in order to minimize thermal degradation of the juice (Tekkanat, 2002).

10.3.4 Ultraclean and Aseptic Packaging

Ultraclean packaging refers to packaging that includes a controlled filling under extreme hygienic conditions and container sterilization to give an extended shelf life compared with pasteurized products. Aseptic packaging is the filling of a commercially sterile product into sterile containers under aseptic conditions and sealing the containers so that reinfection is prevented, that is, so that they are hermetically sealed. The term "aseptic" implies the absence or exclusion of any unwanted organisms from the product, package, or other specific areas, whereas the term "hermetic" (strictly "air tight") is used to indicate suitable mechanical properties to exclude the entrance of microorganisms into a package and the passage of gas or water vapor into or from the package (Robertson, 2006).

Ultraclean and aseptic packaging allows cold or ambient temperature filling of juice, and it is possible to use laminated plastic/alufoil/paperboard cartons and plastic containers (monolayer and multilayer PET bottles, multilayer PP/EVOH/PP thermoformed cups, and multilayer flexible bags of the bag-in-box systems) that do not have heat-set characteristics.

These technologies provide a better final quality of the packed orange juice, as they avoid the large thermal treatment that occurs during hot-filling. If the quality of the raw orange juice is very high, and double or triple thermal treatment (successive pasteurizations) is avoided, by means of aseptic bulk storage and aseptic transfer to the aseptic filler, the quality of the orange juice will be very high (López-Gómez and Barbosa-Cánovas, 2005; Ros-Chumillas et al., 2007).

10.4 ORANGE JUICE QUALITY ATTRIBUTES

The flavor and aroma of freshly squeezed unpasteurized orange juice is the target for the optimum initial quality of pasteurized, packaged orange juice. Consumers in developed countries are becoming more critical and demanding about the food and drink they consume, desiring high-quality foods with fresh flavor, texture, and color. As a result, they are demanding more natural products and fresher foods, with less severe processing and no preservatives, that are safe and easy to prepare (Loureiro and Querol, 1999). This has contributed to the increasing consumption of NFCOJs that have been subjected to a mild pasteurization process, as these meet the requirements of consumers who demand high quality (Esteve et al., 2005).

10.4.1 Color

Color is one of the most characteristic quality parameters of orange juice and has been included in the quality control procedures of the food industries in the European Union (AIJN, 2008). In the United States, the color of citrus juices is one of the parameters evaluated for the commercial classification of the product in relation to its quality, with some studies showing that the color of citrus beverages in general is related to the consumer's perception of flavor, sweetness, and other quality characteristics of these products (Tepper, 1993). Color is also an indicator of the natural transformation resulting from changes that occur during storage or processing.

The color of orange juice is mainly due to carotenoid pigments, a complex mixture of more than 115 natural substances, although not all are precursors of vitamin A (Lee and Coates, 2003; Meléndez-Martínez et al., 2005). Because of the presence of carotenoids and the relatively high consumption of orange juice, it is the most important source of vitamin A carotenoids (β-carotene, α-carotene, and β-cryptoxanthin) and antioxidant carotenoids (β-carotene, β-cryptoxanthin, zeaxanthin, and lutein). These carotenoids have been associated with the reduction of degenerative human diseases, such as heart disease and cancer, because of their antioxidant and free-radical-scavenging properties (Temple, 2000; Sánchez-Moreno et al., 2006).

10.4.2 Flavor

Many research articles have been published about the composition and the effects of process variables on the volatile flavor components of orange juice (Sizer et al., 1988; Pérez-López and Carbonell-Barrachina, 2006; Perez-Cacho and Rouseff, 2008). Moshonas and Shaw (1997) concluded that limonene, myrcene, α-pinene, decanal, octanal, ethyl butanoate, and linalool were important contributors to orange juice flavor. Farnworth et al. (2001) reported that concentrations of acetaldehyde (identified as a major contributor to fresh orange juice flavor), ethyl acetate (a major ester in fresh orange juices, contributing a fruity, solvent-like odor), α-pinene, β-myrcene, limonene, α-terpineol, 1-hexanol, 3-hexen-1-ol, and sabinene concentrations were highest in unpasteurized orange juice.

Excessive heating irreversibly and negatively alters juice flavor so that it no longer has the aroma and character of fresh orange juice. Some processing and packaging developments have resulted in improved flavor because they minimize the application of heat (Braddock, 1999), for example, ultraclean packaging and aseptic processing.

10.5 DETERIORATIVE REACTIONS AND INDICES OF FAILURE FOR ORANGE JUICE

During processing, packaging, and storage, orange juice can suffer several important deteriorative reactions that can result in important quality losses (Ayhan et al., 2001; Polydera et al., 2003; Torregrosa et al., 2006). The five key deteriorative reactions in orange juice are microbiological spoilage, nonenzymic browning, cloud loss, oxidation resulting in loss or degradation of flavor components and nutrients (essentially ascorbic acid), and absorption of flavor compounds by the package (commonly referred to as scalping).

10.5.1 Microbial Spoilage

The major microbial contaminants of unpasteurized juices have generally been recognized as low-heat-resistant microorganisms such as yeasts, molds, and lactic acid bacteria, as these organisms prefer or tolerate the acidic nature (pH < 4) of citrus juices. Although preservatives were commonly added to fruit juices to overcome microbiological problems, recent consumer preference for preservative-free foods has seen their use diminish. Instead, attention to good manufacturing practice in the plant, coupled in many cases with aseptic processing and packaging, has obviated the need for them (Robertson, 2006).

During storage, orange juice may suffer serious problems due to contamination by microorganisms, mainly lactic acid bacteria (*Lactobacillus* spp. and *Leuconostoc* spp.), molds, and yeasts (*Saccharomyces cerevisiae*), which are the main microorganisms of citrus juices because of their low pH. However, spoilage of aseptically packaged apple juice in Germany in 1982 due to an acid-tolerant bacterium with highly heat-resistant spores capable of surviving the usual pasteurization treatments and capable of producing a disagreeable odor presented a new threat to juice manufacturers. The bacterium involved is *Alicyclobacillus acidoterrestris*, which has an optimum temperature for growth of 40–42°C and a reported growth temperature range of 25–60°C. An examination of 75 samples of concentrated orange juice from 11 suppliers found 14.7% to be positive for *Alicyclobacillus* (Eiroa et al., 1999). The flavor taint is due to the formation of 2,6-dibromophenol and 2,6-dichlorophenol, which have taste thresholds at the parts-per-trillion level. There is no evidence that *A. acidoterrestris* poses a human health risk. The ultimate source of this organism is soils, and it likely enters the processing areas on fruit surfaces contaminated with soil during harvesting (Walker and Phillips, 2008). Although the industry is attempting to move away from the use of preservatives and more and more products are being sold without any preservation apart from pasteurization, the latter does not destroy heat-resistant spores such as those of *A. acidoterrestris* (Esteve and Frígola, 2007).

Until recently, microbial spoilage of improperly handled orange juice was wasteful but not deemed particularly dangerous. However, over the past decade fresh juice has increasingly been the source of serious food-poisoning outbreaks. Unpasteurized juice has been implicated in outbreaks of *Salmonella* and emerging pathogens such as *Escherichia coli* O157:H7. These incidents have resulted in much stricter sanitary requirements for commercial fresh juice producers (Bates et al., 2001). In the past, the growth of human pathogens in citrus products was assumed to be avoided because of the acidity of the juice and the heat treatment applied to commercial citrus juices. However, Caggia et al. (2009) have recently observed that cells of *Listeria monocytogenes* adapted to acidic environments can grow in orange juices. They concluded that, from an industrial point of view, the consequences for humans of the survival or acid adaptation of *Listeria* spp. in acidic conditions such as orange-processing environments should be better evaluated.

Most research regarding citrus-processing microbiology has involved detecting and preventing spoilage events due to the growth of fermentative yeasts (mainly *S. cerevisiae*) and lactic acid bacteria. Few reports address filamentous fungal spoilage of citrus juices. Although filamentous fungi are capable of growth in low-pH fruit juices, they have historically not been involved in retail spoilage of citrus juices due to the stability of FCOJ and the inability of molds to compete with other members of the juice microflora during retail shelf life of reconstituted, chilled, single-strength juices. However, recent technological changes in storage and packaging systems used by the citrus industry may allow mold proliferation in pasteurized, chilled, single-strength citrus juices.

Fundamental information regarding growth characteristics at low temperatures of filamentous fungi previously isolated from chilled, pasteurized citrus products has been reported (Wyatt et al.,

1995). Of particular interest were *Penicillium digitatum* and *P. italicum*, because of their frequent occurrence as pathogens of fresh citrus fruits. After packaging in O_2 barrier cartons, the same juice will have a shelf life of 60 days or more in the retail market, where typical storage temperatures may vary from 3°C to 7°C in the United States. Data from this study indicate that opportunistic contamination by filamentous fungi can result in a substantial accumulation of fungal biomass in the product and a reduction in sugar content.

10.5.2 Nonenzymic Browning

The presence of furfural and 5-hydroxymethylfurfural (HMF) in stored orange juice has long been used as an indicator of quality loss; furfural and HMF are related to the browning of the juice and they are also good indicators of excess thermal treatments and lengthy storage times. Consequently, the analysis of these compounds has special importance in the industry. Some inadequate conditions during thermal treatment and storage of the juice are reflected in an increase in the concentration of the different derivatives of furfural, formed by ascorbic acid degradation in the browning pathway (Braddock, 1999).

However, Roig et al. (1999) reported that in freshly produced commercial citrus juice, aseptically filled in laminated plastic/alufoil/paperboard cartons, nonenzymic browning was mainly due to carbonyl compounds formed from L-ascorbic acid degradation. Although formation of 5-HMF was detected in degraded juice samples, its presence could not be used as an index of browning. The compound was found to be unreactive in the browning process in citrus juices and its contribution to browning in this type of products is insignificant, if not negligible. Despite this finding, 5-HMF is still used by many in the citrus industry as an indicator of browning.

10.5.3 Cloud Loss

In orange juice, loss of cloud leads to a decrease in consumer acceptability, as cloud particles impart the characteristic flavor, color, and mouthfeel to orange juice. Cloud is composed of a complex mixture of proteins, pectins, lipids, hemicellulose, cellulose, and other minor components (Baker and Cameron, 1999; Klavons et al., 1991). The enzyme previously discussed in section 10.3.2 PME in orange juice plays an important role in the loss of cloud (Cameron et al., 1997). To maintain the typical turbidity of orange juice, it is necessary to inactivate PME, typically through suitable heat treatment (Ingallinera et al., 2005). In fact, one of the principal reasons for pasteurizing citrus juice is inactivation of the enzyme responsible for loss of cloud, which is a very important quality attribute for consumers (Varsel, 1980).

PME is responsible for the hydrolysis of pectin present in citrus juices, which results in loss of juice cloudiness and gelation of pectin in concentrated juice (Basak and Ramaswamy, 1996). It occurs naturally in oranges and is composed of several isoenzymes. Cameron et al. (1997) isolated four isoenzymes in Valencia oranges and studied the effects of each on juice cloud stability, concluding that the most heat-resistant form, although only 7.9% of the total enzyme had the major influence on juice cloud stability loss at storage conditions of 5–10°C. They also reported that these heat-resistant isoenzymes were located in the albedo and the juice sac membrane. As PME is more heat resistant than the pathogenic and spoilage microorganisms that can be present in orange juice and is responsible for the cloud stability loss, its inactivation is commonly used as an indicator of the adequacy of the pasteurization process (Basak and Ramaswamy, 1996).

A specific indicator of freshness, allowing routine distinction between freshly squeezed orange juices (FSOJs) and FSOJ-like products, was identified by Hirsch et al. (2008). FSOJs were heated at six different temperatures (42–92°C), and the cloud stability and residual activities of PME and POD were monitored during storage at 4°C for up to 62 days, thus replicating the storage conditions of FSOJs in retail markets. The juices processed at temperatures ≥62°C were characterized by minor residual activities. Juices processed at 52°C with a residual PE activity of 33.8% were hardly inferior in terms of cloud stability within the first 14 days compared to juices processed at 62°C. The authors found that the range of approximately 50–60°C is relevant in minimal heat processing for the retention of cloud stability within the short turnover period of FSOJ-like products, with partial PME and POD deactivation being sufficient to distinguish those juices from FSOJs. PME was suggested as an indicator enzyme for the freshness of FSOJs, allowing their unambiguous distinction from minimally heat-processed juices.

10.5.4 Oxidation

10.5.4.1 Flavor

The oil fraction of citrus juices contains many volatiles that have a major impact on citrus aroma and flavor. These oil-based flavor compounds are relatively easily oxidized, resulting in the development of undesirable, terpene-like off-flavors. Removal of O_2 from the juice prior to packaging and avoidance of high pressures during juice extraction so as to limit oil transfer to the juice minimize this form of flavor deterioration, as does using a package that is a good barrier to O_2.

10.5.4.2 Ascorbic Acid Degradation

A major problem associated with the quality of orange juice is the loss of ascorbic acid during heat treatment and storage (Lee and Coates, 1999). Thus, the concentration of ascorbic acid is used to estimate the end of the shelf life of packed natural orange juice, because, according to the Association of Industries of Juices and Nectars from Fruits and Vegetables of the European Union (AIJN), ascorbic acid in orange juice should be greater than 20 mg 100 mL^{-1}.

Although quality and shelf life determination of orange juice are often based on ascorbic acid retention during storage, other quality parameters such as color and flavor are also very important (Lee and Coates, 1999). Ascorbic acid is an essential nutrient for humans, and, because of its high antioxidant activity, it provides protection against the presence of free radicals and thus protects against many diseases.

Ascorbic acid is oxidized and lost during the storage of juice. The rate of degradation of ascorbic acid is highly dependent on the filling and storage conditions, including the efficiency of deaeration, the amount of O_2 in the headspace, the permeation of O_2 through the package into the juice, and the storage temperature (Kabasakalis et al., 2000). Factors affecting ascorbic acid loss in packed orange juice include temperature, DO, and the O_2 barrier provided by the package. Soares and Hotchkiss (1999) showed that the rate of ascorbic acid degradation correlated inversely with the permeation rate for both deaerated and nondeaerated juices regardless of initial DO content. Juices in high-O_2-permeability containers showed a faster decrease in ascorbic acid content, independent of the initial DO content. Ascorbic acid degradation can lead to nonenzymic browning; therefore, not only is ascorbic acid loss important nutritionally, but its degradation is also related to flavor and color changes.

Light appears to have no effect on the stability of ascorbic acid in orange juice. Solomon et al. (1995) did not observe any statistical differences between the ascorbic acid contents of orange juice stored in glass at 8°C under artificial light (200 lux) and of juice stored in darkness. Recently, Berlinet et al. (2008) found no significant differences between the ascorbic acid contents of juices stored under artificial light (750 lux, which is typical of lighting in supermarkets) and in darkness after both 3 and 9 months of storage.

10.5.5 Scalping

The sorption of key aroma and flavor compounds by plastic packaging in contact with juice is referred to as "scalping" (Sajilata et al., 2007). Because of its lipophilic nature, the oil fraction of orange juice will be absorbed by many nonpolar packaging polymers. Orange juice aromas have been demonstrated to be sorbed to different extents, starting with hydrocarbon compounds, which showed the highest affinity to low density polyethylene (LDPE), followed by ketones, esters, aldehydes, and finally alcohols (Nielsen et al., 1992). Factors that affect absorption include the molecular sizes of the aroma compounds and the polarity and solubility properties of both the polymer and the aroma compounds.

The most extensively studied aroma compound with respect to its sorption by polymers is limonene. Limonene is an unsaturated terpene hydrocarbon present in orange juice; it is highly nonpolar and has a high affinity for many polymeric packaging materials. A decrease in limonene content in stored orange juice is attributed to its lipophilic nature and, hence, the ease of its diffusion into the polymer (Nielsen, 1994; Moshonas and Shaw, 1997).

10.6 IMPACT OF PACKAGING ON INDICES OF FAILURE

10.6.1 Microbial Spoilage

The O_2 barrier properties of the package will influence the type of microbial growth that can occur in packaged juice. Most molds and yeasts able to grow in orange juice are aerobic, as are pathogens such as *Salmonella* spp. and *E. coli* O157:H7. The *Lactobacillus* spp. are mostly microaerophilic or anaerobic.

10.6.2 Nonenzymic Browning

The rate of browning and nutrient degradation in fruit juices is largely a function of storage temperature, although the rate is in part dependent on the packaging material. For example, Mannheim et al. (1987) compared the quality of citrus juices aseptically packaged in laminated cartons and glass containers and found that the extent of browning and loss of ascorbic acid was greater in cartons than in glass, presumably because of O_2 permeation into the carton.

10.6.3 Cloud Loss

Packaging does not influence cloud loss in orange juice. However, transparent packages such as those made from glass and plastic provide a visual indication to the consumer as to the stability of the cloud, in contrast to packages made from metal or paperboard, where the juice is not visible.

10.6.4 Oxidation

The O_2 barrier properties of the package will influence the rate of ascorbic acid degradation as well as the oxidation of oil-based flavor compounds, as will the initial DO content, which should, wherever possible, be minimized by deaerating or hot-filling (Tawfik and Huyghebaert, 1998).

10.6.5 Scalping

In a study (Mannheim et al., 1988) comparing the quality of citrus juices aseptically packaged in laminated cartons and glass containers, the D-limonene content of the juices in the cartons was reduced by about 25% within 14 days of storage due to absorption by the polyethylene, and sensory evaluations showed a significant difference after 10–12 weeks between juices packaged in glass and cartons stored at ambient temperatures. In contrast, another study (Pieper et al., 1992) reported that an experienced panel did not distinguish between orange juice stored in glass bottles and juice stored in laminated aseptic cartons. Absorption of up to 50% limonene and other hydrocarbons, small quantities of ketones, and aldehydes had no significant influence on the sensory quality of juice stored at 4°C.

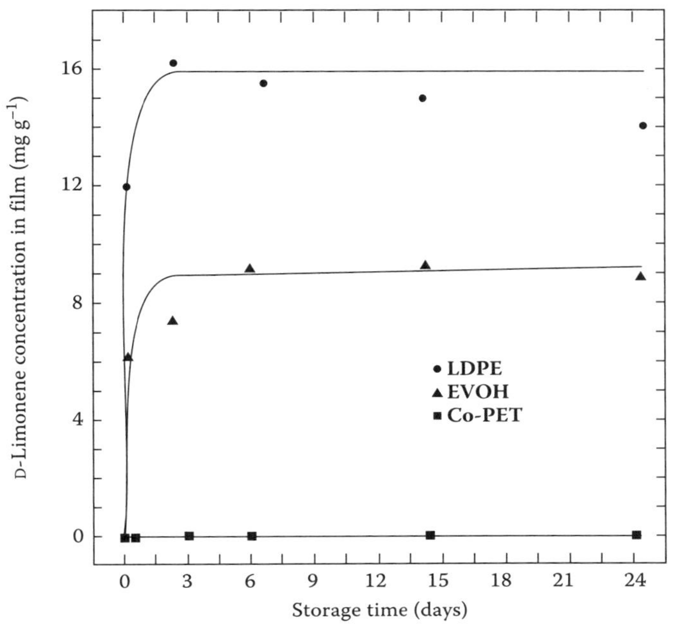

FIGURE 10.3 Sorption of D-limonene by LDPE, EVOH, and Co-PET. (Redrawn from Imai T., Harte B.R., Giacin J.R. 1990. Partition distribution of aroma volatiles from orange juice into selected polymeric sealant films. *Journal of Food Science* 55: 158–161, with permission.)

The presence of juice pulp in orange juice decreased the absorption of volatile compounds into polymeric packaging materials (Yamada et al., 1992). The authors suggested that pulp particles hold flavor compounds such as limonene in equilibrium with the aqueous phase, and this could be responsible for the decreased absorption of these compounds by the plastics.

Another study (Imai et al., 1990) determined the amount of D-limonene sorbed by three different films as a function of storage time, with the amount sorbed varying with the polymer, as shown in Figure 10.3. After 3 days, sorption by LDPE and EVOH plateaued and reached equilibrium, but for Co-PET (a copolyester developmental film) a slow increase was observed for 24 days.

As well as loss of aroma, sorption of organic molecules can affect the mechanical properties of the film and increase its O_2 permeability (Tawfik et al., 1998). In a study involving the sorption of D-limonene by LDPE and ionomer films, rapid absorption was observed, with saturation (around 44% of the initial concentration) being reached after 12 days. There was a reduction in seal and tensile strengths and an increase in O_2 permeability of 2–4 times (Hirose et al., 1988).

Polyesters such as poly(ethylene terephthalate) (PET), poly(ethylene naphthalate) (PEN), and PC have a more polar character than the polyolefins and therefore show less affinity to the common flavor compounds; that is, they absorb fewer flavor compounds. A recent review (Linssen et al., 2003) concluded that although packaging and flavor interactions exist, they do not influence food quality to the extent that they cause insuperable problems in practical situations. This is evident from the fact that packaging materials in which polyolefins are in contact with juices are widely used commercially.

10.7 SHELF LIFE OF ORANGE JUICE IN DIFFERENT PACKAGES

From a packaging point of view, there are three categories of juices: single-strength juices (10–13°Brix), concentrated juices (42 or 65°Brix), and nectars (20–35°Brix). Refrigerated FSOJ has a relatively short shelf life of up to 14 days, on the basis of subjective flavor evaluation. The absence of pasteurization and lack of preservatives allow the growth of bacteria and yeasts, which together with enzyme activity cause off-flavors and oxidation. Staleness can be the primary off-flavor limiting shelf life at refrigerated storage temperatures. Frozen storage of FSOJ results in a longer shelf life than for refrigerated FSOJ. However, once thawed, the orange juice has a refrigerated shelf life of 7–10 days (Lee and Coates, 1999).

10.7.1 Metal Cans

The traditional packaging procedure for single-strength juices involved heating the deaerated juice to around 90–95°C in a tubular or plate heat exchanger, filling the hot juice directly into plain (i.e., unenameled or unlacquered) tinplated steel cans, sealing and inverting the cans, holding them for 10–20 min, and then cooling. This hot-fill/hold/cool process ensured that the juice was commercially sterile and, provided that the seams were of good quality and the juice had been properly deaerated, a shelf life of at least 1–2 years was attainable. Unenameled tinplated steel cans are used because traces of tin dissolve and provide a reducing environment that improves color stability. However, extended storage in such cans, particularly at temperatures over 30°C, must be avoided to prevent the development of off-flavors and excessive metal pickup and corrosion, which threatens can integrity (Hendrix and Redd, 1995). The use of glass containers obviated these problems provided that the container closure (typically metal) was enameled to minimize attack by the juice.

In the United States, the production of FCOJ has become a huge industry. The 42°Brix juice is usually held at −12°C, at which temperature it is still liquid. Typical packaging materials for this product consist of a spiral-wound paperboard tube with aluminum ends or an aluminum can.

10.7.2 Glass Bottles

The use of glass bottles for the packaging of fruit juices is also widespread, although the hot-fill/hold/cool process has to be applied with care to avoid breakage of the glass containers. Glass is still the preferred packaging medium for high-quality fruit juices (Siegmud et al., 2004).

The glass container is being replaced in some markets, and there is a growing tendency to abandon the standard forms and to introduce special forms with a variety of colors. Glassmakers are trying hard to highlight more than before the qualitative virtues of their packaging: inert, hygienic, versatile, hermetic, waterproof, and able to add prestige and image to the product. Faced with the challenge of other materials such as PET, the glass industry has responded by, among other things, developing a new generation of bottles that are lighter and more resilient. It is also carrying out finishing of bottles online. A special lacquer coating powder confers a high degree of protection to the outer surface. At the same time, this coating produces an attractive visual effect similar to frost, which can be carried out with varying intensities and in many colors.

Although it is technologically possible to fill juice into glass bottles aseptically, this packaging technology is not widely used.

10.7.3 Gable-Top Cartons

Gable-top cartons consist of paperboard coated on both sides with polyolefins; occasionally aluminum foil or EVOH may be incorporated into the structure to improve its O_2 barrier, but this is relatively uncommon. The cartons are prefabricated and delivered as blanks to the juice packing facility, where they are erected, filled, and sealed. Although the cartons are handled under nonsterile conditions, steps are taken to avoid recontamination. The filling temperature of the juice is typically 4–5°C to minimize microbial growth, although foaming can be a problem at this low temperature. The cartons are filled to leave a positively controlled headspace, and an inert gas such as N_2 can be injected immediately prior to sealing to remove O_2 from the headspace (Anon., 2004).

According to Wyatt et al. (1995), it is not unusual to observe mold spoilage in single-strength, chilled citrus juices packed in O_2 barrier gable-top cartons. These packages were designed to limit

O_2 permeation, thereby decreasing microbial growth rates and increasing product shelf life. The shelf life of chilled, retail orange juice almost doubled from 35 to 65 days because of this packaging change. Although O_2 permeation across the barrier was virtually eliminated, minor leaks of O_2 into the product can routinely occur along package seams. As neither the cartons nor the filling systems are aseptic, low-level contamination from air, packaging, and surrounding equipment, in combination with the longer shelf life, allow proliferation of molds, especially along package seams. As a result, citrus juice processors have become increasingly aware of filamentous fungi as potential juice spoilage agents. Pasteurized, single-strength orange juice is aseptically stored in bulk at citrus processing facilities for as long as a year at temperatures near 0°C. However, after packaging in O_2 barrier cartons, this juice will only have a shelf life of around 60 days in the retail market, at storage temperatures of 3–7°C. In this case the opportunistic contamination by filamentous fungi can result in the visible presence of fungal biomass in the product.

10.7.4 Aseptically Filled Laminated Cartons

Over the past 30 years an increasing proportion of fruit juices and concentrates have been packaged aseptically, generally into plastic/alufoil/paperboard laminated cartons. In laminated cartons, the aluminum foil is covered by polyolefin coatings (see Figure 10.4). The purpose of the foil is to serve as a barrier to light, O_2, odors, and aromas. These products are then held at room temperature, and the shelf life and nutrient composition are influenced by the interactions of the juice with the carton and by the storage temperature. The end of shelf life is typically at 4–6 months and is related to the extent of nonenzymic browning and the sorption of key aroma and flavor compounds by the plastic in contact with the juice. In a review of aseptically packaged orange juice and concentrate, Graumlich et al. (1986) reported that although aseptic processing produces a higher-quality orange juice than hot-filling, differences in quality may disappear during storage at ambient temperatures. Oxygen dissolved in the product, present in the container headspace, or permeating through the container accelerates the rate of ascorbic acid destruction and nonenzymic browning and reduces shelf life, although these processes will continue in its absence. The most important factor in determining the shelf life of aseptic orange juice and concentrate is the storage temperature.

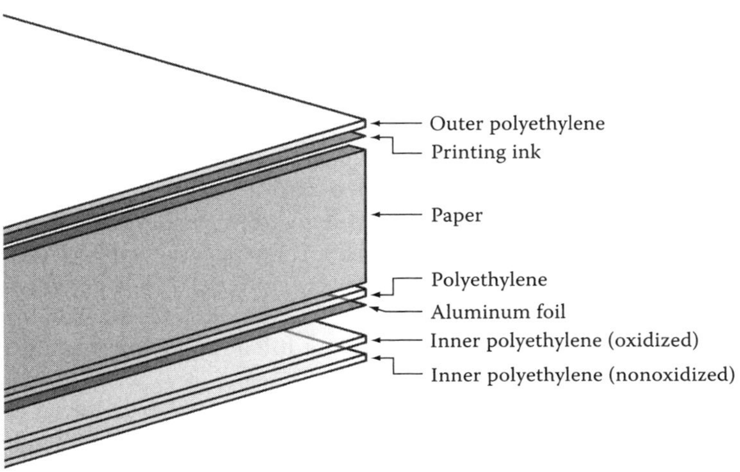

FIGURE 10.4 Typical structure of a paperboard laminate carton for aseptic filling. (From Robertson G.L. 2006. *Food Packaging Principles and Practice*, 2nd edn. Boca Raton, Florida: CRC Press, pp. 457–460, with permission.)

Data on the O_2 permeability of aseptic cartons is limited. Bourque (1985) reported oxygen transmission rates (OTRs) through the flat nonscored area of 1-L laminated cartons to be 30–40 mL m^{-2} day^{-1}; once the material was scored or in other ways flexed, the OTR increased to over 1500 mL m^{-2} day^{-1}, close to a 50-fold increase. The OTR of an empty, finished, sealed 250-mL carton was reported as >5 mL m^{-2} day^{-1}. In contrast, Ahrné et al. (1997) reported OTRs for a similar 1-L laminated carton (surface area not given) of 0.009 mL $pack^{-1}$ day^{-1} at 10°C, 0.014 at 20°C, 0.023 at 30°C, and 0.038 at 40°C, and they noted that O_2 permeation occurs mainly through the seam. Roig et al. (1994) reported OTRs for similar 200-mL laminated cartons of 0.2713 mL after 1 month at 18°C, which corresponded to 0.009 mL of O_2 $pack^{-1}$ day^{-1}. Alves et al. (2001) reported an OTR for a similar 250-mL carton with an inner surface area of 283 cm^2 (0.028 m^2) as <0.105 mL m^{-2} day^{-1} at 25°C in air, which corresponds to 0.003 mL $pack^{-1}$ day^{-1}. Assuming an inner surface area of 495 cm^2 (0.0495 m^2) for the 1-L cartons analyzed by Ahrné et al., the latter's results of 0.009 mL $pack^{-1}$ day^{-1} at 10°C correspond to an OTR of <0.18 mL m^{-2} day^{-1}, 0.28 at 20°C, 0.47 at 30°C, and 0.77 at 40°C.

Recognizing that the O_2 transmission performance of the finished laminated cartons is reasonably poor due to the destruction of the foil in the package manufacturing process, Bourque (1985) concluded that their relatively good shelf life performance can be attributed not to the barrier properties of the packaging material so much as to the lack of O_2 in the package as the result of no headspace. In contrast, Ahrné et al. (1997) concluded that the oxidative reactions in juice packed in laminated cartons were limited by the mass transfer through the package being high enough to maintain a residual O_2 concentration in the juice. The package permeability was much smaller by three orders of magnitude than the oxidative rate constant; the smaller the package, the greater the relative contribution of the seam to the total O_2 uptake.

Consumer concerns about possible migration of aluminum from laminated cartons into orange juice have been shown to be unfounded; analysis of juice stored from 12 hr to 1 year revealed no time-dependent changes in aluminum content (Rodushkin and Magnusson, 2005).

10.7.5 Plastics

The use of materials such as plastics for packaging has grown exponentially in the past few decades owing to their desirable properties, which include high clarity, good mechanical properties, good gas barrier properties, low weight, and ease of recycling (Ophir et al., 2004).

10.7.5.1 Flexible Plastics

Flexible plastic packaging is used for juices and two formats are common. The so-called Doy pack is a stand-up pouch constructed from inside to out using LDPE/alufoil/PET, with a drinking straw attached to the side of the pouch; the sharpened end of the straw is used to pierce a specially prepared area on the pouch. The Cheer pack was developed in Japan during the 1980s and is made up of four panels or sections combined to form a stand-up pack with two side gussets. A variety of laminate constructions are available, but for beverages the most common structure from inside to out is LDPE/PET/alufoil/PET. For specific applications, EVOH, OPA, or PP can be included in the structure. An high density polyethylene (HDPE) neck and "straw" are sealed into the top portion of the pack, which is filled through the neck and then sealed by a tamper evident closure. The packs can be cold- or hot-filled (up to 95°C) and pasteurized after filling if required (Tacchella, 1999).

The stability of fruit juice drinks in aseptic packages constructed from linear low density polyethylene (LLDPE) with either EVOH or PVdC copolymer barrier layers has been investigated (Alves et al., 2001). The OTRs for the films in mL m^{-2} day^{-1} were reported as 1.40 and 2.96 for the films containing EVOH and 13.74 for those with PVdC; the total inner surface area of the 250-mL plastic packs was 309 cm^2. The performance of the packages with EVOH was virtually equivalent to that of the carton packs throughout the storage period studied (90 days).

10.7.5.2 Rigid Plastics

10.7.5.2.1 High Density Polyethylene

Extrusion-blow-molded HDPE bottles have been used for many years to package orange juice. As HDPE is a poor barrier to O_2, such bottles can be used only for chilled juices with a shelf life of up to 3 weeks. The barrier properties can be improved by incorporating a layer of EVOH copolymer or polyamide, permitting shelf lives of up to 6 months at ambient temperatures, depending on the choice and thickness of the barrier layer (Anon., 2004).

Fellers (1988) stored unpasteurized FSOJs in 0.946-L HDPE bottles at temperatures ranging from –1.7°C to 7.8°C. Staleness was the primary off-flavor, limiting shelf life at temperatures of 4.4°C or less, whereas spoilage with diacetyl was primarily responsible at 7.8°C. Microbial counts generally decreased markedly during storage at 4.4°C or less, whereas at 7.8°C an increase was generally noted. Ascorbic acid retention after 2 weeks of storage at 4.4°C or lower was about 91–93%. On the basis of results from experienced panelists, the shelf life ranged from 20–23 days at –1.7°C to 5–8 days at 7.8°C.

Since the early part of this century, orange juice in bag-in-box packaging has successfully used flexible bags of different compositions. This kind of packaging system offers significant cost savings, environmental compliance, product line diversification, packaging differentiation, improved brand recognition, and end-user satisfaction. For example, 3- to 5-L bags offered by Scholle (www.scholle.com; www.boxedjuice.com) for packaging orange juice have a multilayered composition of LLDPE/EVOH/LLDPE or LDPE/MetPET/LLDPE, as can be seen in Figure 10.5, with corresponding OTRs of less than 1.5 and 0.2 mL O_2 m^{-2} day^{-1}, respectively, at 1 atm, 23°C, and 75% RH. According to the data given by the manufacturers, this results in an O_2 ingress of 1.7–4.3 mL L^{-1} in 6 months, which is several times the target maximum O_2 ingress of 0.7 mL L^{-1} in 6 months for an O_2-sensitive product such as orange juice (Brooks, 2002).

10.7.5.2.2 Poly(ethylene Terephthalate)

Since the 1970s PET bottles have increasingly replaced glass as the packaging of carbonated beverages. However, the O_2 barrier properties of PET are insufficient to give a satisfactory shelf life unless the product is kept at chill temperatures. Recent developments in barrier coatings for PET have led to increasing use of PET bottles for fruit juices, and this trend is likely to accelerate as

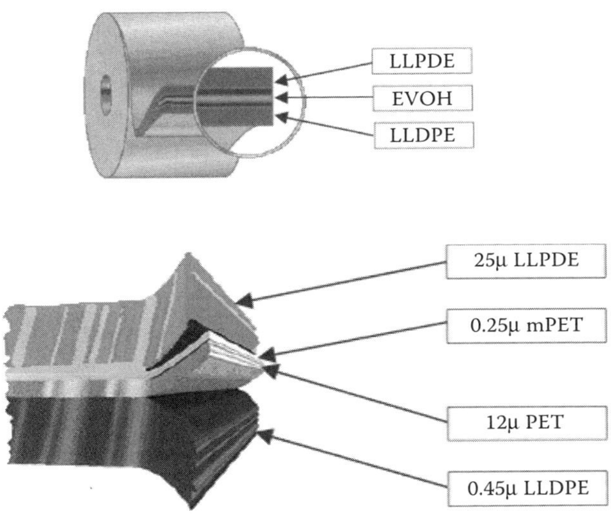

FIGURE 10.5 Different multilayer wall solutions for bag-in-box packaging for orange juice. mPET is metalized PET. (From www.scholle.com, with permission.)

production ramps up and costs come down. Until recently there has been a dearth of independent scientific publications on the performance of the various barrier coatings with respect to gas transfer and shelf life.

Aseptically filled orange juice in multilayer PET bottles has a shelf life at 23°C of 6–12 months (Rodushkin and Magnusson, 2005), whereas ultraclean filled orange juice in monolayer PET bottles and multilayer PP/EVOH/PP thermoformed cups has a shelf life of 30–45 days at 4°C.

According to Ros-Chumillas et al. (2007), orange juice aseptically packaged in monolayer PET bottles has a poor retention of ascorbic acid, and the shelf life is shorter than for juice bottled in glass or a multilayer PET. However, the PET bottling factors considered in their study had an additive effect on ascorbic acid retention such that the shelf life can be extended to that provided by glass and a multilayer PET. If orange juice is packaged in monolayer PET bottles containing an O_2 scavenger, with the addition of a drop of liquid N_2 in the headspace and an aluminum foil seal in the screw cap, the shelf life may exceed 9 months at 4°C and be nearly 8 months at 25°C. Both values are much higher than those actually demanded by the market for juice aseptically packed in glass bottles at 25°C, for which a shelf life of 180 days has been established (Ros-Chumillas et al., 2007). These results are similar to those obtained by Berlinet et al. (2008), who compared orange juice packaged in a monolayer PET with that packaged in a multilayer PET. A multilayer PET with improved O_2 barrier properties showed better ascorbic acid contents and color in orange juice during 9 months of storage.

Berlinet et al. (2005) evaluated three different 330-mL commercial PET bottles: a standard monolayer PET (PET1), a multilayer PET containing an O_2 scavenger and complexed with nylon MXD6 (PET2), and a plasma-treated (internal carbon coating) PET (PET3). The O_2 permeabilities of the PET bottles were 63.21, 5.77, and 5.59 × 10^{-14} mL (STP) cm cm^{-2} s^{-1} (cm Hg)$^{-1}$ for PET1, PET2, and PET3, respectively. Glass bottles (500 mL) were used as the reference packaging. All the bottles were sealed with aluminum foils after filling and the headspace volumes were 20 mL for the PET bottles and 30 mL for the glass bottles. All bottles were stored at 20°C under artificial light. Only limonene and β-myrcene were absorbed, at very low levels, after 5 months of storage, indicating that PET is a satisfactory packaging material to limit flavor absorption from orange juice during long-term storage.

It was also found that the aromatic composition of the stored orange juice samples was controlled by the duration of storage, and not by the packaging material and its O_2 permeability. The levels of volatile components making a positive contribution to orange juice flavor, such as ethyl butanoate, hexanal, octanal, nonanal, and decanal, fell by more than 50%, whereas those of furfural, α-terpineol, β-terpineol, and 4-vinylguaiacol increased during 5 months of storage at 20°C, which could be largely explained by acid-catalyzed reactions within the matrix itself.

Using the same packaging materials, the authors later reported (Berlinet et al., 2006) on ascorbic acid retention in orange juice made from concentrate stored for 9 months at 20°C under artificial light (Figure 10.6). After 9 months of storage, the ascorbic acid contents in orange juice were 310 mg L^{-1} (glass), 132 mg L^{-1} (PET1), 255 mg L^{-1} (PET2), and 230 mg L^{-1} (PET3), respectively; for orange juice, 200 mg L^{-1} ascorbic acid must be guaranteed until the end of the shelf life in the European Union (AIJN, 2008). Thus, if PET1 is used, the ascorbic acid content is lower than the required value after a 9-month storage period. As a consequence, in an industrial setting, the use of a barrier PET technology coupled with juice degassing and headspace nitrogen filling could be a good combination to maintain the ascorbic acid content at the highest possible level. Nevertheless, the PET barrier technologies presented here were not as efficient as glass. Moreover, the increase in O_2 permeability of PET over time would also have to be taken into account. During a 6-month storage period, PET1 O_2 permeability decreased from 63.21 to 52.04 × 10^{-14} mL (STP) cm cm^{-2} s^{-1} (cm Hg)$^{-1}$ and PET2 O_2 permeability remained constant, whereas for PET3 it increased from 5.49 to 12.66 × 10^{-14} mL (STP) cm cm^{-2} s^{-1} (cm Hg)$^{-1}$. The behavior of PET3 was attributed to a possible degradation of the plasma layer during long-term storage.

In a later study, Berlinet et al. (2008) investigated the loss of aroma compounds from orange juice by permeation through the bottle (PET1 and PET2) and the cap. The results showed that permeation

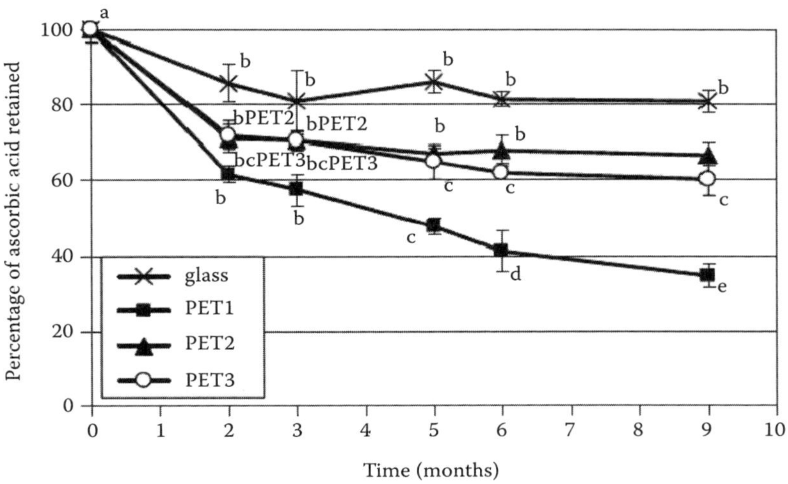

FIGURE 10.6 Percentage of ascorbic acid retained (means ± SD, $n = 3$) in an orange juice from concentrate stored during 9 months at 20°C under artificial light in either glass, PET1, PET2, or PET3. Different letters in the same curve indicate significant differences at $p < 0.05$ (Duncan). (From Berlinet C., Brat P., Brillouet J.-M., Ducruet V. 2006. Ascorbic acid, aroma compounds and browning of orange juices related to PET packaging materials and pH. *Journal of the Science of Food and Agriculture* 86: 2206–2212, with permission.)

mainly took place through the cap. The use of an HDPE multilayer cap with an internal barrier layer of LDPE/EVOH/LDPE considerably limited the permeation of aroma compounds, regardless of which PET bottle was used.

Oxygen scavenger films that effectively reduce the O_2 dissolved in orange juice or initially present in the headspace have been developed on a laboratory scale (Zerdin et al., 2003). The loss of ascorbic acid correlated with an increase in the browning of the juice, with the extent of browning being lower for the juice packed in film containing an O_2 scavenger. The rapid removal of O_2 was found to be an important factor in retaining a higher concentration of ascorbic acid over long storage times.

REFERENCES

Ahrné L.M., Oliveira F.A.R., Manso M.C., Drummond M.C., Öste R., Gekas V. 1997. Modelling dissolved oxygen concentration during storage of packaged liquid foods. *Journal of Food Engineering* 34: 213–224.

AIJN (European Fruit Juice Association). 2008. AIJN Code of Practice for Evaluation of Fruit and Vegetable Juices. http:/www.aijn.org/ (accessed December 18, 2008).

Alves R.M.V., Sarantópoulos C.I.G.L., Saron E.S., Bordin M.R. 2001. Stability of fruit juice drinks in aseptic packages. *Packaging Technology and Science* 14: 79–86.

Anon. 2004. *The Orange Book*, 2nd edn. Lund, Sweden: Tetra Pak Processing Systems AB.

Ashurst P.R. (Ed.). 2005. *Chemistry and Technology of Soft Drinks and Fruit Juices*, 2nd edn. Ames, Iowa: Blackwell Publishing.

Ayhan Z., Yeom H.W., Zhang Q.H., Min D.B. 2001. Flavor, color, and vitamin C retention of pulsed electric field processed orange juice in different packaging materials. *Journal of Agricultural and Food Chemistry* 49: 669–674.

Baker R.A., Cameron R.G. 1999. Clouds and citrus juices and juice drinks. *Food Technology* 53: 64–69.

Barrett D.M., Somogyi L., Ramaswamy H. (Eds). 2005. *Processing Fruits: Science and Technology*, 2nd edn. Boca Raton, Florida: CRC Press.

Basak K., Ramaswamy H.S. 1996. Ultra high pressure treatment of orange juice: a kinetic study on inactivation of pectin methylesterase. *Food Research International* 26: 601–607.

Bates R.P., Morris J.R., Crandall P.G. 2001. Principles and practices of small- and medium-scale fruit juice processing. *FAO Agricultural Services Bulletin* 146: 59–93.

Berlinet C., Ducruet V., Brillouet J.-M., Reynes M., Brat P. 2005. Evolution of aroma compounds from orange juice stored in polyethylene terephthalate (PET). *Food Additives and Contaminants* 22: 185–195.

Berlinet C., Brat P., Brillouet J.-M., Ducruet V. 2006. Ascorbic acid, aroma compounds and browning of orange juices related to PET packaging materials and pH. *Journal of the Science of Food and Agriculture* 86: 2206–2212.

Berlinet C., Brat P., Ducruet V. 2008. Quality of orange juice in barrier packaging material. *Packaging Technology and Science* 21: 279–286.

Bourque R.A. 1985. Shelf life of fruit juices in oxygen permeable packages. *Journal of Plastic Film and Sheeting* 1: 115–121.

Braddock R.J. 1999. *Handbook of Citrus By-Products and Processing Technology*. New York: John Wiley and Sons.

Brooks D.W. 2002. Barrier materials and technology. In: *PET Packaging Technologies*. Brooks D.W., Giles G.A. (Eds). Boca Raton, Florida: CRC Press, pp. 98–115.

Bruemmer J.H., Roe B., Bowen E.R. 1976. Peroxidase reactions and orange juice quality. *Journal of Food Science* 41: 186–189.

Caggia C., Scifò G.O., Restuccia C., Randazzo C.L. 2009. Growth of acid-adapted *Listeria monocytogenes* in orange juice and in minimally processed orange slices. *Food Control* 20: 59–66.

Cameron R.G., Baker R.A., Grohmann K. 1997. Citrus tissue extracts affect juice cloud stability. *Journal of Food Science* 62: 242–245.

Castberg H.B., Osmundsen J.I., Solberg, P. 1995. Packaging systems for fruit juices and non-carbonated beverages. In: *Production and Packaging of Non-Carbonated Fruit Juices and Fruit Beverages*, 2nd edn. Ashurst P.R. (Ed). London, England: Blackie Academic & Professional, pp. 291–309.

Chen C.S., Shaw P.E., Parish M.E. 1993. Orange and tangerine juices. In: *Fruit Juice Processing Technology*. Nagy S., Chen C.S., Shaw P.E. (Eds). Auburndale, Florida: Agscience, pp. 110–165.

Ebbesen A. 1998. Effect of temperature, oxygen and packaging material on orange juice quality during storage. *Fruit Processing* 11: 446–455.

Eiroa M.N.U., Junqueira V.C.A., Schmidt F.L. 1999. *Alicyclobacillus* in orange juice: occurrence and heat resistance of spores. *Journal of Food Protection* 62: 883–886.

Esteve M.J., Frígola A., Rodrigo C., Rodrigo D. 2005. Effect of storage period under variable conditions on the chemical and physical composition and colour of Spanish refrigerated orange juices. *Food and Chemical Toxicology* 43: 1413–1422.

Esteve M.J., Frígola A. 2007. Refrigerated fruit juices: quality and safety issues. *Advances in Food and Nutrition Research* 52: 103–139.

Farnworth E.R., Lagacé M., Couture R., Yaylayan V., Stewart B. 2001. Thermal processing, storage conditions and the composition and physical properties of orange juice. *Food Research International* 34: 25–30.

Fellers P.J. 1988. Shelf life and quality of freshly squeezed, unpasteurized, polyethylene-bottled citrus juice. *Journal of Food Science* 53: 1699–1702.

Graumlich T.R., Marcy J.E., Adams J.P. 1986. Aseptically packaged orange juice and concentrate: a review of the influence of processing and packaging conditions on quality. *Journal of Agricultural and Food Chemistry* 34: 402–405.

Hendrix C.M., Redd J.B. 1995. Chemistry and technology of citrus juices and by-products. In: *Production and Packaging of Non-Carbonated Fruit Juices and Fruit Beverages*, 2nd edn. Ashurst P.R. (Ed). London, England: Blackie Academic & Professional, chapter 2.

Hirose K., Harte B.R., Giacin J.R., Miltz J., Stine C. 1988. Sorption of *d*-limonene by sealant films and effect on mechanical properties. In: *Food and Packaging Interactions*. Hotchkiss J.H. (Ed). ACS Symposium Series #365. Washington DC: American Chemical Society, chapter 3.

Hirsch A.R., Forch K., Neidhart S., Wolf G., Carle R. 2008. Effects of thermal treatments and storage on pectin methylesterase and peroxidase activity in freshly squeezed orange juice. *Journal of Agricultural and Food Chemistry* 56: 5691–5699.

Imai T., Harte B.R., Giacin J.R. 1990. Partition distribution of aroma volatiles from orange juice into selected polymeric sealant films. *Journal of Food Science* 55: 158–161.

Ingallinera B., Barbagallo R.N., Spagna G., Palmeri R., Todazo A. 2005. Effects of thermal treatments on pectinesterase activity determined in blood oranges juices. *Enzyme and Microbial Technology* 3: 258–263.

Jordán M.J., Goodner K.L., Laencina J. 2003. Deaeration and pasteurization effects on the orange juice aromatic fraction. *LWT—Food Science and Technology* 36: 391–396.

Kabasakalis V., Siopidou E., Moshatou E. 2000. Ascorbic acid content of commercial fruit juices and its rate of loss upon storage. *Food Chemistry* 70: 325–328.

Klavons J.A., Bennett R.D., Vannier S.H. 1991. Nature of the protein constituent of commercial orange juice cloud. *Journal of Agricultural and Food Chemistry* 39: 1546–1548.

Lee H.S., Coates G.A. 1999. Vitamin C in frozen, fresh squeezed, unpasteurized, polyethylene-bottled orange juice: a storage study. *Food Chemistry* 65: 165–168.

Lee H.S., Coates G.A. 2003. Effect of thermal pasteurisation on Valencia orange juice color and pigments. *LWT—Food Science and Technology* 36: 153–156.

Lewis M., Heppell N. 2000. *Continuous Thermal Processing of Foods*. Maryland: Aspen Publishers, Inc.

Linssen J.P.H., van Willige R.W.G., Dekker M. 2003. Packaging-flavour interactions. In: *Novel Food Packaging Techniques*. Ahvenainen R. (Ed). Boca Raton, Florida: CRC Press, chapter 8.

López-Gómez A., Barbosa-Cánovas G.V. 2005. *Food Plant Design*. Boca Raton, Florida: CRC Press.

Loureiro V., Querol A. 1999. The prevalence and control of spoilage yeasts in foods and beverages. *Trends in Food Science & Technology* 10: 356–365.

Mannheim C.H., Miltz J., Letzter A. 1987. Interaction between polyethylene laminated cartons and aseptically packed citrus juices. *Journal of Food Science* 52: 737–740.

Mannheim C.H., Miltz J., Passy N. 1988. Interaction between aseptically filled citrus products and laminated structures. In: *Food and Packaging Interactions*. Hotchkiss J.H. (Ed). ACS Symposium Series #365. Washington DC: American Chemical Society, chapter 6.

Meléndez-Martínez A.J., Britton G., Vicario I.M., Heredia F.J. 2005. Color and carotenoid profile of Spanish Valencia late ultrafrozen orange juices. *Food Research International* 38: 931–936.

Moshonas M.G., Shaw P.E. 1997. Flavor and chemical comparison of pasteurized and fresh Valencia orange juices. *Journal of Food Quality* 20: 31–40.

Nielsen T.J., Jägerstad M.I., Oste R.E. 1992. Study of factors affecting the absorption of aroma compounds into low-density polyethylene. *Journal of the Science of Food and Agriculture* 60: 377–381.

Nielsen T.J. 1994. Limonene and myrcene sorption into refillable polyethylene terephthalate bottles, and washing effects on removal of sorbed compounds. *Journal of Food Science* 59: 227–230.

Nisperos-Carriedo M.O., Shaw P.E. 1990. Comparison of volatile flavor components in fresh and processed orange juices. *Journal of Agricultural and Food Chemistry* 38: 1048–1052.

Ophir A., Kenig S., Shai A., Barka'ai Y., Miltz J. 2004. Hot-fillable containers containing PET/PEN copolymers and blends. *Polymer Engineering and Science* 44: 1670–1675.

Perez-Cacho P.R., Rouseff R. 2008. Processing and storage effects on orange juice aroma: a review. *Journal of Agricultural and Food Chemistry* 56: 9785–9796.

Pérez-López A.J., Carbonell-Barrachina A.A. 2006. Volatile odour components and sensory quality of fresh and processed mandarin juices. *Journal of the Science of Food and Agriculture* 86: 2404–2411.

Pieper G., Borgudd L., Ackermann P., Fellers P. 1992. Absorption of aroma compounds of orange juice into laminated carton packages did not affect sensory quality. *Journal of Food Science* 57: 1408–1411.

Pollack S.L., Lin B-H., Allshouse J. 2003. *Characteristics of US orange consumption*. Electronic Outlook Report from the Economic Research Service (www.ers.usda.gov). FTS 305-01, August 2003, pp. 1–17.

Polydera A.C., Stoforos N.G., Taoukis P.S. 2003. Comparative shelf life study and vitamin C loss kinetics in pasteurised and high pressure processed reconstituted orange juice. *Journal of Food Engineering* 60: 21–29.

Robertson G.L. 2006. *Food Packaging Principles and Practice*, 2nd edn. Boca Raton, Florida: CRC Press, pp. 457–460.

Rodushkin I., Magnusson A. 2005. Aluminium migration to orange juice in laminated paperboard packages. *Journal of Food Composition and Analysis* 18: 365–374.

Roig M.G., Bello J.F., Rivera Z.S., Kennedy J.F. 1994. Possible additives for extension of shelf-life of single-strength reconstituted citrus juice aseptically packaged in laminated cartons. *International Journal of Food Sciences and Nutrition* 45: 15–29.

Roig M.G., Bello J.F., Rivera Z.S., Kennedy J.F. 1999. Studies on the occurrence of non-enzymatic browning during storage of citrus juice. *Food Research International* 32: 609–619.

Ros-Chumillas M., Belisario Y., Iguaz A., López-Gómez A. 2007. Quality and shelf life of orange juice aseptically packaged in PET bottles. *Journal of Food Engineering* 79: 234–242.

Sajilata M.G., Savitha K., Singhal R.S., Kanetkar V.R. 2007. Scalping of flavors in packaged foods. *Comprehensive Reviews in Food Science and Food Safety* 6: 17–35.

Sánchez-Moreno C., Plaza L., de Ancos B., Cano P. 2006. Nutritional characterisation of commercial traditional pasteurised tomato juices: carotenoids, vitamin C and radical-scavenging capacity. *Food Chemistry* 98: 749–756.

Schöttler P., Pecoroni S., Günnerwig W. 2002. *Separators, Decanters and Process Line for Citrus Processing.* Technical Scientific Doc. No. 14. 2nd edn. Germany: Westfalia Separator AG, GEA.

Siegmud B., Derler K., Pfannhauser W. 2004. Chemical and sensory effects of glass and laminated carton packages on fruit juice products—still a controversial topic. *LWT—Food Science and Technology* 37: 481–488.

Sizer C.E., Waugh P.L., Edstam S., Ackerman P. 1988. Maintaining flavor and nutrient quality of aseptic orange juice. *Food Technology* 42(6): 152–159.

Sizer C.E., Balasubramaniam V.M. 1999. New intervention processes for minimally processed juices. *Food Technology* 53(9): 64–68.

Soares N.F.F., Hotchkiss J.H. 1999. Comparative effects of de-aeration and package permeability on ascorbic acid loss in refrigerated orange juice. *Packaging Technology and Science* 12: 111–118.

Solomon O., Svanberg U., Sahlström A. 1995. Effect of oxygen and fluorescent light on the quality of orange juice during storage at 8°C. *Food Chemistry* 53: 363–368.

Tachella A. 1999. Packaging of beverages in foil pouches. In: *Handbook of Beverage Packaging*. Giles G.A. (Ed). Boca Raton, Florida: CRC Press, pp. 165–183.

Tawfik M.S., Huyghebaert A. 1998. Effect of storage temperature, time, dissolved oxygen and packaging materials on the quality of aseptically filled orange juice. *Acta Alimentaria*. 27: 231–244.

Tawfik M.S., Devlieghere F., Huyghebaert A. 1998. Influence of *d*-limonene absorption on the physical properties of refillable PET. *Food Chemistry* 61: 157–162.

Tekkanat B. 2002. Hot-fill, heat set, pasteurization and retort technologies. In: *PET Packaging Technologies*. Brooks D.W., Giles G.A. (Eds). Boca Raton, Florida: CRC Press, pp. 292–314.

Temple N.J. 2000. Antioxidants and disease: more questions than answers. *Nutrition Research* 20: 449–459.

Tepper B.J. 1993. Effects of a slight color variation on consumer acceptance of orange juice. *Journal of Sensory Studies* 8: 145–154.

Tomás-Barberán F.A., Espín J.C. 2001. Phenolic compounds and related enzymes as determinants of quality in fruits and vegetables. *Journal of the Science of Food and Agriculture* 81: 853–876.

Torregrosa F., Esteve M.J., Frígola A., Cortés C. 2006. Ascorbic acid stability during refrigerated storage of orange-carrot juice treated by high pulsed electric field and comparison with pasteurized juice. *Journal of Food Engineering* 73: 339–345.

UNCTAD (United Nations Conference on Trade And Development). 2008. http://www.unctad.org/infocomm/anglais/orange/market.htm (accessed December 18, 2008).

USDA. 2008a. *World markets and trade: Orange Juice*. United States Department of Agriculture. Foreign Agricultural Service, Office of Global Analysis, April, 2008.

USDA. 2008b. *Citrus market update: European Union—27*. Foreign Agricultural Service, Office of Global Analysis, April, 2008.

Varsel C. 1980. Citrus juice processing as related to quality and nutrition. In: *Citrus Nutrition and Quality.* Nagy S., Attaway J.A. (Eds), ACS Symposium Series #143. Washington DC: American Chemical Society, pp. 225–271.

Walker M., Phillips C.A. 2008. *Alicyclobacillus acidoterrestris*: an increasing threat to the fruit juice industry? *International Journal of Food Science and Technology* 43: 250–260.

Wyatt M.K., Parish M.E., Widmer W.W., Kimbrough J. 1995. Characterization of mould growth in orange juice. *Food Microbiology* 12: 347–355.

Yamada K., Mita K., Yoshida K., Ishitani T. 1992. A study of the absorption of fruit juice volatiles by the sealant layer in flexible packaging containers. *Packaging Technology and Science* 5: 41–47.

Zerdin K., Rooney M.L., Vermuë J. 2003. The vitamin C content of orange juice packed in an oxygen scavenger material. *Food Chemistry* 82: 387–395.

11 Packaging and the Shelf Life of Coffee

Maria Cristina Nicoli, Lara Manzocco, and Sonia Calligaris
Department of Food Science
University of Udine
Udine, Italy

CONTENTS

11.1 COFFEE QUALITY ATTRIBUTES

The word "coffee" is a general term that comprises roasted coffee (including decaffeinated coffee) and derived beverages, as well as a wide variety of convenience and semimanufactured products such as instant coffee and coffee concentrates. Thus, this term is synonymous with coffee products and encompasses many technological processes responsible for the great compositional complexity of the derived products.

Figure 11.1 shows the key technological steps in obtaining the most important coffee products. Conversion of green coffee beans into a beverage involves three main operations: roasting, grinding, and extraction. Roasting is the key technological operation to transform a green bean, characterized by poor or even unpleasant sensory properties, to the dark one with typical flavor and aroma. Conceptually, roasting is the unit operation in which coffee beans are put in contact with hot surfaces or gases to raise their temperature to 220°C. Exhaustive descriptions of the roasting operation have been given by Clarke (1987a) and Eggers et al. (2001). The complex chemical and physical changes that occur during roasting convert green beans into a very unstable and reactive system. The main reactions occurring in coffee beans during roasting are water removal and carbohydrate

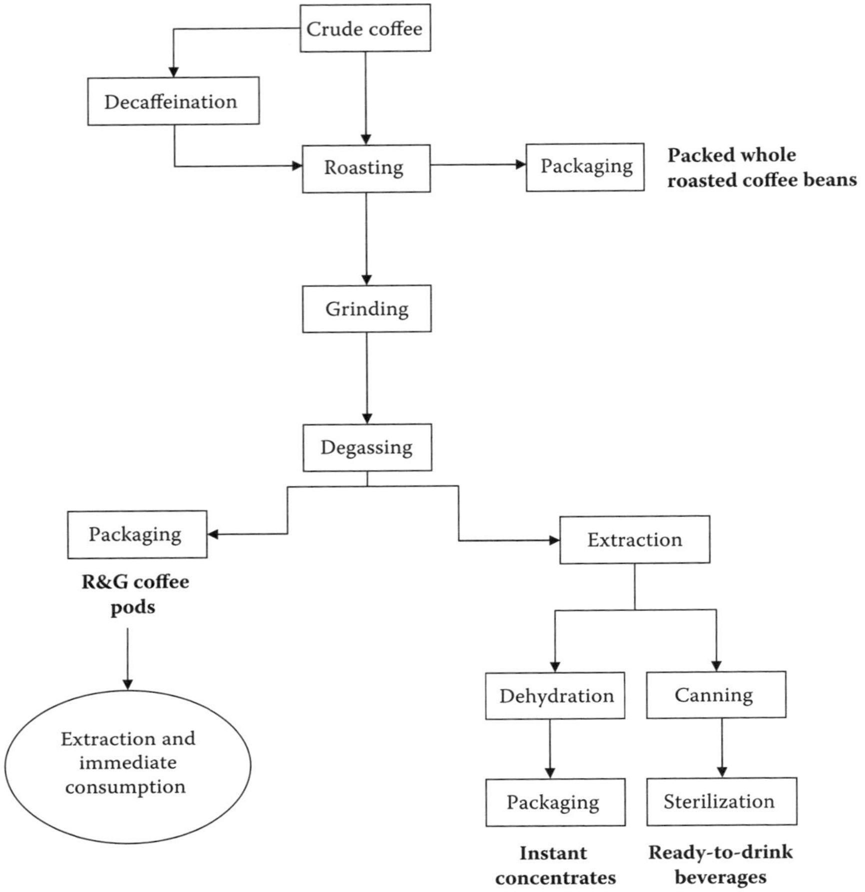

FIGURE 11.1 Key technological steps to obtain the most important coffee products present in the market.

fragmentation and polymerization, via nonenzymic browning (NEB) reactions, and pyrolysis. The chemical composition of the beans is drastically modified, with release of large amounts of CO_2 and the formation of hundreds of substances associated with coffee aroma and taste. There are a large number of volatiles and nonvolatiles such as melanoidins and their precursors among the reaction products. A list of volatile compounds deemed to be of particular importance for coffee flavor is provided by Czerny et al. (1999) and Clarke (2001). Melanoidins are responsible for bean color changes from green-yellow to brown-black. Detailed insight into the roasting chemistry is given by Trugo (1985), Macrae (1985), Clifford (1985), Da Porto et al. (1991), and Homma (2001).

The formation of volatiles and CO_2 during roasting causes the expansion of the beans due to internal buildup of gases, which, along with the high temperatures, allows internal pores and pockets to be formed. However, as the beans become very brittle, they progressively lose their ability to entrap and retain volatiles, which are partially released during storage. For this reason, a degassing step is carried out on ground coffee before packaging in order to avoid the swelling of the packages during storage.

Grinding is the operation that converts whole roasted beans into smaller fragments in order to increase the specific extraction surface area and thus facilitate the transfer of soluble and emulsifiable substances from the coffee matrix to water (Petracco, 2005a, 2005b). Additional technological operations include caffeine removal, which is generally carried out on green coffee beans prior to roasting in order to minimize flavor and aroma losses. Decaffeination procedures include (a) decaffeination with organic solvents such as methylene chloride or ethyl acetate, (b) water decaffeination based on the use of green coffee extract with equilibrium quantities of noncaffeine solubles, and (c) supercritical carbon dioxide extraction (Heilmann, 2001).

Brewing, or, more correctly, solid-liquid extraction, is the operation in which roasted and ground (R&G) coffee is mixed with hot water to give a beverage that may be consumed as such or may constitute a semimanufactured product. Coffee brews prepared at the domestic and catering level are generally consumed immediately after preparation. The extraction methods adopted can vary greatly from country to country, strongly affecting the sensory properties of the cup of coffee (Petracco, 2001, 2005a, 2005b; Pictet, 1987).

Although, in term of volumes sold, roasted whole and ground coffee are the main coffee products present in the market, instant coffee and, more recently, coffee concentrates and ready-to-drink coffee brews represent an interesting and dynamic area. In the past 15 years, the production and sale of instant coffee has increased markedly in most countries, with a noticeable increase in the number of different brands. The success of these products is largely because they come in an attractive and convenient physical form and have a very long shelf life. At present, instant coffee and instant coffee-based products include spray-dried or freeze-dried powder or granules. Instant coffee technology includes extraction and further dehydration carried out by freeze drying or thermal concentration (i.e., spray drying). A comprehensive description of the process and equipment involved in instant coffee manufacture is given by Clarke (2001).

Coffee concentrate technology includes solid-liquid extraction and subsequent partial water removal through thermal or cryo-concentration. Partial dehydration results in concentrates with a solids content of up to 22% and a water activity (a_w) of up to 0.84 (Manzocco and Nicoli, 2007). These products are used as ingredients for the food industry or as semimanufactured products for vending machines and in catering. More recently, ready-to-drink coffee beverages have become very popular, especially in Asian countries, where the tradition of consuming freshly prepared coffee brews is not widespread. They are generally pasteurized and chilled or sterilized and stored at ambient temperature. They include the so-called coffee drinks obtained by using 5 g or more of R&G coffee per cup/can and the soft coffee drinks based on coffee drinks with added sugar, dairy products, emulsifiers, and other additional ingredients (Petracco, 2001).

11.2 DETERIORATIVE REACTIONS AND INDICES OF FAILURE

As previously mentioned, coffee is a term that encompasses a wide number of products that undergo different deteriorative reactions, have different sensitivities to environmental variables, and thus have different shelf lives. For this reason, stability problems relevant to roasted, instant, and liquid coffees will be discussed separately. However, a common element in all these coffee products is their compositional complexity. At the moment, the structures of the majority of the compounds present in roasted coffee, most of which end up suspended or solubilized in the coffee beverage, are unknown. This is the reason for the still inadequate scientific knowledge on the causes and variables affecting coffee product quality loss during storage. In general terms, it is well known that coffee products lose their peculiar sensory properties during storage, but in most cases, very little is known about the chemical and physical events underlying these undesirable changes.

11.2.1 Roasted Coffee

Although generally recognized as a shelf-stable product, roasted coffee undergoes important chemical and physical changes that greatly affect quality and acceptability of the coffee brew. From a sensory point of view, coffee staleness is defined as "a sweet but unpleasant flavor and aroma of roasted coffee which reflects the oxidation of many of the pleasant volatiles and the loss of others"

(Buffo et al., 2004). The most important physical and chemical events involved in roasted coffee staling during storage are as follows:

- Volatile release
- Carbon dioxide release
- Surface oil migration
- Oxidation reactions

Volatiles and CO_2 are easily released into the vapor phase due to diffusion mechanisms enhanced by the fact that the pressure inside the bean pores is greater than atmospheric. The evolution of CO_2 is of particular importance because, although it can be considered an index of the aroma richness of roasted coffee, if it is released after coffee packaging it produces overpressure inside the package, with possible bursting and loss of package integrity. The presence of a pressure gradient between bean pores and outside is also responsible for the migration of lipids, together with their lipophilic solutes, onto the bean surface. As a consequence of the increased exposed surface area, these compounds become more prone to oxidation. Volatile and CO_2 release and oxidation reactions are considered the main causes of coffee staling.

The sensitivity of coffee to oxidation reactions originates from the roasting stage. The majority of compounds formed via NEB during roasting are strong antioxidants able to react both as chain breakers and as oxygen scavengers (Nicoli et al., 1997). It is interesting to note that just after roasting, coffee has a remarkably negative redox potential value, indicating its strong reducing properties (Nicoli et al., 2004). For this reason oxygen availability causes the oxidation of a wide number of coffee compounds. The development of the typical stale flavor of aged coffee is the result of two different reactions: the release of volatiles in the vapor phase and the development of off-flavor as the result of oxidation of some of them (Nicoli and Savonitto, 2005). Steinhardt and Holscher (1991) found that the aroma freshness of whole roasted coffee was mainly determined by certain low-boiling-point components, namely low molecular weight sulfur compounds, Strecker aldehydes, and α-dicarbonyls. Aroma staling is caused by the loss of most of them. The heavier compounds, such as furfuryl mercaptan, remain in the coffee, causing the "aging" note. As coffee is a dry product with a_w values ranging from 0.1 to 0.4, an additional cause of coffee staling is the sorption of moisture. As the moisture content and hence a_w increases, the roasted coffee bean, which is initially brittle and easy to grind, becomes very elastic due to a decrease in the glass transition temperature T_g (Figure 11.2) (Anese et al., 2005; Pittia, 2007). These changes also have a great impact on volatile retention, as shown in Figure 11.3.

Although many of these changes are considered unavoidable, the rate at which they occur mostly depends on the exposed surface area of the coffee (i.e., whether whole or ground) and on environmental variables such as oxygen partial pressure and relative humidity. The greater the surface area exposed, the higher is the degradation rate. This is particularly evident for ground coffee contained in paper pods. Obviously, the influence of these variables is strongly dependent on the packaging solutions adopted. An additional environmental variable that can strongly affect coffee staling is temperature. The temperature dependence of the most important degradation reactions that take place in roasted coffee during storage can be well described by the Arrhenius equation (Manzocco and Nicoli, 2007).

Typical indices of failure (IoFs) for roasted coffee can be the headspace volatiles or CO_2 concentration (Anese et al., 2006; Nicoli and Savonitto, 2005). Among the volatiles, some specific indicators of coffee aroma freshness have been selected: (a) the M/B aroma index, the ratio between methylfuran and 2-butanone (Amstalden, 2001); (b) the flavor quality index, based on five key odorants (hexanal, vinylpyrazine, pyrrol, furfurylmethylketone, and pyridine), which shows an inverse linear relationship with the M/B index and has been used to follow coffee staling in industrial settings; and (c) the MM aroma index, the ratio of methanol to 2-methylfuran (Amstalden, 2001; Buffo et al., 2004; Holsher and Steinhartd, 1992). Steinhardt and Holscher (1991) suggested that the loss of coffee aroma freshness is due to the loss of certain aroma volatiles (mainly methyl mercaptan),

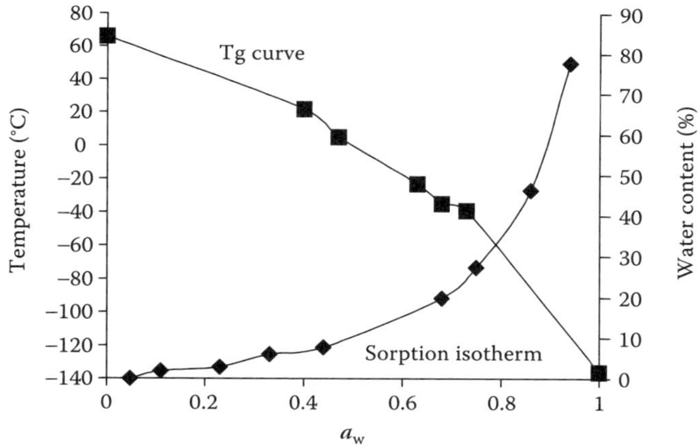

FIGURE 11.2 Modified state diagram of light-roasted coffee. (From Anese A., Manzocco L., Maltini E. 2005. Effect of coffee physical structure on volatile release. *European Food Research and Technology* 221: 434–438, with permission.)

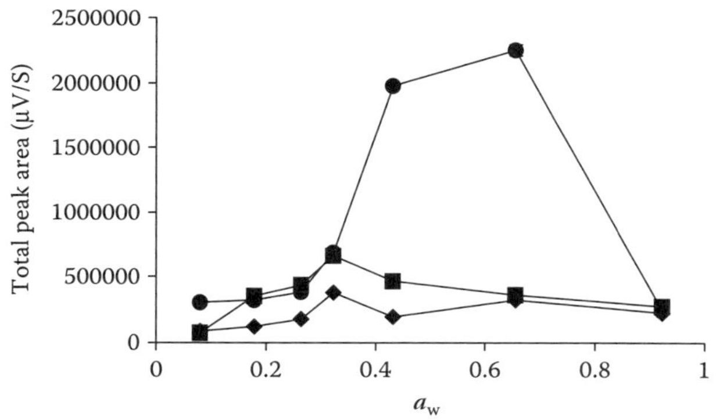

FIGURE 11.3 Total peak area of headspace volatile compounds of light- (◆), medium- (■), and dark roasted (●) soluble coffees as a function of a_w. (From Anese A., Manzocco L., Maltini E. 2005. Effect of coffee physical structure on volatile release. *European Food Research and Technology* 221: 434–438, with permission.)

which can be used as an indicator of freshness. Peroxide and acidity indexes have also been used to follow R&G coffee stability (Chafer et al., 1998). Additional IoFs can be represented by the redox potential value and sensory indicators (Cardelli and Labuza, 2001; Manzocco and Lagazio 2009; Nicoli et al., 2004). No literature data are available on chemical, physical, and sensory IoF values corresponding to acceptability limits for shelf life assessment. Consumer acceptability can also be used as an IoF, and survival analysis can be applied to obtain directly an estimate of the shelf life of the product (Anese et al., 2006).

11.2.2 Instant Coffee

Instant coffee has very long shelf stability if proper packaging solutions are adopted. It is probably for this reason that there are a limited number of studies on the stability and shelf life assessment of instant coffee. In contrast to what happens to R&G coffee, instant coffee does not have the problem of a steady loss of CO_2 during storage. However, volatile release may be responsible for a rapid quality loss of the product, as most of these volatiles are susceptible to oxidation reactions. The extent of oxidation can be enhanced when coffee oils are spread over the granule surface to improve product aroma. Being hygroscopic, instant coffee is also susceptible to the action of ambient moisture, which affects volatile liquid vapor partition. An increase in moisture content to 7–8% is responsible for "caking," where the powder or granules becomes a pasty or solid mass (Clarke, 1987b). Therefore, O_2 and moisture are the environmental variables that must be carefully controlled in selecting suitable packaging. In addition, low storage temperatures should be adopted in order to slow down volatile release rates and keep the product below the glass transition temperature.

Typical IoFs are similar to those previously mentioned for roasted coffee, such as volatile profile or selected volatiles as indicators of the oxidative status of the product, and moisture content. Recently, the Weibull hazard analysis has been applied to sensory acceptability data at different storage times to evaluate the shelf life of spray-dried instant coffee (Ocampo and Giraldo, 2006).

11.2.3 Coffee Concentrates and Drinks

These products, generally pasteurized or sterilized, and thus stable with respect to microbial spoilage, show a very low chemical stability characterized by a change in the flavor profile and an increase in perceived sourness. These changes are accompanied by a pH decrease, corresponding to an increase in titratable acidity. Quality depletion of liquid coffee as well as of coffee concentrates starts immediately after brewing and proceeds at a significant rate, even at subzero temperatures. At present, very little is known about the mechanisms underlying liquid coffee instability. It has been suggested that the decrease in pH could be the consequence of complex reactions, probably related to NEB pathways involving carbohydrates and amino acids. Additional mechanisms involving lactone hydrolysis could also contribute to pH decrease (Balzer, 2001). Similar reactions have been reported for coffee brew aroma staling, which was mainly attributed to interactions between odor-active thiols and melanoidins (Hofmann and Schieberle, 2002; Mueller and Hofmann, 2007). Typical IoFs for this class of product are headspace volatile profile; pH, and, hence H_3O^+ concentration; changes in antioxidant properties; and sensory attributes (Anese and Nicoli, 2003; Dalla Rosa et al., 1990; Nicoli et al., 1989; Perez-Martinez et al., 2008). Consumer acceptability can also be used as an IoF, and survival analysis can be applied to obtain an assessment of the shelf life of the product (Manzocco and Lagazio, 2009).

11.3 HOW PACKAGING IMPACTS INDICES OF FAILURE

11.3.1 Roasted Coffee

The oldest types of packaging used to store and sell roasted whole beans were simple jute or cardboard bags. At that time they perfectly fitted storage and selling, as the requested shelf life of roasted coffee was very short. However, the development of large-scale manufacturers and the increased complexity of the distribution chain demanded a longer shelf life for roasted coffee. Consequently, major attention had to be paid to coffee stability. Therefore, the choice of the packaging materials as well as of the packaging procedures became crucial for delivering the required shelf life. As previously mentioned, the shelf life of roasted coffee is the result of the interaction between the coffee matrix and the packaging, which influences the environmental conditions inside the package. The latter are affected by the permeability of the package to moisture, gases (mainly O_2 and CO_2), and coffee volatiles. In addition, the material should be greaseproof due to the presence of oil on the surface of the coffee beans. Furthermore, the headspace volume and the resistance to increases in internal pressure of the packaging can play a critical role.

Table 11.1 shows the most commonly used materials for the packaging of roasted whole and ground coffee. Tinplate or aluminum cans were the first commercial packages used by the coffee industry (Clarke, 1987b). The use of cans has considerable advantages, such as providing a total barrier to water, gases, and volatile compounds. On the other hand, their management is expensive,

TABLE 11.1
Materials Commonly Used for Packaging of Roasted Coffee

Material	Advantages	Disadvantages
Tinplate	Total barrier Resistant to pressure Recyclable Sturdy Satisfactory ecobalance[a]	Expensive Its rigidity prevents an optimal use of space
Glass	Total barrier Resistant to pressure Recyclable Good ecobalance[a]	Fragile Expensive Heavy Its rigidity prevents an optimal use of space Difficult to sort from other wastes
Aluminum	Total barrier Resistant to pressure Recyclable	Expensive Its rigidity prevents an optimal use of space Difficult to sort from other wastes Poor ecobalance[a]
Combined flexible multiply polymer aluminum	Inexpensive Simple manufacturing technology Large flexibility and optimal use of space Satisfactory ecobalance[a]	Nontight barrier Resistant only to negative pressure Poor strength
Combined multiply aluminum cardboard	Inexpensive Total barrier Satisfactory ecobalance[a]	Nontight barrier Resistant only to negative pressure Poor strength Complex manufacturing technology Difficult to sort from other wastes
Flexible combined multiply polymer	Inexpensive Simple manufacturing technology Satisfactory CO_2 permeability No need to sort from other wastes	

Source: From Nicoli M.C., Savonitto O. 2005. Physical and chemical changes of roasted coffee during storage. In: *Espresso Coffee: the Science of Quality*. Viani R., Illy A. (Eds), 2nd edn. San Diego, California: Elsevier Academic Press, pp. 230–245, with permission.

[a] Ecobalance is defined as the total pollution resulting from energy consumption and that resulting from the product during manufacture, use, and disposal.

because their rigidity and generally cylindrical shape do not allow optimal use of space in the warehouse and during distribution. Similar considerations apply to glass containers, which have the additional disadvantage of fragility. Today, the most commonly used materials are the inexpensive and easy-to-manage flexible polymer–aluminum multiply laminates, which permit both hard and soft packs. These materials ensure an efficient barrier due to the presence of a central layer of aluminum foil. The other layers are a waterproof film on the inner side and a rigid film that gives mechanical strength on the outer side. A typical construction is poly(ethylene terephthalate) (PET)/alufoil/low density polyethylene (LDPE). Other materials used with comparable performance are metalized PET laminated to LDPE or four-ply structures based on biaxially oriented polypropylene (BOPP) or biaxially oriented nylon (BON) in addition to OPET, alufoil, and LDPE (Clarke, 1987b). These packages can be fitted with a one-way valve that opens at a preset pressure to release gases.

A recent innovation in the R&G coffee sector is the coffee pod, which contains the optimal quantity of coffee for a single-cup preparation. The Easy Serving Espresso pod, or ESE pod, is an example. It looks like a small disk and contains 7 g of coffee prepacked in its own paper filter. Coffee pods are generally individually sealed in a high-barrier metalized secondary package flushed with an inert gas (typically N_2) prior to sealing. More recently, coffee capsules have been introduced into the market. Designed for specific espresso machines, capsules have different shapes and may contain different amounts of coffee, ranging from 5.0 to 7.5 g. The capsules are generally made of aluminum coated on the inside with a protective film. Before sealing they are saturated with N_2 to improve shelf life.

11.3.2 Instant Coffee

Instant coffee has been commercialized for many years, and the prevalent package is glass jars or tinplate cans of various shapes. Moisture-proofness is ensured by a paper/plastic laminate or a metal foil sealed to the rim of the container, over which a tight-fitting metal lid is placed in the case of tinplate containers and a plastic or metal screw-cap in the case of jars (Clarke, 1987b). Today flexible materials such as LDPE, BOPP/BOPP, metalized PET/LDPE, and PET/alufoil/LDPE are also employed to package single doses (2 g) (Alves and Bordin, 1998; Robertson, 2006).

11.3.3 Coffee Concentrates and Drinks

Ready-to-drink coffee beverages are available in several packaging formats. Pop-top (easy-open) aluminum cans are a widely popular option, especially in Japan and other Asian countries. Other coffee drinks are available in plastic cups with different shapes. Some of these can be instantaneously heated or cooled due to the presence of an inside cavity containing endothermic or exothermic salts and water separated by a thin layer. By breaking this layer and shaking the container, the salts and water mix, releasing or adsorbing heat. Some manufactures also offer liquid coffee in plastic bottles to those consumers who need to purchase relatively large volumes. Plastic/alufoil/paperboard laminated cartons are also available. Coffee is also available as a liquid concentrate in plastic squeezable bottles or bag-in-box containers for industrial use.

11.4 SHELF LIFE OF COFFEE IN DIFFERENT PACKAGING MATERIALS

11.4.1 Roasted Coffee

11.4.1.1 Effect of Packaging Techniques on Whole and Ground Roasted Coffee Shelf Life

In order to obtain the required shelf life of roasted coffee, the choice of packaging technique is as crucial as the choice of packaging materials. Table 11.2 shows the packaging procedures together with the relevant materials that can be employed for the packaging of both roasted whole and ground coffee. The latter can be packed in multidose containers or in single doses (coffee pods). The table also reports some shelf life data. Unfortunately, very few data on coffee shelf life are available in the literature. In addition, most of the published papers do not specify whether the reported data refer to whole or ground coffee or give clear indications on the methodologies used for shelf life assessment. For this reason the data reported in Table 11.2 only provide an approximate indication of shelf life.

Packaging in air is the easier and older technique. It simply consists of filling and hermetically sealing the package. In this way, coffee is protected against humidity and flavor loss, but the presence of O_2 inside the package greatly shortens the shelf life by inducing the development of oxidative reactions. This technique can be applied to degassed coffee to prevent the swelling and, in extreme cases, the possible bursting of the package. The shelf life of the product is very low, typically around 1 month. Another possibility is fitting a one-way safety valve on the packaging. The valve opens at a preset pressure to release the CO_2 without letting air in. As CO_2 is heavier than air, it tends to stratify at the bottom of the package headspace. This implies that during storage air is first expelled through the valve, followed by O_2 and CO_2. The flushing effect of CO_2 contributes to the

TABLE 11.2
Main Packaging Techniques and Relevant Materials Applied for the Packaging of Roasted Coffee and Reported Shelf Lives

Technique			Materials	Requirements	Shelf Life (Months)
Multidose packaging	Air packaging		Metal cans Flexible laminated materials	Degassing or one-way safety valve	1–3*
	Modified atmosphere packaging	Under vacuum	Metal cans Flexible laminated materials	Degassing	4–6*
		Inert gas	Metal cans Flexible laminated materials	Degassing or one-way safety valve	6–8*
		Active packaging	Metal cans Flexible laminated materials	Degassing or one-way safety valve	n.d.
		Pressurization	Metal cans	One-way safety valve	>18*
Single-dose packaging	Modified atmosphere packaging	Inert gas	Paper pods + flexible laminated materials Plastic capsules	Degassing	>18

Source: * Shelf-life data from Nicoli M.C., Savonitto O. 2005. Physical and chemical changes of roasted coffee during storage. In: *Espresso Coffee: the Science of Quality*. Viani R., Illy A. (Eds), 2nd edn. San Diego, California: Elsevier Academic Press, pp. 230–245, with permission.
n.d. = not determined

control of the internal atmosphere composition, reducing the level of O_2 to 10–12%, thus increasing the shelf life of the product for up to 3 months (Illy and Viani, 1995).

It should be noted that flavors also escape through the valve, leading to beverages with poor sensory quality (Nicoli and Savonitto, 2005). Such problems may be potentially overcome by packing the coffee under modified atmosphere conditions. The conditions may be based on the modification of both the overall pressure and the gas composition inside the container. The simplest way of modifying the atmosphere is based on vacuum application. In such a case, after filling, a high vacuum is created so that the level of O_2 is reduced to 4–6% (Nicoli and Savonitto, 2005). The shelf life of roasted coffee packed under vacuum is prolonged by up to 4–6 months. The vacuum packaging procedure can be applied to both cans and flexible materials. In the latter case, after the vacuum is applied, the package collapses onto the coffee to form the very popular "hard bricks," in which the brick shape is created by the intimate contact between coffee and packaging. To prevent pressure increase inside the container during storage due to volatile release, roasted coffee must be degassed before packaging. It should be noted that the vacuum level that can be reached in rigid containers is lower than that achieved in flexible bags (Robertson, 2006).

The shelf life of roasted coffee could be further improved by replacing the air inside the package with an inert gas such as N_2 or CO_2. Although the latter is not an "inert" gas, it is naturally present in the product. The procedure applied is based on the initial application of vacuum followed by its release with the desired gases. The pressure in the package is in equilibrium with the atmosphere at the moment of sealing. The final percentage of O_2 is expected to be 1–2% and the shelf life up to 6–8 months. Also in this case, coffee should be degassed before packaging to avoid an increase in pressure during storage. Similarly to air packaging, the prevention of increased internal pressure may be obviated by fitting a one-way valve on the package. Inert gas packaging can be performed employing both cans and laminated materials. The packs made with the laminated materials become the so-called soft or pillow packs.

In combination with vacuum and inert gas packaging, further atmosphere modification inside the container can be achieved using active packaging procedures. Although such systems are expected to extend coffee shelf life, to our knowledge no shelf life data are available in the literature. Active packaging is performed to achieve two different goals: (a) reducing the percentage of oxygen residue in the containers and (b) avoiding CO_2 emission from coffee. In the first case, oxygen scavenger systems are represented by sachets inserted into the package or bound by a label to the package wall. The most commonly applied oxygen scavengers in coffee packaging are based on iron powder oxidation (Vermeiren et al., 1999). This system could reduce the level of O_2 to a value lower than that reached by applying inert gas packaging. As an alternative tool to the one-way valve, the inclusion inside the package of sachets containing CO_2 scavengers has recently become quite widespread to avoid the problem of CO_2 release during storage (Vermeiren et al., 1999).

Pressurization is the modified atmosphere packaging that seems to most increase the shelf life of the roasted coffee (Nicoli and Savonitto, 2005). According to the producers, pressurization allows the shelf life of coffee to be extended for up to 18 months. However, no information is available on the criteria adopted to reach such an estimate. Pressurization consists of creating a pressure inside the package that is higher than atmospheric. Such high pressure can be established within the container by packing roasted coffee immediately after air cooling. In this case the pressure passively rises due to the large amount of gas released from coffee upon the degassing process, which, in this case, occurs inside the sealed container. The residual percentage of O_2 is less than 1% due to the relative increase in CO_2 concentration as well as the scavenging effect exerted by coffee melanoidins (Nicoli and Savonitto, 2005). The increase in the pressure inside the containers produces two further effects leading to the quality improvement of the final coffee beverage. As described by Clarke (1987b), the degassing rate decreases under pressurized conditions, reducing oil migration. The oil tends to remain inside the cell being protected against oxidative reaction. In addition, the pressure inside the package becomes higher than the partial pressure of most volatiles, so more of them are dissolved in the lipid phase or bound to melanoidins. Under such conditions the flavors are protected, even when the package is open, and pressurized blends retain their fragrance longer than those packaged using other techniques. However, in order to obtain these effects, the pressurized coffee should be temporarily stored before distribution for at least 15 days, which is the time generally required for the degassing process to reach equilibrium inside the container (Nicoli and Savonitto, 2005).

In contrast, pressure inside the package may also be conventionally increased by sealing the filled containers in the presence of gases at the desired overpressure. In order to withstand the internal pressure, the containers must be made of rigid materials (i.e., tinplate or aluminum), as pressure has been reported to rise to up to 2.2 atm (Illy and Viani, 1995). It should be noted that the packaging techniques reported in Table 11.2 can be used for packaging of both whole and ground roasted coffee. As previously mentioned, the higher exposed surface area of ground coffee compared to the whole bean is responsible for an increase in the release rate of CO_2 and volatiles. For this reason, at the moment of packing, ground coffee is expected to be not only poorer in volatiles, due to their lower concentration and oxidation level, but also more unstable because of the faster dynamics of volatile release. In this regard, Figure 11.4 compares the evolution of headspace volatiles of beans and ground coffee during storage. Thus, it can be inferred that, using the same packaging materials and techniques, ground coffee has a shorter shelf life than coffee beans.

Furthermore, as most CO_2 is released during grinding and degassing operations before packing, O_2 concentration inside the container becomes the most critical variable affecting the shelf life of R&G coffee. Thus, packing under modified atmosphere is required to obtain adequate shelf life. Similar considerations apply to coffee pods, whose large surface area makes ground coffee

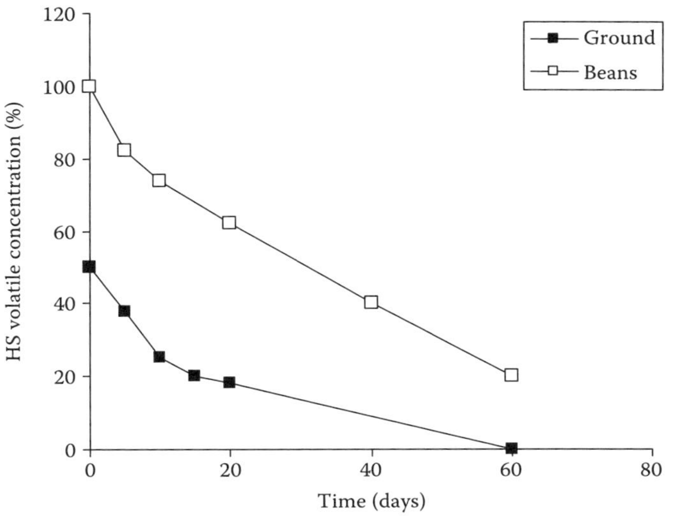

FIGURE 11.4 Headspace volatile concentration of beans and ground coffee as a function of storage time at 25°C. (From Nicoli M.C., Innocente N., Pittia P., Lerici C.R. 1993. *Staling of roasted coffee: volatile release and oxidation reactions during storage*. Proceedings of the 15th International Scientific Colloquium on Coffee, Montpellier, France: pp. 557–566, with permission.)

more prone to oxidation. Pods in modified atmospheres are generally attributed by their producers a shelf life higher than 18 months, not significantly different from that of R&G multidose coffee. However, such indications seem to contradict the shelf life data reported for bulk coffee packed under inert gas. It can be hypothesized that shelf life data reported for bulk coffee could be underestimated because they are based on actual product turnover and not on the assessment of IoFs.

11.4.1.2 Secondary Shelf Life of Coffee

During home or catering use, coffee is almost never consumed immediately after opening the package. When used by consumers, packages of roasted coffee are opened and closed frequently, so coffee degradation may suddenly speed up due to modification of conditions inside the package to those characteristic of an air package. In addition, the moisture content of coffee often increases due to equilibration with the relative humidity of the environment. As a consequence, degradation reactions during consumption proceed at much higher rates than those observed in the originally packed ground coffee, thus contributing to what is known as a secondary shelf life (Cappuccio et al., 2001). Secondary shelf life represents the length of time after opening of the package during which coffee maintains acceptable quality. The end of the secondary shelf life of R&G coffee at 30°C under different a_w conditions was determined by Anese et al. (2006) through consumer rejection methodology, choosing as the acceptability limit a 50% consumer rejection (Table 11.3).

The end of secondary shelf life was almost constant, at around 20 days, at a_w values lower than 0.36. At higher a_w values, the secondary shelf life greatly decreased, to about 13 days, corresponding to an a_w of 0.44. These results were in agreement with the high loss of volatile compounds of coffee equilibrated at the highest a_w. Independent of the hydration degree of the coffee, the end of shelf life was reached when the same total volatile pressure was measured in the coffee headspace. In particular, a loss of about 60% of the volatiles compared to those in the coffee at the time of opening was identified as a proper acceptability limit indicating the end of shelf life.

TABLE 11.3
Secondary Shelf Life of Ground Roasted Coffee as a Function of a_w

a_w	Shelf Life (Days)
0.09	20±3
0.17	23±4
0.23	22±6
0.36	17±6
0.44	13±3

Source: From Anese M., Manzocco L., Nicoli M.C. 2006. Modeling the secondary shelf life of ground roasted coffee. *Journal of Agricultural and Food Chemistry* 54: 5571–5576, with permission.

11.4.2 Instant Coffee

The few shelf life data reported in the literature indicate that instant coffee maintains its original quality for years, provided that moisture content is less than 4–5% w/w to keep the product below its glass transition point, and provided oxygen present in the headspace is low enough to avoid flavor oxidation. The shelf life of unopened instant coffee varies from 18 to 36 months if stored in glass jars or tinplate cans. The Q_{10}'s for coffee staling and moisture absorption are quite low at 1.5–2.0 (Labuza, 1982).

Ocampo and Giraldo (2006) reported that instant coffee packed in glass was rejected by 50% of consumers after approximately 2 years of storage at 18°C. Flexible materials containing aluminum foil (i.e., PET/alufoil/LDPE) as an O_2 and moisture barrier may also be very effective, giving shelf lives of around one year. However, a much lower shelf life is obtained when instant coffee is stored in other plastic materials. Alves and Bordin (1998) investigated the shelf life of portion packs (25 and 50 g) of instant coffee at 30°C and 80% RH in three different plastic structures. The surface areas of the packs were 8 × 11 cm for 25 g and 10 × 16 cm for 50 g. The structure of the packages, their water vapor transmission rate (WVTR), and the shelf life of the coffee at 30°C and 80% RH are shown in Table 11.4. The end of shelf life was when the moisture content had reached 7.8%, at which time powder agglomeration occurred. Data on the shelf life of agglomerated and powdered instant coffee packed in PP and PET pots were presented in a later paper (Alves et al., 2000), with the shelf life in the former being up to 2.5 years, compared to 4 months in the latter.

11.4.3 Coffee Concentrates and Drinks

As previously reported, the extraordinary instability of coffee brews is due to the development of a number of different chemical reactions, leading to rapid off-flavor formation and a decrease in pH. The changes in pH of a dark roasted coffee brew are shown in Figure 11.5 as a function of storage time at different temperatures. The rate of pH decrease is strongly affected by temperature, and other environmental variables, potentially controllable by packaging, do not exert a critical role. Quality depletion is, in this case, strictly related to brew formulation. In other words, in the wide range of coffee derivatives, beverages are the ones whose shelf life is independent of the packaging.

TABLE 11.4
Shelf Life of Instant Coffee in Plastic Packages

Package Structure	WVTR $g\ m^{-2}\ day^{-1}$	Shelf Life (Days)	
		25 g	50 g
LDPE 30 µm	6.1	15	16
BOPP/BOPPP 20 µm/40 µm	1.2	84	94
Met PET/ LDPE 12 µm/70 µm	0.9	108	112

Source: From Alves R.M.V., Bordin M.R. 1998. Shelf life prediction of instant coffee by mathematical method. *Ciencia e Tecnologia de Alimentos* 18: 19–24, with permission.

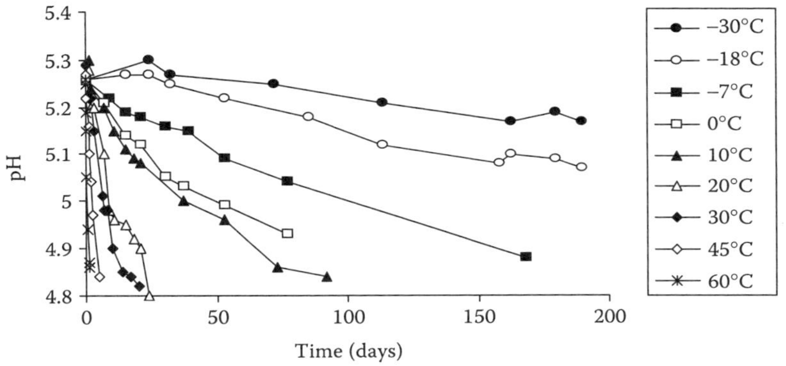

FIGURE 11.5 pH of a dark roasted coffee brew (1.8% w/w solid concentration) as a function of storage time at different temperatures. (From Manzocco L., Nicoli M.C. 2007. Modeling the effect of water activity and storage temperature on chemical stability of coffee brews. *Journal of Agricultural and Food Chemistry* 55: 6521–6526, with permission.)

Manzocco and Lagazio (2009) assessed the shelf life of a dark roasted coffee brew (initial pH 5.26) stored at room temperature by examining the probability of rejection expressed by consumers. Table 11.5 shows the shelf life of the coffee brew as a function of solids concentration and storage temperature. As consumer rejection strongly correlated with H_3O^+ concentration, and thus with pH, the relevant pH values at the end of shelf life were also calculated and are presented in Table 11.6. Such data may be exploited to predict, for instance, that when the coffee brew reaches a pH of 5.16, it is rejected by more that 25% of consumers, and this occurs after 4 ± 1 days of storage.

Coffee brews are also processed to obtain a number of coffee concentrates with different a_w values. Studies carried out on liquid coffee products with different solid content indicate that the rate of H_3O^+ formation was considerably lower for a_w values below 0.5 (0.94% w/w solid concentration). Beyond this critical boundary, the rate increased, reaching a maximum value of 0.8 a_w (78% w/w solids concentration), as shown in Figure 11.6 (Manzocco and Nicoli, 2007). If we consider an acceptability limit for the coffee concentrates as a consumer rejection higher than 25% (pH limit 5.16), the shelf life data can be calculated from the reaction rates reported in Figure 11.6 and are shown in Table 11.6. It can be seen that, at constant temperature, the shelf life of coffee concentrates decreases as solids content increases.

TABLE 11.5
Shelf Life of Liquid Coffee with Increasing Solids Concentration Stored at Different Temperatures

Temperature (°C)	Shelf Life (Days)		
	Solids Concentration (% w/w)		
	1.8	**15**	**78**
0	17	14	12
10	14	9	8
20	4	3	1
30	3	2	1
45	1	1	<1
60	<1	<1	<1

TABLE 11.6
Shelf Life and pH Estimated for Increasing Percentages of Consumers Rejecting the Coffee Brew

Consumers Rejecting the Sample (%)	Shelf Life (Days) ± 95% Confidence Interval	pH ± 95% Confidence Interval
10	2 ± 1	5.22 ± 0.03
25	3 ± 1	5.19 ± 0.03
50	5 ± 1	5.14 ± 0.02
75	6 ± 1	5.12 ± 0.02
90	8 ± 2	5.08 ± 0.04

Source: From Manzocco L., Lagazio C. 2009. Coffee brew shelf life modelling by integration of acceptability and quality data. *Food Quality and Preference* 20: 24–29, with permission.

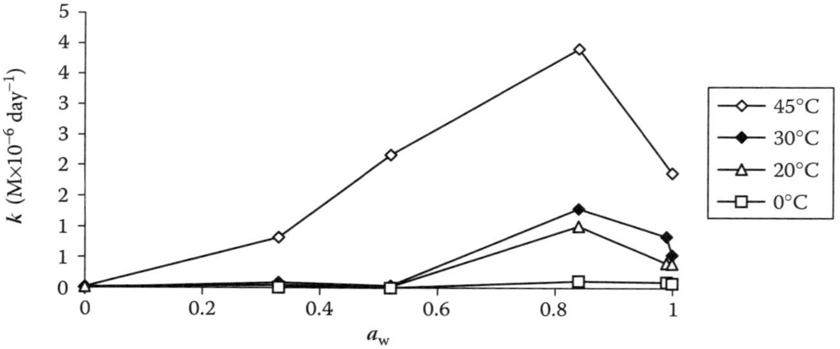

FIGURE 11.6 Apparent zero-order rate constants (k) of H_3O^+ formation as a function of a_w of coffee samples stored at different temperatures. (From Manzocco L., Nicoli M.C. 2007. Modeling the effect of water activity and storage temperature on chemical stability of coffee brews. *Journal of Agricultural and Food Chemistry* 55: 6521–6526, with permission.)

REFERENCES

Alves R.M.V., Bordin M.R. 1998. Shelf life prediction of instant coffee by mathematical method. *Ciencia e Tecnologia de Alimentos* 18: 19–24.

Alves R.M.V., Milanez C.R., Padula, M. 2000. Alternative packaging for instant coffee. *Ciencia e Tecnologia de Alimentos* 20: 204–211.

Amstalden L.C., Leite F. 2001. Staling of ground roasted coffee in punctured commercial packs during storage. *Alimentaria* 319: 103–110.

Anese A., Manzocco L., Maltini E. 2005. Effect of coffee physical structure on volatile release. *European Food Research and Technology* 221: 434–438.

Anese M., Manzocco L., Nicoli M.C. 2006. Modeling the secondary shelf life of ground roasted coffee. *Journal of Agricultural and Food Chemistry* 54: 5571–5576.

Anese M., Nicoli M.C. 2003. Antioxidant properties of ready-to-drink coffee beverages. *Journal of Agricultural and Food Chemistry* 51: 942–946.

Balzer H.H. 2001. Acids in coffee. In: *Coffee: Recent Developments*. Clarke R.J., Vitzthum O.G. (Eds). Oxford, England: Blackwell Science, pp. 18–32.

Buffo R.A., Cardelli-Freire C. 2004. Coffee flavor: an overview. *Flavor and Fragrance Journal* 19: 99–104.

Cappuccio R., Full G., Lonzarich V. Savonitto O. 2001. *Staling of roasted and ground coffee at different temperatures: combining sensory and GC analysis*. Proceedings of the 19th International Scientific Colloquium on Coffee, Trieste, Italy, CD-ROM.

Cardelli C., Labuza T.P. 2001. Application of Weibull Hazard Analysis to the determination of the shelf life of roasted and ground coffee. *LWT—Food Science and Technology* 34: 273–278.

Chafer M., Ortola M.D., Chralt A. 1998. Stability of roast coffee during its shelf life. *Alimentaria* 297: 123–128.

Clarke R.J. 1987a. Roasting and grinding. In: *Coffee*, vol. 2. Clarke R.J., Macrae R. (Eds). New York: Elsevier Science, pp. 73–107.

Clarke R.J. 1987b. Packing of roast and instant coffee. In: *Coffee*, vol. 2. Clarke R.J., Macrae R. (Eds). New York: Elsevier Science, pp. 201–219.

Clarke R.J. 1993. The shelf life of coffee. In: *Shelf Life Studies of Foods and Beverages: Chemical, Biological, Physical and Nutritional Aspects*, (Developments in Food Science, vol. 33). Charalambous G. (Ed). Amsterdam: Elsevier Science, pp. 801–819.

Clarke R.J. 2001. Technology III: Instant coffee. In: *Coffee: Recent Developments*. Clarke R.J., Vitzthum O.G. (Eds). Oxford, England: Blackwell Science, pp. 125–139.

Clifford M.N. 1985. Chlorogenic acids. In: *Coffee*, vol. 1, Clarke R.J., Macrae R. (Eds). New York: Elsevier Science, pp. 153–197.

Czerny M., Mayer F., Grosch W. 1999. Sensory study on the character impact odorants of roasted Arabica coffee. *Journal of Agriculture and Food Chemistry* 47: 695–699.

Dalla Rosa M., Barbanti D., Lerici C.R. 1990. Changes in coffee brews in relation to storage temperature. *Journal of the Science of Food and Agriculture* 50: 227–235.

Da Porto C., Nicoli M.C., Severini C., Sensidoni C., Lerici C.R. 1991. Study on physico-chemical changes of coffee beans during roasting. Note 2. *Italian Journal of Food Science* 3: 197–206.

Eggers R., Pietsch A. 2001. Technology I: Roasting. In: *Coffee: Recent Developments*. Clarke R.J., Vitzthum O.G. (Eds). Oxford, England: Blackwell Science, pp. 90–107.

Heilman W. 2001. Technology II: decaffeination of coffee. In: *Coffee: Recent Developments*. Clarke R.J., Vitzthum O.G. (Eds). Oxford, England: Blackwell Science, pp. 108–124.

Hofmann T., Schieberle P. 2002. Chemical interactions between odor-active thiols and melanoidins involved in the aroma staling of coffee beverages. *Journal of Agriculture and Food Chemistry* 50: 319–326.

Holscher W., Steinhart H. 1992. Investigation of roasted coffee freshness with an improved headspace technique. *Zeitschrift für Lebensmittel-Untersuchung und-Forschung* 195: 35–38.

Homma S. 2001. Chemistry II: non-volatile compounds. Part II. In: *Coffee: Recent Developments*. Clarke R.J., Vitzthum O.G. (Eds). Oxford, England: Blackwell Science, pp. 50–65.

Illy A., Viani R. (Eds). 1995. *Espresso Coffee: the Chemistry of Quality*. San Diego, California: Academic Press.

Labuza, T.P. 1982. Shelf life of coffee and tea. In: *Shelf-life Dating of Foods*. Westport, Connecticut: Food and Nutrition Press, pp. 359–372.

Macrae R. 1985. Nitrogenous components. In: *Coffee*, Vol. 1. Clarke R.J., Macrae R. (Eds). New York: Elsevier Science, pp. 115–149.

Manzocco L., Nicoli M.C. 2007. Modeling the effect of water activity and storage temperature on chemical stability of coffee brews. *Journal of Agricultural and Food Chemistry* 55: 6521–6526.

Manzocco L., Lagazio C. 2009. Coffee brew shelf life modelling by integration of acceptability and quality data. *Food Quality and Preference* 20: 24–29.

Mueller C., Hofmann T. 2007. Quantitative studies on the formation of phenol/2-furfurylthiol conjugates in coffee beverages toward the understanding of the molecular mechanisms of coffee aroma staling. *Journal of Agricultural and Food Chemistry* 55: 4095–4102.

Nicoli M.C., Savonitto O. 2005. Physical and chemical changes of roasted coffee during storage. In: *Espresso Coffee: the Science of Quality*. Viani R., Illy A. (Eds), 2nd edn. San Diego, California: Elsevier Academic Press, pp. 230–245.

Nicoli M.C., Anese M., Manzocco L., Lerici C.R. 1997. Antioxidant properties of coffee brews in relation to the roasting degree. *LWT—Food Science and Technology* 30: 292–297.

Nicoli M.C., Dalla Rosa M., Bonora R., Lerici C.R. 1989. Chemical properties of coffee brew: 3rd note. Kinetics of ageing and influence of some technological operations on the coffee brew stability. *Industrie Alimentari* 6–7: 706–709.

Nicoli M.C., Toniolo R., Anese M. 2004. Relationship between redox potential and chain-breaking activity of model systems and foods. *Food Chemistry* 88: 79–83.

Nicoli M.C., Innocente N., Pittia P., Lerici C.R. 1993. *Staling of roasted coffee: volatile release and oxidation reactions during storage*. Proceedings of the 15th International Scientific Colloquium on Coffee, Montpellier, France: pp. 557–566.

Ocampo J.A., Giraldo G.I. 2006. Application of Weibull hazard analysis to determination of shelf life of spray dried instant coffee. *Alimentacion Equipos y Tecnologia* 25:107–111, 210.

Perez-Martinez M., Sopelana P., Paz de Pena M., Cid C. 2008. Changes in volatile compounds and overall aroma profile during storage of coffee brews at 4 and 25°C. *Journal of Agricultural and Food Chemistry* 56: 3145–3154.

Petracco M. 2001. Technology IV: beverage preparation: brewing trends for the new millennium. In: *Coffee: Recent Developments*. Clarke R.J., Vitzthum O.G. (Eds). Oxford, England: Blackwell Science, pp. 140–164.

Petracco M. 2005a. Grinding. In: *Espresso Coffee: the Science of Quality*. Viani R., Illy A. (Eds), 2nd edn. San Diego, California: Elsevier Academic Press, pp. 215–229.

Petracco M. 2005b. Percolation. In: *Espresso Coffee: the Science of Quality*. Viani R., Illy A. (Eds), 2nd edn. San Diego, California: Elsevier Academic Press, pp. 259–287.

Pictet G., 1987. Home and catering brewing coffee. In: *Coffee*, Vol. 2. Clarke R.J., Macrae R. (Eds). New York: Elsevier Science, pp. 221–252.

Pittia P., Nicoli M.C., Sacchetti G. 2007. Effect of moisture and water activity on textural properties of raw and roasted coffee beans. *Journal of Textural Studies* 38: 116–134.

Robertson G.L. 2006. *Food Packaging Principles and Practice*, 2nd edn. Boca Raton, Florida: CRC Press.

Steinhardt H., Holscher W. 1991. *Storage related changes of low-boiling volatiles in whole beans*. Proceedings of the 14th International Scientific Colloquium on Coffee, San Francisco, California, pp. 156–174.

Trugo L.C. 1985. Carbohydrates. In: *Coffee*, Vol. 1. Clarke R.J., Macrae R. (Eds). New York: Elsevier Science, pp. 83–113.

Vermeiren L., Devlieghere F., Van Beest M., de Kruif N., Debevere J. 1999. Developments in the active packaging of foods. *Trends in Food Science and Technology* 10: 77–86.

12 Packaging and the Shelf Life of Beer

Charles W. Bamforth and John M. Krochta
Department of Food Science and Technology
University of California Davis
Davis, California

CONTENTS

12.1 INTRODUCTION

The vast majority of beers are never better than when they are first filled into their containers. Perhaps some of the more alcoholic brews—those with an alcohol content approaching or even surpassing those of wines (Sherman, 2003)—may register interesting flavor changes with time—for example, the development of flavorsome materials such as ketals through the interaction of alcohols and acids. However, for beers in the customary range of alcohol content, the vector of flavor change is downward, through staling. Furthermore, other quality attributes of beer, such as clarity, color, and foam, are also likely to deteriorate over weeks and months of storage.

The composition of beer is established on the basis of the raw materials used in its manufacture and on the processing procedures employed. From grain and hops growing in the field right way through to closing the final container, there are forces at play that can impact the quality and the ensuing shelf life of the product (Bamforth, 2008b). As this book focuses on packaging issues, only brief reference will be made to the myriad forces prior to the packaging stage that impact beer properties. Rather, the focus will be on the diversity of packaging modes and how they impact beer quality directly and indirectly.

Although this chapter will focus very much on practical issues, it is important to recognize that the package represents the primary interface between the producer and the purchaser. It needs to be attractive, convenient for (and appropriate to) the drinking occasion, and easy to access and store. Over the years a clear ranking has emerged with regard to package sophistication, with glass bottles considered the premier container type. Beer in cans is of lower perceived quality, despite the fact that air ingress into a can will inevitably be less than that into a bottle.

12.2 BEER QUALITY

12.2.1 Beer Production

Figure 12.1 presents a simplified overview of the beer production process. Detailed descriptions of the nuances of brewing can be found in Briggs et al. (2004). Production volumes of beer worldwide are growing, albeit with substantial differences between countries. There is a burgeoning growth in beer production *inter alia* in China, Russia, and Brazil, but a decrease in beer volumes in more traditional beer markets such as the world's largest (per capita)—the Czech Republic, Germany, Ireland, and Denmark (Table 12.1).

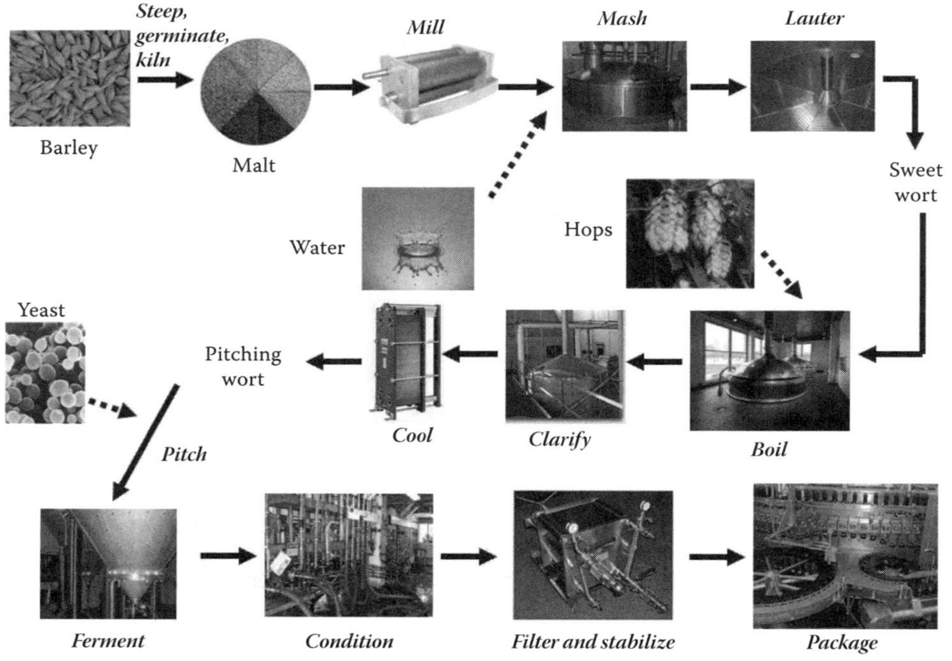

FIGURE 12.1 A simplified overview of the beer production process.

TABLE 12.1
Growth or Decline in Beer Volume (million hL) since 1970

Country	1970	1980	1990	1995	1999	2004
Denmark	7.1	8.2	9.0	10.1	8.0	8.5
France	20.3	21.6	21.4	20.6	19.9	16.8
Germany	103.7	115.9	120.2	117.4	112.8	106.2
Ireland	5.0	6.1	6.4	7.4	8.6	5.2
Netherlands	8.7	15.7	20.0	23.1	24.5	23.8
United Kingdom	55.1	64.8	61.8	56.8	57.9	57.4
Russia	—	—	—	—	44.9	85.2
South Africa	2.5	8.3	22.6	24.5	25.6	24.4
China	1.2	6.0	69.2	154.6	207.4	291.0
Japan	30.0	45.5	66.0	67.3	72.2	66.0
South Korea	0.9	5.8	13.0	17.6	14.9	20.2
Australia	15.5	19.5	19.4	17.9	17.3	17.6
Canada	15.8	21.6	22.6	22.8	23.0	23.1
United States	158.0	227.8	238.9	233.7	236.5	233.2
Brazil	10.3	29.5	58.0	84.0	80.4	82.6
Mexico	14.4	27.3	39.7	44.5	57.3	68.5
World	648.1	938.6	1,166.0	1,249.8	1,367.3	1,554.6

Source: Tighe A. 2006. *British Beer and Pub Association Statistical Handbook*. London, England: Brewing Publications Ltd.

12.2.2 Quality Attributes of Beer

12.2.2.1 Flavor

Although not generally regarded as such by many people, beer is substantially more complex flavorwise than wine (Bamforth, 2008a). The greater diversity of raw materials (grains of various types, hops, water, other novel ingredients as diverse as pumpkins, clamato, limes, and many more), together with greater attention to yeast strain selection and health (Boulton and Quain, 2001), makes for a broader spectrum of flavor-active components. Furthermore, the malting and brewing processes are extremely closely controlled, the maltster and brewer recognizing that achieving consistency at each stage is critical if the beer flavor is to be regularized.

The reality is that the flavor profile of most beers is established by the time the product is filled into the container. One notable exception is those beers whose degree of carbonation is finalized through "natural conditioning," in which residual sugars are converted into CO_2 by a final secondary fermentation in the bottle. Carbon dioxide plays a critical role in flavor delivery, both directly through interaction with the trigeminal sense (Carstens et al., 1998), leading to "tingle" and by driving flavor-active volatiles into the headspace of the product and thence into the naso-olfactory passages. A further exception is the introduction of N_2 gas into the beer during packaging, primarily for the benefits it affords foam, but which impacts the flavor both directly by delivering a smooth mouthfeel to the beer and indirectly by masking certain aromas, notably hoppiness (Carroll, 1979). The science underpinning these effects is unknown.

Otherwise, consideration of the impact of the package on flavor centers on the avoidance of flavor deterioration. Critical factors are the integrity of the inner surface of the container, the amount of O_2 in the package and which can gain access into the package, and the color of any glass or plastic bottle. In addition, the temperature to which the container and therefore the enclosed beer is exposed is of great significance.

Oxidative deterioration of beer in packages is effected through reactions involving reactive species, primarily O_2 radicals, produced with the mediation of transition metal ions such as iron and copper (Bamforth and Parsons, 1985), and also aluminum. The complete coating of the inner surface of a can is therefore critical. Some brewers still add copper to the beer to minimize off-flavors from sulfides (Pfisterer et al., 2004).

Also essential is the achievement of the lowest practical O_2 content in the container, as well as minimizing air ingress into the container. This ingress can occur by permeation through the crown cork closure–bottle interface (Teumac et al., 1990) and, in plastic bottles, through the bottle wall (Oag and Webb, 1984). The former has driven a move back from twist-off to pry-off caps and led to the development of O_2-scavenging inserts into the crowns, and latter-day plastics for beer bottling have greatly improved barrier properties (Huige, 2002).

Regarding bottle color, the marketing vector is ever toward green and clear glass, despite the fact that they are far more transparent than is brown glass to the wavelengths of light that trigger the breakdown of hop bitter acids to generate highly flavor-active molecules generally considered to have a reprehensible skunk-like aroma (Huvaere et al., 2004). One strategy for avoiding such deterioration is to use reduced iso-α-acids that do not yield these breakdown products and that also possess superior foaming properties (Heyerick et al., 2003). Another proposal has been to remove riboflavin from the beer, a molecule that acts to collect the light energy (Duyvis et al., 2002).

The deterioration of beer is minimized for all containers by keeping them away from light and by keeping them cold: as a rule of thumb, the rate of flavor deterioration will be at least halved for every 10°C reduction in storage temperature (although beer should not be frozen if container integrity is to be preserved and the development of sediments is to be avoided). As a general rule, ambient (~20°C) storage of a relatively gently flavored beer will make for a shelf life of approximately 3 months. Researchers have for many years accelerated aging of the beer in the laboratory at either 30°C for 1 month or 60°C for 1 day, in keeping with the Arrhenius prediction of two to three times faster chemical reactions for each 10°C increase in temperature (Bamforth, 2004a).

12.2.2.2 Color

The color of beer is primarily determined by the composition of the grist, that is, the extent to which the malts have been kilned or roasted and to which they are "diluted" by a proportion of sugar (Shellhammer and Bamforth, 2008). There will also be some color development during the boiling of wort. For the paler beers, however, there is a concern about color pickup through enzymic browning, notably involving peroxidases using hydrogen peroxide produced from O_2 in mashing to oxidize polyphenols (Clarkson et al., 1989). For this reason many brewers seek to minimize O_2 levels in the brewhouse as well as in the final package. It may also be that iron- or copper-catalyzed polyphenol oxidation could occur in the final package.

12.2.2.3 Foam

Just as it has a major impact on flavor, the CO_2 content of beer is of much relevance for the foaming properties too. The foam production potential is in proportion to carbonation (Lynch and Bamforth, 2002); however, beer is a supersaturated solution and does not spontaneously generate foam in the absence of nucleation. This can take the form of agitative dispensing devices and procedures, scratches in glasses, and, in the case of some latter-day cans and bottles designed to allow pub-style beer at home, the widget (Browne, 1996). The latter is usually accompanied by the employment of N_2 gas, which leads to greatly enhanced foam stability because of reduced rates of disproportionation (a phenomenon known as *Ostwald ripening*) (Ronteltap et al., 1991).

Foam stability is dependent on the presence of surface-active components in the beer, notably grain-derived polypeptides and hop-derived bitter compounds, but in some markets also foam stabilizers such as propylene-glycol alginate (PGA) (Bamforth, 1993). It is unusual for a beer to be majorly deficient in foam-active molecules, a more frequent problem being foam destabilization through the encroachment of lipids and detergents (Dickie et al., 2001). However, even in the absence of such foam-negatives, the foam stability of beer may decrease with time in the package due to the action of yeast-derived proteinases that progressively degrade polypeptides (Ormrod et al., 1991). Such enzymes are inactivated by pasteurization, and so it is primarily beers that are "sterile-filtered" that display this phenomenon.

12.2.2.4 Clarity

Most beers are "bright" in that they are free from turbidity and are expected to remain that way throughout their shelf life. A few beer styles, however, are intended to be cloudy, notable among them being the Hefeweizens, the turbidity of which is largely due to residual yeast. Even then, it is expected that consistency is achieved in the magnitude of this haziness.

Various forms of insolubility problems can arise in beer (Bamforth, 1999). "Bits" can develop through agitation in the package, such as when a product is shipped long distances, causing foam that leaves residues upon collapse. Bits can also be caused by the interaction of different additions in beer; for example, papain added as a haze-preventive treatment can react with PGA added to stabilize foam. The latter can also react with residual isinglass finings to yield gelatinous precipitates, a process promoted at high ambient temperatures found in some markets. "Invisible hazes" are due to very small particles (<0.1 μm) that are not easy to detect by eye but scatter light powerfully at the 90° angle employed in many haze measuring systems. Visible hazes of the dispersed type in beer have variously been reported as being due to a range of materials that survive less-than-ideal processing in the brewery: protein cross-linking with polyphenols, starch, pentosans, β-glucans, and oxalic acid. The latter is an especial problem in draft beer, where it has a tendency to precipitate out as calcium oxalate in dispensor lines, thereby blocking them. Less common forms of haze arise from dead bacteria and from injudiciously selected can seamer lubricants.

All of the aforementioned are generally grouped into the category of "nonbiological instability." Development of turbidity (as well as flavor problems) in beer due to the growth of microorganisms is called "biological instability" (Flannigan, 2003). Most beers are relatively resistant to microbial contamination because of the presence of antiseptic agents, notably ethanol, bitter acids, and polyphenols; the low pH (usually 4–4.5); the dearth of certain nutrients (notably sugars and amino acids); and the absence of O_2 (although the latter can be a problem for draft beer once the container has been broached, allowing for the growth of the aerobic acetic acid bacteria). Well-run breweries minimize the microbial count in their beer at all stages, thereby preventing spoilage earlier in the process when conditions are more conducive to microbial growth and also ensuring that the count of organisms needed to be removed by either pasteurization or sterile filtration is as low as possible.

All the malting and brewing processes have an impact on the nonbiological instability of beer, with successful elimination of sensitive polypeptides, polysaccharides, and polyphenols variously involving removal during cereal germination, mashing, boiling, fermentation, and cold conditioning. However, most brewers pay especial attention to downstream processing as an exercise in lessening the levels of colloidally sensitive materials. Of particular importance is chilling the beer to as low a temperature as possible without it freezing (Miedl and Bamforth, 2004) and the use of agents to remove polypeptides, such as silica gels, tannic acid, or enzymes (papain, prolyl endopeptidase), or to remove polyphenols (polyvinylpolypyrollidone) (Leiper and Miedl, 2008). Just as for flavor stability, the O_2 and metal ion content of the beer needs to be as low as possible, and the beer needs to be kept cold.

12.2.2.5 Gushing

Gushing is the spontaneous generation of foam on opening a package of beer (Garbe et al., 2008). It can be caused either by solid particles in the beer acting as nucleation sites for bubble release or by stable microbubbles of gas, produced by agitation of a beer and its equilibration to room temperature.

Notable among the nucleation sites that can cause gushing are immensely hydrophobic polypeptides produced by the mould *Fusarium*, which can contaminate barley that encounters damp conditions during growing, and hence gushing can be a prime problem with barley grown in the far north of Europe. Other nucleation agents include calcium oxalate crystals, slithers of glass in inadequately rinsed new glass bottles, rough surfaces on the inside of glass bottles, heavy metals such as nickel, oxidation products in hops, filter aid breaking through filters, haze particles in beer, and excessive carbonation.

12.2.2.6 Wholesomeness

Brewers have long been fastidious in seeking to eliminate undesirable components from their products. Indeed, in many markets there is a strong preference for the avoidance of process additions, such as foam stabilizer and agents that would lengthen flavor life (e.g., sulfur dioxide). Among the materials that can be found in emerge into beer and against which precautions have long been taken are nitrosamines; nitrates and nitrites; chloropropanols; mycotoxins; pesticides and their residues; acrylamide; and lead, cobalt, and other heavy metals (Long, 1999). There is also, of course, concerted attention during packaging to ensure that insoluble "foreign bodies" do not enter into the container.

In recent years brewers have been increasingly prepared to recognize that their products may actually offer some very real benefits to human health when taken in moderation (Bamforth, 2004b).

12.3 IMPACT OF PACKAGING ON INDICES OF FAILURE

12.3.1 Indices of Failure

As indicated in the previous section, there are several quality attributes of beer that are influenced by packaging. These attributes relate to indices of failure (IoFs) associated with quality loss and eventual end of shelf life for beer. Undesirable levels of interaction of beer with O_2 due to the packing/filling operation and the nature of the packaging lead to unacceptable levels of off-flavor, off-color, and haze. Another IoF associated with loss of flavor is loss of carbonation. Additional IoFs that can be associated with packaging include migration of package components to the beer and scalping (absorption) of beer flavor by the packaging. This section discusses the impact of packaging on these IoFs. Table 12.2 summarizes the possible beer–package interactions and the related IoFs for different packages.

TABLE 12.2
Beer–Package Interaction and Related Indices of Failure for Different Beer Packaging

Beer–Package Interaction (BPI)	Indices of Failure (IoFs)	Aluminum Can	Glass Bottle	Plastic Bottle
Light ingress	Color, flavor, nutrient degradation		✓[a]	✓[a]
Oxygen ingress through package body	Color, flavor, etc., oxidation; haze			✓[b,c,d]
Oxygen ingress through package closure lining	Color, flavor, etc., oxidation; haze		✓[b,d]	✓[b,d]
Carbon dioxide egress through package body	Flavor loss			✓[b,c]
Carbon dioxide egress through package closure lining	Flavor loss		✓[b]	✓[b]
Package component migration	Aroma, flavor change	✓[c]		✓[b,c]
Beer component scalping	Aroma, flavor loss			✓[b,c]

[a] Reduced with glass coloring (preferably amber).
[b] Extent depends on material selection.
[c] Reduced with barrier layer or coating.
[d] Reduced with O_2 scavenger.

12.3.2 Off-Flavor and Off-Color

The greatest deterioration problem for beer involves production of off-flavors and off-colors due to oxidation, which is accelerated by light, transition metal ions (e.g., iron, copper, and aluminum), and heat. Oxidation of alcohols, iso-α-acids, and fatty acids in beer produces carbonyl compounds that are detectable in sensory tests at low levels (Kuchel et al., 2006; Narziss et al., 1993). In addition to residual O_2 from the brewing process, O_2 can find its way into packaged beer from three sources: (a) O_2 in the package headspace that is not removed during the filling operation, (b) penetration of O_2 through the package closure, and (c) permeation of O_2 through the packaging material. Even very low levels of O_2 ($\leq$1 ppm) (Brody, 2000) lead to detectable flavor and color problems in beer. Off-flavor in beer resulting from only 0.27 ppm O_2 was detectable by sensory panelists (Table 12.3). Thus, filling methods and packaging that minimize O_2 introduction are highly desirable, as are avoidance of light, metal ions, and high storage temperature.

Packaging used for beer includes cans made from aluminum and bottles made from glass, aluminum, or plastic. Because aluminum is impermeable to gases, vapors, and light, aluminum cans are commonly used for beer. The stronger nature of steel-based cans is not an advantage for beer, because the carbonation in beer stiffens aluminum cans against deformation. Enameling (lacquering) the inside of aluminum cans prevents interaction of aluminum ions with the beer. As the double-seam closure used for aluminum cans prevents transmission of O_2 and CO_2, the only source of O_2 in canned beer is the residual O_2 from the brewing process and filling operation.

The glass bottle is the traditional package for beer and is preferred by many beer consumers. Coloring of the glass, preferably amber, is necessary to limit the effect of light on beer deterioration. The crown or roll-on closures used for glass bottles provide opportunity for O_2 permeation into beer through the closure lining material. Application of the active packaging concept of O_2-scavenging compounds (e.g., ascorbates, iron compounds, and sodium sulfites) in the lining material limits O_2 transmission into the beer (Teumac, 1995; Vermeiren et al., 2003). Aluminum has also been formed into bottles for beer, gaining the advantage of recloseability over aluminum cans. Closure liners for aluminum bottles create the same opportunity for O_2 ingress as closure liners for glass bottles.

The brewing industry is always searching for packaging alternatives. Plastic bottles have been explored for many years, with poly(ethylene terephthalate) (PET) receiving most of the attention. PET bottles have the advantage of looking and feeling like glass bottles, but without the weight, breakability, and complexity of manufacture of glass bottles. PET bottles are commonly used for packaging carbonated soft drinks. This has been possible because carbonated soft drinks can withstand ingress of up to approximately 20 ppm O_2 for citrus-flavored drinks and approximately 40 ppm O_2 for cola drinks before end of shelf life (Robertson, 2006). However, the O_2 permeability of monolayer PET is too large for beer, which can withstand $\leq$1 ppm O_2 before reaching the end of shelf life (Brody, 2000). Similarly, although carbonated soft drinks can withstand up to approximately 20% loss of carbonation, beer can tolerate less than 10% loss of CO_2 (Brody, 2000; Robertson,

TABLE 12.3
Carbonyl Content and Sensory Evaluation of Beer as Affected by Initial O_2 Level

Initial Package Oxygen (ppm)	Carbonyl Content (µg L^{-1})	Aging Note[a]
Fresh beer	74.2	1
0.27	133.8[b]	2.2[b]
5.4	194.4[b]	3.5[b]

Source: Adapted from Narziss L., Miedaner H., Graf H., Eichorn P., Lustig S. 1993. Technological approach to improve flavor stability. *Master Brewers Association of the Americas Technical Quarterly* 30: 48–53.

[a] 1: Fresh; 2: slightly aged; 3: aged; 4: strongly-aged.

[b] After accelerated storage (7 days, 40°C).

2006). Thus, PET has been combined with other materials to increase the barrier properties for use in packaging beer. Unfortunately, the coated, multilayer, blend, copolymer, and active O_2-scavenger-incorporated plastic bottles cannot be easily recycled with single-layer PET bottles. Also, such high-barrier PET bottles do not at present have a price advantage that would motivate beer manufacturers to adopt them widely (Brody, 2001). Nonetheless, PET beer bottles are being used at varying levels in many parts of the world. In locations and applications where long beer shelf life is not necessary, less expensive monolayer PET bottles with no barrier component are being used (Browne, 2006).

The aluminum layer in multilayer plastic/alufoil/paperboard cartons is theoretically a total barrier to O_2. However, the scoring and flexing of the aluminum and heat sealing of the plastic (LDPE or PP) layer that occur in forming cartons create cracks and pinholes in the aluminum and a permeable seal that allows ingress of O_2 and egress of CO_2 (Ahrné et al., 1997; Bourque, 1985). The relatively high O_2 transmission rate and related inability to prevent loss of carbonation has limited consideration of cartons to lightly carbonated beer with short shelf life. The same issues apply for multilayer structures incorporating aluminum layers that are formed into pouches and bags (in boxes). Thus, multilayer cartons, flexible pouches, and bag-in-box packaging commercialized for other beverages have not been useful for beer, because of their inability to sufficiently protect against the ingress of O_2 and egress of CO_2 within the tolerance limits for beer.

12.3.3 Haze

Development of haze in beer is influenced by the presence of O_2 and metal ions (Brody, 2005; Lee et al., 2008). The enamel (lacquer) applied to the inside of aluminum cans prevents interaction of the beer with aluminum ions, which lead to a metallic taste and the formation of haze. Neither glass nor plastic bottles have been associated with haze formation.

12.3.4 Carbonation Loss

Along with being more sensitive to O_2, beer is more sensitive to loss of CO_2 than are carbonated soft drinks (Brody, 1997). When packaged in aluminum cans, loss of CO_2 from beer is prevented by the intimate lid–can seal. Beer packaged in glass, aluminum, or plastic bottles can experience some small loss in carbonation through the lining of the closure. Loss of carbonation becomes a more important issue when considering use of PET bottles. The barrier coatings, layers, and blends being considered and used for PET bottles to prevent ingress of O_2 also minimize egress of CO_2 from beer.

12.3.5 Migration of Packaging Components

The inertness of glass precludes any glass component migration into beer. The closure linings used with glass bottles are a potential source of migration into beer. However, this does not seem to be a problem. Direct interaction of aluminum and beer can result in aluminum ions migrating into the beer, affecting beer flavor and clarity. However, formation of an enamel (lacquer) coating on the interior of the cans prevents such interaction. Depending on the coating material utilized, migration of coating components into the beer is possible and should always be investigated for safety reasons and possible effect on flavor. Plastic bottles also have potential for interaction with beer by transfer of bottle components (e.g., monomers, plasticizers, and antioxidants) into the beer. If multilayer cartons, pouches, or bag-in-box packaging were ever to be considered for beer packaging, migration of components of the interior plastic layer would need investigation. Virgin PET was shown to prevent migration of a number of model contaminants from a core layer of PET spiked with the contaminants, indicating low solubility or diffusivity of the contaminants in PET (Franz and Welle, 2003). Several supercleaning processes have been developed to remove contaminants from recycled postconsumer PET (Franz and Welle, 2003).

12.3.6 Scalping of Beer Components

The lining material used for glass bottle closures can scalp flavor compounds from beer. The extent of sorption depends on the respective compositions of the flavor compound and lining material. Plastic beer bottles also have potential to scalp flavor compounds from beer. If multilayer cartons, pouches, or bag-in-box packaging were ever to be considered for beer packaging, scalping of beer flavor by the interior plastic layer would also need investigation. PET is generally viewed as having low flavor sorption compared to other plastics and, thus, is recommended as both barrier and sealant layer to avoid flavor scalping from fruit juices, alcoholic beverages, and milk (Ishitani, 1995). The solubility of limonene as a model compound was shown to be considerably less in PET than in biaxially oriented polypropylene (BOPP) or acrylic (Ac)-coated BOPP (Franz, 1995). Aroma compounds present in apples were also less soluble in PET than in low density polyethylene (LDPE), linear LDPE, polypropylene (PP), and nylon 6 (PA) (Nielsen and Jagerstad, 1995). Several flavor compounds, including limonene, decanal, hexyl acetate, and 2-nonanone, had a much lower tendency to absorb in PET than in LDPE, PP, and polycarbonate, and they had relatively less effect on PET's O_2 permeability (Linssen et al., 2003). However, acetone as a model compound was shown to have both larger solubility and diffusivity in PET than in meta-xylylene diamine nylon (MXD6) or high density polyethylene (HDPE) (Giacin, 1995).

12.4 PACKAGING EFFECT ON SHELF LIFE OF BEER

The requirements for single-use beer packaging include O_2 ingress <1 ppm (1000 ppb), CO_2 loss <10%, and shelf life ≥60 days at 22°C (Brody, 2000). Beer emerges from the brewing process with approximately 50 ppb O_2 (Robertson, 2006). The amount of O_2 to which the beer is exposed increases due to a combination of the filling operation and ingress into the beer package, with a total exposure to 0.25–1.2 ppm O_2 being the range identified as producing end of shelf life (Robertson, 2006). The result is a shelf life in the range of 80–120 days. The ways in which the exposure occurs and the opportunities for reduced exposure and increased shelf life vary with the type of package, including newer packages such as plastic bottles.

12.4.1 Evolution of Beer Packaging

The functions of food packaging (containment, protection, communication, and convenience) are generally the same for beer packaging. However, the ways in which these packaging functions apply to beer are unique to the specific characteristics of beer and its IoFs.

The earliest beer packages were wooden casks and ceramic containers, meant only to serve the function of containment of the beer for a brief time during short transport and then consumption (Brody, 2000). Beer packaging evolved over many years to provide increasing protection and shelf life to beer, thus allowing it to become an important item of global commerce. Over the past century, manufacture of beer has grown from a small-scale industry, with small breweries distributing beer locally in casks for immediate consumption in pubs, to a large-scale industry, with multinational companies that distribute beer long distances in glass, metal, and plastic containers destined for individual consumers (Dunn, 2006).

For the functions of communication and convenience, beer manufacturers have evolved packaging that suggests product uniqueness and quality and that is easy to use. A distinct package can be associated with a brand and a particular level of quality. Beers imported into the United States are often packaged in green rather than amber glass or plastic containers to communicate the international nature of the beer, which is often associated with unique character. (Ironically, green glass is less effective than amber glass at protecting beer from the oxidation-catalyzing wavelengths of visible and ultraviolet light.) Package labels on beer containers are now more likely to indicate date of filling (or "born on...," as used by one main producer) or date by which it is recommended the beer be used. Such information communicates the beer manufacturer's goal of providing product with greatest freshness and quality. The concern with freshness and quality has also resulted in increased speed of beer movement through the supply chain. With greater globalization, beer manufacturers are challenged to create packaging with more universal appeal or to use different packaging for each distinct market. Within markets, greater choice of product, each in a unique package, is being provided in order to communicate quality and sell to a broader range of consumers. More convenient beer packages have been developed that are easier to open, hold, and reseal, thus providing more attractive features to consumers.

The cost of packaging constitutes the most expensive part of providing beer to the consumer in individual containers, followed by labor costs and then raw materials (Bamforth, 2003, 2006; Dunn, 2006). The costs associated with using glass bottles and aluminum cans are similar (Dunn, 2006). It is critical that packages and the filling/sealing operation provide the greatest quality and shelf life possible for the cost. Specifically, the packages and filling/sealing operation must protect to the highest level possible against the IoFs of beer, including haze, off-flavor, off-color, and carbonation loss.

12.4.2 Filling Method

The goals of the beer filling operation are to (a) avoid product loss, (b) fill containers to the desired content consistently, (c) minimize loss of CO_2 and gain of O_2, and (d) prevent microbial, chemical, or physical contamination (Dunn, 2006). The specific method of filling depends on the nature of the container and the level of O_2 reduction desired, both of which will affect the beer shelf life.

Beer is filled into glass bottles using a rotary carousel filler machine (Bamforth, 2006; Browne, 2006; Dunn, 2006). Bottles are conveyed continuously to the filler, where they move individually under filling heads located around the whole circumference of the carousel. At the proper moment, each bottle is raised to a filler tube. Depending on filler-tube design, an airtight seal can be made between the bottle and tube and air evacuated from the bottle to reduce air content to 10% and thus O_2 content to around 2%. If a second evacuation is applied after filling the bottle with CO_2, the amount of air in the bottle can be reduced to 1% and O_2 content to 0.2%. After the bottle is filled or refilled (if a second evacuation is performed) with CO_2, a premeasured amount of beer is filled into the bottle, and the bottle is finally sealed. Beer comes to the filler from the brewhouse with approximately 50 ppb O_2, and the filling operation can add up to approximately 400 ppb O_2 to the beer (Robertson, 2006). However, use of the double evacuation technique with glass bottles can essentially eliminate any addition of O_2 to beer during the filling operation (Dunn, 2006; Kuchel et al., 2006). Even if the O_2 level in the packaged beer could be reduced to as low as 1 ppb, staling would eventually still occur, but shelf life might be extended to 1 year (Kuchel et al., 2006).

Filling of beer into aluminum cans or PET bottles is similar to filling of glass bottles (Browne, 2003; Dunn, 2006). The difference is that the weakness of the aluminum and PET container walls prevents evacuation of air without collapse of the container. Thus, aluminum cans or PET bottles are only purged with CO_2 before filling with beer. In these cases, a long-tube filler is used, because the tube quickly becomes submerged and O_2 uptake is minimized with the resulting quiet filling. However, residual O_2 in beer filled into aluminum cans or PET bottles may be greater than with glass bottles due to the absence of the air evacuation step(s) in the filler operation.

Each aspect of the filling operation should be part of the beer production process Hazard Analysis and Critical Control Points (HACCP) plan to ensure the proper operation of the filler and quality of the beer. Also, the manufacture, placement, maintenance, cleaning, sanitizing, and operation personnel for the filler should follow good manufacturing practices (GMPs).

Before heat pasteurization to destroy microorganisms, whether before or after filling and sealing of beer into packages, the amount of O_2 dissolved in beer should be less than 0.1 ppm to avoid significant rapid oxidative effect on the beer flavor and color (Bamforth, 2006). This low level of O_2 is also a good target for beer that is cold filtered to remove microorganisms, because it will eventually (albeit more slowly) react to produce off-flavor or off-color.

12.4.3 Aluminum Cans

The total barrier provided by a double-seamed aluminum can prevents ingress of O_2 or egress of CO_2. Any oxidation leading to off-flavors, off-colors, and haze is due to O_2 remaining in the beer after the brewing process and any O_2 added in the filling operation. Thus, the extension of shelf life of beer in cans appears to be dependent on reducing levels of O_2 exposure from these two sources.

12.4.4 Glass Bottles

As with aluminum cans, glass bottles prevent O_2 ingress and CO_2 egress. However, unlike aluminum can double-seam closures, bottle closures provide an opportunity for gas transmission through the closure lining. The amount of O_2 transmission through the closure lining has been identified as 0.6–2.0 μL day^{-1} or 2.0–8.4 ppb day^{-1} for a 355-mL bottle (Robertson, 2006). If no O_2 is added to a bottle of beer during filling, the resulting shelf life for the beer would be 4–13 months for a maximum O_2 ingress of 1 ppm. In order to decrease O_2 ingress through the closure lining, with resulting increase in shelf life, various O_2 scavengers have been developed and commercialized for bottle closures. It is recognized that pry-off crown closures provide a tighter seal than do twist-off crowns.

12.4.5 Plastic Bottles

Because of advantages of low-weight and unbreakability and the potential advantage of lower cost, PET bottles have been explored for many years as containers for beer. Developers of PET bottles for beer have worked to overcome several challenges: (a) rapid development of off-flavors and off-colors in the beer due to rapid ingress of O_2, (b) loss of carbonation due to rapid egress of CO_2, (c) potential scalping of beer flavors by the PET, (d) potential migration of PET components (e.g., monomer, plasticizer) into the beer, with resulting flavor or safety problems, and (e) difficulties with recycling when O_2-barrier coatings or layers or O_2 scavengers are added to PET bottles to overcome high O_2 ingress (Huige, 2002).

PET bottles have the same issues as glass bottles of O_2 ingress and CO_2 egress through the closure lining. However, PET bottles suffer from two other sources of O_2 ingress and CO_2 egress: (a) PET bottles that have come into equilibrium with the atmosphere can absorb surprisingly high levels of O_2 that can desorb relatively quickly into the beer filled into them. The PET bottles can also absorb significant levels of CO_2 as they equilibrate with beer; (b) PET bottles can transmit O_2 from the surrounding air into the beer and CO_2 from the beer into the air at rates that produce large flavor changes and carbonation loss, respectively, over a relatively short time. The shelf life of beer at 21°C in a standard monolayer PET bottle was shown to be only 3 weeks before ingress of 500 ppb O_2 and 4 weeks before 10% loss of CO_2 (Boutroy et al., 2006).

Many concepts have been developed to overcome the O_2 desorption, CO_2 absorption, and O_2 and CO_2 permeation rates of PET bottles, each of which must be tested for scalping and migration potential. These fall into the categories of incorporation into PET bottles of (a) interior layer(s) of high-barrier material, (b) high-barrier coatings formed on the inside or outside of the bottles, and (c) O_2 scavengers that can be incorporated into the PET, high-barrier layers, or high-barrier coatings.

High-barrier interior layers in PET bottles investigated for beer have included ethylene-vinyl alcohol copolymer (EVOH) and MXD6 (Huige, 2002; Robertson, 2006). Because the barrier properties of both of these materials are sensitive to moisture, their performance depends on the relative humidity of distribution, marketing, and storage environments. To overcome this problem, they or the closure linings of such bottles could incorporate an O_2 scavenger (Huige, 2002). Multilayer PET bottles incorporating a PET-compatible O_2-scavenging middle layer have been proposed that are essentially impermeable to O_2, capable of absorbing any residual O_2 from the package headspace, and likely recyclable with other PET bottles (Brody, 2000; Cahill et al., 2003).

External barrier coatings developed include epoxy amine and oxides of silicon (SiO_x), and internal coatings developed include amorphous carbon and SiO_x (Brody, 2000; Boutroy et al., 2006; Huige, 2002; Robertson, 2006). The location of the barrier layer or coating will have a large effect on bottle desorption of O_2 and absorption of CO_2, with interior layers practically eliminating these mechanisms of transfer (Huige, 2002). Theoretically, amorphous carbon and SiO_x coatings should be total barriers, but cracking and flaking can be problems. Coating processes have improved coating durability, leading to commercialization of both SiO_x- and amorphous-carbon-coated PET bottles. Interior diamond-like carbon coatings were shown to reduce O_2 and CO_2 transmission rates of PET bottles by 95% and 90%, respectively, with resulting improvement in sensory evaluation (Table 12.4) (Shirakura et al., 2006). Standard PET bottles with an interior coating of 100-nm hydrogenated amorphous carbon showed a shelf life at 21°C of 25 weeks based on ingress of 500 ppb O_2, or 38 weeks based on loss of 10% of CO_2 (Boutroy et al., 2006). Depending on the barrier layer or coatings and the package surface area:volume ratio, some manufacturers are claiming shelf lives of up to 9 months (Robertson, 2006). Additional advantages of inert ultra-thin barrier coatings such as SiO_x and hydrogenated amorphous carbon are reduced migration and scalping and minimal negative impact on PET bottle recycling (Shirakura et al., 2006).

Another option is PET blended with a barrier polymer such as nylon and an O_2-scavenging compound. Such blends have the advantage that they can be formed into bottles on standard PET injection-molding equipment (Browne, 2006). Poly(ethylene naphthalate) (PEN) has better barrier properties than PET. Thus, it has also been investigated for beer bottles, usually as a copolymer with PET because of its high cost (Lee et al., 2008).

A mathematical model has been developed for assessing the effects of bottle size, layer or coating composition and thickness, beer initial O_2 and CO_2 content, and temperature and relative humidity of the surrounding environment on O_2 ingress and CO_2 egress for plastic bottles (Huige, 2002). The model shows that for a 0.5-L capacity 20-mil (508-μm)-thick PET bottle with middle layer of EVOH comprising 9% of the bottle, the time for ingress (by desorption and permeation) of 1 ppm O_2 would vary from approximately 10 days at 32.2°C and 90% RH to approximately 175 days at 10°C and 50% RH. The model allows assessment of the effect of barrier thickness and the O_2 scavenger requirement for a given bottle composition and desired shelf life. For example, the O_2 scavenger capacity required for a 450-μm monolayer PET bottle at 21.1°C and 70% RH was calculated to be approximately 3.1 mL to limit O_2 ingress to 1 ppm during a targeted shelf life of 120 days. If the bottle were comprised of three 150-μm layers of PET alternating with two 15-μm layers of MXD6,

TABLE 12.4
Bottle O_2 Transmission Rate (OTR), CO_2 Transmission Rate (CO_2TR), and Bottled Beer Sensory Evaluation

	OTR (mL bottle^{-1} day^{-1})		CO_2TR (g bottle^{-1} day^{-1} atm^{-1})		Stale Flavor[a]	
	30°C	**40°C**	**30°C**	**40°C**	**30°C[b]**	**40°C[b]**
PET bottle	0.1132	0.2167	0.0048	0.0071	6.5	7.1
DLC[c]-coated PET bottle	0.0043	0.0061	0.00044	0.00059	4.7	5.0
Glass bottle	0	0	0	0	3.9	5.7

Source: Adapted from Shirakura A., Nakaya M., Koga Y., Kodama H., Hasebe T., Suzuki T. 2006. Diamond-like carbon films for PET bottles and medical applications. *Thin Solid Films*. 494: 84–91.

[a] 0: not perceptible, 9: very strong perceptible.

[b] After storage for 2 weeks.

[c] Diamond-like carbon.

the required O_2 scavenger capacity for 120-day shelf life (i.e., 1 ppm O_2 ingress) would be approximately 1.9 mL. This model has great potential for use in the design of plastic bottles that incorporate barrier layers, coatings, or scavengers, as long as accurate property data on the materials are available. Nonetheless, analytical measurements of O_2 ingress and CO_2 loss and sensory tests of flavor stability and other quality factors are necessary (Huige, 2002).

A mathematical model of the effect of temperature and time on beer flavor oxidation has been used to show the large number of shelf life days that can be lost when beer is transported or stored at high temperature, no matter what packaging is used (Huige, 2004). The model showed that 12 days of transit in a rail car averaging 30°C can result in 36 days of 20°C shelf life lost. The large effect of temperature on beer shelf life demonstrated by this study suggests use of the intelligent packaging concept of time–temperature indicators (Taoukis and Labuza, 2003) on beer packaging, so that manufacturers of beer can ensure that spoiled product is not sold to consumers.

12.5 SUMMARY

Beer is a complex product that has pleasant sensory and desirable wholesome attributes. Beer manufacturers have made great efforts to optimize these attributes in the brewing process. Relatively low levels of the IoFs of off-flavor, off-color, haze, loss of carbonation, migration, and scalping experienced by the beer during transportation, storage, and marketing defeat the brewers' efforts to provide optimum levels of these attributes to consumers. Thus, beer manufacturers have also made great efforts in investigation and selection of filling procedures and packaging to protect the quality of beer. Providers of beer packaging are at the forefront in optimizing traditional packaging and developing new high-barrier and active packaging. Continuing efforts on packaging for beer will ensure that beer is unmatched in the level of protection provided.

REFERENCES

Ahrné L.M., Oliveira F.A.R., Manso M.C., Drummond M.C., Öste R., Gekas V. 1997. Modelling dissolved oxygen concentration during storage of packaged liquid foods. *Journal of Food Engineering* 34: 213–224.

Bamforth C.W. 1993. Recent progress in our understanding of beer foam. *Cerevisiae Biotechnology* 18: 49–53.

Bamforth C.W. 1999. Beer Haze. *Journal of the American Society of Brewing Chemists* 57: 81–90.

Bamforth C.W. 2003. The expensive bit. *Brewers' Guardian* November: 22–24.

Bamforth C.W. 2004a. A critical control point analysis for flavor stability of beer. *Master Brewers Association of the Americas Technical Quarterly* 41: 97–103.

Bamforth C.W. 2004b. *Beer: Health and Nutrition.* Oxford, England: Blackwell Publishing.

Bamforth C.W. 2006. *Scientific Principles of Malting and Brewing.* St. Paul, Minneapolis: American Society of Brewing Chemists.

Bamforth C. 2008a. *Grape versus Grain.* New York: Cambridge University Press.

Bamforth C.W. (Ed). 2008b. *Beer: A Quality Perspective.* Burlington, Massachusetts: Academic Press.

Bamforth C.W., Parsons R. 1985. New procedures to improve the flavor stability of beer. *Journal of the American Society of Brewing Chemists* 43: 197–202.

Boulton C., Quain D. 2001. *Brewing Yeast and Fermentation.* Oxford, England: Blackwell Publishing.

Bourque R.A. 1985. Shelf life of fruit juices in oxygen permeable packages. *Journal of Plastic Film and Sheeting* 1: 115–121.

Boutroy N., Pernel Y., Rius J.M., Auger F., von Bardeleben H.J., Cantin J.L., Abel F., et al. 2006. Hydrogenated amorphous carbon film coating of PET bottles for gas diffusion barriers. *Diamond and Related Materials* 15: 921–927.

Briggs D.E., Boulton C.A., Brookes P.A., Stevens R. 2004. *Brewing: Science and Practice.* Cambridge, England: Woodhead Publishing.

Brody A.L. 1997. Packaging of food. In: *The Wiley Encyclopedia of Packaging.* New York: John Wiley and Sons, pp. 699–704.

Brody A.L. 2000. Beer bottles, old and new. *Food Technology* 54(4): 93–96.

Brody A.L. 2001. Strategies for polyester packaging. *Food Technology* 55(2): 68–69.

Brody A.L. 2005. Micro-oxygen packaging retains quality in beer. *Food Technology* 59(4): 69–71.

Browne J. 2006. Packaging of beer. In: *Brewing—New Technologies.* Bamforth C.W. (Ed). Boca Raton, Florida: CRC Press, pp. 293–305.

Browne J.J.C. 1996. What are widgets? *Brewer* 82: 498–503.

Cahill P.J., Rotter G.E., Chen S.Y. 2003. Zero oxygen permeation plastic bottle for beer and other applications. U.S. Patent 6,558,762.

Carroll T.C.N. 1979. The effect of dissolved nitrogen gas on beer foam and palate. *Master Brewers Association of the Americas Technical Quarterly* 16: 116–119.

Carstens E., Kuenzler N., Handwerker H.O. 1998. Activation of neurons in rat trigeminal subnucleus caudalis by different irritant chemicals applied to oral or ocular mucosa. *Journal of Neurophysiology* 80: 465–492.

Clarkson S.P., Large P.J., Hegarty P.K., Bamforth C.W. 1989. *Oxygen radicals—their influence on process performance and product quality.* Proceedings of the European Brewery Convention Congress, Zurich, pp. 267–274.

Dickie K.H., Cann C., Norman E.C., Bamforth C.W., Muller R.E. 2001. Foam-negative materials. *Journal of the American Society of Brewing Chemists* 59: 17–23.

Dunn A.R. 2006. Packaging technology. In: *Handbook of Brewing.* Priest F.G., Stewart, G.G. (Eds). Boca Raton, Florida: CRC Press, pp. 563–606.

Duyvis M.G., Hilhorst R., Laane C., Evans D.J., Schmedding D.J.M. 2002. Role of riboflavin in beer flavor instability: determination of levels of riboflavin and its origin in beer by fluorometric apoprotein titration. *Journal of Agricultural and Food Chemistry* 50: 1548–1552.

Flannigan B. 2003. The microbiota of barley and malt. In: *Brewing Microbiology.* Priest F.G., Campbell I. (Eds), 3rd edn. New York: Kluwer Academic/Plenum Publishers, pp. 113–180.

Franz R. 1995. Permeation of flavour compounds across conventional as well as biodegradable polymer films. In: *Foods and Packaging Materials—Chemical Interactions.* Ackermann P., Jagerstad M., Ohlsson T. (Eds). Cambridge, England: The Royal Society of Chemistry, pp. 45–50.

Franz R., Welle F. 2003. Recycling packaging materials. In: *Novel Food Packaging Techniques.* Ahvenainen R. (Ed). Boca Raton, Florida: CRC Press, pp. 497–518.

Garbe L.-A., Schwarz P., Ehmer A. 2008. Beer gushing. In: *Beer: A Quality Perspective.* Bamforth C.W. (Ed). Burlington, Massachusetts: Academic Press, pp. 185–212.

Giacin J.R. 1995. Factors affecting permeation, sorption, and migration processes in package-product systems. In: *Foods and Packaging Materials—Chemical Interactions.* Ackermann P., Jagerstad M., Ohlsson T. (Eds). Cambridge, England: The Royal Society of Chemistry, pp. 12–22.

Heyerick A., Zhao Y., Sandra P., Huvaere K., Roelens F., De Keukeleire D. 2003. Photolysis of hop-derived trans-iso-alpha-acids and trans-tetrahydroiso-alpha-acids: product identification in relation to the lightstruck flavor of beer. *Photochemistry and Photobiology Science* 2: 306–314.

Huige N.J. 2002. Evaluating barrier-enhancing and scavenger technologies for plastic beer bottles. *Master Brewers Association of the Americas Technical Quarterly* 39: 218–230.

Huige N.J. 2004. Chemical engineering for quality brewing. *Master Brewers Association of the Americas Technical Quarterly* 41: 9–17.

Huvaere K., Sinnaeve B., Van Bocxlaer J., De Keukeleire D. 2004. Photooxidative degradation of beer bittering principles: product analysis with respect to lightstruck flavor formation. *Photochemistry and Photobiology Science* 3: 854–858.

Ishitani T. 1995. Active packaging for food quality preservation in Japan. In: *Foods and Packaging Materials—Chemical Interactions.* Ackermann P., Jagerstad M., Ohlsson T. (Eds). Cambridge, England: The Royal Society of Chemistry, pp. 177–188.

Kuchel L., Brody A.L., Wicker L. 2006. Oxygen and its reactions in beer. *Packaging Technology and Science* 19: 25–32.

Lee D.S., Yam K.L., Piergiovanni L. 2008. *Food Packaging Science and Technology.* Boca Raton, Florida: CRC Press.

Leiper K.A., Miedl M. 2008. Colloidal stability of beer. In: *Beer: A Quality Perspective.* Bamforth C.W. (Ed). Burlington, Massachusetts: Academic Press, pp. 111–161.

Linssen J.P.H., van Willige R.W.G., Dekker M. 2003. Packaging-flavour interactions. In: *Novel Food Packaging Techniques.* Ahvenainen R. (Ed). Boca Raton, Florida: CRC Press, pp. 144–171.

Long D.E. 1999. From cobalt to chloropropanol: de tribulationibus aptis cervisiis imbibendis. *Journal of the Institute of Brewing* 105: 79–84.

Lynch D.M., Bamforth C.W. 2002. Measurement and characterization of bubble nucleation in beer. *Journal of Food Science* 67: 2696–2701.

Miedl M., Bamforth C.W. 2004. The relative importance of temperature and time in the cold conditioning of beer. *Journal of the American Society of Brewing Chemists* 62: 75–78.

Narziss L., Miedaner H., Graf H., Eichorn P., Lustig S. 1993. Technological approach to improve flavor stability. *Master Brewers Association of the Americas Technical Quarterly* 30: 48–53.

Nielsen T.J., Jagerstad M. 1995. Supercritical fluid extraction (SFE) coupled to capillary gas chromatography for the analysis of aroma compounds sorbed by food packaging material. In: *Foods and Packaging Materials—Chemical Interactions*. Ackermann P., Jagerstad M., Ohlsson T. (Eds). Cambridge, England: The Royal Society of Chemistry, pp. 59–64.

Oag I.D.M., Webb T.J.B. 1984. Practical experiences of packaging beer in PET bottles. *Master Brewers Association of the Americas Technical Quarterly* 21: 131–139.

Ormrod I.H., Lalor E.F., Sharpe F.R. 1991. The release of yeast proteolytic enzymes into beer. *Journal of the Institute of Brewing* 97: 441–443.

Pfisterer E., Richardson I., Soti A. 2004. Control of hydrogen sulfide in beer with a copper electrolysis system. *Master Brewers Association of the Americas Technical Quarterly* 41: 50–52.

Robertson G.L. 2006. *Food Packaging Principles and Practice*, 2nd edn. Boca Raton, Florida: CRC Press.

Ronteltap A.D., Hollemans M., Bisperink C.G.J., Prins A. 1991. Beer foam physics. *Master Brewers Association of the Americas Technical Quarterly* 28: 25–32.

Shellhammer T., Bamforth C.W. 2008. Assessing color quality of beer. In: *Color Quality of Fresh and Preserved Foods*. Culver C.A., Wrolstad R.E. (Eds). ACS Symposium Series #983. Washington, DC: American Chemical Society, pp. 192–202.

Sherman S. 2003. http://www.sptimes.com/2003/06/18/Taste/For_beer_lovers__utop.shtml. Accessed November 2008.

Shirakura A., Nakaya M., Koga Y., Kodama H., Hasebe T., Suzuki T. 2006. Diamond-like carbon films for PET bottles and medical applications. *Thin Solid Films* 494: 84–91.

Taoukis P.S., Labuza T.P. 2003. Time-temperature indicators (TTIs). In: *Novel Food Packaging Techniques*. Ahvenainen R. (Ed). Boca Raton, Florida: CRC Press, pp. 103–126.

Teumac F.N. 1995. The history of oxygen scavenger bottle closures. In: *Active Food Packaging*. Rooney M.L. (Ed). London, England: Blackie Academic & Professional, pp. 193–202.

Teumac F.N., Ross B.A., Rassouli M.R. 1990. Air ingress through bottle crowns. *Master Brewers Association of the Americas Technical Quarterly* 27: 122–126.

Tighe A. 2006. *British Beer and Pub Association Statistical Handbook*. London, England: Brewing Publications Ltd.

Vermeiren L., Heirlings L., Devlieghere F., Debevere J. 2003. Oxygen, ethylene and other scavengers. In: *Novel Food Packaging Techniques*. Ahvenainen R. (Ed). Boca Raton, FL: CRC Press, pp. 22–49.

13 Packaging and the Shelf Life of Wine

Malcolm J. Reeves
Faculty of Science and Technology
Eastern Institute of Technology
Taradale, New Zealand

CONTENTS

13.1 INTRODUCTION

Despite the fact that less than 10% of wines are bought to age, with most wines being consumed within 48 hr of purchase (Lockshin, 2008, personal communication), consumers expect wine to improve with age. Yet this change is neither universal nor everlasting, is variable from wine to wine, and the expectation varies from consumer to consumer. Thus, the required shelf life can be a few weeks to many years.

Broadbent (2002) wrote of the highly prized Chateau Latour 1961 that, despite its extraordinary sweet, nose-filling bouquet, all the component parts were excessively represented and yet it had another half-century of life. But not all top-rated wines are expected to age. New Zealand's highly rated Sauvignon Blancs, often said to be the best in the world (Waugh, 2008), are, in the opinion of expert reviewers, best drunk young (Vogel, 2008).

Wine quality is a multidimensional construct and consumers engage with it depending on their varying involvement with the product (Charters and Pettigrew, 2007). It is a mix of extrinsic and intrinsic factors. With prices ranging from as little as $2 to more than $1000 per bottle, it is understandable that there is a significant variation in consumer's quality expectations.

The packaging of such a range will not surprisingly be diverse and be required to fulfill not just common roles such as physical protection, containment, and presentation but also provision of a shelf life appropriate to the distribution system and the consumer's expectations.

13.2 WINE QUALITY ATTRIBUTES

13.2.1 Initial Wine Flavors and Aromas

13.2.1.1 Flavor Compounds from Grapes

Originating from the grapes, yeast, and bacterial fermentations, from oak in some wines, and from chemical reactions occurring during maturation, there are some 600–800 compounds in wine that possess aroma properties (Rapp, 1998). The concentration of these compounds is influenced by viticultural and winemaking practices, and they have highly different stabilities over time in wine.

Methoxypyrazines, present in the grapes at about 10 times their sensory threshold levels, contribute the herbaceous and bell pepper notes characteristic of the flavor of Sauvignon Blanc and Cabernet Sauvignon wines. Some of these compounds, in particular 2-methoxy-3-isobutylpyrazine (IBMP), are considered to be stable, especially to oxidation (Allen and Lacey, 1999). Others are not stable over time; they are either lost or converted into other flavor compounds, frequently with lower sensory impacts. Acid and enzymic hydrolysis can release additional quantities of some, such as terpenes (found in Riesling varieties) and thiols from nonodorous precursors during fermentation and maturation (Clarke and Bakker, 2004; Lee et al., 2008).

13.2.1.2 Flavor Compounds from Fermenting Yeast

Yeasts produce variable levels of compounds such as higher alcohols (including isoamyl alcohol and propanol); fatty acids and their esters, ranging from the simple ethyl acetate to the larger ethyl decanoate; aldehydes (such as methional); and ketones, which influence the sensory profile of a newly fermented wine. These undergo further changes as a result of hydrolysis and esterification during post-bottling storage.

Acceptably low levels of acetic acid and ethyl acetate are normally produced simultaneously during fermentation and by spoilage bacteria after fermentation. Typically, acetic acid levels do not alter significantly once the wine is packaged. Isoamyl acetate, found at up to 8 mg L^{-1}, contributes to the fruity aroma of white wines.

13.2.1.3 Oak-Derived Flavor Compounds

Oak can make a pronounced contribution to the flavor and aroma of wine. The seasoning and, in particular, the heating of the oak affects the range and level of the resulting compounds such as vanillin and "oak lactone".

13.2.2 Sensory Changes during Storage

That a wine changes during storage/aging is one of its most fascinating properties. Regrettably, most wines improve only for a few months to a few years before showing irreversible loss in quality (Jackson, 2000). The timing and extent of changes are difficult to predict reliably.

13.2.2.1 Aroma and Flavor

The sensory properties of wine change during maturation before packing and continue once packaged; they are influenced by the wine's environment. Although some changes are beneficial, ultimately over time all changes become unacceptable. "Maturation bouquet" may be considered as post-fermentation but prepacking changes, whereas "aging bouquet" may be considered as "in-package" development. The loss of existing fruit-derived aromas is frequently but not always perceived as a quality reduction. Simultaneously, the appearance of new compounds, generically termed "bottle bouquet," may be appreciated and considered to be a quality improvement.

Fruitiness in young wines is modified during storage by ester hydrolysis, especially of acetate esters. Higher temperatures and lower pHs favor this process; the equilibrium position depends on the relative rates of esterification and hydrolysis. Ethyl decanoate and other volatile esters trend downward early (Boulton et al., 1999).

The compounds responsible for bottle bouquet include 1,1,6-trimethyl-1,2 dihydronaphthalene (TDN), methional, sotolon, eugenol, and phenylacetaldehyde (Escudero et al., 2000). TDN is found in many wines but particularly in Riesling-type varieties from warmer countries such as Australia. It has a kerosene-like character, unacceptable to some, rising in some wines to 10 times its 20 μg L^{-1} threshold in 10 to 12 years (Winterhalter, 1991). Its occurrence coincides with changes in esters and terpenes, which are prime varietal aroma contributors that are converted to vitispirane, linalool oxide, and nerol oxide (Simpson, 1978; Simpson et al., 1977) although these may be below their taste and odor thresholds. Sotolon has aromas reminiscent of roasted and baked products and maple syrup, while phenylacetaldehyde has a honey, roasted-nut-like character.

In common with a number of sulfur compounds, methional has a cooked-vegetable character. Dimethyl sulfide often increases with bottle age and has been described as being like canned corn, cabbage, and quince (Vidal and Aagaard, 2008). By itself it smells like boiled corn, which is quite unlike desirable bottle bouquet, but with other compounds it contributes to the bottle bouquet.

13.2.2.2 Acidity

The acid taste of wine softens with age due to esterification of tartaric acid by ethanol. It takes several years and is favored in extent by low temperatures and in rate by high temperatures. The formation of as much as 1.5 g L^{-1} of the ester reduces the acid impact and can make acid wines more acceptable (Edwards et al., 1985).

13.2.2.3 Astringency and Bitterness

These properties of red wines, in particular, change due to the much higher levels of phenolics. Most winemakers fine their wine before bottling to eliminate excessive astringency and bitterness, but many young reds still have a pronounced astringency so that a softening of their aggressiveness is frequently sought through in-bottle phenolic polymerization during aging. This is not a shelf life issue until the polymers become so large, frequently precipitating, that the wine loses mouthfeel, color, and too much astringency. This is typically accompanied by an excessively browned color, and the generation of dull oxidized flavors and excessive precipitated phenolic polymers.

The time for this process varies greatly between reds such that it is impossible to predict precisely when a red wine will reach its limit of acceptability. The indices of "chemical age" developed by Somers and Evans (1977) give methods for assessing the degree of polymerization, but they do not provide a shelf life prediction model.

Because the polymerization can be influenced by the ingress of O_2 (Vidal and Aagaard, 2008) as well as the compositional factors discussed in the earlier text, the packaging system plays an important part in determining the character and shelf life of a red wine. Vidal and Aagaard suggested that some red varieties such as Syrah would benefit from a closure that provides controlled ingress of O_2.

13.2.2.4 Influence of Oxygen

Wine can pick up O_2 during handling at many stages of processing, and some red wines are even given small, controlled doses of O_2 during maturation in a process termed *microoxygenation*. However, all wine can be exposed to too much O_2, resulting in dull, flat, "oxidized" flavors. White wines with their lower levels of phenolics are more prone to exhibiting loss of freshness, fruitiness, and an increase in brownness due to O_2. Odorants formed during wine oxidation include acetaldehyde, 1-octen-3-ol, trans-2-nonenal, furfural, benzaldehyde, 2-butoxyethanol, acetovanillone, and a dioxolane isomer.

There is considerable debate on closures and their permeability to O_2; it is clear that the closure and packaging system can allow or contribute varying amounts of O_2 into the wine. This can have noticeable effects on the aroma of a wine, as graphically portrayed by Figure 13.1.

13.2.2.5 Effect of Inadequate Oxygen

The amount of O_2 required for optimal development varies from wine to wine. Some wines become reductive with insufficient O_2 (Vidal and Aagaard, 2008). The ingress of small amounts of O_2 can result in the production of quinones from phenolics, leading to the loss of odorous sulfur compounds; SO_2 is also intimately involved in this situation.

Wines such as Syrah are prone to the development of sulfur-like odors (SLOs) that can be formed from precursors in the wine such as *S*-methyl methionine (Vidal and Aagaard, 2008) either

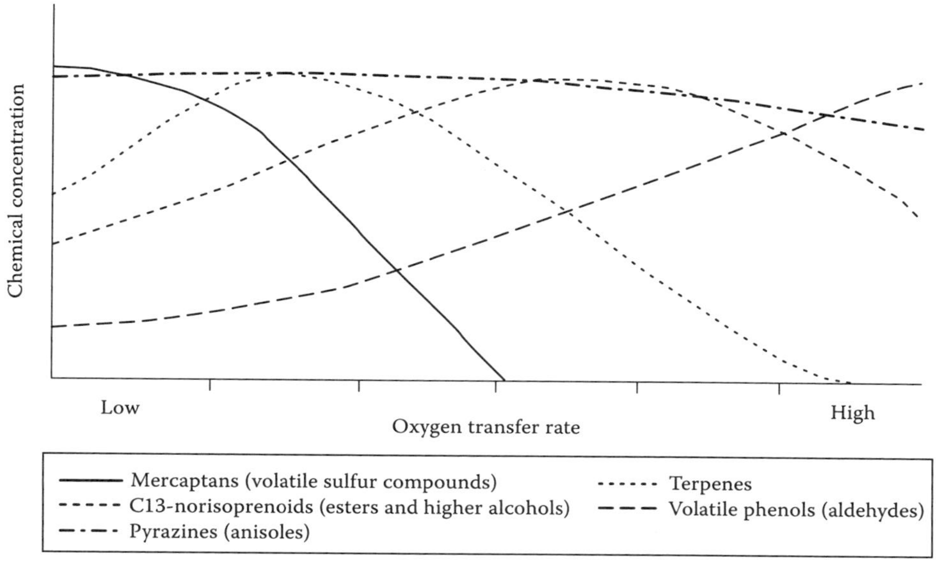

FIGURE 13.1 Oxygen sensitivity of the main aroma groups expressed as the amount of oxygen ingress into the wine. (From Vidal S., Aagaard O. 2008. Oxygen management during vinification and storage of Shiraz wine. *Australian Wine Industry Journal* 23(5): 21–23, with permission.)

by hydrolysis or under anaerobic conditions during fermentation, barrel aging, and bottle aging (Limmer, 2006a). Some SLOs can make a positive contribution to wine quality but at high levels are considered a fault.

13.3 OXYGEN, SULFUR DIOXIDE, AND PHENOLICS

Three of the major factors that determine the shelf life of wine are O_2, SO_2, and phenolics. There is an intricate interaction between these components in wine, and they cannot be considered in isolation. Oxygen is associated with both beneficial and detrimental changes to the sensory properties of wine. The antioxidant SO_2 has been generally found to protect the sensory qualities, but it can also have a negative impact on the color of red wine. Phenolics contribute to the sensory properties in a positive and negative fashion and also react with both O_2 and SO_2. Such interactions occur from the crushing of the grapes at the start of winemaking and continue until the packaged product is consumed. For many wines the end products from the reactions of O_2, SO_2, and phenolics will determine the acceptability and hence shelf life of a wine.

Despite considerable knowledge about the reactions and their products, it is not yet possible to construct a single definitive model for predicting the shelf life of wine. The considerable variety of wine styles, compositional variations within a given style, range of winemaking techniques coupled with the packaging used, as well as differences in personal perceptions of what is desirable and undesirable in wine makes such a model impossible. Nonetheless, it is possible to manipulate conditions so that the desired shelf life of the packaged wine is achieved in most cases when certain parameters are specified. An example for packing wine in flexible bag-in-box packs (Section 13.5.1.6) has been given (Casey, 1989b).

13.3.1 Role of Oxygen

According to Lopes et al. (2005), O_2 is one of the most important factors determining the aging potential of bottled wine. Depending on its phenolic content, a liter of wine can absorb 60–600 mL of O_2, but in doing so, constituents of the wine are oxidized and the character of the wine is altered or even lost (Singleton, 1987). For red wines some O_2 can improve color stability and soften astringency due to polymer formation. As discussed in the preceding text, O_2 in white wines reduces desirable sensory properties; therefore O_2 can be a major factor in determining a wine's shelf life.

Although prepacking operations are important, O_2 levels at packing and subsequent ingress into the package are fundamental in determining subsequent shelf life. At the time of packing, most winemakers seek to get the level of dissolved O_2 (DO) below 1 mg L^{-1}, and depending on wine type, prefer even less than 0.4 mg L^{-1}. This compares with levels of 2–3 mg L^{-1} that may result from deliberate exposure during various prior operations such as transfer to and from barrels. Once present in the wine, O_2 can undergo a range of competing reactions, many of which produce intermediates that cause further, frequently deleterious, reactions.

According to Casey (1988, 1989a) the eventual decline in wine quality is to a large extent caused by the cumulative effects of the very gradual ingress of atmospheric O_2 as the permeability of the traditional natural cork stopper changes with time. The sidewall force reduces as a result of loss of cork elasticity, allowing O_2 to enter via the glass–cork interface. Also, as the wine penetrates the cork, there is a suggestion of changes in cork permeability although data on this is unavailable. This situation is not limited to wines packed in glass bottles with cork stoppers as O_2 can permeate into nearly all packaged wines regardless of closure type or packaging material, albeit at different rates.

13.3.2 Role of Sulfur Dioxide

During winemaking, SO_2 inhibits some microorganisms and enzymes, including polyphenoloxidases (PPO) and laccase, an oxidative enzyme produced by *Botrytis cinerea*, the fungus responsible for noble rot on grapes used to produce distinctive sweet dessert wines such as Sauternes. In white wines the retention of fruitiness and freshness, together with prevention of the development of oxidized flavors, correlated strongly, even if not directly, with the level of SO_2 in the wine (Ough, 1987). The principal function of SO_2 in packaged white wines is to bind the major carbonyls as bisulfite addition compounds, particularly acetaldehyde and chromophoric carbonyl groups (Casey, 2002).

13.3.2.1 Chemistry of Sulfur Dioxide in Wine

In wine SO_2 is considered to exist in two forms, so-called "free" and "bound," which together make up the "total." In reality SO_2 is not sharply partitioned as this simple concept of free and bound might suggest; experience shows varying levels of binding.

When there are elevated levels of SO_2 (e.g., total >150 mg L^{-1}, with say 40 mg L^{-1} being free and the remaining 110 mg L^{-1} being bound), it is possible to decrease both the free and bound by the addition of a strong oxidizer such as hydrogen peroxide (H_2O_2). These decreases are more pronounced at even higher levels of SO_2, especially in sweet white wines where the high residual glucose can loosely bind SO_2 and act as a reservoir when the free is depleted by the H_2O_2. However, some SO_2 is tightly bound; for example, the binding by acetaldehyde is more than 4×10^5 times greater than that by glucose and is unaffected by H_2O_2. Despite this, in the accepted analytical aspiration method for SO_2 determination, such glucose-bound SO_2 is determined as bound.

This liberation is also seen in red wines where the colorless anthocyanin–bisulfite addition compounds are labile, with color being restored once the SO_2 is stripped off. Addition compounds are also formed by α-keto acids such as pyruvic acid. These have dissociation constants that are considerably greater than that for acetaldehyde.

This apparent availability of SO_2 has implications for packaged wine. Depending on the composition of the wine and the level of total and bound SO_2, there may be more antioxidant capacity than is simply represented by the free SO_2 at the time of packing. This situation is discussed by Casey (1989a), Boulton et al. (1999), and Rotter (2008).

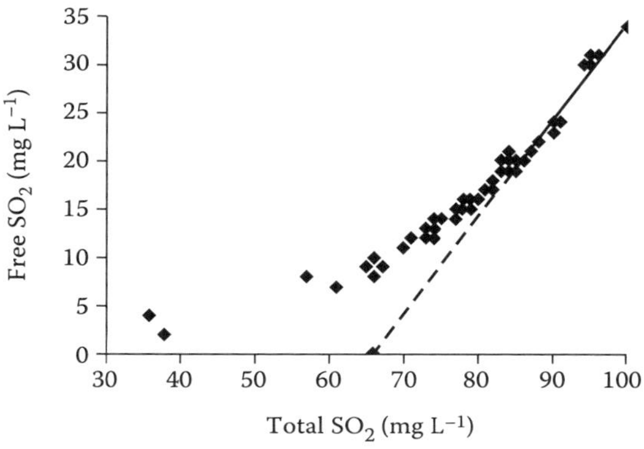

FIGURE 13.2 Relationship between free and total SO_2 for a Semillon wine. (Data from Godden P., Francis L., Field J., Gishen M., Coulter A., Valente P., Høj P., Robinson E. 2001. Wine bottle closures: physical characteristics and effect on composition and sensory properties of a Semillon wine. 1. Performance up to 20 months post-bottling. *Australian Journal of Grape and Wine Research* 7: 64–105 and plotted by Casey J. 2002. A commentary on the AWRI closure report. *Australian and New Zealand Grapegrower and Winemaker* 457: 65–69.)

Casey (1989b, 1996) presented graphs of free versus total SO_2 for hypothetical white and red wines showing, in effect, three states: free, which is lost first during storage; labile, which tests as bound but is so loosely bound that it replenishes the free; and permanently bound. Initially the free is likely to decrease with no great change in the labile, but when the free reaches a low level, the labile begins to dissociate. With time just the permanently bound will remain. It is at this point that oxidative-induced color changes will begin to appear. The wine type and composition will determine the level of permanently bound SO_2 at which this occurs, and it is this that influences the shelf life.

Casey (2002) applied this approach to the closure trial SO_2 results of Godden et al. (2001) involving Semillon wine stored in bottles closed with a wide selection of closures. It was concluded that a better indication of shelf life could be gained by measuring the free and labile SO_2 (taking into account the O_2 permeation into the package) rather than just free SO_2 (see Figure 13.2).

13.3.2.2 Equilibrium Forms of Free SO_2

In wine, free SO_2 exists as three species: $SO_2.H_2O$, HSO_3^-, and $SO_3^=$. All three species are in a pH-dependent equilibrium (Boulton et al., 1999). As the three species have different antioxidant behavior, with reaction kinetics dependent on the concentration of the specific SO_2 species involved, the rate of loss of SO_2 due to oxidation will depend on pH. However, the ultimate antioxidant capacity will be much the same regardless of pH through equilibrium maintenance as determined by the equilibrium constants.

13.3.2.3 Sulfite Oxidation Stoichiometry and Processes

Stoichiometry indicates that it requires 4 mg of O_2 to reduce the measured SO_2 content of wine by 1 mg, the reaction occurring via $SO_3^=$; in practice it is found to vary between 2 and 4 mg. At the pH of wine, the reaction between $SO_3^=$ and O_2 is slow, with a half time of about 30 days (Boulton et al., 1999). Moreover, the increase in $SO_4^=$ in wine is normally less noticeable than the corresponding reduction in total SO_2, indicating that the direct reaction is not the only mechanism operating. Indeed Casey (2008, personal communication) reported an approximate one-to-one conversion when a small number of red wines were examined but reiterated that this process involves intermediate oxidants and that the direct process is virtually nonexistent.

13.3.3 Role of Phenolics

Danilewicz (2008), Waterhouse (2008), and du Toit et al. (2006) discussed a possible interlinked sequence of reactions involving the conversion of phenolics to corresponding quinones, and the formation of peroxide ($O_2^{=}$), and radical formation including phenolic radicals and a superoxide radical ($O_2^{\bullet-}$), which are better oxidants than molecular O_2. Iron and copper act as catalysts in this scheme. It is this sequence that explains variable relationships between O_2 ingress and loss of SO_2.

According to Boulton et al. (1999), the reaction between 30 mg L^{-1} of $SO_3^{=}$ and O_2 in a model wine with pH 3.6 has a half time of about 30 days for the loss of O_2. This is in contrast to ascorbate (sometimes used as an antioxidant in wine at 100 mg L^{-1}), where the half time was considerably shorter at 35 min, reflecting the far greater reactivity of ascorbate with O_2. This indicates that the $SO_3^{=}$ is not a key protector of wine from oxidation by molecular O_2. Waterhouse (2008) reported that the reaction between O_2 and phenolics is the quickest and that the H_2O_2 formed is responsible for the loss of SO_2.

Once the available SO_2 has been depleted to a low level, the peroxide oxidizes flavor compounds and phenolics to brown-colored compounds. In combination with the variable effects of temperature, it is readily understood that each wine will behave differently at its own rate, explaining why accurate shelf life prediction is difficult.

13.4 GLASS PACKAGING AND OXYGEN

13.4.1 Initial Dissolved Oxygen

Today many winemakers target DO levels of 0.2–0.4 mg L^{-1} in wine before bottling but some tolerate as much as 1 mg L^{-1}. From the preceding discussion it would seem self-evident that DO levels should be minimized as much as practicable.

13.4.2 Dissolved Oxygen Increase during Bottling

Variables such as closure type, flushing of empty bottles with inert gas, filler style, equipment maintenance, filler speed, and preclosure headspace flushing affect the DO level in bottled wine. Where empty bottle flushing is practiced, dosing with 1–2 volumes of inert gas (CO_2, N_2, or even Ar if cost is not an issue) may reduce the potential O_2 contribution from the bottle to 1–3 mg L^{-1} (Boulton et al., 1999). Bottling with mildly elevated wine CO_2 levels can also help to purge the bottle and keep the final DO levels to less than 0.5 mg L^{-1} (Brajkovich M., 2008, personal communication). DO pickup can also be influenced by wine temperature (commonly 16°C at filling) and turbulence during the filling process itself. In the case of cylindrical closures (i.e., corks and their analogues as against screw caps), a vacuum is often applied to the bottle to ensure that the headspace gas pressure is kept to about ±20 kPa after insertion. Vacuum screw cappers are also now available (Stelzer, 2005).

Depending on bottling procedure, the approximately 3.5 mL of headspace in a cork closed bottle may contain from 0.2% to 7% O_2, which could potentially increase the DO in a 750-mL bottle of wine by 0.47 mg L^{-1}. The potential DO contribution to wine DO under screw cap would be about three times greater based on the approximate 8 mL of headspace (Stelzer, 2005), and yet bottlers report typical final DO levels of <0.5 mg L^{-1} (Brajkovich M., 2008, personal communication; Nowell-Usticke T., 2008, personal communication). Such increases are rather lower than the 1.1 and 2.8 mg L^{-1} for 750-mL bottles closed with corks and screw caps, respectively, as quoted by Gibson (2005). Some winemakers relax the target levels of DO for red wines because of their phenolic content.

13.4.3 The Package

Once closed, the bottle contains O_2 from the original bulk wine, with the headspace O_2 remaining after any gas flushing or vacuum application at the time of closure application, and any O_2 within a cellular structured closure if such was used. This latter O_2 will move with time, with some escaping from the bottle to its surroundings and some passing into the interior. Over time steady state ingress may be established. The O_2, SO_2, and CO_2 transmission rates of the package including closures will impact on the shelf life. With O_2 playing a pivotal role in the development and longevity of bottled wine, the O_2 level in the newly packaged wine and the quantity that subsequently ingresses into the product is a key determinant of shelf life.

13.4.3.1 Glass Bottles and Glass Bottle Closures

> The wine industry as a whole has not yet arrived at the perfect closure but what seems evident is that there will no longer be a single closure.
>
> **Heald and Heald, 2008**

Glass wine bottles have been closed with traditional bark cork (obtained from the oak tree *Quercus suber*) for some 300 years. Cork has performed reasonably well although it can present problems such as cork dust, leakage and off flavors including cork taint. The occurrence of cork taint (see Section 13.6.2) in wines closed with natural corks, variously estimated as 3–9% of wines by Oeneo Closures USA (2008) and 1–5% by Sefton & Simpson (2005), has been a major motivation for the development of agglomerated corks and other alternative closures.

The various so-called "technical corks" have been developed from reformed, frequently treated comminuted cork into more uniform products such as Twin Top® and more recently Diam® closures. The agglomerate cork, originally developed as a closure for sparkling wine, consists of small pieces or granules of clean, natural cork bound together with resin or a chemical binder into a single stopper, frequently with one or more thin discs of intact natural cork stuck on the end intended to be in contact with the wine. Reduced levels of cork taint have resulted from a disinfecting or deodorizing process which extracts volatile components from the cork material. This, it is claimed, eliminates the possibility of contaminants being retained inside the lenticels or pores through which gases are exchanged between the atmosphere and plant tissues (Robertson, 2006a).

The past 40 years has seen a proliferation of other non-cork closures, including Stelvin®, Zork®, Nomacorc®, and the "Vino-Lok" glass closure, designed specifically for use with table wine. These can offer reduced cost, easy of application and removal, uniform performance, and lower O_2 ingress.

The opportunity to develop a technically superior wine closure to the cork was recognized by a French closure manufacturing company, Le Bouchage Mecanique (LBM), which began research in the late 1950s on a metal closure to replace cork closures. Their Stelcap closure had already gained widespread acceptance in use over aperitifs, spirits, and liqueurs. LBM aimed to modify the Stelcap and develop a quality table wine closure that would completely replace cork. By the late 1960s LBM had developed the Stelvin; it was made of aluminum, was corrosion resistant, and had a treated and chemically inert wad facing that was completely compatible with wine. The wad consists of three components: an expanded 2-mm LDPE foam substrate to provide controlled and uniform compressibility, a 20-μm layer of tin to provide a gas barrier, and a 19-μm PVdC copolymer facing that isolated the tin film from the product and provided an additional O_2 barrier (Robertson, 2006a; Stelzer, 2005).

13.4.3.2 Oxygen Ingress Measurement

Natural cork has given variable protection to wine from oxidation. Despite the assertion of some researchers that corks do not breathe (Limmer, 2006b, 2006c; Petersen, 2008a, 2008b), there is much data confirming the ingress of O_2 through closure systems. Correlated observations were among the early methods of assessment of the ingress of O_2. In particular, wine properties such as the SO_2 status (Casey, 1994) and browning as determined by the A_{420} (absorbance at 420 nm) have been used (Skouroumounis et al., 2005).

Among the common methods for oxygen transmission rate (OTR) measurement are coulometric instruments such as the MoCon OX-TRAN, changes in optical properties of O_2-sensitive materials such as dyes (OxyDot developed by Oxysense) and bis-9,10-anthracene-(4-trimethylphenylammonium) dichloride (BPAA) used by Skouroumounis and Waters (2007), and indigo carmine (Lopes et al., 2005). Oxygen pressure decay over a period of 18–72 hr has also been used (Aracil, 2004).

Most methods have been criticized. Changes in SO_2 and A_{420} are only very loosely correlated with O_2 ingress and changes in sensory performance. Their changes can be influenced by wine composition and the presence of other antioxidants such as ascorbic acid. Skouroumounis and Waters (2007) noted that the MoCon method does not reflect normal cork usage where wine is typically stored horizontally and the cork is in contact with wine, and not the headspace gas. Yet under the MoCon test setup there is gas either side of the closure. Research has shown that corks in particular perform differently when the bottle is stored horizontally (or inverted) compared to upright (Gibson, 2005; Lopes et al., 2006).

Some of the dye oxidation methods have been criticized because it is considered that other wine components may interfere with the reaction, and so give a false result. Skouroumounis and Waters (2007) consider the fact that the OxyDot reagent does not consume O_2 could be a disadvantage. They also note that very careful technique is needed with the indigo carmine method.

13.4.3.3 Oxygen Transmission Rate Results

Different bases for expressing the OTR are found in the literature. Here OTR values are expressed as mL day^{-1} per closure with air on one side. Casey (1994) reported OTR values for natural cork equivalent to 0.002 mL day^{-1} based on SO_2 losses in white wine stored horizontally for 21 months. It was assumed that the loss of SO_2 was entirely due to permeation of O_2 into the wine.

Table 13.1 gives literature OTR values for a variety of wine closures. Not only are there differences between closures but also between researchers for the same type of closure, especially natural cork. Peck (2007) noted that with screw caps not only can the liner type and layer thicknesses affect the OTR but also cap application variables such as top pressure, reform depth, thread roller pressure, and glass finish.

Lopes et al. (2005) reported an OTR range of 0.24–0.50 mg L^{-1} month^{-1} for natural corks for the 2- to 12-month stage of storage. Godden (2004) found a range of 0.0001–0.1227 mL day^{-1} and Gibson (2005) quotes Southcorp results of <0.001 to >1.00 mL day^{-1}, a difference in magnitude in excess of 1000 within the trials but also a difference of 10 between the trials. It has been noted that, in some cork studies where "selected" corks were used, the OTR values found were both lower and more consistent than where the corks were said to have been taken randomly.

13.4.3.4 Gas Permeation through Closures

Robertson (2006b) provides detailed treatment of the mechanisms by which gases pass though polymeric materials. Two effects, a pore effect and a solution-diffusion effect, are discussed. The pore effect is gas transfer through microscopic pores, pinholes, and cracks in the material, whereas the solution-diffusion effect involves gases dissolving in a material, diffusing through it under the influence of a concentration gradient and evaporating at the other surface. This latter process is "true permeability" and varies inversely with thickness.

Although porosity falls very sharply with increasing material thickness, in the case of natural corks it may still be a factor as cork has pores and cracks by virtue of the natural growth process. The extent of these in a given line of corks may be influenced by the visual grading processes employed during manufacture. It is therefore quite conceivable that both pore and solution-diffusion processes will occur in natural corks, whereas only the solution-diffusion process will be found in technical and synthetic cork analogues, most of which should be pore-free by virtue of the manufacturing process and their length.

A third process, transfer of gas through the interface between the glass and the closure material itself where the two surfaces may have parted microscopically, may also be present, but to an extent that is controlled by the individual closure characteristics such as closure diameter, elasticity, dimensional stability with time, and ability to bond with the glass surface; variations in the internal diameter of glass bottle necks is also an important variable, but no published data has been found on this. O'Brien (2005) and Lopes et al. (2007) portrayed the three mechanisms.

TABLE 13.1
Reported OTR Values for Wine Bottle Closures

Closure	OTR min–max	Average	Method	Reference	Other Details
Natural*		0.002	SO_2 loss	Casey (1994)	Horizontally stored white wine
Natural	0.0001–0.1227	0.0179	MoCon	Godden et al. (2005)	$N = 12$; 44 × 24 mm Bottles stored upright
Natural	<0.001–>1.00	0.013	MoCon	Hart and Kleinig (2005)	$N = 35$; 44 × 24 mm
Natural	0.078–0.083		Unstated	Silva et al. (2003)	
Natural	0.004–0.009		MoCon	Gibson (2005)	Stored white wine OK
Natural	0.064–0.192		MoCon	Gibson (2005)	Stored white wine oxidized
Natural	0.002–0.004		Indigo carmine	Lopes et al. (2005)	2–12 months after bottling 1st grade diameters of 22–26 mm
Natural	0.0017–0.0061		Indigo carmine	Lopes et al. (2006)	Horizontal storage between 2 and12 months storage
Natural	0.0001–0.0023		Indigo carmine	Lopes et al. (2006)	Horizontal storage between 12 and 24 months storage
Natural	0.0005–0.0044		Indigo carmine	Lopes et al. (2006)	Vertical storage between 2 and 12 months storage
Natural	0.0001–0.0027		Indigo carmine	Lopes et al. (2006)	Vertical storage between12 and 24 months storage
Natural		0.006	Coulometric	Ortiz et al. (2004)	
Twin Top	0.0002		Indigo carmine	Lopes et al. (2005)	
Technical	0.001–0.006		Indigo carmine	Lopes et al. (2006)	Horizontal storage, >2 months
Synthetic expanded LDPE	0.030–0.038		Unstated	Silva et al. (2003)	Horizontal 24 months Supplied by manufacturer
Synthetic with diaphragm	0.015–0.017		Unstated	Silva et al. (2003)	Horizontal 24 months Supplied by manufacturer
Nukorc	0.007–0.009		MoCon	Gibson (2005)	
Nucork	0.015		Coulometric	Ortiz et al. (2004)	
Integra	0.014		Coulometric	Ortiz et al. (2004)	
Neocork	0.018		Coulometric	Ortiz et al. (2004)	
Nomacorc		0.0308	MoCon	Gibson (2005)	http://www.us.nomacorc.com/products_data.php
Nomacorc	0.006		Indigo carmine	Lopes et al. (2006)	Horizontal storage, >2 months
Nomacorq Classic	0.011		MoCon	Ortiz et al. (2004)	
Superemcorq 2	0.006		MoCon	Ortiz et al. (2004)	
Supremecorq original	0.011		MoCon	Ortiz et al. (2004)	
Supremecorq	0.011–0.015		Indigo carmine	Lopes et al. (2006)	Horizontal storage, >2 months
Synthetics	0.016–0.0308		MoCon	Gibson (2005)	http://www.us.nomacorc.com/products_data.php
Zork		0.0078	MoCon	Leske (2007)	
Altec	0.007–0.0013	0.001	MoCon	Godden (2004)	$n = 6$

(Continued)

TABLE 13.1 (*Continued*)

Closure	OTR min–max	Average	Method	Reference	Other Details
Screwcap	0.0003–0.0007		Indigo carmine	Lopes et al. (2006)	Horizontal storage, >2 months
Screw cap foil lined	<0.001		MoCon	Gibson (2005)	
Screwcap foil		0.0002	MoCon	Peck (2005)	0.002 std dev
Screwcap Saran		0.001		Peck (2005)	0.0003 std dev
Screwcap PVC		0.004	MoCon	Peck (2005)	0.0006 std dev
Screwcap LDPE		0.09	MoCon	Peck (2005)	0.062 std dev
Screwcap	0.0002–0.0008	0.0005		Godden et al. (2005)	n =6
Injection A	0.008		Coulometric	Ortiz et al. (2004)	Used a multicell test rig
Extruded A	0.010		Coulometric	Ortiz et al. (2004)	Used a multicell test rig
Injection B	0.005		Coulometric	Ortiz et al. (2004)	Used a multicell test rig
Extruded B	0.009		Coulometric	Ortiz et al. (2004)	Used a multicell test rig

* Natural = natural cork.

13.4.4 Dissolved Oxygen Behavior after Bottling

The pattern of change in DO after bottling depends on initial DO levels, remaining headspace O_2, the O_2 in the closure, and subsequent O_2 ingress, with the last two being dependent on the closure system. As Table 13.1 shows, O_2 ingress varies considerably from one closure system to the other and also according to how the bottles are stored.

For natural cork, in particular, and to some extent synthetic analogues, mathematical modeling of the processes that occur after insertion is difficult. On the basis that the closure is essentially cellular and that the cells are intact, the increase in internal gas pressure for a typical 45 mm long and 24 mm-diameter natural cork, compressed into a bottle neck of diameter 18.5 ± 0.5 mm, would be 70%. Driven by this pressure increase, cellular gas will gradually permeate out at each end of the cork. At the outside end, this will continue until the internal closure pressure equals that of the atmosphere. The inside end process is more complex, being affected by factors such as closure internal pressure; gas flushing, if used; the extent of pre-insertion vacuum application; and wine composition. In addition, at the inside end the solution-diffusion process will occur, controlled by compositional differences between the closure gas and the headspace gas composition.

DO in the wine will be consumed more quickly than it will be supplied by the various processes due to the reactivity of phenolics (Vidal and Moutounet, 2007). Depending on temperature and wine composition, the initial DO, headspace O_2, and O_2 released from compression of the cork will be essentially consumed within a few days. The pressure migration process will probably be continuing albeit at a decreasing rate for perhaps some weeks. The rate of the solution-diffusion process will also decrease as the overall concentration of O_2 in the closure decreases. The depletion will occur at the surface adjacent to the wine; in the cork body the concentration will be that of the original air. However, as the surface layer O_2 depletes (the N_2 will tend to stay as it is not soluble), the O_2 from the adjacent inner layers will move toward the wine end. Gradually the depletion zone will progress through the cork until it reaches the outer surface. Any depletion will then be made up from the atmosphere. As the diffusion through cork is slow, the establishment of this gradient will take quite some time. Some but not all research shows that the overall transfer process is influenced by the storage position of the bottle (Gibson, 2005; Lopes et al., 2006).

Petersen (2008b) quotes Ribereau-Gayon (1933) that

> 0.1–0.38 mL of O_2 diffused out of the inner end of a cork into the wine bottle in the first 3 weeks after bottling, but less than 0.07 mL over the next 4 months. Clearly, the initial air has come out of the compressed cork cells near the inside end of the cork and not 'through' the cork. As this compressed air escapes (from both ends of the cork) the pressure inside the cork cells diminishes, returning to near normal in 2, 3, or 4 weeks. It probably tails off to extremely low levels for a few months more before stopping entirely, but that wasn't measured.

Once the closure internal pressure equals the headspace pressure, pressure-driven transfer will cease. However, a steady state situation with regard to O_2 ingress will not have been reached due to O_2 removal from the headspace by reaction with wine components. The O_2 concentration at the inside end of the closure will be lower than at the outside end, which will be at atmospheric concentration. Driven by a concentration difference, O_2 will continue to diffuse out of the closure into the wine. The zone of internal O_2 depletion in the cork will gradually progress toward the outside end, and the rate of transfer into the wine will decrease as the magnitude of the concentration gradient in the closure decreases.

Depending on the closure characteristics, storage position, and temperature, and presuming that the wine has the capacity to react with all the transferred O_2, there could be a substantial time period before the steady state is reached for the solution-diffusion process.

Skouroumounis and Waters (2007) followed the cumulative O_2 ingress through a synthetic closure using BPAA and identified three stages, as shown in Figure 13.3. An initial rapid loss in BPAA occurred for about 10 days followed by a period of about 25 days after which there was a generally linear increase in total loss. The latter would represent the steady state solution-diffusion stage. A natural cork could display much the same behavior apart from any contribution from pore diffusion. The rates for the three phases are likely to be different due to the fundamental structural differences. Both closure types may have permitted O_2 to enter as a result of closure/glass interface diffusion.

For bottles stored horizontally, Lopes et al. (2007) found that O_2 in an inserted natural cork continued to be released into the bottle slowly over a period of 12 months, and only a small amount diffused through the cork; the synthetic plastic corks tested (Nomacorc) were permeable to atmospheric O_2 after the first month. They did not discuss possible pore diffusion (a process that will vary from cork to cork) nor transfer at the closure–glass interface.

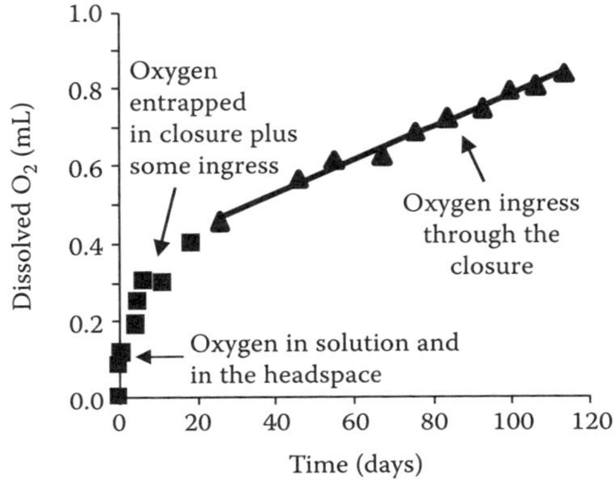

FIGURE 13.3 Ingress of oxygen into a 375-mL wine bottle sealed with a synthetic closure during storage at 18°C. Oxygen content was determined, nondestructively, from the loss of BPAA. (Reproduced from Skouroumounis G., Waters E. 2007. Oxygen ingress into bottled wines. Technical Review No. 170. Australian Wine Research Institute, Adelaide, Australia, with permission. Copyright The Australian Wine Research Institute 2007.)

13.4.4.1 Glass–Cork Interface

Shortly after insertion a cork exerts a sidewall pressure of between 1.5 and 3 kg cm^{-2}; over time this lessens as the cell walls begin to collapse and the unrestrained volume of the cork decreases (Casey, 1988). Depending on magnitude, fluctuations in temperature can encourage the egress of wine and ingress of air. This will be minimized where such fluctuations are seasonal rather than diurnal.

Keenan et al. (1999), after examining the structure of paraffin wax coatings used on corks, calculated that O_2 permeation along the length of paraffin coating on a cork closure would not allow ingress of sufficient O_2 to cause oxidation of wine during storage.

Waters et al. (2001) measured the ingress of O_2 (as A_{420} values) at the closure–glass interface in wines with natural and synthetic closures partly and totally covered with epoxy resin and compared them with uncovered controls and bottles closed with Stelvin closures. Their results indicated that there was some O_2 ingress along the glass–cork interface for natural corks but little through the cork body; for synthetics the reverse applied, with the body being a significant ingress route. The Stelvin absorbance values were the same as for the 100% covered synthetic closure and the 70% and 100% covered natural cork. However, the authors warned that data interpretation must be made with caution due to the small sample size, which ranged from just 4 to 60 bottles for the different closures.

Lopes et al. (2007) investigated ingress routes by similarly covering zones of the outside cork surface and assessing O_2 ingress using indigo-carmine dye. They concluded that technical corks were essentially impermeable for the first 24 months, while natural corks showed low permeability for the first 12 months, with O_2 penetrating in very tiny amounts through the glass–cork interface thereafter. They found that Nomacorc synthetic closures were permeable to atmospheric O_2, especially after the first month of storage.

Differences in O_2 ingress caused by bottle storage position may be explained by wine penetration into the cork and the closure–glass interface. Skouroumounis et al. (2005) found significant differences after just 6 months in the extent of this latter penetration for natural cork from two different sources. For one line of corks, the penetration remained constant, after 6 months, at about 10% of the length of the cork for bottles stored in either inverted or upright positions. For the second line of corks, the penetration increased to 70% for bottles stored upright for 5 years and 90% for the inverted bottles. This difference in wine penetration may influence O_2 ingress via the same path.

13.4.4.2 Movement through Channels

Horizontally stored bottles closed with natural corks frequently show a very small percentage with wine leakage through continuous longitudinal channels. Such channels can facilitate the egress of wine and ingress of O_2, even under relatively stable storage temperatures. This is largely influenced by pressure due to wine volume changes with temperature. For a 750-mL bottle of 12% v/v alcohol wine, the pressure increase due to the thermal expansion alone when the temperature is increased from 20°C to 25°C will be about 11%.

Such channels could be present to a greater or lesser extent depending on cork grading. Poorer, lower-density corks may be affected to a greater extent. Together with a variable glass–cork interface performance, this could contribute to the higher OTR values observed in trials where corks have been randomly selected, such as the Godden (2004) and Gibson (2005) trials, which both reported a 1000-fold OTR range.

After completion of an OTR trial, Gibson (2005) investigated the natural corks for leakage. Of the 64 corks tested, 45% showed bubbling, ranging from small (16%) to gross (3%). Although the pressures used are not encountered in normal everyday storage, the test does support the wide range of OTR values reported by Hart and Kleinig (2005). Of the gas leakers, 33% leaked through the body and 66% leaked at the glass–cork interface. About 70% of the leakers (32% overall) had OTR values > 0.1 mL day^{-1} compared with just 30% of the nonleakers (17% overall) having OTR values >0.1 mL day^{-1}. The level of 48% of the corks with OTR values >0.1 mL day^{-1} is less than the 80% in the 35 random cork trial reported by Hart and Kleinig. To a winemaker, variability in performance is every bit as important as averages in terms of determining shelf life and product acceptability.

However, Casey (2008, personal communication) expressed doubts as to the real contribution of leakers to high OTR values and hence wine shelf life reduction.

13.4.4.3 Effect of Storage Position

As shown in Table 13.1, O_2 ingress has been assessed for bottles stored in upright, horizontal, and inverted positions. Lopes et al. (2006) concluded that storage position had little effect on O_2 ingress in natural corks and its analogues over 20 months. However, Gibson (2005) found some noticeable differences with upright storage resulting in more rapid wine development with some closures. Gibson discussed the possible effect of the O_2–wine interface actually being located in the interior of the cork as the result of wine penetration over time into the cork.

13.4.5 Relationship between Oxygen Transmission Rate and Wine Parameters

Despite some less than perfect aspects to their measurement, OTR values should provide guidance on shelf life and packaging decisions. For bottled wine, Gibson (2005) found a broad correlation between OTR and oxidized character. Three white wines that were judged as being "OK" had mean cork permeabilities of 0.004, 0.007, and 0.009 mL day^{-1}, whereas bottles of the same wines judged to be oxidized had OTR values of 0.192, 0.179, and 0.064 mL day^{-1}. Those judged most oxidized had significantly higher OTR values.

With knowledge of the impact of closure type on SO_2 retention, SO_2 levels at bottling can be appropriately specified. Francis et al. (2003) found that roll-on tamper-evident (ROTE) or screw-cap closures such as Stelvin provided a seal that allowed the SO_2 concentration in a white wine to remain higher during storage than all of the other closures. Also, wines closed with screw caps had the lowest A_{420} values and were consistently rated as high in fruit descriptors without significant oxidation aroma.

Nicolau (2005) observed that screw caps are better at preserving the fruity bouquet of New Zealand's Sauvignon Blanc. This is in particular due to the lower OTR of screw caps giving better protection to the oxidation susceptible thiols, major contributors to the variety's distinctive tropical bouquet.

The relationships between O_2 ingress, free SO_2, freshness of flavor, and color development are not simple. In some cases the observed differences do not reflect the difference in magnitude in closure OTR values.

After 4 years' storage, the SO_2 in a Chardonnay closure trial had dropped from an initial 27 mg L^{-1} free and 64 mg L^{-1} total to 3 and 32 mg L^{-1} for natural cork, 5 and 37 mg L^{-1} for screw cap, and 5 and 39 mg L^{-1} in a glass ampoule (Godden, 2004). Clearly changes in SO_2 are not totally dependent on O_2 ingress. Figure 13.4 shows there are greater differences in color development than the SO_2 figures would suggest. However, the order of browning is generally in line with the reported OTR values. It should be noted that there was greater color development in the bottles stored upright. As was the case with total SO_2, Gibson (2005) found a correlation between A_{420} in wine and OTR values when bottles were stored upright but not when stored inverted.

The imperfect relationship between SO_2 loss and O_2 ingress is supported by Waterhouse (2005), who stated that one cannot determine the amount of O_2 that comes in by looking at the loss of SO_2. Although it is possible to use stoichiometry to predict changes in SO_2 from O_2 ingress, much of that O_2 goes into reaction products that do not react, ultimately, with SO_2. Thus, stoichiometric SO_2 loss underestimates total O_2 ingress.

Despite this, Zoecklein (2005) stated that SO_2 and A_{420} are perhaps the best predictive measures of how wines will hold up after bottling, regardless of closure type. For example, there is a nonlinear relationship between oxidized character development and the concentration of free SO_2. The cutoff appears to be about 13 mg L^{-1} free SO_2 for white wines. As wines develop in the bottle and retain a level of free SO_2 exceeding 13 mg L^{-1}, the likelihood of developing oxidized aroma/flavors is minimized and the wine will usually remain sound. When the free SO_2 level drops to less than about 13 mg L^{-1}, perceptible developed and/or oxidative aromas can be expected. This is in line with Godden et al. (2001), who stated that wines are substantially affected by oxidation at <10 mg L^{-1}

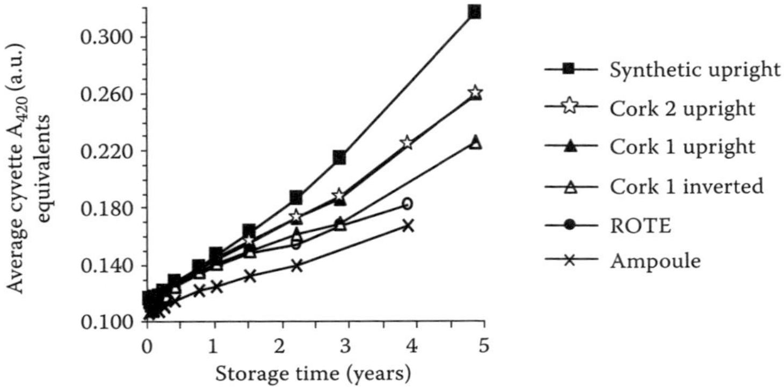

FIGURE 13.4 Effect of closure type and storage position on A_{420} of stored wine. Mean A_{420} values over 5 years of storage for a Chardonnay wine to which ascorbic acid was added at bottling, sealed with various closures, and stored either upright or inverted. (From Skouroumounis G., Kwiatkowski M., Francis I., Oakey H., Capone D., Peng Z., Duncan B., Sefton M., Waters E. 2005. The impact of closure type and storage conditions on the composition, colour and flavour properties of a Riesling and wooded Chardonnay wine during five years' storage. *Australian Journal of Grape and Wine Research* 11: 369–377, with permission, Wiley-Blackwell, Oxford, UK.)

free SO_2. Zoecklein also stated that it is likely that critical loss of antioxidant levels relate, at least in part, to O_2 permeability via closures that will relate to the development of oxidized aroma and flavors, and the variation among wines bottled with different closure types.

13.4.6 Implications for Shelf Life

This was well summarized by Godden (2004), who stated that in closure trials, whenever wines were bottled under different closures, they were changed so radically that they could effectively be thought of as different wines as they aged not only at different rates but in different ways. Godden et al. (2005) added that the ability to link such variables to wine development after bottling creates the possibility of reliably predicting and, therefore, optimizing wine development in the bottle. Thus better knowledge of the wine composition and closure parameters would improve shelf life prediction.

This opens the debate as to what is the optimum condition for a given wine and what defines the IoFs for wine and hence shelf life. Nonetheless, the greater the exposure to O_2, either by the initial DO and/or greater ingress, the shorter will be the shelf life of any wine. Ultimately there will be unacceptable deterioration.

13.5 OTHER PACKAGING FORMATS

Besides the traditional glass bottle, an increasing proportion of the world's wine production is packed in other containers, including bag-in-box (BIB), poly(ethylene terephthalate) (PET) bottles, paperboard laminates, and aluminum cans.

As with bottles, the major influences on shelf life of these alternative packaging formats are temperature, initial DO before packing, pickup during filling, package headspace, and subsequent O_2 ingress. To some extent these are controllable by the winery filling line operator and related to equipment design and setup as well as choice of container material.

13.5.1 Bag-in-Box

Bag-in-box (BIB) wines became popular in Australasia more than 30 years ago, and today about 50% of wines are packed in this format. There has been increasing international acceptance of BIB wines with a 9% share by value in France and the United Kingdom, 42% in Norway, 33% in

Sweden, 25% in Finland, 12% in Denmark, and 6% in the United States (El Amin, 2007). The BIB package offers a high level of convenience, a shelf life of 6 months or more, volumetric efficiency, and a low packaging-to-contents weight ratio.

Today's bags consist of a five-layer coextrusion bag that includes LDPE and nylon with ethylene-vinyl alcohol (EVOH) as the O_2 barrier layer (Lingle, 2004). Metalized PET (mPET) can also be used to improve O_2 barrier properties but it has lower flex resistance. Multilayer metalized laminates can be produced with O_2 permeabilities as low as 0.02 mL m^{-2} day^{-1}.

13.5.1.1 Performance

Typically wines packed in BIB are marked with a "best before" date of some 6–9 months after packing. Apart from variable consumer acceptance standards, this shelf life is highly dependent on the following:

a. Rate of O_2 ingress, which in turn depends on film composition and valve type, and the presence of air between multilayers in the case of bags consisting of two or more webs
b. Temperature of storage
c. Relative humidity
d. Wine composition, including wine type, white versus red, level of antioxidants (SO_2 and ascorbic acid), and DO at the time of bottling
e. Bag size and surface area:volume ratio
f. Film fracturing as a result of transportation and handling
g. Rate of use

13.5.1.2 Oxygen at Filling

Besides the normal DO in wine as it is prepared for bottling, some additional air is normally trapped in the bag as it is filled. Thus, a 4-L bag can contain the equivalent of 4–6 mL of O_2 immediately after filling. Where bags are multilayer as distinct from laminated, O_2 can also be trapped between the layers. In Casey's experience (1989b) there did not appear to be any difficulty in maintaining interlayer air volumes below 20–30 mL, although, unfortunately and inexplicably, they frequently exceeded 40 mL and occasionally 60–70 mL.

13.5.1.3 Oxygen Permeability

13.5.1.3.1 Bag

This depends very much on the film and valve composition. Casey (1989b) stated that shelf life based on O_2 permeabilities currently used were less than 1.0 mL m^{-2} atm^{-1} day^{-1}, with some film combinations going as low as 0.02 mL m^{-2} atm^{-1} day^{-1}. He also noted that shelf life based on these figures can be optimistic as the permeabilities are based on pristine (i.e., unconverted) films. By using a gel with an O_2-sensitive dye, the seams on heat-sealable plastic bags have been demonstrated to be zones of greater O_2 ingress than the plain panels of the bag (Anon, 2004).

13.5.1.3.2 Taps

Taps used in BIBs are now typically multicomponent types made from a variety of compounds including polypropylene (PP), low density polyethylene (LDPE), and thermoplastic elastomer resin (TPE). Davis (1978) and later Doyon et al. (2005) reported that up to 60% of total O_2 ingress in BIB is through the tap, an element overlooked by testing standards based on film material performance. Doyon found that "10–16% of tap permeation was due to the neck and valve-neck interface, and the snugness of the fit was particularly sensitive to temperature."

Published values for OTRs of tap systems are in the range 0.04–0.2 mL tap^{-1} day^{-1}. Scholle (2005) claim the equivalent of 0.08 mL tap^{-1} day^{-1}. Some bags are sealed with a barrier membrane in the tap fitting, and, before puncturing on first use, their OTRs are in the region of 0.01 mL tap^{-1} day^{-1} (Casey, 1989b).

13.5.1.4 Sulfur Dioxide

The overall rate of SO_2 loss is variable in BIB wines. The greatest difference occurs in the initial phase due to the variable initial level of O_2 in the package. Thereafter SO_2 loss in an undisturbed 4-L cask at 20°C takes place at essentially a steady rate of 0.15–0.20 mg L^{-1} day^{-1}. No significant differences between red and white wines were noted (Casey, 1989b). As Davis (1978) showed by some elegant calculations, SO_2 egress is not likely to be significant for BIB wines.

As with white wines in other forms of packaging, once the free SO_2 level drops below 10 mg L^{-1}, the wine shows increasing oxidation and has effectively reached the end of its shelf life. The shelf life of BIB reds can be considered acceptable down to zero free SO_2 (Casey, 1989b).

13.5.1.5 Loss of Carbon Dioxide

This is of interest but is not quite as crucial as SO_2. Dissolved CO_2 affects the sensory balance of wine, enhancing fruitiness in young wines, and increasing the perception of freshness. Its loss may therefore reduce the quality and acceptability. Typically BIB wines are packed with a mildly elevated CO_2 content as high as 1.1 g L^{-1} (Casey, 2008; personal communication). Casey (1989b) gives losses of 1–2 g of CO_2 in 200 days, but more modern bags are likely to perform better.

13.5.1.6 Shelf Life Calculation

Casey (1989b) gives the following simplistic formula to calculate the shelf life of BIB white wine based on a drop of 50 mg L^{-1} of SO_2, resulting in an oxidized wine:

$$t = \frac{50 \times V_w - 5.3(O_h + O_i)}{0.21 \times 5.3(P_t + P_f A)} \tag{13.1}$$

where

t = shelf life (days)
O_h = volume of O_2 in the headspace (mL)
O_i = volume of O_2 between the interlayers (mL)
V_w = volume of wine (L)
P_t = O_2 permeability of the tap (mL tap^{-1} atm^{-1} day^{-1})
P_f = O_2 permeability of the barrier film (mL m^{-2} atm^{-1} day^{-1})
A = surface area of barrier film (m^2)

13.5.1.7 Effect of Transport

Flexing of bags during transport and distribution can result in increased O_2 ingress and, in extreme cases, in leaking. Sundell et al. (1992) found the shelf life of BIB wines was rather longer than their theoretical figures of 130–450 days for new bags of mPET but only 30–130 days for worn (transported) bags. For mPET Doyon et al. (2005) found transport-induced damage occurred mainly near the edges and around the tap. OTRs ranged from 0.5 to 1.0 mL m^{-2} day^{-1} after transporting for between 414 and 809 km, giving shelf lives of 3–8 months.

13.5.2 Poly(ethylene Terephthalate) Bottles

Wine packaged in PET bottles is becoming increasingly accepted (Carter, 2007). It offers considerable weight and not insignificant reductions in total volume for the same content volume (Newhouse, 2008).

13.5.2.1 Performance

With an O_2 permeability of around 3×10^{-10} mL cm cm^{-2} s^{-1} (cm Hg)$^{-1}$ at 25°C, monolayer PET does not have an acceptable barrier for highly O_2-sensitive products such as wine. Its performance for wine can be improved by the use of a barrier layer (MXD6 nylon) between two PET layers and the addition of O_2 scavengers could give 2–4 times greater shelf life, with further improvement given by closures containing O_2 scavengers. Although different bottle volumes are involved, a comparison can be made with the O_2 ingress values for various closures reported by Godden (2004) and Gibson (2005).

A potential shelf life of 1 year is being claimed for 187-mL barrier-coated PET bottles fitted with aluminum caps (Linkplas, 2008). According to Birkby (2006), a scavenger layer can result in less than 10 mg of O_2 permeating into a 2-L PET bottle in a year. New coatings such as oxides of silicon are also claimed to significantly reduce O_2 permeation (Mans, 2008). Atmospheric ingress through the closure should also be controlled so that the headspace O_2 is scavenged preferentially.

13.5.2.2 Loss of Carbon Dioxide

For wines packed in PET, the CO_2 permeation rate is 4–5 times higher than O_2. A typical 1.5-L PET bottle can lose as much as 40 mg day^{-1} with an internal pressure of 300 kPa; for still wines, this loss rate could be reduced to 2 mg day^{-1}.

13.5.2.3 Gases Dissolved in Poly(ethylene Terephthalate)

Dissolved O_2 in new PET bottles at the time of filling will have a negligible impact based on a 1.5-L bottle being about 40–50 g and the saturation level of O_2 being between 0.008 and 0.012 mL g^{-1}.

The 10–20 ng g^{-1} of acetaldehyde that can be found in PET is not a problem for wine, because not only will any free acetaldehyde rapidly react with free SO_2, rendering it nonodorous, but the resulting reduction in free SO_2 (about 0.003 mg L^{-1}) will also be insignificant. Moreover, acetaldehyde bound to SO_2 is found in wine as a result of normal fermentation.

13.5.3 Laminated Paperboard Containers

Semi-rigid paperboard containers laminated with alufoil and LDPE and aseptically filled have been used for many years for wine. Despite offering weight and space savings and easy disposal once empty, they have not been accepted universally but have become extremely popular in certain parts of the world, including Latin America and Scandinavia.

13.5.3.1 Performance

Foil in laminates provides a good O_2 barrier (about 0.02 mL O_2 m^{-2} day^{-1}), providing about 12 months' shelf life even with packs as small as 200 mL (Casey, 1989b). As with all other forms of wine packaging, the DO and headspace O_2 would be a significant contributor to loss of initial free SO_2 and hence influence shelf life. Brick-type packs can be formed with zero headspace, and so with good wine handling and filling technology, there is no reason why the initial DO levels in the packed wine would be significantly different to bottled wine.

Buiatti et al. (1997) found that wine in laminated paperboard cartons could be stored for up to 24 months, as measured by several quality-related properties, and that this was longer than for PET and BIB packaging. Recent improvements to the barrier and scavenging properties of PET have improved the performance of this type of package.

13.5.3.2 Environmental Profile

A Life Cycle Inventory (LCI) evaluated three types of container systems for wine: laminated paperboard containers, glass bottles, and PET bottles (Anon, 2006). The paperboard containers ranged in capacity from 200 mL to 1 L; the glass bottles were 187 mL and 750 mL, as were the PET bottles. The paperboard containers had the lowest weight per delivered volume of wine and the lowest total energy requirements; the glass bottles had the highest. The production of container materials accounts for the largest share of total energy for all container systems. The glass bottles had significantly higher transportation requirements than the paperboard containers or PET bottles.

13.5.4 Metal Cans

Although widely accepted for many beverages, cans have not been especially well received as packaging for wine. Even when coated with enamel (lacquer), corrosion of the can caused by sulfites and anions such as chlorides has been a major hurdle. Recently, a new product has been launched (Barokes, 2008) which claims success with an epoxy internal coating that can withstand 35 mg L^{-1} free SO_2 and up to 250 mg L^{-1} total SO_2. Sidewall rigidity is achieved by carbonating the wine to an internal pressure of about 170 kPa.

Wine in metal cans may from time to time show problems with SLOs (see Section 13.6.1).

13.6 QUALITY FAULTS ASSOCIATED WITH PACKAGING

13.6.1 Sulfur-like Odors

Odorous sulfur compounds or "reduced characters," particularly hydrogen sulfides, dimethyl sulphide, and thiols, have long been recognized as a fault in wines. At low levels they can add some complexity and breadth to the flavor profile, but when they become identifiable and dominate the fruit and maturation characters in wine, their presence is a fault. The sensory thresholds for the various SLOs (CQC, 2007) are of particular interest. For example, dimethyl sulfide has a threshold that is some 20 times greater than those of hydrogen sulfide and methanethiol.

Not all wines are equally prone to developing SLOs. It was noted (Coulter, 2008; Godden, 2006; Godden et al., 2005; Limmer, 2006b, 2006c) that some wines are predisposed to the development of SLOs when bottled with low-OTR closures; in higher OTR packages the SLOs are oxidized and so are not noticed. Attention to winemaking practices has decreased the predisposition to their development.

There has been some poorly written comment about closures and SLOs in the wine media. This has provoked responses from the scientific community (Pretorius, 2006). Limmer (2005b) correctly summed up the situation by stating that much of what has been written has little origin in science and more reflects popular beliefs.

The chemistry of the SLOs and their interconversion has been examined at length by Limmer (2005a). Under oxidizing conditions, thiols can be converted to disulfides, thereby decreasing their sensory intensity. Winemakers have practiced aerating wines to reduce the SLO impact. However, under reducing conditions such as may exist in a bottle closed with a low-OTR closure, the low-impact disulfides at bottling can be reduced back to the more pungent thiols during storage, a process that make take 2 years (Limmer, 2006b). This explains the recurrence of SLOs in wines that were thought to be "corrected."

Skouroumounis et al. (2005) demonstrated that this process is mediated by reducing conditions by storing the same wine in glass ampoules, as well as bottles closed with a variety of closures. The ampoule and low-OTR closed wines developed "reductive" characters, as shown in Figure 13.5. Gibson (2005) noted that results from at least six published trials showed consistent presence of SLOs in wine in bottles sealed with low-OTR closures; the same wine in bottles sealed with higher OTR closures did not show SLOs. New screw caps with modified liners that will permit known rates of O_2 ingress are being developed.

Limitation of storage life through oxidation and the generation of SLOs under certain reducing conditions is thus a balance that involves closure type and wine composition. Recommendations have been made as to how to achieve this balance (Godden et al., 2005).

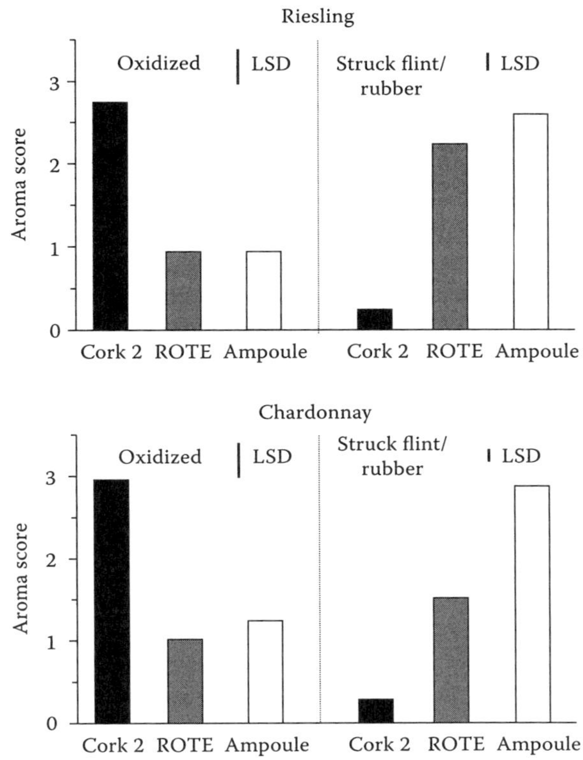

FIGURE 13.5 The effect of closure type on the struck flint/rubber and oxidized aroma score after 4 years (LSD = least significant difference). (From Skouroumounis G., Kwiatkowski M., Francis I., Oakey H., Capone D., Peng Z., Duncan B., Sefton M., Waters E. 2005. The impact of closure type and storage conditions on the composition, colour and flavour properties of a Riesling and wooded Chardonnay wine during five years' storage. *Australian Journal of Grape and Wine Research* 11: 369–377, with permission, Wiley-Blackwell, Oxford, UK.)

13.6.2 Organohalogens

Organohalogens have low sensory thresholds and can impart taints to foods at concentrations in the parts per billion (ppb) or parts per trillion (ppt) range. A musty or corked character in wines (referred to as cork taint) has long been associated with 2,4,6-trichloroanisole (TCA), with 1–5% of corked wines being affected (Sefton and Simpson, 2005, and references therein). TCA was first identified as the major cause of cork taint in bottled wine by Buser et al. (1982), who reported concentrations of 2,4,6-TCA from 20 to 370 ppt (ng L^{-1}) in a series of red and white wines with this distinct off-flavor that is perceivable in wine at concentrations as low as 10 ppt. It was presumed that 2,4,6-TCA and other related chlorinated compounds originated from chlorination of lignin-related substances during the chlorine bleaching used in the processing of cork and that these compounds later migrated into the wine. About 12 other compounds have been implicated, although their role has not received adequate attention and is incompletely understood (Sefton and Simpson, 2005). However, a review by researchers from Portugal, the world's major producer of cork (Pereira et al., 2000), concluded that faulty corks were responsible for only a minority of contaminated wines and that targeting TCA contamination of cork closures as the major cause of wine taint was not justified.

Chloro- and bromoanisoles are produced by fungal methylation of the parent halophenols under humid and warm conditions (xerophilic fungi require a moisture content between 12% and 16% and a temperature between 25°C and 35°C for germination and growth).

A survey of 2400 wines from northern hemisphere wineries found that many wines exhibiting cork taint had low or undetectable concentrations of 2,4,6-TCA (Soleas et al., 2002). The authors concluded that trichlorophenol (TCP) and TCA are, at most, minor components of cork taint in commercial wines. Their critical examination of the primary data on which the original assertions are based revealed small sample numbers, a tendency to ignore anomalous results, and the absence of any statistical proof for the asserted relationship. Their study has been criticized by Sefton and Simpson (2005) on the grounds that the analysis was conducted without an internal standard, had no means of ensuring peak homogeneity, and was based on a standard addition curve with concentrations three orders of magnitude greater than those determined in real wine samples. They also criticized the fact that wine was categorized as tainted if assigned as such by as few as one of an unknown number of assessors.

An important form of cork taint described as "fungal must" and considered to be second only to TCA in importance has recently been identified as 2-methoxy-3,5-dimethylpyrazine (Simpson et al., 2004). It has an aroma threshold in white wine of 2.1 ppt and was first reported as being responsible for an obnoxious odor (musty, foul drains, or sour dishcloths) present in certain machine-cutting emulsions used in engineering workshops; it was produced by an aerobic, gram-negative bacterium (Mottram et al., 1984). The origin of this compound in corks is unknown but, unlike TCA, it is not necessarily bacterial; the cutting punches that produce the cork cylinders could be one possible source (Simpson et al., 2004).

French researchers identified 2,4,6-tribromoanisole (TBA) in wines causing pungent, musty odors in the absence of chloroanisoles (Chatonnet et al., 2004). The TBA came from the precursor 2,4,6-tribromophenol (TBP), and both derived mainly from environmental pollution in wineries where the atmosphere was contaminated with TBA coming from TBP used recently to treat wood or originating from much older structural elements of the winery or from used wooden containers. In certain cases, although the initial source had been eliminated, residual pollution adsorbed on walls or found in old barrels could be sufficient to make a building unsuitable for storing wines or sensitive materials (including corks and glass bottles) intended for direct contact with wine. When wine is tainted before bottling, the taint is usually detected by the winemaker and the tainted wine should not then reach the consumer.

Champagne corks packed in polyethylene bags inside fiberboard cartons and loaded into a shipping container were contaminated during transport to Australia (Simpson and Lee, 1990). The concentration of 2,4,6-TCA in the corks was 1–8 ppb and in the fiberboard cartons, 25 ppb. It was deduced that the floor of the shipping container had been treated with the fungicide 2,4,6-TCP.

13.6.3 Flavor Scalping

This is not a major problem with wine packaging. Godden et al. (2001) noted that although the loss of fruit aroma intensity is likely to be directly caused by oxidative damage to flavor compounds, it could also be due to a masking effect of low levels of aldehydes being formed; an alternate possibility is the absorption of flavor compounds by the closure material.

With respect to BIB, Casey (1989b) reported that despite the ability of some inner liners such as LDPE to absorb flavors, problems of flavor loss were not encountered. A study by Capone et al. (1999) demonstrated chloroanisoles, and, in particular, TCA and 2,3,4,6-tetrachloroanisole (TeCA) were absorbed by such packaging, thereby improving the sensory quality of wines so affected.

Pollnitz et al. (2002) found that 75% of TDN in a BIB test wine was absorbed within 24 hr and 97% overall after 32 days. The losses for β-ionone and rose oxide, both significant wine aroma compounds, were about 35% greater than the reduction for the same wine stored in a glass ampoule. Another hydrocarbon closely resembling TDN, 2,3,6-trimethylphenylbutadiene (TPB), a contributor to the bottle age character in Semillon, may also be absorbed by corks, where its level was found to be about 70% of that in ampoule-stored wines. However, given that aged whites wines, which have higher TDN, are not normally packed and sold in BIB formats, this is not significant. The loss for β-damascenone, about 40% in glass, was about 45% in the BIB pack.

Just as TDN was absorbed by bags, it was found that after 2 years up to 98% was absorbed by some synthetic closures and that natural corks, Twin Tops, and screw caps removed 50%, 70%, and 0% respectively, as shown in Figure 13.6 (Capone et al., 2005). Overall, wines closed with screw caps could not be differentiated from wines stored in glass ampoules.

The situation with esters is complicated by compositional and equilibrium considerations. Capone and Sefton (2003) and Capone et al. (2003, 2005) reported that in glass-packed wine some synthetic closures scalped flavors over a period of 2 years. It was found that less-polar, long-chain esters such as ethyl decanoate were affected, with a final level in the bag being just 40% of the glass-stored wine. The effect was slightly less for ethyl octanoate and hexanoate: 50% and 70% respectively, as shown in Figure 13.7. There was no absorption of monoterpenes, with the changes in all trial wines being effectively the same.

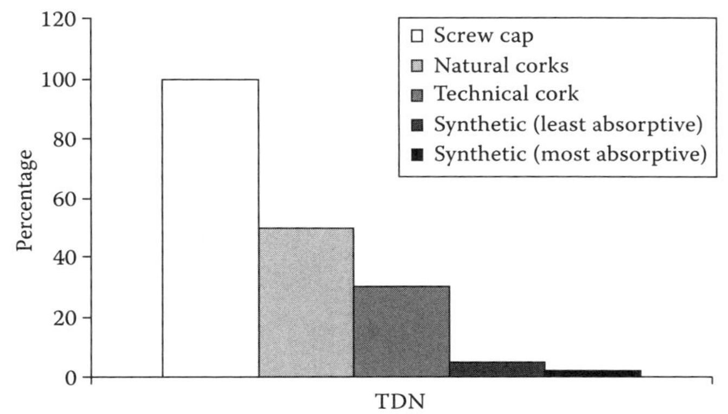

FIGURE 13.6 Percentage of trimethyldihydronaphthalene (TDN) remaining in white wine stored horizontally after 2 years storage of wine in glass bottles closed with various closures compared to wine stored in ampoules. (From Capone D., Simpson R., Cox A., Duncan B., Skouroumounis G., Sefton M. 2005. New insights into wine bottle closure performance—flavour 'scalping' and cork taint. *Proceedings of the 12th Australian Wine Industry Technical Conference*. Blair R.J., Williams P.J., Pretorius I.S. (Eds). July 24–29, 2004, Australian Wine Industry Technical Conference Inc., Urrbrae, South Australia, pp. 215–218, with permission.)

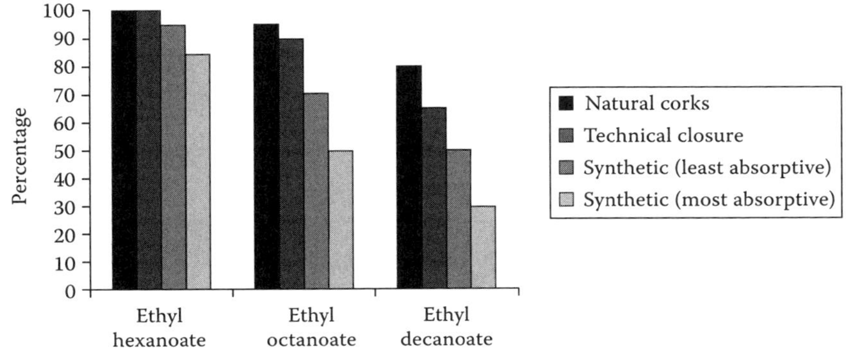

FIGURE 13.7 Concentration of ethyl esters in bottled wine as a percentage of wines stored in glass ampoules after storage for 2 years. (From Capone D., Simpson R., Cox A., Duncan B., Skouroumounis G., Sefton M. 2005. New insights into wine bottle closure performance—flavour 'scalping' and cork taint. *Proceedings of the 12th Australian Wine Industry Technical Conference*. Blair R.J., Williams P.J., Pretorius I.S. (Eds). July 24–29, 2004, Australian Wine Industry Technical Conference Inc., Urrbrae, South Australia, pp. 215–218, with permission.)

Brajkovich et al. (2005) and Herbst et al. (2008) investigated the impact of closures on volatile thiols, in particular mercaptohexyl acetate (MHA), which has a passion fruit aroma, and mercaptohexanol (MH), with a fruity, grapefruit aroma, both being important contributors to Sauvignon Blanc flavor. MHA can be hydrolyzed, or in the presence of phenols oxidized, to MH, which has a 15-fold higher sensory threshold. Closures with low OTR and SO_2 are thought to protect the MHA. Brajkovich et al. (2005) found that wines stored for 2 years under corks had between 18% and 23% lower levels of volatile thiols than those closed with screw caps. As there were no signs of oxidation, it was suggested that scalping may have been a possible cause, although initial higher DO in bottles closed with corks as well as possible OTR differences may have been the cause. In contrast, Herbst et al. (2008) found that, although MHA is the least stable of the volatile thiols (it had disappeared from most of the wines after 12 months of storage), the type of closure was not found to have a major impact on the level of aromas and antioxidants under investigation.

REFERENCES

Allen M.S., Lacey M.J. 1999. Methoxypyrazines of grapes and wines. In: *Chemistry of Wine Flavor*. Waterhouse A.L., Ebeler S.E. (Eds). ACS Symposium Series #714. Washington, DC: American Chemical Society, pp. 31–38.

Anon. 2004. Novel Approach to Package Evaluation: GFTC's Oxygen Sensitive Indicator. http://www.gftc.ca/tech/profiles/osi.cfm. Accessed September 2008.

Anon. 2006. *Life Cycle Inventory of Container Systems for Wine*. Prairie Village, Kansas: Franklin Associates, a Division of ERG. www.tetrapak.ca/pics/winepack-report-EN.pdf. Accessed 30th January 2009.

Aracil J.-M. 2004. Oxygen permeability control in technological closures. http://www.tcafreecorks.com/downloads_docs/rd_AWITC_JMA_permeability_technologicals_closures_07_2004_GB.pdf. Accessed November 2008.

Barokes Pty Ltd. 2008. Wine in a can. http://wineinacan.com/. Accessed June 2008.

Birkby D. 2006. Plastic Packaging—the New Choice of the Wine Industry. http://aux.myplasticsrecruiter.com/aux/newsletter/articles/article046.html. Accessed June 2008.

Boulton R.B., Singleton V.L., Bisson L.F., Kunkee R.E. 1999. *The Principles and Practice of Wine Making*. New York: Springer Elsevier Science.

Brajkovich M., Tibbits N., Peron G., Lund C., Dykes S., Kilmartin P., Nicolau L. 2005. Effect of screw cap and cork closures on SO_2 levels and aromas in a sauvignon blanc wine. *Journal of Agricultural and Food Chemistry* 53: 10006–10011.

Broadbent M. 2002. *Vintage Wine. Fifty Years of Tasting over Three Centuries of Wine*. Brettenham House, Lancaster Place, London: Little, Brown an imprint of Time Warner Books.

Buiatti S., Celotti E., Ferrarini R., Zironi R. 1997. Wine packaging for market in containers other than glass. *Journal of Agricultural and Food Chemistry* 45: 2081–2084.

Buser H.R., Zanier C., Tanner H. 1982. Identification of 2,4,6-trichloroanisole as a potent compound causing cork taint in wine. *Journal Agricultural and Food Chemistry* 30: 359–362.

Capone D., Sefton M. 2003. Flavour 'scalping' by wine bottle closures. *Technical Review 144*. Australian Wine Research Institute, Adelaide, Australia, pp. 6–9.

Capone D., Sefton M., Pretorius I.S., Høj P.B. 2003. Flavour 'scalping' by wine bottle closures. *Australian and New Zealand Wine Industry Journal* 18(5): 16–20.

Capone D., Simpson R., Cox A., Duncan B., Skouroumounis G., Sefton M. 2005. New insights into wine bottle closure performance—flavour 'scalping' and cork taint. *Proceedings of the 12th Australian Wine Industry Technical Conference*. Blair R.J., Williams P.J., Pretorius I.S. (Eds). July 24–29, 2004, Australian Wine Industry Technical Conference Inc., Urrbrae, South Australia, pp. 215–218.

Capone D., Skouroumounis G., Barker D., McLean H., Pollnitz A., Sefton M.1999. Absorption of chloroanisoles from wine by corks and by other materials. *Australian Journal of Grape and Wine Research* 5: 91–98.

Carter M. 2007. Sainbury's to sell wine in PET bottles. http://www.seriousaboutwine.co.za/?m=200707. Accessed July 2008.

Casey J. 1988. Wine and Beverage Closures. Address to the Australian Institute of Packaging Seminar on Closures. Sydney, Australia, November 16, 1988.

Casey J. 1989a. Closures for wine bottles—a user's viewpoint. *Australian Grapegrower and Winemaker* 304: 99–107.

Casey J. 1989b. Flexible packages for wine. *Australian Grapegrower and Winemaker* 304: 57–62.
Casey J. 1994. Is cork a good seal for wine? *Australian Grapegrower and Winemaker* 372: 37, 39–41.
Casey J. 1996. Oxidation, serendipity and sulfur dioxide. *Australian Grapegrower and Winemaker. Annual Technical Issue* 438: 42–52.
Casey J. 2002. A commentary on the AWRI closure report. *Australian and New Zealand Grapegrower and Winemaker* 460: 65–69.
Casey J. 2008. Personal communication.
Charters S., Pettigrew S. 2007. The dimensions of wine quality. *Food Quality and Preference* 18: 997–1007.
Chatonnet P., Bonnet S., Boutou S., Labadie M.D. 2004. Identification and responsibility of 2,4,6-tribromoanisole in musty, corked odors in wine. *Journal Agricultural and Food Chemistry* 52: 1255–1262.
Clarke R., Bakker J. 2004. *Wine Flavour Chemistry*. Oxford, England: Blackwell Publishing, pp. 173–174.
Coulter A. 2008. Closure issues. Unified Wine and Grape Symposium, Sacramento, California: January 30, 2008. http://www.foodsci.purdue.edu/research/labs/enology/Coulter%20Unified%20Symp%20Reduction%20The%20Closure%20Issues.pdf. Accessed October 2008.
CQC. 2007. Internal Report. Sulfide Analysis of Screw Cap Wines. Cork Quality Council. http://www.corkqc.com/currentresearch/Micro-Oxidation/SulfideReportWeb.pdf. Accessed October 2008.
Danilewicz J. 2008. The interaction of polyphenols, oxygen and sulfite in wine. Address to the Annual Meeting of the American Society for Enology and Viticulture, Portland, Oregon, June 18, 2008.
Davis E.G. 1978. The performance of liners for retail wine casks. *Journal of Food Technology* 13: 235–241.
Doyon G., Clément A., Ribéreau S., Morin G., 2005. Canadian bag-in-box wine under distribution channel abuse: material fatigue, flexing simulation and total closure/spout leakage investigation. *Packaging Technology and Science* 18: 97–106.
du Toit W., Marais J., Pretorius I., du Toit M. 2006. Oxygen in must and wine: A review. *South African Journal of Enology and Viticulture* 27: 76–94.
Edwards T.L., Singleton V.L., Boulton R.B. 1985. Formation of ethyl esters of tartaric acid during wine aging: chemical and sensory effects. *American Journal of Enology and Viticulture* 36: 118–124.
El Amin A. 2007. Traditional French begin switch to bag-in-box wine. *Food Production Daily* February 15th 2007. http://www.foodproductiondaily.com/Packaging/Traditional-French-begin-switch-to-bag-in-box-wine. Accessed August 2008.
Escudero A., Cacho J., Ferreira V. 2000. Isolation and identification of odorants generated in wine during its oxidation: a gas chromatography-olfactometric study. *European Food Research and Technology* 211: 105–110.
Francis L., Lattey K., Smyth H. 2003. Reduced aroma in screw-cap bottled white wines. *Technical Review* 142. Australian Wine Research Institute, Adelaide, Australia, pp. 10, 51–53.
Gibson R. 2005. How good is your seal? Closure oxygen transmission and wine quality. Rutherglen Wine Show seminar, Victoria, Australia. http://www.scorpex.net/gibsonoxtransrutherglen2005.pdf. Accessed June 2008.
Godden P. 2004. Results of the AWRI trial of the technical performance of various wine bottle closures up to 63 months post-bottling, and an examination of factors related to 'reductive' aroma in bottled wine. 1st International Screw Cap Closure Symposium, Marlborough, New Zealand.
Godden P. 2006. The Oeneo Closures Debate 2006. London International Wine and Spirits Fair. http://www.wineanorak.com/oeneo_closures_debate_2006.pdf. Accessed August 2008.
Godden P., Francis L., Field J., Gishen M., Coulter A., Valente P., Høj P., Robinson E. 2001. Wine bottle closures: physical characteristics and effect on composition and sensory properties of a Semillon wine. 1. Performance up to 20 months post-bottling. *Australian Journal of Grape and Wine Research* 7: 64–105.
Godden P., Lattey K., Francis L., Gishen M., Cowey G., Holdstock M., Robinson E., et al. 2005. Towards offering wine to the consumer in optimal condition—the wine, the closures and other packaging variables: a review of AWRI research examining the changes that occur in wine after bottling. *Australian and New Zealand Wine Industry Journal* 20(4): 20–30.
Hart A., Kleinig A. 2005. The role of oxygen in the ageing of bottled wine. *Australian and New Zealand Wine Industry Journal* 20(2): 46–50.
Heald E., Heald R. 2008. Getting Closure. The continuing search for the best way to seal the bottle. http://wine.appellationamerica.com:80/wine-review/532/Nomacorc-research.html. Accessed September 2008.
Herbst M., Kilmartin P., Nicolau L. 2008. Stability of varietal aromas in Marlborough Sauvignon Blanc wines. Poster summary In: Proceedings of the 13th Australian Wine Industry Technical Conference. Blair R.J., Williams P.J., Pretorius I.S., (Eds). AWITC Inc, Urrbrae South Australia.
Jackson R.S. 2000. *Wine Science: Principles Practice and Perception*, 2nd edn. San Diego, California: Academic Press, Elsevier.

Keenan C., Gözükara M., Christie G., Heyes D. 1999. Oxygen permeability of microcrystalline paraffin wax and relevance to wax coatings on natural corks used as wine bottle closures. *Australian Journal of Grape and Wine Research* 5: 66–70.

Lee S., Rick F., Dobson J., Reeves M., Clark H., Thomson M., Gardner R. 2008. Grape juice is the major influence on volatile thiol aromas in Sauvignon blanc. *Australian and New Zealand Grapegrower and Winemaker* 533: 78–86.

Leske P. 2007. ZORK V5 Provisor Trial at 38 Months. http://www.zorkusa.com/pdfs/38_month_test_summary.pdf. Accessed November 2008.

Limmer A. 2005a. Do corks breathe? Or the origin of SLO? *Australian and New Zealand Grapegrower and Winemaker* 497: 89–98.

Limmer A. 2005b. The chemistry of post–bottling sulfides in wine. *Chemistry in New Zealand* September 69(3): 2–5.

Limmer A. 2006a. Cork as a closure—post-bottling reduction, and 'permeability' performance. Address to 'A closer look at cork closures' at the Portuguese Cork Association Symposium. Napa, California: June 23, 2006.

Limmer A. 2006b. Part 1. Do corks breathe? The origin of post-bottling sulfides. *Practical Winery and Vineyard* 28(1): 21–24, 26–28.

Limmer A. 2006c. Part 2. Do corks breathe? The origin of post-bottling sulfides. *Practical Winery and Vineyard* 28(2): 71–73, 88–89.

Lingle R. 2004. Boxed wine aims high. Packaging World Magazine. April p. 47 http://www.packworld.com/print.php?id=17365. Accessed August 2008.

Linkplas. 2008. 'Pettle' Wine Bottle. http://www.linkplas.com/product/pettle.htm. Accessed June 2008.

Lopes P., Saucier C., Glories Y. 2005. Nondestructive colorimetric method to determine the oxygen diffusion rate through closures used in winemaking. *Journal Agricultural and Food Chemistry* 53: 6967–6973.

Lopes P., Saucier C., Teissedre P., Glories Y. 2006. Impact of storage position on oxygen ingress through different closures into wine bottles. *Journal Agricultural and Food Chemistry* 54: 6741–6746.

Lopes P., Saucier C., Teissedre P.L., Glories Y. 2007. Main routes of oxygen ingress through different closures into wine bottles. *Journal Agricultural and Food Chemistry* 55: 5167–5170.

Mans J. 2008. Plasma coating protects wine in PET bottle. *Packaging Digest.* http://www.packagingdigest.com/article/CA6602568.html. Accessed November 2008.

Mottram D.S., Patterson R.L.S., Warrilow E. 1984. 2,6-Dimethyl-3-methoxypyrazine: a microbiologically-produced compound with an obnoxious musty odour. *Chemistry and Industry* 12: 448–449.

Newhouse D. 2008. PET wine breakthrough to replace glass on airlines. http://www.trend-news.com/default.asp?newsid=5525. Accessed November 2008.

Nicolau L. 2005. Screw caps extends Sauvignon shelf life. http://www.corkInformation/Screw caps extend sauvignon shelf life.htm. Accessed September 2008.

O'Brien V. 2005. Closures—An Independent View. http://www.provisor.com.au/uploads/documents/winetech_20_7_05.pdf. Accessed July 2008.

Oeneo Closures USA. 2008. The truth about TCA. http://www.tcafreecorks.com/. Accessed November 2008.

Ortiz R., Hill E., DeLassus P. 2004. Permeation protocols for synthetic wine closures. Presentation at Pack Expo 2004 http://www.pmmi.org/ms/peconf/w11.pdf. Accessed June 2008.

Ough C., 1987. Use of PET bottles for wine. *American Journal of Enology and Viticulture* 38: 100–103.

Peck J. 2005. Oxygen transmission (OTR) measurement and variability. Presentation to The Science of Closures, 56th Annual Meeting of the American Society for Enology and Viticulture. Seattle, Washington, June 24, 2005.

Peck J. 2007. Beam Wine Estates Technical conference, Clois du Bois Winery, Healdsburg, California, July, 2007.

Pereira C.S., Marques J.J.F., San Romão M.V. 2000. Cork taint in wine: scientific knowledge and public perception—a critical review. *Critical Reviews in Microbiology* 26: 147–162.

Petersen R. 2008a. Please stop telling people that corks 'breathe'. http://www.wine.appellationamerica.com/wine-review/548/Cork-Breath.html. Accessed March 2008.

Petersen R. 2008b. Corks do not breathe—part 2. http://www.wine.appellationamerica.com/wine-review/636/Cork-Breath-Part-2.html. Accessed October 2008.

Pollnitz A., Capone D., Campbell J., Franke S., McLean H., Skouroumounis G., Sefton M. 2002. Some applications of analyses of volatile flavor compounds to wine. In: *Proceedings of the 11th Australian Wine Industry Technical Conference.* Blair R., Williams P., Høj P. (Eds). October 7–11, 2001. Adelaide, Australia: Australian Wine Research Institute, pp. 162–164.

Pretorius S. 2006. The Australian Wine Research Institute refutes Paul White's 'Scientifically Speaking' feature in Harpers' Closures supplement in December 2005. *Harpers Wine and Spirit*, August 31, 2006.

Rapp A. 1998. Volatile flavor of wine: correlation between instrumental analysis and sensory perception. *Nahrung/Food* 42: 351–363.

Robertson G. 2006a. *Food Packaging Principles and Practice*. Boca Raton, Florida: CRC Press, pp. 55–78.

Robertson G. 2006b. *Food Packaging Principles and Practice*. Boca Raton, Florida: CRC Press, pp. 171–172.

Rotter B. 2008. Sulphur dioxide. http://www.brsquared.org/wine. Accessed July 2008.

Scholle. 2005. FlexTap® Push-Button Dispensing. http://www.scholle.com/PDF/wine/FlexTap_sellsheet_ENG_US_1.06.pdf. Accessed August 2008.

Sefton M.A., Simpson R.F. 2005. Compounds causing taint and the factors affecting their transfer from natural cork closures to wine—a review. *Australian Journal of Grape and Wine Research* 11: 226–240.

Silva A., Lambri M., De Faveri M. 2003. Evaluation of the performances of synthetic and cork stoppers up to 24 months post-bottling. *European Food Research and Technology* 216: 529–534.

Simpson R. 1978. 1,1,6-Trimethyl-1,2-dihydronaphthalene: an important contributor to the bottle aged bouquet of wine. *Chemistry and Industry* 1: 37.

Simpson R., Strauss C., Williams P. 1977. Vitispirane: a C13 spiro-ether in the aroma volatiles of grape juice, wines and distilled grape spirits. *Chemistry and Industry* 15: 663–664.

Simpson R.F., Lee T.H. 1990. The microbiology and taints of cork and oak. In: *Proceedings 9th International Oenological Symposium*, May 24–26, 1990, Cascais, Portugal. Breisach, West Germany: International Association for Modern Winery Technology and Management, pp. 653–667.

Simpson R.F., Capone, D.L., Sefton M.A. 2004. Isolation and identification of 2-methoxy-3,5-dimethylpyrazine, a potent musty compound from wine corks. *Journal Agricultural and Food Chemistry* 52: 5425–5430.

Singleton V. 1987. Oxygen with phenols and related reactions in musts, wines, and model systems: observations and practical implications. *American Journal of Enology and Viticulture* 38: 69–77.

Skouroumounis G., Kwiatkowski M., Francis I., Oakey H., Capone D., Peng Z., Duncan B., Sefton M., Waters E. 2005. The impact of closure type and storage conditions on the composition, colour and flavour properties of a Riesling and wooded Chardonnay wine during five years' storage. *Australian Journal of Grape and Wine Research* 11: 369–384.

Skouroumounis G., Waters E. 2007. Oxygen ingress into bottled wines. *Technical Review* No. 170. Australian Wine Research Institute, Adelaide, Australia.

Soleas G.J., Yan J., Seaver T., Goldberg D.M. 2002. Method for gas chromatographic assay with mass selective detection of trichlorocompounds in corks and wines applied to elucidate the potential cause of cork taint. *Journal Agricultural and Food Chemistry* 50: 1032–1039.

Somers T., Evans M. 1977. Spectral evaluation of young red wines: anthocyanin equilibria, total phenolics, free and molecular SO2 "chemical age." *Journal of the Science of Food and Agriculture* 28: 279–287.

Stelzer T. 2005. *Taming the Screw: A Manual for Winemaking with Screw Caps*. Brisbane, Australia: Wine Press.

Sundell H., Holen B., Nicolaysen F., Hilton C., Løkkeberg O. 1992. Bag-in-box packaging for wine: analysis of transport stress in barrier films. *Packaging Technology and Science* 5: 321–329.

Vidal J., Moutounet M. 2007. Suivi de l'oxygène au cours du conditionnement: nouveau critère de l'assurance qualite www.oiv2007.hu/documents/viniculture/181_oiv_juin_07_vidal_jc_fran_ais_full_text.pdf. Accessed October 2008.

Vidal S., Aagaard O. 2008. Oxygen management during vinification and storage of Shiraz wine. *Australian Wine Industry Journal* 23(5): 21–23.

Vogel M. 2008. http://www.foodreference.com/html/a-other-white-wine-403.html. Accessed June 2008.

Waterhouse A. 2005. Recent Research: The Science of Closures. http://www.winebusiness.com/Winemaking/webarticle.cfm?dataId=40000. Accessed October 2008.

Waterhouse A. 2008. Wine oxidation mechanisms. Address to the Auckland University Red Wine Oxidation Workshop held at the Eastern Institute of Technology, Taradale, New Zealand: December 11, 2008.

Waters E., Skouroumounis G., Sefton M., Francis L., Peng Z., Kwiatkowski M. 2001. Team Report on Random Oxidation, Australian Wine Research Institute Annual Report. Adelaide, Australia: Australian Wine Research Institute.

Waugh A. 2008. http://www.nzwine.com/winestyle/blanc.php. Accessed August 2008.

Winterhalter P. 1991. 1,1,6-Trimethyl-1,2-dihydronaphthalene (TDN) formation in wine. I: Studies on the hydrolysis of 2,6,10,10-tetramethyl-1-oxaspiro[4.5]dec-6-ene-2,8-diol rationalizing the origin of TDN and related C13 norisoprenoids in Riesling wine. *Journal of Agricultural and Food Chemistry* 39: 1825–1829.

Zoecklein B. 2005. Enology Notes #98. http://www.vtwines.info/. Accessed August 2008.

14 Packaging and the Shelf Life of Fresh Red and Poultry Meats

Alex O. Gill
Bureau of Microbial Hazards
Health Canada
Sir F. G. Banting Research Centre
Ottawa, Ontario
Canada

Colin O. Gill
Agriculture and Agri-Food Canada
Lacombe Research Centre
Lacombe, Alberta, Canada

CONTENTS

14.1 INTRODUCTION

The adoption of preservative packaging for raw meats has led to major changes in the processing and marketing of such products. As a result of the widespread adoption of vacuum packaging for primal cuts of red meats, trade in red meat carcasses has declined to trivial proportions in many developed countries, and the international trade in chilled raw meats has greatly increased, with a consequent decline in trading of frozen meats. The enhanced stability of vacuum-packaged products has facilitated consolidation of meat-packing facilities. Consequently, in many countries, abattoirs of intermediate size have been replaced by relatively few, large facilities, with small abattoirs providing services for niche markets. The use of preservative packaging with retail cuts has facilitated the central preparation of retail-ready product, with progressive displacement of fresh meat retailing from traditional butchers' shops to supermarkets.

In this chapter, the effects on retail-ready raw meats of the various forms of protective or preservative packaging that are used with them will be discussed. Preservative packaging of primal cuts or bulk meats will be considered only in relation to the effects on the stability of retail-ready items of prior storage in such packaging.

14.2 FOOD QUALITY ATTRIBUTES AND INDICES OF FAILURE

14.2.1 Intrinsic Organoleptic Qualities

14.2.1.1 Color

Appearance greatly influences the decisions of consumers on whether or not to purchase raw meats offered for retail sale (Cornforth, 1994). The colors of meat that are generally regarded as acceptable reflect how consumers expect fresh meat to appear. For red meats, bright red muscle tissue and bone marrow, if the latter is exposed, are much preferred, and white fat is preferred to yellow. For poultry meat the general preference is for flesh and skin that are white and bright.

The color of muscle tissue is primarily determined by the pigmentary compounds myoglobin and hemoglobin (Ledward, 1984). Cytochromes, flavins, and catalase also provide pigmentation, but as they are present in only relatively small amounts, they have only minor effects upon muscle color (Ledward, 1984). In the tissues of mature red meat animals, myoglobin can comprise more than 90% of the total pigments (Warris and Rhodes, 1977). In pale muscle tissues the red blood cell pigment hemoglobin can form a substantial part of the total pigments, depending on the amount of blood that is retained in the tissue. Thus, in chicken breast meat, most of the pigment present may be hemoglobin (Rhee and Ziprin, 1987).

Both myoglobin and hemoglobin are heme proteins that function physiologically to bind and store O_2. Their reactions with O_2 and other ligands are similar, and result in comparable color changes. Therefore, to describe the effects of various conditions on meat color, only the reactions of myoglobin need to be considered (Govindarajan, 1973).

The colors of hemoglobin and myoglobin depend upon the reaction of the iron atom bound within the heme prosthetic group with O_2 and other ligands (Mancini and Hunt, 2005). In their physiologically functional states, the globin molecules are folded into structures that place the heme group in a hydrophobic environment and limit the size of the molecules that can access and react with the heme iron (Cornforth and Jayasingh, 2004). Consequently, physiologically functional hemoglobin and myoglobin can react with only small ligands such as O_2.

In the absence of O_2, myoglobin is present in muscle tissue as deoxymyoglobin, which is a dull, purple color (Figure 14.1). On exposure to O_2, myoglobin is oxygenated to form bright red oxymyoglobin. The interconversion of deoxy- and oxymyoglobin is rapid and reversible, and dependent on the partial pressure of O_2 to which myoglobin is exposed (Giddings, 1977). At physiological O_2 concentrations, small changes in O_2 concentration result in relatively large changes in the ratio of oxy- to deoxymyoglobin. At the O_2 concentration in air, oxymyoglobin predominates. When the atmosphere contains carbon monoxide, stable, cherry red carboxymyoglobin is formed.

The heme iron of myoglobin can be oxidized from the ferrous to the ferric state, with the formation of physiologically inactive, brown metmyoglobin (Wallace et al., 1982). Oxymyoglobin is more resistant to oxidation than is deoxymyoglobin. Consequently, metmyoglobin is formed more rapidly at low concentrations of O_2, at which deoxymyoglobin predominates, than at high O_2 concentrations. Metmyoglobin is converted slowly back to deoxymyoglobin by enzymic reactions collectively termed *metmyoglobin reduction activity* (MRA). The activities of the MRA enzymes decline, and the pool of the reduced coenzyme (NADH) that is required for MRA is depleted postmortem. Consequently, during storage the ability of muscle tissue to regenerate deoxymyoglobin from

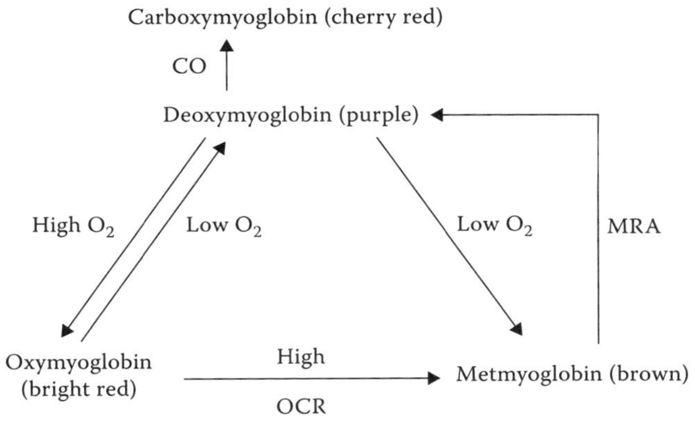

FIGURE 14.1 Reactions of myoglobin with oxygen and carbon monoxide and the effects on the state of myoglobin of the oxygen consumption rate (OCR) and metmyoglobin reduction activity (MRA) of muscle.

metmyoglobin progressively declines (Echevarne et al., 1990). Grinding of meat greatly accelerates enzyme inactivation and depletion of NADH and so results in loss of MRA.

When muscle tissue is cut, the newly exposed surfaces develop the bright red color of oxymyoglobin, which consumers greatly prefer. However, the O_2 concentration declines with distance below the tissue surface, and a band of metmyoglobin will form a few millimeters below the surface, where the O_2 partial pressure is less than 1% of the atmospheric pressure. The tissue below that is anoxic and myoglobin is in the deoxy form. The band of metmyoglobin will progressively expand toward the meat surface as the myoglobin in the oxygenated tissues oxidizes (Madhavi and Carpenter, 1993). Eventually, the presence of metmyoglobin at the meat surface will be obvious, but before that the perceived color of the meat will be affected by metmyoglobin formed beneath the surface, because light is reflected from within as well as from the surface of muscle tissue (Swatland, 2004).

In fresh meat, metmyoglobin formation is affected by the enzymic activities of the muscle tissue. Tissues with a high oxygen consumption rate (OCR) will tend to form metmyoglobin at a relatively shallow depth below the meat surface and discolor rapidly (Millar et al., 1994), whereas MRA will reconvert metmyoglobin to deoxymyoglobin and retard discoloration (Echevarne et al., 1990). Consequently, the rate at which discoloration of fresh muscle tissues develops will depend upon the relative rates of O_2 consumption by, and metmyoglobin reduction in, the tissue. These activities differ between muscles, resulting in wide variation in their intrinsic color stabilities. Thus, in chops, the main *longissimus dorsi* muscle can retain a red color for up to three times longer than the smaller *psoas major* muscle (O'Keeffe and Hood, 1982). However, the OCR as well as MRA decays during the storage of meat, so the color stabilities of muscle in cuts of meat that have been stored in vacuum packs tend to become similar and relatively low.

Exposed bone marrow in bone-in retail cuts will darken during display, and can become black in cuts prepared from vacuum-packaged meat. The darkening is due to the formation of methemoglobin from hemoglobin released from red blood cells in the bone marrow as they disintegrate during storage. Progressive accumulation of methemoglobin, at cut marrow surfaces, appears to account for bone marrow blackening in cuts from stored meats (Gill, 1996).

The accumulation of metmyoglobin and methemoglobin reduces the acceptability of red meats as it dulls and darkens muscle tissues and cut bone surfaces. The display life of such meats is usually terminated by extensive brown discoloration of exposed muscle tissue surfaces, which is unacceptable to consumers.

Poultry meat contains little myoglobin and has a high OCR. Consequently, fresh poultry meat exposed to air contains little oxymyoglobin, and instead of a bright red color it has the duller tones imparted by deoxy- and metmyoglobin (Millar et al., 1994). As the surface color of fresh poultry meat is not dependent upon the presence of the oxygenated forms of the muscle and blood pigments, there is little change in the color of poultry meat during storage. Consequently, the storage life of poultry meat is not usually terminated by color deterioration.

14.2.1.2 Odor, Flavor, Tenderness, and Exudate

Oxidation of lipids gives rise to rancid odors and flavors in meats (Campo et al., 2006). The susceptibility of muscle tissue lipids to oxidation is highly dependent on their fatty acid compositions, as rates of oxidation increase with greater amounts of unsaturated fatty acids and the degree to which those acids are unsaturated (Kanner, 1994). Lipid oxidation is catalyzed by ferrous iron and heme compounds. Thus, lipid oxidation is accelerated by grinding or other treatments that disrupt muscle tissue and release these compounds (Gray et al., 1996). Discoloration of muscle tissue and lipid oxidation are closely linked, with the development of oxidative rancidity being more rapid in color-unstable than in color-stable muscles (McKenna et al., 2005). Vacuum or other packaging that greatly restricts the exposure of meat to O_2 can delay or prevent the development of oxidative rancidity (Knox et al., 2008). Modified atmospheres (MAs) enriched with O_2 may accelerate oxidative rancidity, although the difference in the rates at which oxidative rancidity develops in meat stored in such atmospheres or in air may be small (Gatellier et al., 2001; Ordenez and Ledward, 1977).

Loss of exudate from meat is unavoidable. In contrast to the situation with vacuum-packaged primal cuts that will be subject to further fabrication, exudate loss from retail packaged cuts does not result in a loss of salable weight, because the cost to the customer is based on the cut weight at the time of packing. However, exudate in retail packs is undesirable because of its adverse effects on the appearance of the packed product and the problems it creates for handling of the product during its preparation for consumption. Exudate losses are lower from meat of pH > 6 than from meat of lower, normal pH, but exudate losses from all meats are increased by cutting large pieces into smaller ones and by pressure on the product (Offer and Knight, 1988). Uniform pressure from the packaging on a meat cut may reduce exudate loss for a time (Barros-Velazquez et al., 2003), but, in general, exudate in retail packs must be dealt with by including in-pack absorbent pads of sufficient capacity to absorb all the exudate that may be released.

Most consumers consider tenderness to be the main factor determining the eating quality of meat (Jeremiah et al., 1993). Tenderness tends to increase with the holding time of meat in carcass form or in vacuum packs. However, the rate of tenderization declines exponentially with time, with little further tenderization occurring after about 2 weeks of storage at chiller temperatures (Channon et al., 2004; Dransfield et al., 1994). Thus, unless retail cuts are prepared from carcasses soon after chilling, retail-ready meat is not likely to increase substantially in tenderness during distribution and display. Indeed, storage under MAs rich in O_2 may lead to toughening of the meat because of oxidation and cross-linking of muscle structural proteins (Lund et al., 2007).

14.2.2 Bacterial Spoilage

The spoilage flora of chilled meats is composed of psychrotrophic bacteria with minimum temperatures for growth of –3°C or below (Gill, 2006). The minimum temperature at which muscle tissue can be held indefinitely without freezing is –1.5°C ± 0.5°C (Gill and Phillips, 1993). Therefore, spoilage bacteria will always grow on, and ultimately spoil, chilled raw meat. When the numbers of bacteria on the meat at the time it is packaged are relatively small, the flora that develops will be increasingly dominated by the organisms with the fastest growth rate in the medium provided by the meat and under the atmosphere imposed by the packaging (Gill, 1986). However, spoilage bacteria that grow relatively slowly may still form a substantial, although diminished, fraction of the spoilage flora at the time of the onset of spoilage if they were initially present in high numbers.

Meat is spoiled when its appearance, odor, or flavor is adversely affected by activities of spoilage bacteria (Gill, 1981). The number of bacteria that are needed to produce such changes vary with the intrinsic qualities of the meat tissues and with the types of bacteria that are present in the spoilage flora (Table 14.1). Intact muscle tissue does not usually contain bacteria, and so spoilage bacteria grow only on the surfaces of meats. When meat is ground, or otherwise comminuted, the surface area is increased and bacteria are spread throughout the meat mass. Because of the reactions between meat and O_2, the tissues a few millimeters below the surface of a ground meat mass will be anoxic irrespective of the surrounding atmosphere. As aerobic flora usually grows faster, and often contains more potent spoilage organisms than anaerobic flora, the spoilage of ground meat can occur at the surface, rather than throughout the meat mass, unless conditions at the surface are anaerobic.

Fresh meats stored in air usually develop flora in which species of *Pseudomonas* predominate. These strictly aerobic organisms preferentially utilize glucose and strongly repress their utilization of other substrates present in meat while glucose is available (Gill and Newton, 1977). When glucose diffusing from within the meat cannot adequately supply the needs of the bacteria on the surface, they commence utilization of amino acids. The products of glucose metabolism are inoffensive, but the breakdown of amino acids results in the release of by-products such as ammonia, amines, and organic sulfides, which confer objectionable odors and flavors on the meat when they are present in only small quantities (Nychas et al., 1988). On muscle tissue of normal pH (between 5.4 and 5.8), aerobic bacteria will usually reach numbers of about 10^8 colony-forming units (cfu) per cm^2 of meat surface before glucose is exhausted. However, muscle tissue of pH 6.0 or above may contain little or no glucose, and little glucose is available on fat tissue surfaces that are not bathed with exudates, whatever the pH of the muscle tissue in the meat (Newton and Gill, 1978). On such tissues the pseudomonads will commence utilization of amino acids when their numbers are still low, and objectionable odors can then be detected when the numbers of the spoilage flora reach about 10^6 cfu cm^{-2} (Gill and Newton, 1980).

As the pseudomonads are strictly aerobic, they cannot grow under anaerobic conditions, such as can exist in vacuum packs. Low partial pressures of O_2 will slow this growth, but maximum growth rates can be achieved with O_2 partial pressures at or above 1% of atmospheric pressure (Clark and Burki, 1972). The flora that develop on normal pH meat under anaerobic conditions are usually dominated by lactic acid bacteria, which include organisms of the genera *Lactobacillus*, *Carnobacterium*, and *Leuconostoc*. The compositions of lactic flora can change substantially during storage as the numbers increase. Carnobacteria are often initially dominant before being displaced by the more aciduric lactobacilli and leuconostocs (Jones, 2004). These organisms ferment only glucose and a few other substrates present in meat at low concentrations. When those substrates are depleted growth of the flora stops, with maximum numbers on the surfaces of cuts at about 10^8 cfu cm^{-2}. The metabolic by-products of the lactic acid bacteria have a low sensory impact as they impart

TABLE 14.1
Principle Spoilage Bacteria of Fresh Meats

Organism	Gram Reaction	Oxygen Requirement	Spoilage Products	Growth Characteristics
Brochothrix thermosphacta	+	Facultative anaerobe	Diacetyl, and acetic, isovaleric, and isobutyric acids	No anaerobic growth below pH 5.8
Enterobacteriaceae	–	Facultative anaerobe	Amines, sulfides	No anaerobic growth below pH 5.8
Lactic acid bacteria	+	Aerotolerant anaerobe	Lactic acid and ethanol	Ferment a restricted range of substrates
Pseudomonas	–	Aerobe	Amines, ethyl esters	Aerobic growth only
Shewanella putrefaciens	–	Facultative anaerobe	Sulfides	No anaerobic growth below pH 6.0

Source: Spoilage characteristics from Whitfield, F.B. 1998. Microbiology of food taints. *International Journal of Food Science and Technology* 33: 31–51.

mild acidic or dairy flavors to meat only some time after the flora has attained maximum numbers (Dainty et al., 1979). However, if the pH of the meat is high, facultative anaerobes of high spoilage potential such as *Brochothrix thermosphacta*, *Shewenella putrefaciens*, and psychrotrophic enterobacteria may grow and impart stale or putrid odors and flavors to the meat as the flora approach the maximum numbers (Blickstade, 1983). Vacuum-packaged beef can also be spoiled by psychrotolerant clostridia that produce large volumes of gas and cause gross swelling of packs (Broda et al., 2003). However, this form of spoilage has not been reported to occur with retail-ready raw meats.

The growth of pseudomonads can be inhibited, but not prevented, by CO_2 when O_2 is present. Inhibition increases with increasing CO_2 concentrations in pack atmospheres, up to about 20%. At that CO_2 concentration growth rates are about half those in air (Gill and Tan, 1980). Further increases in the concentration of CO_2 have a minimal effect on growth rates, provided the atmosphere remains aerobic. When pseudomonads are inhibited by CO_2, the aerobic flora is usually dominated by lactic acid bacteria accompanied by facultative anaerobes, particularly *B. thermosphacta* and enterobacteria (Gill and Jones, 1996). The lactic acid bacteria may continue to predominate in the spoilage flora when meat that has been stored in a vacuum pack or under a MA is subsequently exposed to air during display. However, on exposure to air, spoilage will ultimately be due to strict aerobes and facultative anaerobes, which will form a progressively larger fraction of the flora as its numbers increase (Gill and Jones, 1996).

14.3 DISTRIBUTION AND DISPLAY OF RETAIL-READY MEAT

In traditional practice, retail cuts of meat are prepared from carcasses or vacuum-packaged primal cuts at butchers' shops or butchering facilities at retail food stores. However, in most developed countries, meat is increasingly prepared and packaged for retail display at central cutting facilities or meat packing plants. Because of the many economic advantages of central preparation (Quigley, 2002), most of the meat offered for retail sale in several countries is now prepared in that manner (Crews, 2007; Dietrich, 2005). Various systems for packaging retail-ready meats to address the wide range of marketing requirements have been developed (Belcher, 2006).

The time for which raw meat will remain acceptable during distribution and display is highly dependent upon the temperatures experienced by the products. The rate at which muscle tissue will discolor increases linearly with increasing temperature, with the rate of increase varying for different muscle types. Thus, in aerobic atmospheres, the color-stable *longissimus dorsi* will discolor twice as fast at 10°C than at 0°C, whereas the color-unstable *psoas major* will discolor five times faster at 10°C than at 0°C (Hood, 1980). Under anaerobic conditions the rate at which color stability on exposure to air is lost also increases with temperature, to about twice and four times the rate at 0°C than at 5°C and 10°C, respectively (O'Keeffe and Hood, 1982). The rates of lipid oxidation and exudate loss also increase with increasing temperature.

The growth rates of spoilage bacteria at any given temperature within their growth temperature range differ widely, but the growth rates of all organisms increase rapidly with small increases in temperature above –1.5°C, the optimum temperature for storage of chilled meat (Gill et al., 1988). For meat stored in air, in a MA, or under vacuum, the proportional reductions with increasing temperature of the time to onset of microbial spoilage are similar (Gill and Jones, 1992). Thus, whatever the packaging, the time before onset of microbial spoilage will be substantially reduced by storage at temperatures above –1.5°C (Figure 14.2).

Regulations in various countries require that fresh meats be stored, transported, and displayed at temperatures less than 4°C or 2°C (CFIA, 1998; Pastors, 2005). At those temperatures the storage life of chilled meats is only about 35% or 50%, respectively, of the storage life at –1.5°C (Gill, 1995). Although product may be stored at such temperatures before it is prepared and packaged as consumer portions, product temperatures may rise during processing and during assembly of consignments at packing plants or central cutting facilities. Distribution of multiple small consignments from a single load of a refrigerated vehicle, such as commonly occurs with deliveries from central

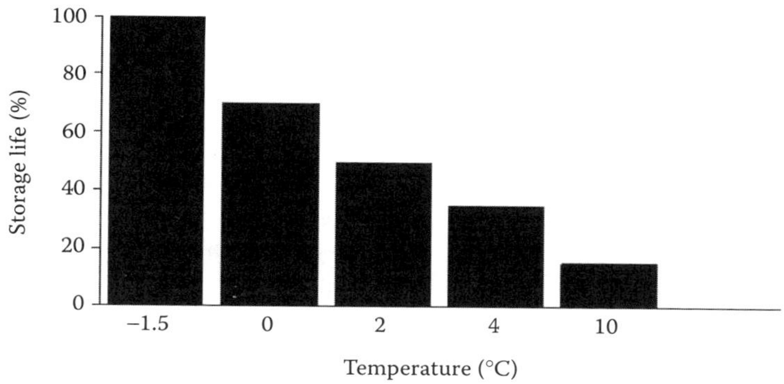

FIGURE 14.2 Effect of storage temperature on the storage life of chilled meat as determined by the onset of microbial spoilage.

cutting facilities or central warehouses, can result in progressive warming of product because of the ingress of ambient air during each delivery (Malton, 1978). Such loss of temperature control is countered in some distribution systems by operating vehicle refrigeration equipment at subzero temperatures as low as −5°C (Gill et al., 1995, 2002a). At retail outlets, raw meat storage facilities are likely to operate at temperatures less than 4°C (Gill et al., 1995, 2002b), but temperature control during retail display may be relatively poor.

In display cases, cold air is usually blown over the product from the back of the case, and a curtain of cold air descending from the top of the case at the front of a vertical case, or blown above the product from the back of a horizontal case, prevents the ingress of ambient air (James and James, 2002). Product is generally warmer at the front than at the back of a case, and at higher than lower positions on a shelf. Inappropriate product stacking can result in blocking of the flow of cold air over the product, or disruption of the air curtain, with consequent ingress of ambient air (Faramarzi et al., 2003). The case refrigeration equipment must be periodically defrosted, usually two, three, or four times a day, and during defrosting the temperature of the air blown through the case will rise toward the ambient temperature (Brolls, 1986). Because of these and other factors, the mean temperatures of product in many meat display cases are above 4°C, and the maximum temperatures experienced by displayed product may exceed 10°C (Gill et al., 2002c, 2003; Torstveit and Magnussen, 1998). Growth of pathogens and rapid spoilage of product must be expected at such temperatures.

14.4 EFFECTS OF RETAIL PACKAGING ON INDICES OF FAILURE

14.4.1 Overwrapped Trays

Raw meats prepared for retail display continue to be packaged mainly in expanded polystyrene (PS) trays overwrapped with a stretchable poly(vinyl chloride) (PVC) film. The inner surfaces of the trays may have a low density polyethylene (LDPE) or PVC coating, and the product is usually placed on a soaker pad to absorb exudate. The overwrapping film, which is usually heat-sealed on the underside of the tray, must have a high O_2 transmission rate (OTR) if discoloration of red meats is not to be accelerated by low O_2 concentration in areas of contact between the meat and the film. An OTR of about 1700 mL m^{-2} day^{-1} at 23°C/50% RH is adequate, as the color stability of beef steaks is not increased when overwrapping films with OTRs up to four times higher are used (Petersen et al., 2004).

From the retailer's point of view, overwrapped trays have the advantages of being familiar to consumers and inexpensive. Therefore, much centrally prepared retail-ready meat is packaged in that manner. The limited display life of the product is, in many systems, largely addressed by frequent, often daily, delivery of product from central cutting facilities. Thus, most retail packs of fresh meat are preferably sold within 2 days of being prepared at a retail or central cutting facility (Gill et al., 2002c). In addition, much centrally prepared meat in overwrapped trays is master packaged for distribution under MAs in bags of low gas permeability.

Master packs containing overwrapped trays are usually prepared using equipment in which the mouth of a filled bag is sealed around two flattened tubes, termed *snorkels*, that extend into the bag. Air is withdrawn and gas is pumped into the bag through the snorkels. When a gas atmosphere has been established, the snorkels are retracted and the master pack is heat sealed.

The drawing of a vacuum within a master pack would result in crushing of the overwrapped trays. Therefore, the air initially present in a bag is usually evacuated for a set time that will inevitably leave substantial residual volumes of air within packs. As the initial volume of the master pack is not controlled, the amount of residual air is variable. Repeated flushing of the master pack will move the composition of the pack atmosphere toward that of the input gas, but usually a master pack is flushed only once, or is not flushed at all. Thus, the gas atmospheres of master packs formed using snorkel equipment can be highly variable.

Gas mixtures used when master packing red meats commonly contain O_2, CO_2, and N_2 at concentrations of about 60:20:20. The high O_2 concentration will tend to enhance and maintain a desirable red color for the meat, whereas the CO_2 will reduce the rates of growth of aerobic spoilage bacteria. The inclusion of N_2 to buffer against pack collapse arising from absorption of CO_2 by the meat is unnecessary in these circumstances because of the high gas volume:meat weight ratio. If the meat color is to be maintained, the trays must be arranged to expose all overwrapped meat surfaces to the pack atmosphere. Alternatively, a porous layer, such as several sheets of absorbent paper, can be placed between the overwrapped meat and other surfaces to allow access of the atmosphere to the meat. As meat is often retained for only short times in such crudely formed packs, mechanical protection of the filled trays during their distribution is probably equally or more important than any effect of the pack atmosphere on meat color or development of the spoilage flora.

With poultry meats, master packs are prepared using an input gas containing CO_2 at concentrations between 50% and 80%, with the balance N_2 (Sarantopoulos et al., 1998). The input gas is sometimes supplemented with 5% O_2 in the mistaken belief that the pack atmosphere would otherwise be anaerobic, and that O_2 in the atmosphere is necessary to suppress the growth of *Clostridium botulinum* (Lambert et al., 1991). However, *Cl. botulinum* has never been reported to be a hazard in MA packaged chicken, and whether or not the input gas is supplemented with O_2, the pack atmosphere will contain some low concentration of O_2. As red meats discolor more rapidly at low O_2 concentrations than in air, such atmospheres are wholly unsuitable for use with red meats.

However, atmospheres of CO_2 and N_2 that contain residual O_2 can be used with red meats if the input gas is supplemented with carbon monoxide (CO) so that cherry red carboxymyoglobin is formed (Lanier et al., 1978). Thus, raw meats exposed to CO can develop a persistent red color, but the use of CO with master packaged overwrapped meats has as yet been only experimental. The colors of overwrapped pork chops, beef cuts, and ground beef master packaged using input gases that contained 0.4% CO but no O_2 were maintained or enhanced by the pack atmosphere (Hunt et al., 2004; Wilkinson et al., 2006). Even so, the color stability of some beef muscles and ground beef was reduced by storage in such an atmosphere.

Another method proposed for delaying microbial spoilage of meat in overwrapped trays is the incorporation of chemicals that generate CO_2 on exposure to water in absorbent pads used in the trays. Moistening of the pads by exudate from the meat will then generate a CO_2-enriched atmosphere within the pack (Brody, 2008). Although such packaging might be of practical use with poultry meats, it would seem to have little application for red meats, which usually become unsalable due to discoloration before the onset of microbial spoilage. The compositions of the atmospheres generated within packs are likely to vary widely because of variations in both the amounts of exudate produced by retail portions and the loss of CO_2 from packs. The inhibition of bacterial growth is then likely to be equally variable, whereas reductions of the O_2 concentration in some pack atmospheres might result in the acceleration of red meat discoloration.

14.4.2 Lidded Trays

Despite the rapid turnover of much retail packaged meat, storage plus display lives of 7 or 10 days are considered to be desirable for centrally prepared meat distributed locally or regionally, respectively (Brody, 2007). To attain such storage lives for meat, the product is commonly packaged in lidded trays containing MAs. A large number of systems for preparing lidded trays with MAs are available. The trays may be preformed or formed from role stock immediately before filling. Trays may be composed of PS, polypropylene (PP), or PVC laminated with a film having good gas barrier properties, such as one containing an ethylene-vinyl alcohol (EVOH) copolymer. The OTRs of trays and lidding films can be expected to be less than 0.42 and less than 4.2 mL m^{-2} day^{-1}, respectively, although negligible OTRs are technically possible (Brooks et al., 2008; Ho et al., 2003).

Input gas mixtures used with lidded trays are typically composed of 60–70% O_2, 20–30% CO_2, and 10–20% N_2 (Anon., 2008), or 30–70% of both CO_2 and N_2 without or with 0.3–0.5% CO.

Atmospheres rich in O_2 are widely used with red meats to enhance and maintain a desirable color. However, the display life of the meat is still usually limited by discoloration of the muscle tissue and by blackening of the cut surfaces of bones in bone-in cuts (Grobbel et al., 2006), rather than by bacterial spoilage. This is because the growth of aerobic spoilage flora is inhibited by the CO_2. N_2 is included in the atmosphere to guard against pack collapse because of absorption of CO_2. CO_2 is highly soluble in both muscle and fat tissues, and will be absorbed into the meat from the pack atmosphere in amounts that vary with muscle tissue pH, type of fat, temperature (Figure 14.3), and the fraction of CO_2 in the pack atmosphere (Gill, 1988). The pack atmosphere will change over time, initially because of rapid dissolution of CO_2 in the meat and subsequently because of O_2 consumption by muscle tissue and bacteria, and gas exchange across the lidding film. After an initial decrease in CO_2 concentration, the concentration of CO_2 may change little, whereas the concentration of N_2 rises and that of O_2 falls (Nortje and Shaw, 1989). Therefore, to ensure that the O_2 concentration remains at levels that preserve meat color, the volume of gas in the pack should be at least twice that of the meat it contains (Kennedy et al., 2004). However, packs that are oversized for the product they contain add cost, reduce the amount of product displayed, and may not be attractive to consumers. Consequently, lidded packs with high-O_2 atmospheres are often less than the optimal size for product preservation and the life of the product is reduced.

Lidded trays for red meats that are not oversized can be used with input gases of CO_2 and N_2 supplemented with CO. The shelf life of the product is then limited by bacterial spoilage rather than discoloration of meat or bone. CO is highly and cumulatively poisonous, so packing operations in

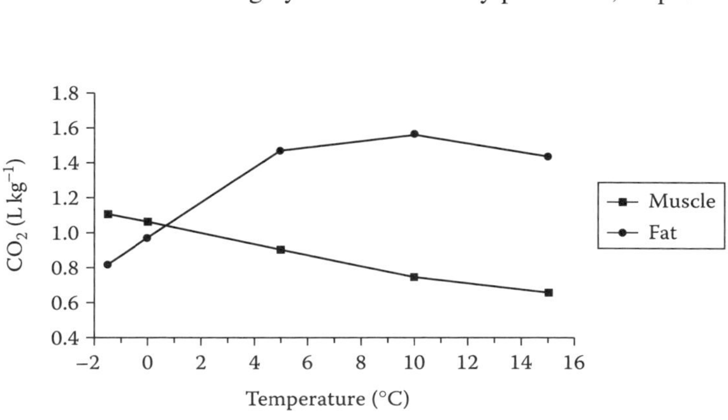

FIGURE 14.3 Effects of temperature on the solubility of CO_2 (L kg^{-1} at STP) in muscle tissue, pH 5.8, and beef fat tissue.

which gas mixtures containing it are used must be well controlled. However, the risks to consumers from the small quantities of CO in meat exposed to atmospheres containing the gas are trivial (Sørheim et al., 1997). Atmospheres containing CO were used with meat in Norway for a number of years, but the practice has been discontinued to accord with EU regulations. Recently, the use with meats of atmospheres containing CO has been permitted in the United States (FDA, 2004). Even so, the use of CO with red meats continues to be contentious, mainly because of concern that the stable red color imparted by CO could result in customers being sold meat that is close to being or has been spoiled by bacteria (Bjerklie, 2007).

Poultry meats can and are packaged in lidded trays using mixtures of CO_2 and N_2 without O_2 as input gases. The atmospheres of lidded trays formed with input gases without O_2 typically contain about 1% residual O_2 (Smiddy et al., 2002). Red meats would rapidly discolor due to metmyoglobin formation in such pack atmospheres but, as discussed earlier, the colors imparted to poultry meat by metmyoglobin and methemoglobin are usual and acceptable for fresh poultry meat. Use of lidded films of very low OTR, inclusion of CO in the pack atmosphere, or relatively high O_2 concentrations in the initial atmosphere are unlikely to substantially affect the time to microbial spoilage of chicken in lidded trays with low O_2 atmospheres (Pettersen et al., 2004).

14.4.3 Vacuum Packs

Although consumers certainly prefer raw meats of bright red color, vacuum-packaged beef and pork primal cuts with dull, purple colors imparted by deoxymyoglobin are now commonly offered for retail sale in at least some countries. Vacuum skin packaging (VSP), in which a heated film is tightly applied around meat portions and sealed to a bottom web beneath the meat, has been used extensively with poultry meats (Jenkins and Harrington, 1991; Spaulding, 1994). As the color of poultry meat in anoxic packaging is acceptable, VSP using upper and lower films of low OTRs (typically less than 2 mL m^{-2} day^{-1}) can be used to extend the storage life by inhibition of aerobic spoilage flora (Kartika et al., 2003).

VSP has also been used with red meats, with an O_2-permeable film being cast over the product and subsequent packaging of the VSP product under a MA (Coventry et al., 1998). Meat can also be packaged in VSP, where a film of high O_2 permeability is first applied to the meat and then a film of low O_2 permeability is applied over that (Brody 1996). The vacuum skin pack can be removed from the MA, or the outer vacuum skin film can be peeled from the inner film, to allow the product in an O_2-permeable vacuum skin pack to bloom to a bright red color when displayed in air. Typical polymers used for the packaging of chilled meat are presented in Table 14.2. A red color can be maintained by meat in vacuum packs if the meat is exposed to an atmosphere containing CO to allow extensive formation of carboxymyoglobin before the meat is packaged (Aspé et al., 2008). However, there is no published information as to whether or not packaging systems of any of these types are currently used commercially.

It has been claimed that the bacterial loads on meat surfaces are substantially reduced by the application of hot films during VSP (Vázquez et al., 2004). These claims can be dismissed, because heating sufficient to produce some substantial pasteurizing of the meat surface would inevitably result in discoloration of the meat (Gill and Badoni, 2002).

14.4.4 Oxygen-Depleted-Atmosphere Packaging

Red meats that are packaged in pouches of O_2-impermeable film under an atmosphere of CO_2 that contains little residual O_2 at the time of pack closure will bloom to a bright red color when exposed to air after long periods of storage (Gill, 1989). Packaging of this type is used with primal cuts, and with lamb and turkey carcasses. Use of such packaging with retail packaged red meats has been extensively investigated, although the use of O_2-depleted atmospheres with retail-ready product in commercial practice has been limited. Most studies have involved master packaging of retail cuts

TABLE 14.2
Typical Materials Used for Packaging Chilled Meat

Pack Type	Bottom Web Materials	Top Web Materials (Where Applicable)
Flexible vacuum pack	PA-LDPE, coextruded as 5-layer film	
Flexible MAP pack	PA-LDPE	OPA-LDPE
	PA-EVOH-LDPE	
	PA-EVOH-PA-LDPE	PET-PVdC-LDPE
	PP-EVOH-LDPE	
	LDPE-EVOH-LDPE	
Rigid vacuum pack	APET	OPA-LDPE
	PVC or PVC-LDPE	PET-PVdC-LDPE
	PS-EVOH-LDPE	OPA-LDPE-EVOH-LDPE
		PET-LDPE-EVOH-LDPE
Rigid MAP pack	PVC	OPA-LDPE
	PVC-LDPE or PVC-EVOH-LDPE	PET-PVdC-LDPE
	APET	OPA-LDPE-EVOH-LDPE
	APET-LDPE or APET-EVOH-PE	PET-LDPE-EVOH-LDPE
	PS-EVOH-LDPE	
Skin packs	PVC-LDPE	Several combinations of up to seven or more layers but incorporating EVOH as gas barrier
	PS-EVOH-LDPE	
	APET	
	APET-LDPE	

Source: From Mondry H. 1996. Packaging systems for processed meat. In: *Meat Quality and Meat Packaging.* Taylor A.A., Raimundo A., Severini M., Smulders F.J.M. (Eds). Utrecht, Holland: ECCEAMST (European Consortium for Continuing Education in Advanced Meat Science and Technology), pp. 323–356, with permission.

APET: amorphous poly(ethylene terephthalate); EVOH: ethylene-vinyl alcohol copolymer; LDPE: low density polyethylene; OPA: oriented polyamide; PA: polyamide; PET: poly(ethylene terephthalate); PP: polypropylene; PS: polystyrene; PVC: poly(vinyl chloride); PVdC: poly(vinylidene chloride).

of meat in overwrapped or lidded trays, but a system for packaging individual retail cuts under an O_2-depleted atmosphere has been commercially available.

For master packaging in O_2-depleted atmospheres, cuts must be retail packed on trays made of solid plastic rather than expanded polystyrene (EPS), because O_2 from air entrained within EPS will contaminate the master pack atmosphere and discolor the meat. Air may also be trapped in soaking pads beneath the meat, and so trays should be constructed with ridged bases that allow free movement of gases beneath the meat; also the pads used in them should not have absorbent material sealed within perforated plastic films that can entrap air. The trays must be overwrapped or lidded with film of high O_2 permeability, and the films should be perforated to allow rapid change of the atmospheres within the retail packs (Tewari et al., 2002).

Master pack pouches are composed of metalized or EVOH laminates, with OTRs of about 0.1 or 1 mL m^{-2} day^{-1}, respectively (Bell et al., 1996; Tewari et al., 2002). Master pack atmospheres may be established by repeated flushing of packs using conventional snorkel equipment, or by use of equipment in which air is evacuated from a hood placed over the bag while the bag is evacuated, flushed, and filled through snorkels (Gill and McGinnis, 1995; Isdell et al., 2003). Simultaneous evacuation of air from within the hood and the bag avoids crushing of the bag contents, and partial breaking of the vacuum in the hood when flushing and filling of the bag commences reduces the residual volume of air in the bag to a minimum. Residual O_2 in the atmospheres established in packs with such equipment can be reduced to a few ppm but not to zero. The residual O_2 in the pack will be scavenged by the meat and will discolor it. However, provided the meat has not been stored for an extended period before packing under an O_2-depleted atmosphere, the discoloration will be reversed over 1 or 2 days by the MRA of the muscle tissue (Gill and Jones, 1994). The amount of residual O_2 in the atmosphere that can be tolerated can be increased by inclusion in the retail packs of O_2 scavengers that function in the presence of CO_2.

The initial volume of the master pack atmosphere must be sufficient to allow for absorption of CO_2 by the meat. With an atmosphere of only CO_2, the meat will absorb approximately its own volume of gas (Gill, 1988; Jakobsen and Bertelsen, 2004). Possible difficulties with pack collapse of individual retail trays under an O_2-depleted atmosphere can be avoided by establishing an atmosphere of 30% CO_2:70% N_2 in packs with a rigid plastic dome fitted over the lidded tray, with the dome being removed for display of the product (Zhao et al., 1994).

14.5 MICROBIOLOGICAL SAFETY

Muscle tissue from healthy animals is free of bacterial or viral pathogens. As with spoilage organisms, pathogens are deposited on meat surfaces during processing and handling of the carcass. Three potential sources of pathogen contamination may be identified. Animal-associated pathogens may be transferred to meat from the hide, skin, or feathers and the intestinal tract of the animal during carcass processing. Human-associated pathogens may be transferred from personnel during handling of product. Both human- and animal-associated pathogens may contaminate processing equipment and tools, which if inadequately cleaned and sanitized may act not only as vehicles for pathogen transfer but also as sources of contamination.

The major bacterial pathogens associated with fresh meats are *Aeromonas hydrophila*, *Campylobacter* spp., *Cl. botulinum*, verotoxigenic *E. coli*, *Salmonella*, *Shigella* spp., *Listeria monocytogenes*, and *Yersinia enterocolitica*. These organisms are facultative or obligate anaerobes, or are microaerophilic (Table 14.3). Consequently, concerns have been raised about the safety of fresh meats in preservative packaging, particularly under microaerobic or anaerobic conditions in which pathogens could potentially increase in numbers or produce toxin before the onset of overt spoilage (Hintlian and Hotchkiss, 1986).

Consideration of the growth requirements for the major pathogens associated with meat indicates that there are limited possibilities for growth of these organisms on meat in preservative packaging at chiller temperatures. With the exceptions of *A. hydrophila*, *L. monocytogenes*, and *Y. enterocolitica*, the major pathogens associated with meat cannot grow at temperatures below 3°C. On meat of pH < 6 the growth of these three cold-tolerant pathogens is inhibited under anaerobic

TABLE 14.3
Major Bacterial Pathogens Associated with Fresh Meats

Organism	Gram Reaction	Oxygen Requirement	Minimum Temperature	Associated Meats
Aeromonas hydrophila	–	Facultative anaerobe	0°C	Red meats, poultry
Campylobacter	–	Microaerophile	28°C	Red meats, poultry
Clostridium botulinum, Proteolytic A, B, F	+	Obligate anaerobe	10°C	Red meats
Clostridium botulinum, Nonproteolytic B, E, F	+	Obligate anaerobe	3°C	Red meats
Verotoxigenic *E. coli*	–	Facultative anaerobe	7–10°C	Red meats
Salmonella typhimurium	–	Facultative anaerobe	8°C	Red meats, poultry
Shigella spp.	–	Facultative anaerobe	10°C	Red meats, poultry
Yersinia enterocolitica	–	Facultative anaerobe	–1°C	Pork
Listeria monocytogenes	+	Facultative anaerobe	0°C	Red meats

conditions (Grau and Vanderlinde, 1990; Palumbo, 1988), but under those conditions they can potentially grow on dark, firm, dry (DFD) meat, moist fat surfaces, or chicken leg muscle or skin, all of which have pH values > 6. However, the growth of these organisms is slowed by high CO_2 concentrations (Garcia de Fernando et al., 1995) and their growth is completely inhibited under a high-CO_2, anaerobic atmosphere at 5°C or below (Gill and Reichel, 1989). In addition to inhibiting the growth of cold tolerant pathogens at chiller temperatures, high CO_2 concentrations (30% CO_2:70% O_2 or 0.4% CO:60% CO_2:39.6% N_2) have been reported to inhibit growth of *E. coli* O157:H7 at the abusive storage temperature of 10°C (Nissen et al., 2000). At chiller temperatures of 0–2°C, high CO_2 concentrations (20% CO_2:80% O_2 or 0.4% CO:30% CO_2:69.6% N_2) have been shown to produce a 1 log reduction in the numbers of *E. coli* O157:H7 and *Salmonella typhimurium* (Brooks et al., 2008).

It has been suggested that preservative packaging could potentially promote pathogen growth by the suppression of the growth of competitive background flora (Farber, 1991). However, this is unlikely to impact the safety of raw meats because, on nutrient-rich substrates like meat, competition between bacteria for the available nutrients does not affect rates of growth until the flora approaches maximum numbers (Gill, 1986). In addition, fresh meats under an anaerobic atmosphere develop flora dominated by lactic acid bacteria, some species of which, when present in high numbers, have been demonstrated to be antagonistic to pathogens on fresh meats. *Lactobacillus reuteri* has been reported to reduce the viability of *E. coli* O157:H7 on refrigerated ground beef packaged under nitrogen (Muthukumarasamy et al., 2003). On raw chicken at 5°C, the presence of 10^8 *Lactobacillus delbrueckii* subsp. *lactis* resulted in a 1 log reduction in the numbers of *E. coli* O157:H7 compared to controls (Brashears et al., 1998). *Lb. delbrueckii* subsp. *lactis* has been reported to produce similar small declines in *E. coli* O157:H7 and *S. typhimurium* on refrigerated pork and beef (Senne and Gilliland, 2003). Inhibition of *L. monocytogenes* and *C. jejuni* by a variety of lactic acid bacteria has been reported *in vitro* (Jones et al., 2008), but has not been demonstrated with those bacteria on raw meats. Although the reported antagonistic effects of lactic acid bacteria on pathogens are not sufficient to substantially increase the safety of contaminated product, they do indicate that packaging that promotes the growth of lactic acid bacteria will likely not increase the risks from pathogens on meat.

Concerns have been raised over the potential risk in anaerobic packaging systems of growth and toxin production by the strict anaerobe *Cl. botulinum*. Proteolytic *Cl. botulinum*, types A and B, cannot grow at chiller temperatures and so pose no risk when proper temperatures for chilled meats are maintained (Hauschild et al., 1985). However, *Cl. botulinum* types E and F and nonproteolytic type B do pose a potential risk of toxin production in chilled foods as they can grow at temperature of about 3°C (Hauschild, 1989). The inclusion of O_2 in pack atmospheres to inhibit the potential growth of *Cl. botulinum* has been proposed but has a negligible impact on improving safety, as raw meats can provide anaerobic niches whether or not O_2 is present in the atmosphere (Lambert et al., 1991). Although there seems to be a possibility of *Cl. botulinum* growth and toxin production on packaged fresh meats, acquisition of botulism from such a source is unknown. This may be in part due to the fact that growth of *Cl. botulinum* in packs of meat that have been temperature abused will be accompanied by rapid development of gross spoilage of the product (Hauschild et al., 1985). Also, botulinum toxin is heat liable and it will be denatured by cooking. Woodburn et al. (1979) demonstrated that rapid and exponential inactivation of botulinum toxin occurs at 85°C.

There is, then, no real possibility of increase in the risks from pathogens on red meat of normal pH as a result of storage of product under anaerobic or low-O_2 atmospheres, provided temperatures below 2°C are maintained. Though the growth of *A. hydrophila*, *L. monocytogenes*, and *Y. enterocolitica* may occur on high-pH product such as DFD red meat or poultry skin, these pathogens may be inhibited by high CO_2 concentrations in pack atmospheres. High CO_2 concentrations may also provide a barrier to pathogen growth if temperature abuse occurs during storage or display.

14.6 FUTURE DEVELOPMENTS

Central preparation of retail packaged meat offers a number of economic advantages (Eilert, 2005). Meat-cutting operations can be mechanized to reduce requirements for skilled workers, who are increasingly unavailable in many developed countries, and to enhance control over portioning of meat. Meat-cutting facilities at retail outlets can be reduced in size or eliminated, freeing costly floor space for retailing rather than product preparation. Control of product hygiene is easier to achieve at a few central preparation facilities than at multiple facilities at retail outlets. In addition, waste tissues and packaging from primal cuts or hanging meat do not have to be transported to and from retail outlets.

Despite these advantages a number of factors militate against the total displacement of retail-level meat-processing facilities by centrally prepared product (McMillin, 2008). Customers may often prefer to buy meat that is evidently freshly cut in retail level facilities and may be cut to their requirements on request. Moreover, the types and amounts of products prepared at retail-level facilities can be rapidly adjusted to meet with changing or unanticipated patterns of sale. In addition to these marketing matters, two technical factors seem to be of major importance. One is the uncertain control over and generally relatively high temperatures of product in retail cases. Poor temperature control reduces the time that red meats exposed to O_2 retain a desirable red color and greatly reduces the time before the onset of microbial spoilage. The other technical factor is the limited extension of red meat display life obtainable with high-O_2 MAs and the inability of such atmospheres to reduce the large differences in color stability not only between different muscles but also between muscle tissues, bone marrow, and exudate-stained fat.

As the control over display case temperatures does not seem to have greatly improved during the past 30 years, it seems unlikely that display case temperatures will be substantially reduced in the foreseeable future. However, with good processing hygiene, the development of bacteria spoilage could be adequately controlled at current case temperatures if product were retail packaged under vacuum or in an anaerobic atmosphere to inhibit the growth of aerobic spoilage organisms. Packaging with CO would still be required to impart a bright red color to red meats.

Thus, if consumers generally come to accept either or both the anoxic color of red meats in product for retail sale or the use of CO to obtain a red color for anoxic meats, then the technical difficulties that currently restrict central preparation systems for retail-ready meats would largely disappear. However, if neither of these options proves to be acceptable, then technical constraints will prevent central preparation systems displacing retail-level cutting facilities in many regions.

REFERENCES

Anon. 2008. Fresh meat in modified atmosphere packages—status report. *Fleischwirtschaft International* 2008(4): 46–48.

Aspé E., Roeckel M., Marti M.C., Jiménez R. 2008. Effect of pre-treatment with carbon monoxide and film properties on the quality of vacuum packaging of beef chops. *Packaging Technology and Science* 21: 395–404.

Barros-Velazquez J., Carreira L., Franco C., Vazquez B.I., Fente C., Cepeda A. 2003. Microbiological and physicochemical properties of fresh retail cuts of beef packaged under an advanced vacuum skin system and stored at 4°C. *Journal of Food Protection* 66: 2085–2092.

Belcher J.N. 2006. Industrial packaging developments for the global meat market. *Meat Science* 74: 143–148.

Bell R.G., Penney N., Moorhead S.M. 1996. The retail display life of steaks prepared from chill stored vacuum and carbon dioxide-packed sub-primal beef cuts. *Meat Science* 42: 165–178.

Bjerklie S. 2007. MAP stepping on the gas. *Meat and Poultry* 53:102–104.

Blickstade E. 1983. Growth and end product formation of two psychrotrophic *Lactobacillus* spp. and *Brochothrix thermosphacta* ATCC 11509T at different pH values and temperatures. *Applied and Environmental Microbiology* 46: 1345–1350.

Brashears M.M., Reilly S.S., Gilliland S.E. 1998. Antagonistic action of cells of *Lactobacillus lactis* toward *Escherichia coli* O157:H7 on refrigerated raw chicken meat. *Journal of Food Protection* 61: 166–170.

Broda D.M., Boerema J.A., Bell R.G. 2003. PCR detection of psychrophilic *Clostridium* spp. causing 'blown pack' spoilage of vacuum-packed chilled meats. *Journal of Applied Microbiology* 94: 515–522.

Brody A. 2008. Shelf life assurance. Active packaging pushes the envelope. *Meat and Poultry* 54(3): 86–91.

Brody A.L. 1996. Integrating aseptic and modified atmosphere packaging to fulfill a vision of tomorrow. *Food Technology* 50(4): 56–66.

Brody A.L. 2007. Case-ready packaging for red meat. *Food Technology* 61(3): 70–72.

Brolls E.K. 1986. Factors affecting retail display cabinets. In: *Meat Chilling*. Paris: International Institute of Refrigeration, pp. 325–333.

Brooks J.C., Alvarado M., Stephens T.P., Kellermeier J.D., Tittor A.W., Miller M.F., Brashears M.M. 2008. Spoilage and safety characteristics of ground beef packaged in traditional and modified atmosphere packages. *Journal of Food Protection* 71: 293–301.

Campo M.M., Nute G.R., Hughes S.I., Enser M., Wood J.D., Richardson R.I. 2006. Flavor perception of oxidation in beef. *Meat Science* 72: 303–311.

CFIA. 1998. Shipment of meat products from registered establishments. In: *Meat hygiene manual of procedures*, section 8.6. Ottawa: Canadian Food Inspection Agency.

Channon H.A., Kerr M.G., Walker P.J. 2004. Effect of Duroc content, sex and ageing period on meat and eating quality attributes of pork loin. *Meat Science* 66: 881–888.

Clark D.S., Burki T. 1972. O_2 requirements of strains of *Pseudomonas* and *Achromobacter. Canadian Journal of Microbiology* 18: 321–326.

Cornforth D. 1994. Color—its basis and importance. In: *Quality Attributes and Their Measurement in Meat, Poultry and Fish Products. Advances in Meat Research*, vol. 9. Pearson A.M., Dutson T.R. (Eds). London: Blackie Academic and Professional, pp. 34–78.

Cornforth D.P., Jaysingh P. 2004. Color and pigment. In: *Encyclopedia of Meat Sciences*. Jensen W.K., Devine C., Dikeman M. (Eds). Amsterdam: Elsevier, pp. 249–256.

Coventry M.J., Hickey M.W., Mawson R., Drew P., Wan J., Krause D., Boghossian V. 1998. The comparative effects of nitrogen and oxygen on the microflora of beef steaks in carbon dioxide-containing modified atmosphere vacuum skin-packaging (MA-VSP) systems. *Letters in Applied Microbiology* 26: 427–431.

Crews J. 2007. A closer look. The 2007 national meat case study sheds new light on retail protein offerings. *Meat and Poultry* 53(10): 42–52.

Dainty R.M., Shaw B.G., Harding C.O., Michanie S. 1979. The spoilage of vacuum packed beef by cold tolerant bacteria. In: *Cold Tolerant Microbes in Spoilage and the Environment*. Russel A.D., Fuller R. (Eds). London, UK: Academic Press, pp. 83–100.

Dietrich A. 2005. A difficult choice. Pros and cons of packaging materials for case ready applications. *Fleischwirtschaft International* 1: 24–26.

Dransfield E. 1994. Tenderness of meat, poultry, and fish. In: *Quality Attributes and Their Measurement in Meat, Poultry and Fish Products*. Advances in Meat Research vol. 9. Pearson A.M., Dutson T.R. (Eds). London: Blackie Academic and Professional, pp. 287–315.

Echevarne C., Renerre M., Labas R. 1990. Metmyoglobin reductase activity in bovine muscles. *Meat Science* 27: 161–172.

Eilert S.J. 2005. New packaging technologies for the 21st century. *Meat Science* 71: 122–127.

Farber J.M. 1991. Microbiological aspects of modified-atmosphere packaging technology—a review. *Journal of Food Protection* 54: 58–70.

Faramarzi R., Coburn B., Sarhadian R. 2003. Effects of improper product loading on the performance of an open vertical display case. *ASHRAE Transactions* 109: 267–272.

FDA. 2004. US Food and Drug Administration. Agency Response Letter, GRAS Notice No. GRN000143. http://www.cfsan.fda.gov/~rdb/opa-g143.html. Accessed June 2008.

Garcia de Fernando G.D., Mano S.B, Lopez D., Ordóñez J.A. 1995. Effect of modified atmospheres on the growth of pathogenic psychrotrophic microorganisms on proteinaceous foods. *Microbiologia* 11: 7–22.

Gatellier P., Hamelin C., Durand Y., Renerre M. 2001. Effect of a dietary vitamin E supplementation on color stability and lipid oxidation of air- and modified atmosphere-packaged beef. *Meat Science* 59: 133–140.

Giddings G.G. 1977. The basis of color in muscle foods. *CRC Critical Reviews in Food Science and Nutrition* 9: 81–114.

Gill C.O. 1981. Meat spoilage and evaluation of the potential storage life of fresh meat. *Journal of Food Protection* 46: 444–452.

Gill C.O. 1986. The control of microbial spoilage in fresh meats. In: *Advances in Meat Research: Meat and Poultry Microbiology*, Advances in Meat Research, vol 2. Pearson A.M., Dutson T.R. (Eds). London: Blackie Academic and Professional, pp. 49–88.

Gill C.O. 1988. The solubility of carbon dioxide in meat. *Meat Science* 22: 65–71.

Gill C.O. 1989. Packaging meat for prolonged chilled storage. The Captech process. *British Food Journal* 91: 11–15.

Gill C.O. 1995. MAP and CAP of fresh, red meats, poultry and offals. In: *Principles of Modified Atmosphere and Sous Vide Product Packaging*. Farber J.M., Dodds K.L. (Eds). Lancaster, PA: Technomic Publishing, pp. 105–136.

Gill C.O. 1996. Extending the storage life of raw chilled meats. *Meat Science* 43: S99–S109.

Gill C.O. 2006. Microbiology of frozen foods. In: *Handbook of Frozen Food Processing and Packaging*. D.-W. Sun (Ed). Boca Raton, Florida: CRC Press, pp. 85–100.

Gill C.O., Badoni M. 2002. Microbiological and organoleptic qualities of vacuum-packaged ground beef prepared from pasteurized manufacturing beef. *International Journal of Food Microbiology* 74: 111–118.

Gill C.O., Friske M., Tong A.K.W., McGinnis J.C. 1995. Assessment of the hygienic characteristics of a process for the distribution of processed meats, and of storage conditions at retail outlets. *Food Research International* 28: 131–138.

Gill C.O., Jones S.D.M. 1992. Efficiency of a commercial process for the storage and distribution of vacuum-packaged beef. *Journal of Food Protection* 55: 880–887.

Gill C.O., Jones T. 1994. The display life of retail-packaged beef steaks after their storage in master packs under various atmospheres. *Meat Science* 38: 385–396.

Gill C.O., Jones T. 1996. The display life of retail packaged pork chops after their storage in master packs under atmospheres of N_2, CO_2 or $O_2 + CO_2$. *Meat Science* 42: 203–213.

Gill C.O., Jones T., Rahn K., Houde A., McGinnis J.C., Campbell S., Holley R.A., LeBlanc D.I. 2002a. Control of product temperatures during the distribution of retail ready beef to stores and vacuum packaged beef to restaurants. *Dairy, Food and Environmental Sanitation* 22: 422–428.

Gill C.O., Jones T., Rahn K., Campbell S., LeBlanc D.I., Holley R.A., Stark R. 2002b. Temperatures and ages of boxed beef packed and distributed in Canada. *Meat Science* 60: 401–410.

Gill C.O., Jones T., Leblanc D.I., Rahn K., Campbell S., Holley R.A., Stark R., Houde A. 2002c. Temperatures and ages of packs of beef displayed at stores in Canada. *Meat Science* 60: 139–144.

Gill C.O., Jones T., Houde A., LeBlanc D.I., Rahn K., Holley R.A., Starke R. 2003. The temperatures and ages of packs of beef displayed in multi-shelf retail cabinets. *Food Control* 14: 145–151.

Gill C.O., McGinnis J.C. 1995. The use of oxygen scavengers to prevent the transient discoloration of ground beef packaged under controlled, oxygen-depleted atmospheres. *Meat Science* 41: 19–27.

Gill C.O., Newton K.G. 1977. The development of aerobic spoilage flora on meat stored at chill temperatures. *Journal of Applied Bacteriology* 43: 189–195.

Gill C.O., Newton K.G. 1980. Development of bacterial spoilage at adipose tissue surfaces of fresh meat. *Applied and Environmental Microbiology* 39: 1076–1077.

Gill C.O., Phillips D.M. 1993. The efficiency of storage during distant continental transportation of beef sides and quarters. *Food Research International* 26: 239–245.

Gill C.O., Phillips D.M., Harrison J.C.L. 1988. Product temperature criteria for shipment of chilled meats to distant markets. In: *Refrigeration for Food and People*. Paris: International Institute Refrigeration, pp. 40–47.

Gill C.O., Reichel M.P. 1989. Growth of the cold-tolerant pathogens *Yersinia enterocolitica*, *Aeromonas hydrophila* and *Listeria monocytogenes* on high-pH beef packaged under vacuum or carbon dioxide. *Food Microbiology* 6: 223–230.

Gill C.O., Tan K.H. 1980. Effect of carbon dioxide on growth of meat spoilage bacteria. *Applied and Environmental Microbiology* 39: 317–319.

Govindarajan S. 1973. Fresh meat color. *CRC Critical Reviews in Food Technology* 4: 117–140.

Grau F.H., Vanderlinde P.B. 1990. Growth of *Listeria monocytogenes* on vacuum packaged beef. *Journal of Food Protection* 53: 739–741.

Gray J.I., Gomaa E.A., Buckley D.J. 1996. Oxidative quality and shelf life of meats. *Meat Science* 43: S111–S123.

Grobbel J.P., Dikeman M.E., Smith J.S., Kropf D.H., Milliken G.A. 2006. Effects of polyvinyl chloride overwrap film, high-oxygen modified atmosphere packaging, or ultra-low-oxygen modified atmosphere packaging on bone marrow discoloration in beef humerus, rib, thoracic vertebra, and scapula. *Journal of Animal Science* 84: 694–701.

Hauschild A.H.W. 1989. Clostridium botulinium. In: *Foodborne Bacterial Pathogens*. Doyle M.P. (Ed). New York: Marcel Dekker, pp. 111–189.

Hauschild A.H.W., Poste L.M., Hilsheimer R. 1985. Toxin production by *Clostridium botulinum* and organoleptic changes in vacuum-packaged raw beef. *Journal of Food Protection* 48: 712–716.

Hintlian C.B., Hotchkiss J.H. 1986. The safety of modified atmosphere packaging: a review. *Food Technology* 40(12): 70–76.

Ho C.P., Huang N.Y., McMillin K.W. 2003. Microflora and color of ground beef in gas exchange modified atmosphere packaging with abusive display temperatures. *Journal of Food Science* 68: 1771–1776.

Hood D.E. 1980. Factors affecting the rate of metmyoglobin accumulation in pre-packaged beef. *Meat Science* 4: 247–265.

Hunt M.C., Mancini R.A., Hachmeister K.A., Kropf D.H., Merriman M., Delduca G., Milliken G. 2004. Carbon monoxide in modified atmosphere packaging affects color, shelf life, and microorganisms of beef steaks and ground beef. *Journal of Food Science* 69: FCT45–FCT52.

Isdell E., Allen P., Doherty A., Butler F. 2003. Effect of packaging cycle on the color stability of six beef muscles stored in a modified atmosphere mother pack system with oxygen scavengers. *International Journal of Food Science & Technology* 38: 623–632.

Jakobsen M., Bertelsen G. 2004. Predicting the amount of carbon dioxide absorbed in meat. *Meat Science* 68: 603–610.

James S.J., James C. 2002. Chilled and frozen retail display. In: *Meat Refrigeration*, Cambridge, UK: Woodhead Publishing, pp. 231–250.

Jenkins W.A., Harrington J.P. 1991. Fresh meat and poultry. In: *Packaging Foods with Plastics,* Lancaster, PA: Technomic Publishing, pp. 109–131.

Jeremiah L.E., Tong A.K.W., Jones S.D.M., McDonell C. 1993. A survey of Canadian consumer perceptions of beef in relation to general perceptions regarding foods. *Journal of Consumer Studies and Home Economics* 17: 13–37.

Jones R.J. 2004. Observations on the succession dynamics of lactic acid bacteria populations in chill-stored vacuum-packaged beef. *International Journal of Food Microbiology* 90: 273–282.

Jones R.J., Hussein H.M., Zagorec M., Brightwell G., Tagg J.R. 2008. Isolation of lactic acid bacteria with inhibitory activity against pathogens and spoilage organisms associated with fresh meat. *Food Microbiology.* 25: 228–234.

Kanner J. 1994. Oxidative processes in meat and meat products: quality implications. *Meat Science* 36: 169–189.

Kartika S., Candogan K., Grimes L.W., Acton J.C. 2003. Rinse treatment and oxygen barrier properties of films for improving quality retention in vacuum-skin packaged fresh chicken. *Journal of Food Science* 68: 1762–1765.

Kennedy C., Buckley D.J., Kerry J.P. 2004. Display life of sheep meats retail packaged under atmospheres of various volumes and compositions. *Meat Science* 68: 649–658.

Knox B.L., van Lack R.L J.M., Davidson P.M. 2008. Relationships between ultimate pH and microbial, chemical, and physical characteristics of vacuum-packaged pork loins. *Journal of Food Science* 73: M104–M110.

Lambert A.D., Smith J.P., Dodds K.L. 1991. Effect of initial O_2 and CO_2 and low-dose irradiation on toxin production by *Clostridium botulinum* in MAP fresh pork. *Journal of Food Protection* 54: 939–944.

Lanier T.C., Carpenter J.A., Toledo R.T., Reagan J.O. 1978. Metmyoglobin reduction in beef systems as affected by aerobic, anaerobic and carbon monoxide-containing environments. *Journal of Food Science* 43: 1788–1792.

Ledward D.A. 1984. Haemproteins in meat and meat products. In: *Developments in Food Proteins*, vol. 3. Hudson B.J.F. (Ed). New York: Elsevier Applied Science, pp. 33–68.

Lund M.N., Lametsch R., Hviid M.S., Jensen O.N., Skibsted L.H. 2007. High-oxygen packaging atmosphere influences protein oxidation and tenderness of porcine *longissimus dorsi* during chill storage. *Meat Science* 77: 295–303.

Madhavi D.L., Carpenter C.E. 1993. Aging and processing affect color, metmyoglobin reductase and oxygen consumption of beef muscles. *Journal of Food Science* 58: 939–942.

Malton R. 1978. Refrigerated transport. *Meat Science* 51(7): 21–26.

Mancini R.A., Hunt M.C. 2005. Current research in meat color. *Meat Science* 71:100–121.

McKenna D.R., Mies P.D., Baird B.E., Pfeiffer K.D., Ellebracht J.W., Savell J.W. 2005. Biochemical and physical factors affecting discoloration characteristics of 19 bovine muscles. *Meat Science* 70: 665–682.

McMillin K.W. 2008. Where is MAP going? A review and future potential of modified atmosphere packaging for meat. *Meat Science* 80: 43–65.

Millar S., Wilson R., Moss B.W., Ledward D.A. 1994. Oxymyoglobin formation in meat and poultry. *Meat Science* 36: 397–406.

Mondry H. 1996. Packaging systems for processed meat. In: *Meat Quality and Meat Packaging*. Taylor A.A., Raimundo A., Severini M., Smulders F.J.M. (Eds). Utrecht, Holland: ECCEAMST (European Consortium for Continuing Education in Advanced Meat Science and Technology), pp. 323–356.

Muthukumarasamy P., Han J.H., Holley R.A. 2003. Bactericidal effects of *Lactobacillus reuteri* and allyl isothiocyanate on *Escherichia coli* O157:H7 in refrigerated ground beef. *Journal of Food Protection* 66: 2038–2044.

Newton K.G., Gill C.O. 1978. Storage quality of dark, firm, dry meat. *Applied and Environmental Microbiology* 36: 375–376.

Nissen H., Alvseike O., Bredholt S., Holck A., Nesbakken T. 2000. Comparison between the growth of *Yersinia enterocolitica, Listeria monocytogenes, Escherichia coli* O157:H7 and *Salmonella* spp. in ground beef packed by three commercially used packaging techniques. *International Journal of Food Microbiology* 59: 211–220.

Nortje G.L., Shaw B.G. 1989. The effect of ageing treatment on the microbiology and storage characteristics of beef in modified atmosphere packs containing 25% CO_2 + 75% O_2. *Meat Science* 25: 43–58.

Nychas G.J., Dillon V.M., Board R.G. 1988. Glucose, the key substrate in the microbiological changes occurring in meat and certain meat products. *Biotechnology and Applied Biochemistry* 10: 203–231.

Offer G., Knight P. 1988. The structural basis of water holding in meat: part 2, drip loss. In: *Developments in Meat Science*, vol. 4. Lawrie R.A. (Ed). London: Elsevier, pp. 173–243.

O'Keeffe M., Hood D.E. 1982. Biochemical factors influencing metmyoglobin formation on beef from muscles of differing color stability. *Meat Science* 7: 209–228.

Ordonez J.A., Ledward D.A. 1977. Lipid and myoglobin oxidation in pork stored in oxygen- and carbon dioxide-enriched atmospheres. *Meat Science* 1: 41–48.

Palumbo S.A. 1988. The growth of *Aeromonas hydrophila* K144 in ground pork at 5°C. *International Journal of Food Microbiology* 7: 41–48.

Pastors P.M. 2005. The optimal logistic window. *Fleischwirtschaft International* 2005(3): 22–25.

Petersen J.H., Togeskov P., Hallas J., Olsen M.B., Jorgensen B., Jakobsen M. 2004. Evaluation of retail fresh meat packagings covered with stretch films of plasticized PVC and non-PVC alternatives. *Packaging Technology and Science* 17: 53–66.

Pettersen M.K., Nissen T.E., Nilsson A. 2004. Effect of packaging materials and storage conditions on bacterial growth, off-odour, pH and colour in chicken breast fillets. *Packaging Technology and Science* 17: 165–174.

Quigley L. 2002. Canadian style case ready. *Meat and Poultry* 48(2): 30–36.

Rhee K.S., Ziprin Y.A. 1987. Lipid oxidation in retail beef, pork and chicken muscles as affected by concentrations of heme pigments and nonheme iron and microsomal enzymic lipid peroxidation activity. *Journal of Food Biochemistry* 11: 1–15.

Sarantopoulos C.I.G.L., Alves R.M.V., Contreras C.J.C., Galvao M.T.E.L., Gomez T.C. 1998. Use of a modified atmosphere masterpack for extending the shelf life of chicken cuts. *Packaging Technology and Science* 11: 217–229.

Senne M.M., Gilliland S.E. 2003. Antagonistic action of cells of *Lactobacillus delbrueckii* subsp. *lactis* against pathogenic and spoilage microorganisms in fresh meat systems. *Journal of Food Protection* 66: 418–425.

Smiddy M., Fitzgerald M., Kerry J.P., Papkovsky D.B., O'Sullivan C.K., Guilbeaut G.G. 2002. Use of oxygen sensors for the non-destructive measurement of the oxygen content in modified atmosphere and vacuum packs of cooked chicken patties: impact of oxygen content on lipid oxidation. *Meat Science* 61: 285–290.

Sørheim O., Aune T., Nesbakken T. 1997. Technological, hygiene and toxicological aspects of carbon monoxide use in modified-atmosphere packaging of meat. *Trends in Food Science and Technology* 8: 307–312.

Spaulding M. 1994. Vacuum skin-packs add shelf life triple sales. *Packaging* 39(8): 23–24.

Swatland H.J. 2004. Progress in understanding the paleness of meat with low pH. *South African Journal of Animal Science* 34: 1–7.

Tewari G., Jeremiah L.E., Jayas D.S., Holley R.A. 2002. Improved use of oxygen scavengers to stabilize the color of retail-ready meat cuts stored in modified atmospheres. *International Journal of Food Science & Technology* 37: 199–207.

Torstveit A.K., Magnussen O.M. 1998. Temperature conditions in refrigerated counters in Vest Agder, Norway. In: *Hygiene, Quality and Safety in the Cold Chain and Air Conditioning*. Paris: International Institute of Refrigeration, pp. 237–241.

Vázquez B.I., Carreira L., Franco C., Fente C., Cepeda A., Barros-Velázquez J. 2004. Shelf life extension of beef retail cuts subjected to an advanced vacuum skin packaging system. *European Food Research and Technology* 218: 118–122.

Wallace W.J., Houtchens R.A., Maxwell J.C., Caughey W.S. 1982. Mechanism of autoxidation for haemoglobins and myoglobin. *Journal of Biological Chemistry* 257: 4966–4969.

Warris P.D., Rodes D.N. 1977. Haemoglobin concentration in beef. *Journal of the Science of Food and Agriculture* 28: 931–934.

Whitfield F.B. 1998. Microbiology of food taints. *International Journal of Food Science and Technology* 33: 31–51.

Wilkinson B.H.P., Janz J.A.M., Morel P.C.H., Purchas R.W., Hendriks W.H. 2006. The effect of modified atmosphere packaging with carbon monoxide on the storage quality of master-packaged fresh pork. *Meat Science* 73: 605–610.

Woodburn M.J., Somers E., Rodriguez J., Schantz E.J. 1979. Heat inactivation rates of botulinum toxins A, B, E and F in some foods and buffers. *Journal of Food Science* 44: 1658–1661.

Zhao Y., Wells J.H., McMillin K.W. 1994. Application of dynamic modified atmosphere packaging systems for fresh red meats: review. *Journal of Muscle Foods* 5: 299–328.

15 Packaging and the Shelf Life of Fish

Steve Slattery
Innovative Food Technologies
Emerging Technologies
Primary Industries and Fisheries
Queensland Department of Employment,
Economic Development and Innovation
Hamilton, Australia

CONTENTS

15.1 INTRODUCTION

Fish is a rapidly growing component of the modern diet. It has been promoted by health specialists for its health benefits as well as its taste and texture. Unfortunately, fish flesh is highly perishable and has a much shorter shelf life than terrestrial meats. Limiting the loss of quality is dependent on the rapid and effective handling and processing conditions utilized by the fisher after harvest. Packaging can play a key role in limiting any loss of fish quality. Most fish is sold either as fillets or as a prepared product. Modern customers expect convenient forms of packaging for their fish purchases and now look for packs that require minimal preparation or contain ready-to-cook meals.

15.2 QUALITY ATTRIBUTES OF FISH

15.2.1 Nutritional Content

Fish has similar amounts of protein to red meats, pork, and poultry but lower levels of bioavailable iron and zinc; the vitamin B_{12} content is similar to or even higher than that of red meat. Fat content of fish is variable and can range from 1% to 10% of the flesh weight. Oily fish are a particularly good source of omega-3 fatty acids and at levels much higher than that in plants. Fish is a good source of phosphorous and magnesium as well as important mineral salts such as sodium, potassium, calcium, iron, chlorine, and iodine. The latter is an important dietary component especially in areas distant from the sea where soils are low in iodine. Carbohydrate levels in fish are less than 1.0%.

15.2.2 Structure of Fish Flesh

The muscle and connective tissues of fish, unlike those of terrestrial meats, are arranged in a fixed segmented structure that gives fish fillets their flaky appearance when cooked. Many fish species, such as orange roughy (*Hoplostethus atlanticus*), do not have a strong flavor, and so texture plays a major role in consumer acceptability (Manju et al., 2007). Fish is known for its tenderness.

15.3 DETERIORATIVE REACTIONS AND INDICES OF FAILURE

15.3.1 Chemical Changes

The problems of cold shock and thaw rigor that are associated with rapid processing after capture should not be an issue when packaging fresh fish, as the progression of rigor should be fully resolved by the time it is ready to be packed; therefore, these ATP-dependent defects will not be discussed here. The stress induced during capture will result in elevated metabolic rates that persist for a considerable time, resulting in impacts on the quality of fish as food such as gaping of the flesh (a condition where the muscle tissue separates from the connective tissue) and bitterness.

The texture of cooked fish flesh can be unacceptably tough at low pH (6.4) or sloppy at neutral pH, and gaping becomes progressively worse as the postmortem pH falls (Love, 1988). The fall in muscle pH has been associated with an increase in lactic acid that happens when fish struggle during landing using up ATP (Regenstein and Regenstein, 1991).

The different chemistry of the two major muscle types goes beyond metabolism of glycogen, with some activity of the red muscle being similar to that of the liver. Hultin (1992) discussed the possibility that the muscle tissue of species that break down trimethylamine oxide (TMAO) to dimethylamine (DMA) and formaldehyde, via enzymic processes, toughens on storage due to cross-linking of the proteins by formaldehyde. After denaturation, interaction of the proteins via hydrophobic forces induces toughening. The rate of formation of trimethylamine (TMA) and DMA is species dependent, and lower temperatures during storage result in less production of DMA. Phospholipids are associated with fish fats and are rich in TMA. Bacteria and enzymes present in the fish split TMA from the phospholipids, causing the strong, characteristic fishy odor encountered by consumers. The ready oxidation of fish fats adds rancid off-odors to this fishiness as the product deteriorates, and thus the nutritional benefits of the omega-3 fat are also lost over time.

There are a range of protease enzymes present in the flesh of fish (Shahidi and Janak Kamil, 2001). The presence of endogenous collagenolytic activity in the muscle of fish can result in the breakdown of collagen during storage, resulting in gaping. Other enzymes attack the muscle protein and connective tissue, causing the muscle bundles to weaken and separate during handling, which results in the gaping that can be observed on fish fillets.

Rapid objective methods for measuring the quality of fish are becoming a requirement of industry due to quality and traceability issues. Nesvadba (2003) reviewed the multisensor approach for monitoring the quality of fish. This included textural analysis, visible light spectrometry, image analysis, electronic noses, analysis of electrical properties, and color measurement, and the Hyldig et al. (2007) sensory-based quality index method (QIM) was used as the reference method.

15.3.2 Microbiological Changes

Theoretically the flesh of fish is sterile, but, as the surface slime and digestive tract harbor large numbers of many types of bacteria, they rapidly contaminate the flesh after death and during processing. Trawl-caught fish undergo a lot of surface abrasion during capture and so their flesh is more rapidly infiltrated. Large pelagic fish have reduced scales and thicker skin than demersal fish. Line-caught fish encounter less damage; Spanish mackerel stored uncut under ice had flesh microbial total counts of log 1 after 2 days and less than log 6 after 18 days (Slattery, 1998), much less than trawled fish such as gutted cod, which had flesh counts of log 4 on landing and log 6 after 5 days (Villemure et al., 1986). Fish struggle when caught, using up most of the glycogen in their muscles, so little remains to be converted to lactic acid after death. Thus, the preservative action of excess lactic acid in the tissue is low during storage.

Bacteria are the main cause of off-odors of packaged fish. The majority of the bad-smelling compound TMA is produced from bacterial breakdown of TMAO, as the intrinsic enzymes of fish can produce only a small amount by themselves (Pedrosa-Menabrito and Regenstein, 1990).

Temperate marine fish microflora are dominated by psychrotrophic gram-negative bacteria belonging to the genera *Pseudomonas*, *Moraxella*, *Acinetobacter*, *Shewanella*, *Vibrionaceae*, *Clostridium*, *Lactobacillus*, and *Corynebacterium* (Giatrakou et al., 2008). Spoilage patterns of seafood are different for tropical species than for colder water species. As tropical fish come from warm waters, when they are chilled after landing, the lower temperatures inhibit the growth of heterotrophic bacteria present on their surface. It takes some time for the psychrotrophic bacteria to colonize the surface of the fish, but when it occurs, they show similar changes to temperate fish.

Shewan (1970) reported that for cod there was a lag phase of 2–3 days before a logarithmic increase in total bacterial numbers, so that by day 10 counts were as high as 10^8 colony-forming units per gram (cfu g^{-1}). During the first phase (0–6 days), no marked spoilage of cod stored on ice occurred; during the second phase (7–10 days), there was a lack of fresh odor, followed by the development of sour or slightly sweet to fruity odors during the third phase (11–14 days). During the fourth phase (>14 days), sulfur odors were apparent. Hydrogen sulfide–producing bacteria, such as *Pseudomonas* and *Alteromonas*, become the predominant genera in stored fish, thus explaining the strong off-odors that develop.

Histamine is a product of seafood decomposition caused by the growth of certain bacteria such as *Morganella psychrotolerans* and *Photobacterium phosphoreum* (Emborg et al., 2005). It is formed by the breakdown of the amino acid histidine and is a common occurrence in fish such as tuna, mahi-mahi, carangids, bluefish, mackerel, herring, anchovies, and sardines. The amount of amine that forms is a function of the bacterial species present and the temperature and time of exposure, and may exceed 1000 mg kg^{-1}. Consumption of fish with high levels of histamine, usually in excess of 200 mg kg^{-1}, causes a condition commonly referred to as "scombroid poisoning." Good-quality fish contain less than 10 mg kg^{-1} histamine, a level of 30 mg kg^{-1} indicates significant deterioration, and 50 mg kg^{-1} is the defect action level (DAL) at which regulatory actions are taken (Federal Register, 1995).

Another toxin that can be produced by the fish itself is tetrodotoxin. This is limited to a particular family of fish known as the puffer fish. There are also toxins produced by marine plants, such as ciguatoxin, or others associated with algal blooms (red tides) that the fish ingest or take up through respiration and absorption. These toxins tend to be limited to known species of fish, and so the best way to avoid these contaminants is to exclude these species from any packaging lines. Parasites can be killed by normal cooking procedures and by freezing (e.g., the round worm *Anisakis*); however, if eaten raw or semicooked, some parasites can lead to serious illness. The identification and then removal of contaminated fish or portions of the fillet from raw product lines through visualization with strong lights is the best safety procedure. Molina-Garcia and Sanz (2002) found that *Anisakis* larva were also killed by high hydrostatic pressure processing (200 MPa for 10 min).

15.3.2.1 Specific Spoilage Organisms

Although there is a great variation in species of microorganisms present on fresh fish, only a select few—the so-called specific spoilage organisms (SSOs)—are responsible for the offensive off-flavors associated with seafood spoilage (Gram and Dalgaard, 2002). The microflora changes according to the preservation conditions, with gram-negative fermentative bacteria (such as *Vibrionaceae*) spoiling unpreserved fish, whereas psychrotolerant gram-negative bacteria (*Pseudomonas* spp. and *Shewanella* spp.) grow on chilled fish (Gram and Huss, 2000).

It may be difficult to predict initial bacterial loads before packing and thus what shelf life may be achievable. Regular testing of the raw material supply being used during packaging will assist in determining suitable times for end of shelf life labeling. There are predictive tools available on the Internet that can provide this service for a range of packaging types and conditions. The Danish Institute of Fisheries Research has released, free of charge, the Seafood Spoilage and Safety Predictor (SSSP) software (Dalgaard, 2005), which predicts shelf life and growth of bacteria in different fresh and lightly preserved seafood by integrating physical data such as the product temperature profiles recorded by data loggers.

15.3.3 Prepackaging Treatments to Improve Fish Shelf Life

The high-quality shelf life (HQL) of most seafood in chill storage is relatively short, being only a few days. This short period does not allow sufficient time from reception through to distribution and display to ensure the restaurateur or consumer can obtain seafood at its best. Many of the high-quality characteristics are depleted in the marketing and distribution chain, indicating that only fish of the best quality should be packaged. This issue will be discussed in detail in the section on modified atmosphere packaging (MAP).

The concept of introducing a number of hurdles or barriers to the processing and packaging environment has been the core of systems such as hazard analysis and critical control point (HACCP), good manufacturing practice (GMP), and total quality management (TQM). This is especially important when dealing with fish. Microorganisms may be inhibited by chilling, freezing, water activity reduction, nutrient restriction, acidification, modification of packaging atmosphere, fermentation, or addition of antimicrobial compounds.

The best treatment for fresh fish is to keep the temperature close to that of the freezing point of the water inside the fish flesh. This slows down most enzymic processes and inhibits many bacteria. Holding fish at or below 3°C (37°F) inhibits the growth of *Clostridium botulinum*, preventing the production of botulism toxin (Regenstein and Regenstein, 1991). Because of the salt content of fish flesh, it does not start to freeze until a few degrees below 0°C. Reducing the temperature to between 0°C and when ice is finally formed is called "superchilling" (Regenstein and Regenstein, 1991).

The UK Advisory Committee on the Microbiological Safety of Food recommended, in addition to chill temperatures, a number of hurdles to use for vacuum packaged (VP) and MAP foods to prevent growth and toxin production by nonproteolytic *C. botulinum* in chilled foods with a shelf life of more than 10 days (ACMSF, 2006). These included the following:

- A heat treatment of 90°C for 10 min or equivalent lethality
- A pH of 5 or less throughout the food and throughout all components of complex foods
- A minimum salt level of 3.5% in the aqueous phase throughout the food and throughout all components of complex foods
- A water activity of 0.97 or less throughout the food and throughout all components of complex foods
- A combination of heat and preservative factors that can be shown consistently to prevent growth and toxin production by nonproteolytic *C. botulinum*

These treatments, however, alter the fish from being a fresh product, and so other procedures may be needed. The use of chemical additives should include only those compounds approved for fish products as permitted by the Codex Alimentarius (FAO/WHO, 2008) and the food standards of countries where the product is being packaged and exported to.

Dipping fish in chemical solutions to reduce bacterial loads is seen by industry as the easiest way of treating fish. The amount of uptake by the fish is significantly lower than the dip concentration because the exposure time is usually short and the dip is diluted by the liquid content of the fish (Regenstein and Regenstein, 1991). Potassium sorbate (3% w/v for 10 or 30 sec) was found effective in reducing the growth of *Alteromonas putrefaciens* and limited TMA values on Atlantic cod fillets, although it did not reduce the total aerobic count (Shaw et al., 1983). Phosphates have been used for processing fish for some time. The ability of polyphosphates to chelate metal ions appears to play an important role in their antimicrobial activity (Davidson, 1997). Gram-positive bacteria are more susceptible to phosphates than are gram-negative bacteria. The main effect of polyphosphates is to reduce the amount of drip loss from frozen fish products on thawing. There are a number of commercial products on the market that claim to improve shelf life. Some of the other compounds used in conjunction with phosphates are citric acid, potassium sorbate, ascorbic acid, and salt (Regenstein and Regenstein, 1991). Another mixture, containing sodium benzoate and fumaric acid, has also been used on fish. The main effectiveness of these chemicals is that they are acidic in nature. Many countries limit the amount of phosphate that can be added to seafood.

Other chemicals, such as sodium acetate, extend shelf life of fish through the control of pH. Kim et al. (1995) found that the use of 0.75% and 1% sodium acetate lowered initial aerobic counts by 0.6–0.7 log units and resulted in 6 days' longer shelf life. Manju et al. (2007) found a 2% (w/v) sodium acetate dip of 30 min resulted in longer shelf life in VP. Lactic acid (0.5–2.0% for 10 min) has been shown to help remove large numbers (one or more log counts) of microbes and even reduce levels of *Listeria monocytogenes* and *Edwardsiella tarda* (an important cause of hemorrhagic septicemia in fish and fish handlers) from the skin of catfish, both cleaned and when mucus is present (Kim and Marshall, 2001).

Chlorine is regularly used in the fish-processing environment for equipment and surfaces, but it is limited in effectiveness. Sykes (1970) claimed that a 100 mg kg^{-1} hypochlorite solution buffered at pH 7.0 used as a 5-min dip will kill 10^4 *Bacillus subtilis* var *niger*. Chlorine dioxide solutions sprayed at concentrations between 7.5 and 25 mg kg^{-1} were more effective than aqueous chlorine in killing *Escherischia coli* and *L. monocytogenes* inoculated on fish cubes (Lin et al., 1996). Kim et al. (1999) showed that dipping in 100 and 200 mg kg^{-1} chlorine dioxide solution for 5 min was effective in reducing the bacterial load on a range of seafood but was only successful in reducing the initial load by one log count when used on whole fish; similar results were reported by Slattery et al. (1998) when spraying at 500 mg kg^{-1}.

Sodium and potassium nitrite have specialized use in cured products to inhibit bacteria such as *C. botulinum*, *C. perfringens*, *E. coli*, *Achromobacter*, *Enterobacter*, *Flavobacterium*, *Micrococcus*, and *Pseudomonas*. *Listeria monocytogenes* growth on smoked salmon is also inhibited by 200 mg kg^{-1} of sodium nitrite (Pelroy et al., 1994). It is a requirement in some countries for there to be an addition of sodium erythorbate or isoascorbate to products containing nitrites as a cure accelerator and as an inhibitor to the formation of nitrosamines, which are carcinogenic compounds formed by reactions of nitrite with secondary or tertiary amines (Davidson, 1997). These compounds, such as TMA, are prevalent in fish and without precautions could put the consumer at risk when nitrites are used.

Ozone generators are becoming common in fish-processing plants. The insertion of activated O_2 in the wash water not only sanitizes the water but can also reduce the surface bacterial load. Campos et al. (2006) found that an ozone–slurry ice combined refrigerated system doubled the shelf life of farmed turbot, although ozone may accelerate lipid oxidation (Regenstein and Regenstein, 1991). Electrolyzed water is a possible alternative to ozonated water for fish shelf life extension (Park et al., 2002) as both have been demonstrated to reduce cattle hide contamination by several log counts (Bosilevac et al., 2005).

Hypobaric treatment is now becoming a viable method of sterilizing food, but there can be visual changes present that may detract from the final product. Pressures of several thousand atmospheres (500–600 MPa) are used to kill microorganisms but have little effect on spores or enzymes (Tucker,

2003). Unfortunately this treatment can alter the appearance of raw fish so that the muscle appears grilled or boiled (Borderias, 1996).

Irradiation of foods with ionizing radiation (gamma rays) has been seen as a viable process, and many countries have installed plants that can treat food. A dose of 2 kGy was effective in reducing initial populations of bacteria and improved shelf life of VP trout (Savvaidis et al., 2002). Although it can limit microbial growth on the product, biochemical changes can continue, causing defects during storage.

Spices and herbs contain essential oils (EOs) that have varying degrees of antimicrobial activity. The strongest plant antimicrobials come from cloves, cinnamon, oregano, thyme, and, to a lesser extent, sage and rosemary (Davidson, 1997). The review by Burt (2004) of EOs reported that levels between 0.5 and 20 $\mu L\ g^{-1}$ were required to obtain significant reductions in microbial counts on fish. Extensive research on sources and activities of natural antimicrobials is still under way.

Thyme and oregano oils added at 0.05% (v/v) to the inside of a polyethylene bag that contained an individual sea bass (*Lates calcarifer*) were effective in reducing bacterial counts during storage (Harpaz et al., 2003). When wild thyme was used as a hydrosol incorporated into ice, there was at least 15–20 days' increase in shelf life of whole and gutted barb (Oral et al., 2008). Oil-soluble inolens from rosemary offer full protection against rancidity, taste change, and color alteration in meat and meat products, fish, ready-to-eat meals, bakery and confectionary products, nut and seed mixes, and snacks and various savory applications (Anon., 2008). Tea polyphenols (0.2% w/v dip for 90 min) have been found to extend the shelf life of silver carp due to suppression of both bacterial and chemical degradation (Fan et al., 2008).

One technique previously used to reduce initial bacterial loads was the use of antibiotic dips; this practice is now banned in most countries because of the risk of developing antibiotic-resistant strains that would be detrimental to the health of workers and consumers. Natural preservatives using bacterial metabolites and live bacteria are a more acceptable way of inhibiting microbial growth (Delves-Broughton et al., 2007). Some lactic acid bacteria (LABs) produce antimicrobial proteins called bacteriocins that inhibit spoilage and pathogenic bacteria without changing the physical-chemical nature of the food being preserved (Montville and Winkowski, 1997). The LAB *Carnobacterium piscicola* has been found to have activity against the pathogen *L. monocytogenes* present on lightly preserved fish (Alves et al., 2005). Kim and Hearnsberger (1994) found that lactic acid cultures in combination with sodium acetate and potassium sorbate effectively inhibited gram-negative bacterial growth on catfish fillets during refrigerated storage.

Nisin is the best-known LAB bacteriocin and has been used on smoked and fresh fish as an adjunct to MAP. It increases the shelf life and delays toxin production by type E botulinal strains in fresh fish packaged in a CO_2 atmosphere, although toxin can sometimes be detected before samples are obviously spoiled (Taylor et al., 1990). One issue with nisin is that resistance can develop (Montville and Winkowski, 1997).

15.4 HOW PACKAGING IMPACTS INDICES OF FAILURE

The shelf life of food can be extended by storage under VP or MAP (Robertson, 2006). For such a highly perishable food as fish, packaging tends to delay the impact of many of the indices of failure. The main mechanism that operates in packaged fish is the modification from an aerobic psychrotrophic population of bacteria (*Pseudomonas*, *Moraxella*, and *Acinetobacter* spp.) to one consisting predominately of LABs and *Bromothrix thermosphacta*; there is also inhibition of oxidative reactions (Mullan and McDowell, 2003). The downside is that the anaerobic environment encourages the growth of *C. botulinum*, if present. A good example of how packaging affects the SSOs of prawns is provided by Bremner (2004). Under aerobic conditions (permeable materials present) the prime spoiler is *Pseudomonas fragi*, whereas under anaerobic conditions (impermeable materials present) *Shewanella putrefaciens* becomes the prime SSO.

Loss of water from high-moisture-content product such as fish while in VP or MAP can be considerably higher than is seen in fresh-air-stored fish. The barrier film used on these packs must not only have lower water vapor transmission rates (WVTRs) but for MAP there is also the need for antifogging protection so that buyers can readily see the fish in the pack in the retail outlet refrigerated cabinet.

CO_2 increases cell wall permeability. It is absorbed by the packaged fish, inducing increased drip and reducing the internal pressure of the MAP (Borderias, 1996). Discoloration can occur via bleaching of cut surfaces due to low pH precipitation of sarcoplasmic proteins (Statham and Bremner, 1989). Organoleptic changes due to very high concentrations of CO_2 may also be unfavorable. The reduction of bound water during excessive drip loss can result in a coarsening of the texture (Tiffnety and Mills, 1982). Packaging also does not directly inhibit the fish's own proteolytic enzymes that may contribute to weak texture.

Spoilage of fish results in low molecular weight volatile compounds being generated, which directly impacts the type of barrier properties required of the packaging films (Mullan and McDowell, 2003). The barrier film, while preventing the tainting of fish from strong odors of closely stored commodities in the same container or trailer, can retain any off-odors released by bacteria present on the fish. In the early stages of storage this is not an issue, but when bacterial numbers become quite large, the odors can be off-putting to consumers. In packs containing fish still safe to consume (total microbial count of $<10^6$ cfu g^{-1}), there can be a noticeable amount of off-odor, which would dissipate rapidly after opening, but it would be difficult to convince the person present at the opening that this fish would be pleasant to eat.

For this reason shelf life labeling should be based on the HQL of the fish rather than purely on the legislated total microbial counts that are acceptable. A number of systems to help determine the end of shelf life for packaged fish experiments include QIM appraisal scoring (Hyldig et al., 2007) at the time of pack opening; the counts of total, H_2S-producing, and pathogenic bacteria; and the sensory scores of cooked samples. Rejection of packaged fish by sensory panelists can occur due to the H_2S-producing bacteria when they represent a high proportion of the flora, and usually it will be several days later when the product (containing only low numbers of H_2S-producing bacteria) is rejected due to high total bacterial count (Slattery et al., 1998).

15.5 ACTIVE PACKAGING

Active packaging has been defined as "packaging in which subsidiary constituents have been deliberately included in or on either the packaging material or the package headspace to enhance the performance of the package system" (Robertson, 2006, p. 288). Oxygen scavenging is a widely used active-packaging technology, whereby, for example, iron-based pouches or sachets are inserted into individual food packages to retard oxidation and spoilage. There are also CO_2 scavengers or emitters; ethylene scavengers, ethanol emitters, and other preservative releasers; and moisture, flavor, or odor absorbers. The use of active packaging technologies such as O_2 scavengers and CO_2 emitters does not improve microbial safety above that obtained by traditional MAP, and gives little or no additional shelf life to fresh seafood products compared to MAP and VP (Sivertsvik et al., 2002).

Hansen et al. (2007) found that CO_2 emitters allowed a reduction of pack volume for MAP farmed cod; the presence of O_2 also prevented formation of TMA during storage. Debevere and Boskou (1996) found that increased levels of O_2 in MAP contributed to a slightly lower production of TMA and considered *P. phosphoreum* to be the culprit rather than H_2S-producing bacteria. Absorbent pads are a necessity because of the increased drip loss induced by VP and MAP. The inclusion of antimicrobial ingredients into these sachets and pack liners can provide a dual outcome. Colloidal silver has been used effectively for treating domestic and processed water supplies and equipment. There are, however, side-effects from excessive silver exposure such as the skin condition argyria (Glenn and Walker, 2002). Silver-ion-based treatment incorporated into a paper sheet intended for fish storage applications is known to be effective against bacteria (MicrobeGuard Corporation, 2004).

A zirconian phosphate–based ceramic ion exchange resin made with silver has been cleared by the FDA's list of food contact substances as an acceptable fish contact polymer (FDA, 2008). There are also films that will provide the controlled release of volatile chlorine dioxide that have received generally recognized as safe (GRAS) status for fish use in the United States.

The concept of intelligent packaging that can evaluate real-time freshness of fish and seafood while in the package is also being developed. Pacquit et al. (2007) evaluated a packaging film that contained a pH-sensitive dye within the polymer matrix that responded to fish spoilage volatiles by changing color. A systematic approach for modeling fish shelf life and time–temperature integrator selection in order to plan and apply an effective quality monitoring scheme for fish chill chains was developed by Taoukis et al. (1999), who modeled the growth of the SSOs of the Mediterranean fish boque (*Boops boops*).

15.6 SHELF LIFE OF FISH IN DIFFERENT PACKAGES

15.6.1 Overwrapping

Although overwrapping is used to prevent further contamination of foods, it does not prevent deterioration of products that are already contaminated. A cling-wrap film made of plasticized PVC with high WVTR and low oxygen permeability is commonly used.

Özogul et al. (2007) achieved only 8 days of shelf life for sea bream wrapped in a PVC cling-film or aluminum foil at $2 \pm 1°C$, whereas those stored in ice lasted 18 days. Özogul et al. (2005) found similar reduced shelf life for wild sea bass using these packaging materials. This reduced shelf life occurs because melting ice assists in the removal of bacteria from the surface of fish, whereas the films not only prevent washing but also act as an insulator to some extent. Tiffnety and Mills (1982) found that the presence of overwrap reduced shelf life of fish stored under MAP. Gimenez et al. (2002) found that sea bream fillets had the shortest shelf life when overwrapped; VP provided an improvement, and MAP resulted in the longest shelf life.

15.6.2 Vacuum Packaging

Vacuum packaging tends to use films with low oxygen permeability. Vacuum skin packaging (VSP) is used for delicate products such as smoked or pickled fish, shucked scallops, and soft-shell crabs. A softened film is placed over the product and vacuum applied. The film may be heat-molded over the product. If the VSP is placed in an MAP master pack (Slattery et al., 1998), then the film has to be permeable.

Vacuum packaging causes product compression followed by drip release. Borderias (1996) thought that the appearance of retail products in VP was inadequate. Manju et al. (2007) found that the success of VP is completely dependent on the initial quality of the fish and on adequate temperature control throughout storage. Statham and Bremner (1983) found that VP in conjunction with 0.1% potassium sorbate resulted in minimal extension of shelf life for blue grenadier (*Macruronus novaezelandiae*). Sallam (2008) treated Pacific saury (*Cololabis saira*) with marinating solutions before VP and obtained more than 90 days' shelf life, compared with 60 days for brined fillets.

The conditions produced by smoking fish, adding low levels of NaCl, slight acidification, and chill storage in VP result in LAB (*Lactobacillus* and *Carnobacterium*) dominating in association with gram-negative fermentative bacteria (*P. phosphoreum*) and psychrotrophic Enterobacteriaceae (Truelstrup Hansen et al., 1995). Penney et al. (1994) found that smoked blue cod had much longer shelf life in MAP than in VP. Dalgaard stated that MAP is more effective than VP in preventing histamine production during storage (Wray, 2007).

The anaerobic conditions of VP and MAP foods encourage the growth of *C. botulinum*, which can then produce a harmful toxin. Therefore, the UK Food Standards Agency has provided nonbinding guidance for industry recommending a maximum 10-day shelf life for VP and MAP ready-to-eat and raw foods stored at 3°C to less than 8°C when other specified controlling factors (discussed in Section 15.3) could not be identified (Callaghan, 2008). One of the reasons for recommending a limited shelf life is the practice of repackaging VP and MAP foods after processing. It was considered that the shelf life given to repacked product should not exceed the shelf life given to the original product.

15.6.3 Modified Atmosphere Packaging

Using MAP technology for seafood is not new, with the first extensive research on seafood stored in CO_2 being reported in the early 1930s (Robertson, 2006). A 1991 review of food stored in MAP listed 12 authors who experimented on many different seafood species (Farber, 1991), and Sivertsvik et al. (2002) listed 37 authors. MAP has been commercially available in the United Kingdom and France since the 1970s and is now well established in Europe. The adoption of this technology by the seafood industry has been a lot slower in the United States because of strict control over certification by the National Maritime Fisheries Service.

The active component in MAP is CO_2, which is readily soluble in water, with a small fraction being hydrated to carbonic acid, acting as a microbial inhibitor (Statham, 1984) and selecting for *P. phosphoreum* and LAB (Dalgaard, 2000). High levels of CO_2 cause the walls of gas-tight semi-rigid packs to cave in over time as a result of CO_2 absorption by the fish; eventually the upper and lower surfaces of the packs become compressed onto the fish. Increasing the initial CO_2 level from 40% to 60% extended the cooked flavor storage life for most fish species, but the higher CO_2 level collapsed the pack before the fish were considered unacceptable (Tiffnety and Mills, 1982).

This effect can be minimized by reducing the product weight:headspace volume ratio; ratios of two parts atmosphere to one part product should be the minimum used. Rotabakk et al. (2008) found that exposing fish to a CO_2 atmosphere prior to MAP (soluble gas stabilization) reduced the risk of pack collapse. Increasing concentrations of CO_2 reduced the growth rate and delayed histamine formation by *M. psychrotolerans* and 40% CO_2:60% O_2 had a strong inhibitory effect (Emborg and Dalgaard, 2008). Fatty fish are less of a problem as the higher lipid content means there is less water to dissolve the CO_2 (Urch, 2003).

The use of MAP (CO_2:N_2 60:40) in conjunction with "partial freezing" (reducing the temperature of the fish to 1–2°C below its freezing point) or superchilling was compared with storage in air at normal chilled (4°C) and superchilled (–2°C) temperatures (Sivertsvik et al., 2003). The shelf life of MAP salmon at –2°C (24 days) was 2.5 times that of MAP salmon at 4°C and 3.5 times that of salmon at 4°C stored under air. It was suggested that the synergistic effect of MA and temperature may be influenced by increased solubility of CO_2 at the superchilled temperature.

The following materials have been recommended by Air Products (2008) for MAP of fish: PVC/PE, APET/PE, HDPE, and EPS/EVOH/PE for trays; PET/PE–EVOH–PE, OPA/PE–EVOH–PE, OPP/PE–EVOH–PE, and PET/PVdC/PE for lidding film; and PA/PE and PA/EVOH/PE for bulk packaging such as Master packs and bag-in-box.

FDA (2002) import alert #16-125 specifies a minimum OTR at 24°C of 10,000 mL m^{-2} day^{-1} for hermetically sealed seafood. According to Gnanaraj et al. (2005), very few films are available commercially that meet this high-OTR specification. They reported OTRs for four commercial films at various temperatures between 10°C and 35°C; the highest OTR at 23°C was 8620 mL m^{-2} day^{-1} for a mixed-density polyethylene film. Generally the OTR at 10°C was less than half that at 23°C. They noted that the FDA specification does not take into consideration the specific design and geometry of the package; however, their results showed little variation in spore outgrowth behavior between small and large bags.

Sivertsvik et al. (2002) believed that storage of products in MAP may not increase the risks from *Salmonella*, *Staphylococcus*, *C. perfringens*, *Yersinia*, *Campylobacter*, *Vibrio parahemolyticus*, and *Enterococcus* above those of air-stored products. One important consideration is that when the internal temperature becomes higher than that of a chiller (>3.3°C), the conditions present in MAP can encourage botulism. Worse is that toxin has been detected in packaged fish prior to the product being considered spoiled, and the inclusion of O_2 does nothing to prevent the growth of this organism. Anaerobic conditions may develop on the surface of fish due to respiration of the tissue and aerobic bacteria, allowing spore germination and outgrowth. Botulism toxin may be produced in atmospheres containing 100% O_2 (Huss et al., 1980). Because of this potential health risk, the initial quality of the product needs to be of a high standard with low microbiological loads.

One question always asked by industry when discussing MAP is "How many days of shelf life can be achieved?" Outcomes are dependent on the microbial quality of the seafood being packed, and shelf life trials are required to determine reliable use-by labeling dates. The author has found that the number of days seafood remains acceptable in air can usually be doubled using MAP (Slattery et al., 1998). There is debate over the cost benefits of using MAP for fish, as sometimes only small increases in safe shelf life occur. The amount of extended shelf life needed for retail sales will obviously limit what may be cost-effective to pack.

To highlight the importance of having high-quality starting material, a brief review of 21 of the author's own MAP storage trials follows. The pertinent methods applied to a range of fish species, including Atlantic salmon, rainbow trout, yellowtail kingfish, and broadbill swordfish, were a gas mixture of 60% CO_2:40% N_2 and a storage temperature of 4°C. Alternate gas conditions will be discussed later in the chapter.

The various trial data are grouped according to the closest integer total bacterial aerobic log count obtained at the start of storage. The representative group regression lines and 95% confidence limits, which help show the statistical difference between groups of data when they do not cross, for the trials of Slattery et al. (1998) are shown in Figure 15.1. The intercepts at day 0 of the trials with starting counts of log 2 and 3 were significantly different ($p<.01$) from the intercepts of trials starting at log 4 and 5. Intersection of the 95% confidence limits occurred only at the start of storage between trials starting at logs 2 and 3 and between logs 4 and 5, showing that the data can be separated based on initial aerobic log count.

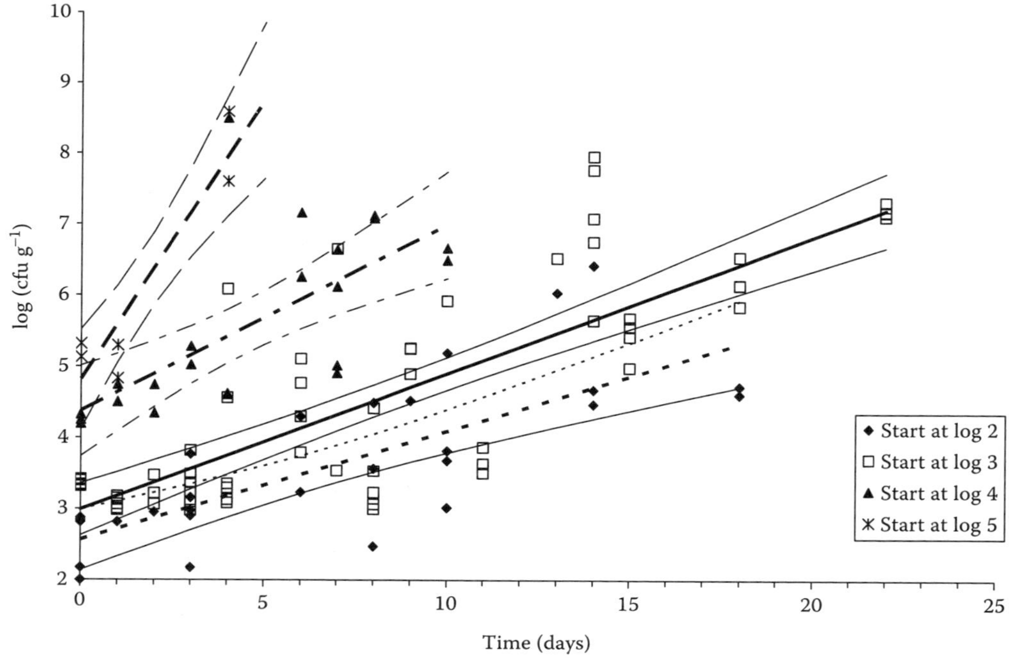

FIGURE 15.1 Total microbial counts and 95% confidence limits recorded by Slattery et al. (1998) for seafood packed in modified atmospheres during storage at 4°C.

The total log aerobic count increased consistently during storage in all trials. The slopes of the regression lines in Figure 15.1 were significantly different ($p<.01$) for the four groups. The increased growth rate that occurred as the initial count increased indicates that bacteria better adapted to the storage conditions are present when there are higher levels of contamination. The average end of shelf life (>log 6) for the respective groups was 23, 16, 7, and 2 days. The data show that the shelf life is very limited for fish that have log counts of 4 or greater at the time of packing. Most supermarkets like to sell product that has at least 7 days' shelf life in the store. This outcome shows the limitation of packing and shipping seafood in MAP with moderate bacterial loads.

To improve the universality of this outcome, trials from others authors were combined with the data discussed here. Some researchers include moderate levels of O_2 in their MAP to further reduce the risk of growth of *Clostridia* (Mullan and McDowell, 2003), some use 100% CO_2, some store product at higher than 5°C, and some use surface tissue sampling for the total microbial count. These conditions were considered too different, and so only those trials that used mainly anaerobic conditions and microbial test methods on pieces of flesh have been included. The most common features of the many trials conducted by other researchers were package atmospheres ranging from 40% to 75% CO_2 with the remainder made up with N_2, and in one case also 6% O_2, and storage temperatures that ranged between 0°C and 5°C. The extra trials are summarized in Table 15.1.

The trials were grouped as described earlier. The representative group regression lines and 95% confidence limits for all the trials are shown in Figure 15.2. The slopes of the group regression lines from trials conducted by other researchers were similar to those of Slattery et al. (1998).

The R^2 values of all trial regression lines analyzed were greater than 0.85. There was no significant difference for intercept or slope between the trials with similar starting microbial counts that used only CO_2 and N_2 when compared with the one that also included a small amount of O_2 in the gas mixture. Statistical analysis of the grouped regression lines found no significant difference between the gradients of the four groups. Although there was no significant difference between trials that started with total aerobic counts of log 2 and log 3 for day 0 intercept, and thus these data could have been combined, the intercept was significantly ($p<.01$) larger for trials that started at a log count of 4, which were in turn significantly different when the initial count was log 5.

Intersection of some of the 95% confidence limits at the start of storage occurred only between the upper limit of the trials starting at log 4 and the lower limit of trials starting at log 5. There was also some intersection from the confidence limits for trials starting at log 4 as the total count approached the end of shelf life (log 6).

TABLE 15.1
Modified Atmospheres and Storage Conditions Used by Various Authors

Authors	Gas Mixture ($\%CO_2$:$\%N_2$:$\%O_2$)	Storage Temperature (°C)	Product
Cann et al., 1984	60:40:0	5	Atlantic salmon steaks
Cann et al., 1984	60:40:0	5	Gutted trout
Cann et al., 1984	60:40:0	0	Atlantic salmon steaks
Cann et al., 1984	60:40:0	0	Gutted trout
Stier et al., 1981	60:40:0	4.4	Atlantic salmon fillets
Reddy et al., 1992	75:25:0	4	Catfish fillets
Reddy et al., 1995	75:25:0	4	Tilapia fillets
Brown et al., 1980	40:60:0	4	Silver salmon steaks
Brown et al., 1980	40:60:0	4	Rockfish fillets
Poli et al., 2006	40:60:0	2	Sea bass fillets
Woyewoda et al., 1984	60:40:0	1	Cod fillets
Handumrongkul and Silva, 1994	60:34:6	2	Striped bass strips

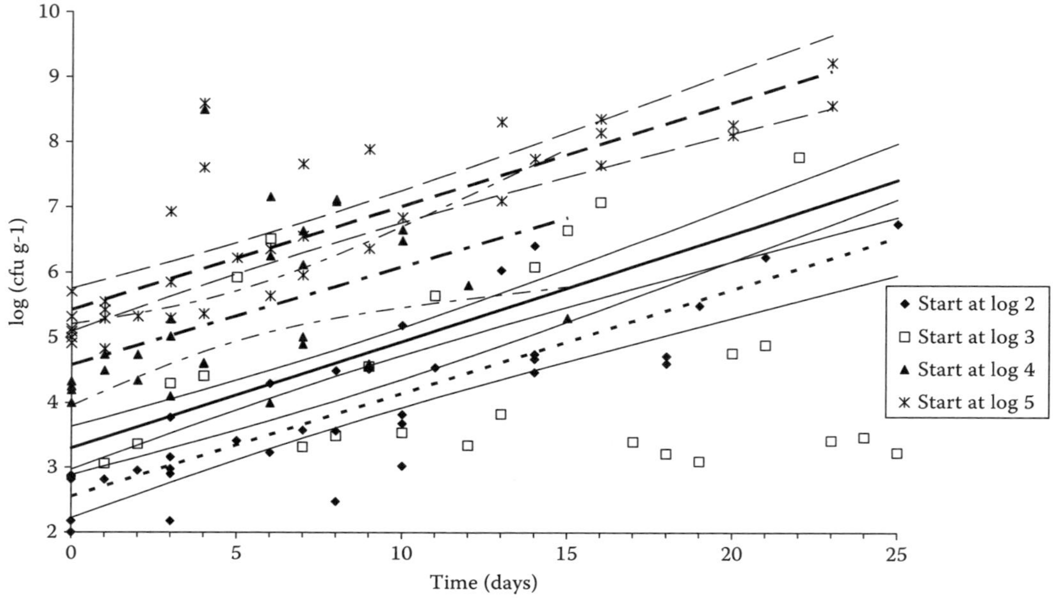

FIGURE 15.2 Total microbial count and 95% confidence limits recorded by authors listed in Table 15.1 and Slattery et al. (1998) for seafood packed in modified atmospheres during storage at 0–5°C.

The average shelf life identified from the regression lines dropped from 22–16 to 10–4 days, respectively, as the starting log count increased from log 2 to log 5. It is recommended that producers not package seafood in MAP for export or domestic markets with long distribution times when the total microbial count is log 4 or greater, as only limited shelf life will be attained (Slattery et al., 1998).

Gas mixtures of 30% O_2:40% CO_2:30% N_2 are used for white nonprocessed nonfatty fish (Mullan and McDowell, 2003). Tiffnety and Mills (1982) found that O_2 (30%) actually increased shelf life of white fish but excluded its use for fatty and cured products. Arkoudelos et al. (2007) found that eel had a sensory acceptability of 18 days, compared with only 11 days for VP and air, when stored under this atmosphere. Pantazi et al. (2008) obtained 11–12 days' shelf life for swordfish packed under 30% O_2:40% CO_2:30% N_2, whereas VP and air resulted in 9 and 7 days respectively. After testing many combinations of O_2, CO_2, and N_2, Sivertsvik (2007) found 63% O_2 and 37% CO_2 to be the optimum gas mixture for MAP of farmed cod. However, Gimenez et al. (2002) found that atmospheres without O_2 provided better shelf life extension for gilt-head sea bream.

There are online tools available to help food processors decide which gas mixture to use (Air Products, 2008). Fish species with a high proportion of red muscle tissue require the presence of O_2 to maintain good color. This is especially important for high-value species, such as tuna, that are also prone to histamine formation. Fortunately there was a strong inhibitory effect on formation of this compound in an atmosphere of 40% CO_2:60% O_2 recommended for tuna (Emborg and Dalgaard, 2008).

The use of carbon monoxide (CO) has been a topic for MAP application to fish for some time, with even a book devoted solely to this issue (Otwell et al., 2006). Many authors have found that CO improved fish shelf life but there were some risks to its use. The FDA has found that many samples of CO-treated tuna have high histamine levels, which could lead to sickness in the consumer (Cruz et al., 2002). Commercial application has been limited in many countries because of the gas toxicity and the formation of potentially explosive mixtures in air (Mullan and McDowell, 2003). Carbon monoxide also masks the changes that occur during normal degradation in the red muscle tissue of fish, which are used by buyers as indicators of freshness, resulting in product being kept too long before sale. Consumers could select and eat this product unaware that it could contain toxins.

The O_2-holding compounds within an animal are myoglobin and hemoglobin. During respiration myoglobin within the muscle tissue takes up O_2, forming red-colored oxymyoglobin. As fish deteriorates, oxymyoglobin is oxidized, resulting in the bright red muscle tissue turning brown due to the formation of metmyoglobin (Sørheim et al., 1997). This occurs more rapidly in fish than in meat. Meat is more stable as its muscle pigment can rebloom (return to a red color) after anaerobic storage such as VP, which is less likely to happen with fish. Carbon monoxide combines with myoglobin to form carboxymyoglobin, which is very stable and more resistant to oxidation than oxymyoglobin (Wolfe, 1980). This cherry red pigment can be quite different to the natural red pigment normally encountered in fish and can be used by buyers as an indicator to avoid CO-gas-flushed fish.

Instead of being applied directly as a component of the modified atmosphere of the pack, CO has been known to be applied earlier in the processing chain through a filtered smoking procedure called "tasteless smoke" or as a gas before any packaging. The first process is the only one accepted in the United States under GRAS Notice No. GRN 000083 (FDA, 2002), and it requires labeling of the treatment. Although the EU considers there is no health concern over the use of CO for meat (EU, 2001), it bans imports of fish treated with this gas (EU, 2004). Although some individual member states such as the Netherlands accept "clear smoked" product, Italy has banned the import of tuna treated by either method (Anon., 2003). There are analytical methods available to test food products to determine whether CO has been used (Anderson and Wu, 2005; Wolfe et al., 1978).

Freeze-chilling has been used as a way of obtaining much longer shelf life times in MAP rather than just holding the product chilled. This allows many months of frozen storage so that supply can be assured year round, but the thawed product will be quite different from a chilled-only one. Also, the chilled shelf life after thawing may be limited. Fagan et al. (2004) obtained 5–7 days' chilled shelf life in MAP for whiting, mackerel, and salmon after thawing.

REFERENCES

ACMSF. 2006. Annual Report 2006. *Advisory Committee on the Microbiological Safety of Food*, paragraphs 53–56. Food Standards Agency, London, UK. FSA/1191/0807. www.food.gov.uk/multimedia/pdfs/publication/asmsfar07.pdf. Accessed December 2008.

Air Products. 2008. Freshline MAP gas selector. www.airproducts.com/Food/content/MAP/IE/MAP-gas-selector/default.asp. Accessed December 2008.

Alves V.F., de Martinis C.P., Destro M.T., Vogel B.F., Gram L. 2005. Antilisterial activity of a *Carnobacterium piscicola* isolated from Brazilian smoked fish (Surubim [*Pseudoplatystoma* sp.]) and its activity against a persistent strain of *Listeria monocytogenes* isolated from Surubim. *Journal of Food Protection* 68: 2068–2077.

Anderson C.R., Wu W.-H. 2005. Analysis of carbon monoxide in commercially treated tuna (*Thunnus* spp.) and mahi-mahi (*Coryphaena hippurus*) by gas chromatography/mass spectrometry. *Journal of Agricultural and Food Chemistry* 53: 7019–7923.

Anon. 2003. Carbon monoxide treatment banned. *Seafood International* August: 4.

Anon. 2008. Expanding the organic rosemary portfolio. *FoodProcessing.Com* www.foodprocessing.com/articles/2008/302.html. Accessed December 2008.

Arkoudelos J., Stamatis N., Samaras F. 2007. Quality attributes of farmed eel (*Anguilla anguilla*) stored under air, vacuum and modified atmosphere packaging at 0°C. *Food Microbiology* 24: 728–735.

Borderias A.J. 1996. New complementary technologies for extending shelf life of chilled seafood products. *Refrigeration and Aquaculture*, Proceedings of the conference of IIR Commission C2, Bordeaux, France, 20–22 March: 265–274.

Bosilevac J.M., Shackelford S.D., Brichta D.M., Koohmaraie M. 2005. Efficacy of ozonated and electrolyzed oxidative waters to decontaminate hides of cattle before slaughter. *Journal of Food Protection* 68: 1393–1398.

Bremner H.A. 2004. Quality issues and more efficient utilization of the fish catch. *Developments in Food Science* 24: 275–287.

Brown W.D., Albright M., Watts D.A., Heyer B, Spruce B., Price R.J. 1980. Modified atmosphere storage of rockfish (*Sebastes miniatus*) and silver salmon (*Oncorhynchus kisutch*). *Journal of Food Science* 45: 93–96.

Burt S. 2004. Essential oils: their antibacterial properties and potential applications in foods: a review. *International Journal of Food Microbiology* 94: 223–253.

Callaghan K. 2008. Food Standards Agency guidance on the safety and shelf-life of vacuum and modified atmosphere packed chilled foods with respect to non-proteolytic *Clostridium botulinum*. Food Standards Agency July, London, UK. www.food.gov.uk/multimedia/pdfs/publication/vacpacguide.pdf Accessed February 12, 2008.

Campos C.A., Losada V., Rodriguez O., Aubourg S., Barros-Velazquez J. 2006. Evaluation of an ozone-slurry ice combined refrigeration system for the storage of farmed turbot (*Psetta maxima*). *Food Chemistry* 97: 223–230.

Cann D.C., Houston N.C., Taylor L.Y., Smith G.L., Thomson A.B., Craig A. 1984. *Studies of salmonids packed and stored under a modified atmosphere*. Aberdeen, Scotland: Torry Research Station, Technical Report.

Cruz L., Chin K., Yu E., Lee L. 2002. Carbon monoxide-treated tuna with high histamine levels. *2002 FDA Science Forum Poster Abstract* www.cfsan.fda.gov/~frf/forum02/a18212.htm. Accessed December 2008.

Dalgaard P. 2000. Fresh and lightly preserved seafood. In: *Shelf-life Evaluation of Foods*. Man C.M.D., Jones A.A. (Eds). London, England: Aspen Publishers Inc., pp. 110–139.

Dalgaard P. 2005. The Seafood Spoilage and Safety Predictor (SSSP) software v. 2.0—multi-language version. Danish Institute of Fisheries Research. www.difres.dk/micro/sssp/Home/Home.aspx. Accessed November 2008.

Davidson M.P. 1997. Chemical preservatives and natural antimicrobial compounds. In: *Food Microbiology Fundamentals and Frontiers*. Doyle M.P., Beuchat L.R., Montville T.J. (Eds). Washington, DC: ASM Press, pp. 520–556.

Debevere J., Boskou G. 1996. Effect of modified atmosphere packaging on the TVB/TMA-producing microflora of cod fillets. *International Journal of Food Microbiology* 31: 221–229.

Delves-Broughton J., Weber G., Elsser D. 2007. Natural preservation of foods using bacterial metabolites and live addition of bacteria. *Food Science Central*. www.foodsciencecentral.com./fsc/ixid14740. Accessed March 2007.

Emborg J., Dalgaard P. 2008. Modelling the effect of temperature, carbon dioxide, water activity and pH on growth and histamine formation by *Morganella psychrotolerans*. *International Journal of Food Microbiology* 128: 226–233.

Emborg J., Laursen B.G., Dalgaard P. 2005. Significant histamine formation in tuna (*Thunnus albacares*) at 2°C—effect of vacuum- and modified atmosphere-packaging on psychrotolerant bacteria. *International Journal of Food Microbiology* 101: 263–279.

EU. 2001. Opinion of the Scientific Committee on food on the use of carbon monoxide as component of packaging gases in modified atmosphere packaging for fresh meat. Brussels, Belgium. http://ec.europa.eu/food/fs/sc/scf/out112_en.pdf. Accessed December 2008.

EU. 2004. Final report of a mission carried out in Indonesia from 15 to 23 April 2004 assessing the conditions of production of fishery products intended for export to the European Union (Article 11 of Council Directive 91/493/EEC). http://ec.europa.eu/food/fs/inspections/vi/reports/indonesia/vi_rep_indo_7148-2004_en.pdf. Accessed December 2008.

Fagan J.D., Gormley T.R., Ui Mhuircheartaigh M.M. 2004. Effect of modified atmosphere packaging with freeze-chilling on some quality parameters of raw whiting, mackerel and salmon portions. *Innovative Food Science and Emerging Technologies* 5: 205–214.

Fan W., Chi Y., Zhang S. 2008. The use of a tea polyphenol dip to extend the shelf life of silver carp (*Hypophthalmicthys molitrix*) during ice storage. *Food Chemistry* 108: 148–153.

FAO/WHO. 2008. Code of practice for fish and fishery products (CAC/RCP 52-2003), Codex Alimentarius. www.codexalimentarius.net/download/standards/10273/CXP_052e.pdf. Accessed December 2008.

Farber J.M. 1991. Microbiological aspects of modified-atmosphere packaging technology—a review. *Journal of Food Protection* 54: 58–70.

Federal Register. 1995. Decomposition and histamine—raw, frozen tuna and mahi-mahi; canned tuna; and related species; *Compliance Policy Guideline # 540.525*.

FDA. 2002. Agency response letter GRAS Notice No. GRN 000083. U.S. Food and Drug Administration, Rockville, MD. www.cfsan.fda.gov/~rdb/opa-g083.html. Accessed October 25, 2002.

FDA. 2002. Import Alert # 16–125. Washington, DC: FDA.

FDA. 2008. Inventory of Effective Food Contact Substance (FCS) Notifications. U.S. Food and Drug Administration, Rockville, Maryland. www.cfsan.fda.gov/~dms/opa-fcn.html. Accessed December 2008.

Giatrakou V., Kykkidou S., Papavergou A., Kontominas M.G., Savvaidis I.N. 2008. Potential of oregano essential oil and MAP to extend the shelf life of fresh swordfish: a comparative study with ice storage. *Journal of Food Science* 73: M167–M173.

Gimenez B., Sanchez-Escalente A., Torrescano G., Roncales P., Beltran J.A. 2002. Different packaging conditions to improve shelf life of filleted gild-head sea bream (*Sparus aurata*). *Journal of Aquatic Food Product Technology* 11(3/4): 275–286.

Glenn J.B., Walker A.N. 2002. Argyria in an elderly man. *The Internet Journal of Dermatology* 1(2). www.ispub.com/ostia/index.php?xmlFilePath=journals/ijd/vol1n2/argyria.xml. Accessed December 2008.

Gnanaraj J., Welt B.A., Otwell W.S. and Kristinsson H.G. 2005. Influence of oxygen transmission rate of packaging film on outgrowth of anaerobic bacterial spores. *Journal of Aquatic Food Product Technology* 14(4): 51–69.

Gram L., Dalgaard P. 2002. Fish spoilage bacteria—problems and solutions. *Current Opinion in Biotechnology* 13: 262–266.

Gram L., Huss H.H. 2000. Fresh and processed fish and shellfish. In: *The Microbiological Safety and Quality of Foods*. Lund B.M., Baird-Parker A.C., Gould G.W. (Eds). London, England: Chapman and Hall, pp. 472–506.

Handumrongkul C., Silva J.L. 1994. Aerobic counts, color and adenine nucleotide changes in CO_2 packed refrigerated striped bass strips. *Journal of Food Science* 59: 67–69.

Hansen A.Å., Mørkøre T., Rudi K., Olsen E., Eie T. 2007. Quality changes during refrigerated storage of MA-packaged pre-*rigor* fillets of farmed Atlantic cod (*Gadus morhua* L.) using traditional MAP, CO_2 emitter, and vacuum. *Journal of Food Science* 72: M423–M430.

Harpaz S., Glatman L., Drabkin V., Gelman D.A. 2003. Effects of herbal essential oils used to extend the shelf life of freshwater-reared Asian Sea Bass fish (*Lates calcarifer*). *Journal of Food Protection* 66: 410–417.

Hultin H.O. 1992. Trimethylamine-*N*-oxide (TMAO) demethylation and protein denaturation in fish muscle. In: *Advances in Seafood Biochemistry*. Flick G.J., Martin R.E. (Eds). Lancaster, Pennsylvania: Technomic Publishing Company, pp. 25–41.

Huss H.H., Schoeffer I., Pedersen A., Jepsen A. 1980. Toxin production by *Clostridium botulinum* type E in smoked fish in relation to the measured oxidation reduction potential (E_h), packaging method and associated microflora. In: *Advances in Fish Science and Technology*. Connell J.J. (Ed). Farnham: Fishing News Books, pp. 476–479.

Hyldig G., Bremer A., Martinsdottir E., Schevis R. 2007. Quality Index Methods, In: *Handbook of Meat, Poultry and Seafood Quality*. Nollet L.M.L. (Ed). Oxford, England: Blackwell Publishing, pp. 529–549.

Kim C.R., Hearnsberger J.O. 1994. Gram negative bacteria inhibition by lactic acid culture and food preservatives on catfish fillets during refrigerated storage. *Journal of Food Science* 59: 513–516.

Kim J., Marshall D.L. 2001. Effect of lactic acid on of *Listeria monocytogenes* and *Edwardsiella tarda* attached to catfish skin. *Food Microbiology* 18: 589–596.

Kim C.R., Hearnsberger J.O., Vickery A.P., White C.H., Marshall D.L. 1995. Extending shelf life of refrigerated catfish fillets using sodium acetate and monopotassium phosphate. *Journal of Food Protection* 58: 664–647.

Kim J.M., Huang T.-S., Marshal M.R., Wei C.-I. 1999. Chlorine dioxide treatment of seafoods to reduce bacterial loads. *Food Microbiology and Safety* 64: 1089–1093.

Lin W.-F., Huang T.-S., Cornell J.A. Lin C.-M., Wei C.-I. 1996. Bactericidal activity of aqueous chlorine and chlorine dioxide solutions in a fish model system. *Journal of Food Science* 61: 1030–1034.

Love R.M. 1988. *The Food Fishes: Their Intrinsic Variation and Practical Implications*. London, England: Farrand Press.

Manju S., Leema J., Srinivasa Gopal T.K., Ravishankar C.N., Lalitha K.V. 2007. Effects of sodium acetate dip treatment and vacuum-packaging on chemical, microbiological, textural and sensory changes of Pearlspot (*Etroplus suratensis*) during chill storage. *Food Chemistry* 102: 27–35.

MicrobeGuard Corporation. 2004. Antimicrobial paper products. www.microbeguard.com. Accessed December 2008.

Molina-Garcia A.D., Sanz P.D. 2002. *Anisakis simplex* larva killed by high-hydrostatic-pressure processing. *Journal of Food Protection* 65: 383–388.

Montville T.J., Winkowski K. 1997. Biologically based preservation systems and probiotic bacteria. In: *Food Microbiology, Fundamentals and Frontiers*. Doyle M.P., Beuchat L.R., Montville T.J. (Eds). Washington DC: ASM Press, pp. 83–100.

Mullan M., McDowell D. 2003. Modified atmosphere packaging. In: *Food Packaging Technology.* Coles R., McDowell D., Kirwan M. (Eds). Oxford, England: Blackwell Publishing, pp. 303–339.

Nesvadba P. 2003. Introduction to and outcome of the project "Multi-sensor techniques for monitoring the quality of fish". In: *Quality of Fish from Catch to Consumer: Labeling, Monitoring and Traceability.* Luten J.B., Oehlenschlager J., Olafsdottir G. (Eds). Wageningen, Netherlands: Wageningen Academic Publishers, pp. 175–187.

Oral N., Gülmez M., Vatansever L., Güven A. 2008. Application of antimicrobial ice for extending shelf life of fish. *Journal of Food Protection* 71: 218–222.

Otwell W.S., Kristinsson H.G., Balaban M.O. (Eds). 2006. *Modified Atmospheric Processing and Packaging of Fish: Filtered Smokes, Carbon Monoxide, and Reduced Oxygen Packaging.* New York: Wiley-Blackwell.

Özogul F., Kuley E., Özogul Y. 2007. Sensory, chemical and microbiological quality parameters in sea bream (*Sparus aurata*) stored in ice or wrapped in cling film or in aluminium foil at 2±1°C. *International Journal of Food Science and Technology* 42: 903–909.

Özogul F., Gökbulut C., Özyurt G., Özogul Y., Dural M. 2005. Quality assessment of gutted wild sea bass (*Dicentrarchus labrax*) stored in ice, cling film and aluminium foil. *European Food Research and Technology* 220: 292–298.

Pacquit A., Frisby J., Diamond D., Lau K.T., Farrell A., Quilty B., Diamond D. 2007. Development of smart packaging for the monitoring of fish spoilage. *Food Chemistry* 102: 466–470.

Pantazi D., Papavergou A., Pournis N., Kontominas M.G., Savvaidis I.N. 2008. Shelf-life of chilled fresh Mediterranean swordfish (*Xiphias gladius*) stored under various packaging conditions: microbiological, biochemical and sensory attributes. *Food Microbiology* 25: 136–143.

Park H., Hung Y.-C., Kim C. 2002. Effectiveness of electrolyzed water as a sanitizer for treating different surfaces. *Journal of Food Protection* 65: 1276–1280.

Pedrosa-Menabrito A., Regenstein J.M. 1990. Shelf-life extension of fresh fish—a review. Part III—fish quality and methods of assessment. *Journal of Food Quality* 13: 209–223.

Pelroy G.A., Peterson M.E., Holland P.J., Eklund M.W. 1994. Inhibition of *Listeria monocytogenes* in cold-process (smoked) salmon by sodium nitrite and packaging method. *Journal of Food Protection* 57: 114–119.

Penney N., Bell R.G., Cummings T.L. 1994. Extension of the chilled storage life of smoked blue cod (*Parapercis colias*) by carbon dioxide packaging. *International Journal of Food Science and Technology* 29: 167–178.

Poli B.M., Messini A., Parisi G., Scappini F., Vigiani V., Giorgi G., Vincenzini M. 2006. Sensory, physical, chemical and microbiological changes in European sea bass (*Dicentrarchus labrax*) fillets packed under modified atmosphere/air or prepared from whole fish stored in ice. *International Journal of Food Science and Technology* 41: 444–454.

Reddy N.R., Armstrong D.J. 1992. Shelf life extension and safety concerns about fresh fishery products packaged under modified atmosphere: a review. *Journal of Food Safety* 12: 87–118.

Reddy N.R., Villanueva M., Kautter D.A. 1995. Shelf life of modified-atmosphere-packaged fresh tilapia fillets stored under refrigeration and temperature-abuse conditions. *Journal of Food Protection* 58: 908–914.

Regenstein J.M., Regenstein C.E. 1991. *Introduction to Fish Technology.* New York: Van Nostrand Reinhold.

Robertson G.L. 2006. *Food Packaging Principles and Practice.* Boca Raton, Florida: CRC Press.

Rotabakk, B.T., Birkeland, S., Lekang, O.I., Sivertsvik, M. 2008. Enhancement of modified atmosphere packaged farmed Atlantic halibut (*Hippoglossus hippoglossus*) fillet quality by soluble gas stabilization. *Food Science and Technology International* 14: 179–186.

Sallam K. 2008. Effect of marinating process on the microbiological quality of Pacific saury (*Cololabis saira*) during vacuum-packaged storage at 4°C. *International Journal of Food Science and Technology* 43: 220–228.

Savvaides I.N., Skandamis P., Riganakos K.A., Panagiotakis N., Kontominas M. 2002. Control of natural flora and *Listeria monocytogenes* in vacuum-packaged trout at 4 and 10°C using irradiation. *Journal of Food Protection* 65: 515–522.

Shahidi F., Janak Kamil Y.V.A. 2001. Enzymes from fish and aquatic invertebrates and their application in the food industry. *Trends in Food Science and Technology* 12: 435–464.

Shaw S.J., Bligh E.G., Woyewoda A.D. 1983. Effect of potassium sorbate application on shelf life of Atlantic cod (*Gadus morhua*). *Journal of the Canadian Institute of Food Science and Technology* 16: 237–241.

Shewan J.M. 1970. Bacterial standards for fish and fishery products. *Chemical Indicators* 6: 193–199.

Sivertsvik M., Jeksrud W.K., Rosnes J.T. 2002. A review of modified atmosphere packaging of fish and fishery products: significance of microbial growth, activities and safety. *International Journal of Food Science and Technology* 37: 107–127.

Sivertsvik M., Rosnes J.T., Kleiberg G.H. 2003. Effect of modified atmosphere packaging and superchilled storage on the microbial and sensory quality of Atlantic salmon (*Salmo salar*) fillets. *Journal of Food Science* 68: 1467–1472.

Sivertsvik M. 2007. The optimized modified atmosphere for packaging of pre-*rigor* filleted farmed cod (*Gadus morhua*) is 63 ml/100 ml oxygen and 37 ml/100 ml carbon dioxide. *LWT—Food Science and Technology* 40: 430–438.

Slattery S.L. 1998. Shelf-life of Spanish mackerel (*Scomberomorus commerson*) from northern Australian waters. *Journal of Aquatic Food Product Technology* 7(4): 63–79.

Slattery S.L., Reeves R., Warfield B. 1998. Extending the high quality shelf life of seafood products. *FRDC Final Report Project No 96/338*. www.frdc.com.au/shop/merchant.mvc?Screen=PROD&Store_Code=B&Product_Code=1996-338-DLD.pdf&Category_Code=. Accessed November 2008.

Sørheim O., Aune T., Nesbakken T. 1997. Technological, hygienic and toxicological aspects of carbon monoxide used in modified-atmosphere packaging of meat. *Trends in Food Science and Technology* 8: 307–312.

Statham J.A. 1984. Modified atmosphere storage of fisheries products: the state of the art. *Food Technology in Australia* 36: 233–239.

Statham J.A., Bremner H.A. 1983. Effect of potassium sorbate on spoilage of blue grenadier (*Macruronus novaezelandiae*) as assessed by microbiology and sensory profiles. *Journal of Food Protection* 46: 1084–1091.

Statham J.A., Bremner H.A. 1989. Shelf-life extension of packaged seafoods: a summary of a research approach. *Food Australia* 41: 614–620.

Stier R.F., Bell L., Ito K.A., Shafer B.D., Brown L.A., Seeger M.L., Allen B.H., Porcuna M.N. and Lerke P.A. 1981. Effect of modified atmosphere storage on *C. botulinum* toxigenesis and the spoilage microflora of salmon fillets. *Journal of Food Science* 46: 1639–1642.

Sykes G. 1970. The sporicial properties of chemical disinfectants. *Journal of Applied Bacteriology* 33: 147–156.

Taoukis P.S., Koutsoumanis K., Nychas G.J.E. 1999. Use of time-temperature integrators and predictive modeling for shelf life control of chilled fish under dynamic storage conditions. *International Journal of Food Microbiology* 53: 21–31.

Taylor L.Y., Cann D.D., Welch B.J. 1990. Antibotulinal properties of nisin in fresh fish packed in an atmosphere of carbon dioxide. *Journal of Food Protection* 53: 953–957.

Tiffnety P., Mills, A. 1982. Storage trials of controlled atmosphere packaged fish products. *Sea Fish Industry Authority Technical Report* No. 191. Reprinted July 2006. www.seafish.org/pdf.pl?file=seafish/Documents/SR191.pdf. Accessed December 2008.

Truelstrup Hansen L., Gill T., Huss H.H. 1995. Effects of salt and storage temperature on chemical, microbiological and sensory changes in cold smoked salmon. *Food Research International* 28: 123–130.

Tucker G.S. 2003. Food biodeterioration and methods of preservation. In: *Food Packaging Technology*. Coles R., McDowell D., Kirwan M. (Eds). Oxford, England: Blackwell Publishing, pp. 32–64.

Urch M. 2003. Extending the shelf-life of fresh fish. *Seafood International* May: 42–44.

Villemure G., Simard R.E., Picard G. 1986. Bulk storage of cod fillets and gutted cod (*Gadus morhua*) under carbon dioxide atmosphere. *Journal of Food Science* 51: 317–320.

Wolfe S.K. 1980. Use of CO- and CO_2-enriched atmospheres for meat, fish, and produce. *Food Technology* 34(2): 55–63.

Wolfe S.K., Watts D.A., Brown W.D. 1978. Analysis of myoglobin derivatives in meat and fish samples using absorption spectrophotometry. *Journal of Agricultural and Food Chemistry* 26: 217–219.

Woyewoda A.D., Bligh E.G., Shaw S.J. 1984. Controlled and modified atmosphere storage of cod fillets. *Journal of the Canadian Institute of Food Science and Technology* 17: 24–27.

Wray T. 2007. MAP helps prevent histamine fish poison. *Seafood Processor* 26: 20–21.

16 Packaging and the Shelf Life of Fruits and Vegetables

Nathalie Gontard and Carole Guillaume
Agropolymers Engineering and Emerging Technologies
University of Montpellier II, Montpellier, France

CONTENTS

16.1 INTRODUCTION

Increasing demand for a wide range of harvested fruits or vegetables (raw and fresh-cut) has led to dynamic growth in sales and new market opportunities for the fresh produce sector. However, their preservation still constitutes one of the most challenging applications for the food industry as they share the ability to respire with only a few other foods such as shellfish and ripened cheese. As indicated in Figure 16.1, in addition to the common physicochemical (dehydration, oxidation) and microbiological deteriorative reactions that occur in all foods, they are also subjected to physiological degradations caused by respiration, transpiration, and ripening phenomena. These time-dependent reactions are influenced by processing (peeling, cutting, slicing, destoning) and storage/distribution conditions, leading to changes that inevitably lower the initial quality of the product. If properly handled, temperature and packaging can act favorably in preserving fruits and vegetables. Chill temperatures slow down reaction rates, and packaging may avoid dehydration as well as water vapor saturation, reduce respiration rate without asphyxiating tissues, slow down ripening, and minimize microbiological changes, depending on its mass transfer properties.

The structure of the polymeric matrix (blend of polymers, grafted polymers, microperforations, nanofillers, etc.) determines the gas and vapor permeation properties of packaging materials. Addition of active agents can result in absorption or release of gases, vapors, or solute by or from the packaging; these agents can be placed into pads or sachets that are added to the package. However, these foreign objects do not always meet with consumer acceptance, and thus active agents are increasingly incorporated directly into the packaging material. These properties are useful in

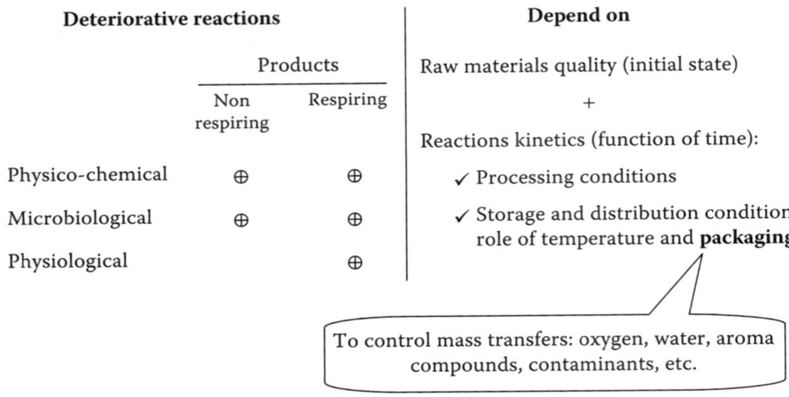

FIGURE 16.1 Factors affecting deterioration of food products.

modified atmosphere packaging (MAP) in which the headspace composition is different from that of the normal atmosphere (air). MAP can be achieved (a) in a passive way when considering only gas or vapor permeation properties of the material or (b) in an active way when also acting on gas or vapor absorption/releasing and/or gas flushing. In both cases, it reduces the need for preservatives or chemical additives within and in contact with the food. Although this technology is already applied in the preservation of fresh fruits and vegetables (both raw and fresh-cut), it is often the result of many time- and money-consuming experiences to determine a packaging material that suits the food requirements. The use of modeling and prediction tools is then of great interest to select an adequate combination (food/packaging system) prior to trials and ultimately commercialization.

The first part of this chapter gives an overview of the deteriorative reactions and external factors that affect the quality of fresh fruits and vegetables. In the second part, quality is related to MAP by focusing on modeling and development of agro-based materials. Polymers from agricultural resources offer unique functional properties that open new applications in the field of biodegradable passive and active MAP. The third part discusses passive and active MAP, and the chapter concludes with a discussion of future trends.

16.2 QUALITY CHANGES IN FRUITS AND VEGETABLES

Quality is a term frequently used by researchers, producers, handlers, marketers, and consumers, but its meaning varies depending on the user. Shewfelt (1999) distinguished "product-oriented quality" from "consumer-oriented quality." In the former definition, it corresponds to a series of attributes that can be accurately measured to evaluate the effect of cultural managements, cultivar selections, handling techniques, or postharvest treatments. The latter definition of quality is linked to consumer attitudes and acceptability of the product that vary according to cultural and demographic perspectives. For instance, pears can be consumed firm or soft, depending on the consumer preference. Quality covers several points indicated in Figure 16.2.

Sensory attributes are time-dependent and mainly refer to texture, flavor and odor, color, and visual development of produce, all of which evolve with maturation and aging of produce. Softening of fresh produce is due to the solubilization and depolymerization of pectins by the action of pectinolytic enzymes, and dehydration of produce with loss of turgidity can lead to a floury texture. Starch degradation and decarboxylation of organic acids induce modification of flavor, where the ratio between sweetness and acidity is important. Aromas are synthesized from alcohol and acids with the enzymic activity increasing according to the stage of maturity. Change in color is attributed to the degradation (or formation of chlorophyll in the case of endives for instance) as well as

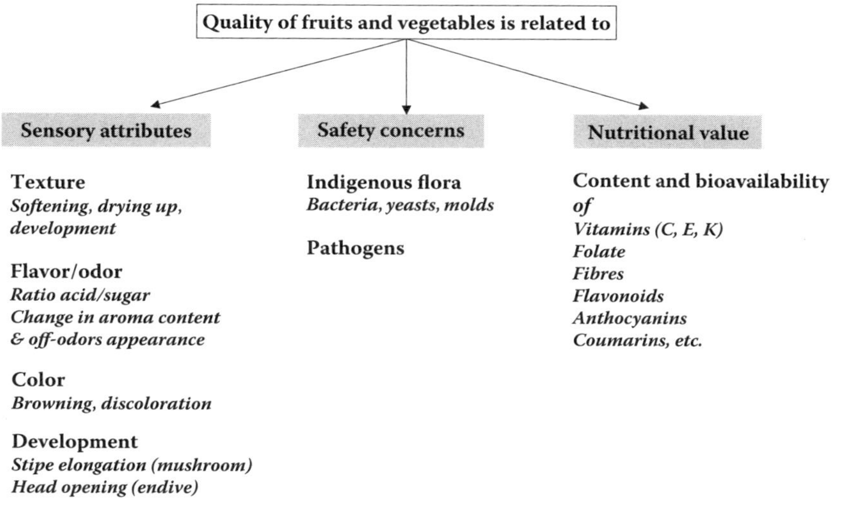

FIGURE 16.2 Quality of fresh fruits and vegetables.

synthesis of colored compounds such as anthocyanins or carotenes, or enzymic browning. The importance of these changes can vary depending on the particular produce. For instance, green leaves, browning of basal parts, and bitterness are commonly unacceptable for endives; common mushrooms are of excellent quality if they are white and firm. It should be stressed that minimal processing greatly affects these time-dependent reactions; for example, at the wounded surface, oxidation is favored and enzymic reactions may occur.

Microbiological quality of fresh fruits and vegetables is of great importance from a safety point of view. Spoilage microorganisms are involved in produce degradation but generally do not directly affect human health (fungi mycotoxin is an exception). Although fruits have generally been considered safe from pathogens, some might be found sporadically in vegetables; recent outbreaks of foodborne illness have been reported (Warriner, 2005; Zhao, 2005). Although sanitation methods are commonly employed, they are not totally effective as microorganisms can locate in subsurface structures of produce and survive biocidal washes (Burnett and Beuchat, 2001; Takeuchi and Frank, 2000).

The nutritional value of fresh fruits and vegetables is also part of their quality and depends on the variety of the produce and its age/maturity. The initial content of phytonutrients in fresh produce is important, but their bioavailability depends on the extent of release of phytonutrients during digestion and subsequent absorption (Southon and Faulks, 2002). Instead of quality, shelf life is often mentioned in the literature and is mainly related to sensory and microbiological aspects, despite the importance of the nutritional value of produce. All these parameters are subjected to physiological changes occurring in harvested organs that are listed below; they depend on temperature and atmosphere composition.

16.2.1 Respiration and Ripening

Fruits and vegetables are living produce that obtain all the nutrients they need for their growth when connected to the parent plant. Once detached from it, these harvested organs have to draw on their own reserves to achieve aerobic respiration and maintain their cellular integrity. During respiration, stored carbohydrates are broken down into glucose, which is oxidized into CO_2, water, and energy via several enzymic steps. As soon as these substrates become unavailable, other carbonated resources that might be essential, such as constitutive protein or membrane lipids, are used, leading to the death of the harvested organ. Thus, the potential shelf life of fruits and vegetables is closely related to their respiration rate, expressed as the quantity of O_2 consumed (or CO_2 released) per time and per mass of produce. A hyperbolic relationship has been demonstrated for vegetables (Marcellin, 1975) and exotic fruits (Paull, 1994): the lower the respiration rate, the longer the potential shelf life.

The respiration rate is specific to a particular harvested organ, in terms of species and varieties, but may differ depending on its maturity, mainly in climacteric fruits. Fruits are classified as climacteric (peaches, apples, kiwis, etc.) or nonclimacteric (citrus fruits, small red fruits, etc.) according to their ability to synthesize the growth hormone ethylene in an autocatalytic way or not, respectively (Robertson, 2006). For climacteric fruits, an increase in respiration is associated with an increase in ethylene: when the respiration rate reaches a maximum value (the so-called respiration crisis) ethylene production also reaches its maximum (the so-called climacteric crisis). During the biosynthesis of ethylene, the produce ripens and several biochemical and structural changes occur, such as softening due to the degradation of pectins (Soda et al., 1987) or internal browning caused by the enzymic oxidation of phenolic compounds (Mayer, 1987).

Minimal processing such as peeling, slicing, and destoning cause mechanical stresses that result in an increase in respiration activity from 2- to 8-fold (Brecht, 1995) and stimulate ethylene production from 5- to 20-fold (Pech et al., 1994). By increasing the wounded surface area, mechanical stresses also result in enzymic alterations in fresh commodities such as enzymic browning or pectin degradation, and increased proliferation of microbial spoilage (Brecht, 1995; Nguyenthe and Carlin, 1994; Watada and Qi, 1999).

16.2.2 Dehydration

In association with aerobic respiration, moisture vapor is produced and natural dehydration occurs, caused by diffusion of moisture vapor from the high-concentration compartments in fresh product to the low concentration in the surrounding environment. Marketing of items that have suffered significant water loss becomes difficult. As fruits and vegetables contain over 90% water, a loss of 5% or more water is visually noticeable, lowering the grade of the produce and resulting in a decrease in its commercial value. Major effects of water loss are a reduction in weight and a wilted appearance; there is also a reduction in nutritional value as the amount of water-soluble components decreases when water is released, a loss in aroma and flavor, and an enhanced sensitivity to chilling injuries (Maguire et al., 2006; Paull, 1999). In the case of climacteric fruits, water loss can accelerate the climacteric crisis as observed in avocados (Adato and Gazit, 1974). Although excessive dehydration is not recommended whatever the product, moderate dehydration may be beneficial to some vegetables such as bulbs or common mushrooms (Barron et al., 2002; Roy et al., 1996).

16.2.3 Temperature Effect

It is well known that temperature is one of the major factors affecting shelf life of fresh produce and needs to be positively controlled during handling and marketing of such commodities (Brecht et al., 2003). Respiration of raw fruits and vegetables increases 2- to 3-fold for every 10°C rise in temperature within the range of temperature usually encountered in the distribution and marketing chain (4–30°C) (Exama et al., 1993; Varoquaux and Ozdemir, 2005; Zagory and Kader, 1988). In the case of fresh-cut produce, this factor may increase by 3.4- to 8.3-fold (Watada and Qi, 1999). By decreasing temperature, the rate of enzymic reactions and respiration is reduced according to the Arrhenius relationship. For climacteric fruits, postharvest storage at low temperatures might be necessary to allow ripening, such as for some pears (Lelièvre et al., 1997; Morin et al., 1985) or might accelerate their ability to produce ethylene, such as for kiwifruit (Jobling et al., 1991; Knee et al., 1983). Although refrigeration appears to be the most appropriate method for fresh produce preservation, it is not always easy to achieve in retail distribution. In addition, it does not suit all commodities as observed by the development of chilling injuries (e.g., core flush or soft scald observed in most exotic fruits) that are often combined with a rise in respiration rate (Marcellin, 1992).

16.2.4 Gas Composition Effect

Another factor affecting respiration and consequently the overall quality of fresh produce is gas composition and is well presented by Varoquaux and Ozdemir (2005). Usually, lowering the O_2 level is really effective in reducing respiration, but anoxia (a switch to anaerobic catabolism and growth of anaerobic flora that produce undesirable off-flavors and off-odors) should be avoided (Nguyenthe and Carlin, 1994; Varoquaux and Ozdemir, 2005). High CO_2 levels (more than 10%) might also reduce the respiration of several commodities such as onions, strawberries, or cucumbers (Mathooko, 1996; Varoquaux and Ozdemir, 2005) and can limit the production of ethylene as observed in kiwifruit (Rothan and Nicolas, 1994) or inhibit its production, as observed in tomatoes (Rothan et al., 1997). High CO_2 levels induce a bacteriostatic effect on aerobic bacteria but might lead to the development of anaerobic flora.

Injuries might occur when a fresh product is exposed to a level of CO_2 above its tolerance limit: formation of brown spots on lettuce or yellowing of mushrooms are common visual degradations caused by a high CO_2 content (Lopez Briones et al., 1992; Zagory and Kader, 1988). Together with CO_2, 1-methylcyclopropene (1-MCP) is also considered as a competitive inhibitor of ethylene action and can be used to delay ethylene production and the respiration crisis (Fan et al., 2000; Hershkovitz et al., 2005). As the optimal combination of O_2 and CO_2 greatly depends on the respiratory activity of the product (values range from less than 35 up to 300 mg O_2 kg^{-1} h^{-1}) and its sensitivity to CO_2, there is no unique atmosphere composition that could be applied to all fresh commodities. Critical concentrations of O_2 and CO_2 exist for each fruit or vegetable and have been published in a convenient form by Kader et al. (1998) or in table form by Bishop and Hanney (2008). Packaging materials must be properly chosen for each product; otherwise the headspace atmosphere may be inefficient or detrimental (Al-Ati and Hotchkiss, 2003; Exama et al., 1993).

16.3 MODIFIED ATMOSPHERE PACKAGING

As long ago as the 1930s, controlled atmospheres were used in storage rooms during shipment and transportation to preserve the freshness of fruits. The gas machinery allowed the maintenance of low O_2 and moderate to high CO_2 levels. That is why MAP historically refers to a ratio of O_2 and CO_2 in the headspace of packaging. It can be and is increasingly extended to other gases (e.g., argon and even xenon; see Zhang et al., 2008) or vapors (water vapor, ethylene, aroma compounds, etc.). In passive MAP, gas and vapor exchanges occur between the produce and its surroundings as well as through the packaging material, as indicated in Figure 16.3. If there is microbial contamination, respiration pathways of microorganisms (aerobic or anaerobic) are also involved in gas exchanges. Initially the headspace composition is air and then, after a transient period, it reaches a steady state when gas and vapor permeation through the material balances gas and vapor consumption and production from the produce. This steady state atmosphere must be as close as possible to the optimal recommended atmosphere; otherwise it might be detrimental to the quality of the commodity (Floros and Matsos, 2005).

Thus, it is essential to carefully select a film with suitable gas and vapor permeabilities. It should be stressed that most synthetic polymer films exhibit too low a permeability to gases and vapors and most often need to be perforated (micro-holes) to allow sufficient gas exchange. Whatever the packaging material used, the produce can be exposed to unsuitable gas compositions during the transient period, thus preventing the positive effects of the steady state atmosphere. For instance, Barron et al. (2002) showed that, in mushroom packages, the transient CO_2 peak induced browning.

Active MAP has been developed to overcome this drawback and suppress or reduce the transient period (Guillaume et al., 2008). Gas flushing with an inert gas mixture is used to quickly reach a

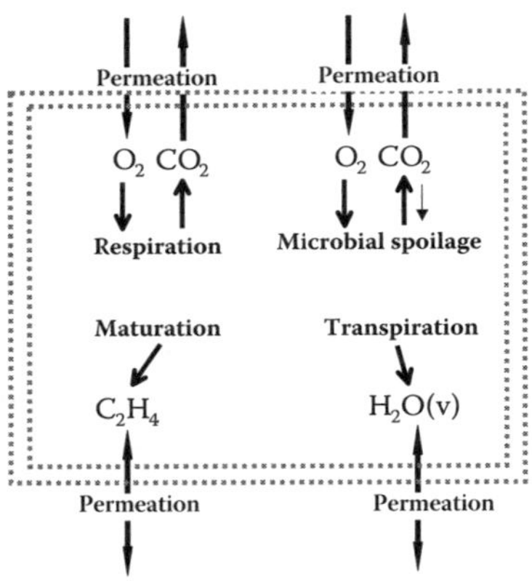

FIGURE 16.3 Relations between physiological phenomena and packaging permeation in passive MAP.

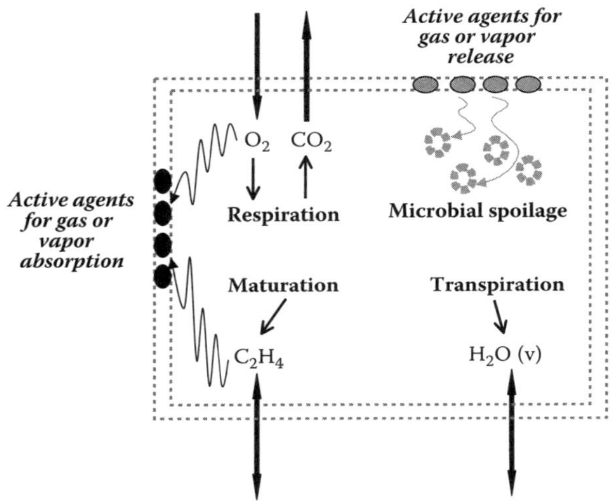

FIGURE 16.4 Relations between physiological phenomena and packaging permeation and absorption/releasing properties in active MAP.

desired ratio of O_2 and CO_2 and suppress the transient period of passive MAP so that the initial headspace composition is different from normal air. As shown in Figure 16.4, active MAP can also be achieved by the release or absorption of gases or vapors from or by active agents in combination with gas/vapor permeation. This opens up new possibilities for packaging such as the controlled release of antimicrobial agents in case of temperature abuse and/or the removal of unsuitable substances such as ethylene when produced. Such systems are commercially available for MAP of fresh produce.

Kontroll films (Kontek SRL, Italy) are multilayered and fiber-based materials that release sulfite under high-moisture conditions and are used for the preservation of table grapes. Wasaouro (Mitsubishi Kagaku Foods Co., Japan) are microcapsules of cyclodextrins containing allyl isothiocyanate (AITC) that can be printed onto a plastic or paper support and release AITC in case of temperature abuse and/or high relative humidity. The use of AITC as well as other aroma compounds constitutes alternative strategies to control postharvest spoilage of fresh fruits and vegetables (Ayala-Zavala et al., 2008; Chalier et al., 2009; Utto et al., 2008; Valero et al., 2006; Valverde et al., 2005) but should not alter the aroma perception of the fresh produce by consumers. Ethylene scavengers are already commercialized to delay ripening of climacteric fruits and are mainly based on ethylene oxidation by potassium permanganate (Frisspack, Dunpack Ltd., Hungary) or the incorporation of mineral clays such as zeolite within the packaging materials (Green bag, Evert-Fresh Corp., USA). The effect of mineral clay was attributed to its adsorption of ethylene and maybe to the formation of pores within the packaging material that could modify ethylene permeability of the film (Brody et al., 2001).

16.3.1 Modeling Aspects

Designing passive MAP for living produce is a complicated task. Many factors should be taken into account such as film characteristics (O_2, CO_2, and N_2 permeability, thickness, surface area, etc.), free volume inside the package, release/absorption capacity and kinetics of emitters/scavengers, product weight, and respiratory parameters. Several models are available in the literature to describe change in gas composition in the headspace of packaging as a function of time (Cameron et al., 1989; Edmond et al., 1991; Fishman et al., 1995; Henig and Gilbert 1975; Mannapperuma et al., 1989). Basically these models use the principles of O_2 and CO_2 mass balances to define the interactions among product respiration, film permeability, and the environment:

- Respiratory activity is described by a Michaëlis-Menten-type equation with a noncompetitive CO_2 inhibition (Fonseca et al., 2002; Lee et al., 1991).
- Gas exchanges through the plastic film are represented by the classical permeability equation based on Fick's first law of diffusion for thin and infinite films (Crank and Park, 1968).

The models can be improved by also considering the interaction with active agents such as O_2 scavengers. In this case, the absorption kinetics of individual gas scavenger sachets is used according to an apparent first order reaction (Charles et al., 2006; Tewari et al., 2002). Whatever model is used, validation experiments must be performed to compare computed and measured gas concentrations and estimate the fitting of predictions (Charles et al., 2003, 2005a).

Figure 16.5 shows an example of experimental and predicted gas changes in the headspace of passive MAP (Figure 16.5A) or active MAP (Figure 16.5B) for endives storage at 20°C. Predicted values correctly fit the experimental data during the transient period as well as during the steady state. However, after almost 8 days, the experiment indicated a metabolic deviation that was attributed to the growth of microorganisms, mainly yeasts and molds (Charles et al., 2005b), that was not predicted by modeling. Oxygen scavengers did not modify the gas composition at the steady state with O_2 and CO_2 partial pressures at about 3 and 5 kPa in passive and active MAP (Figure 16.5A and 16.5B respectively). These values are close to those recommended by Mannapperuma et al. (1989) to maintain acceptable visual appearance of endive. Oxygen scavengers reduced the transient period duration by half (50 hr compared to 100 hr without scavengers) and consequently accelerated the setting up of the optimal modified atmosphere. This led to an important delay in head opening, leaf greening, and browning of the basal part of endives, markers of produce aging (Charles et al., 2008).

The effect of temperature on the gas flows in MAP can also be predicted by using the Arrhenius relationship for respiratory activity, gas permeability, and the absorption kinetics of the gas scavengers:

$$R_{O_2} = R_{O_2 0} \exp\left(-\frac{E_a}{RT}\right) \quad (16.1)$$

$$P = P_0 \exp\left(-\frac{E_a}{RT}\right) \quad (16.2)$$

$$k = A \exp\left(-\frac{E_a}{RT}\right) \quad (16.3)$$

where R_{O_2} is the respiration rate; $R_{O_2 0}$ is a temperature-dependent factor for respiration; P is the gas permeability of the material; P_0 is a temperature-dependent factor for permeability; k is the absorption rate constant; A is the temperature-dependent factor for absorption; E_a the energy of activation for respiration, permeation, or absorption; R is the gas constant; and T is the absolute temperature.

It should be stressed that the Q_{10} value (the factor by which the respiration rate increases for a rise in temperature of 10°C) is widely used by physiologists, but it is increasingly being replaced by E_a to simplify the global equation in coupling models in which gas diffusion through materials is also considered. The relationship between E_a and Q_{10} is

$$E_a = R\frac{T_1 T_2}{10}\ln Q_{10} \tag{16.4}$$

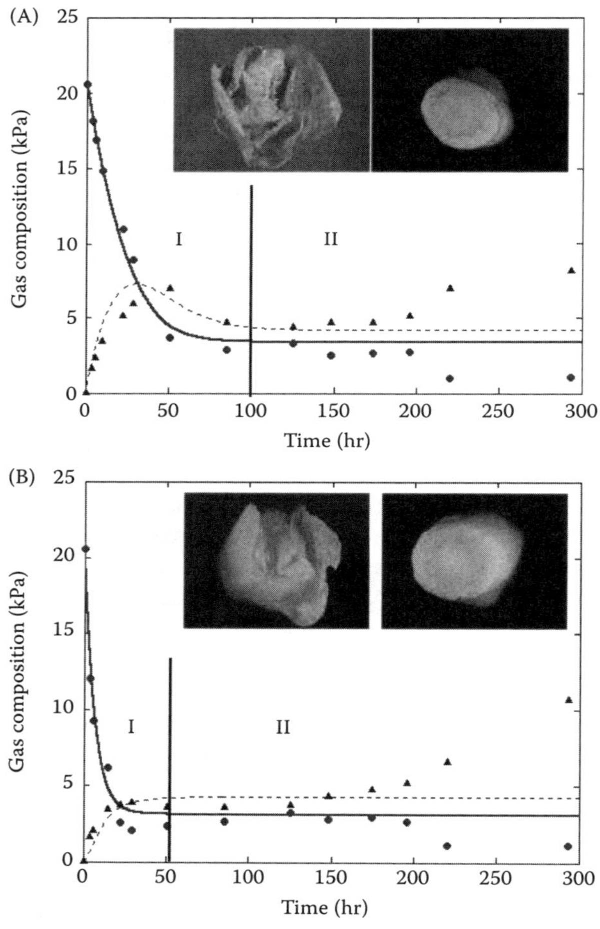

FIGURE 16.5 Experimental and predicted O_2 (•, —) and carbon dioxide (▲, – –) partial pressure changes in LDPE package (V = 1.8 L) containing around 500 g of endive and with (B) or without (A) individual O_2 scavenger sachet (ATCO LH-100) sealed under air at 20°C. Photograph of leaves and basal part of endives after 7 days (168 hr) of storage. I: Transient period, II: Steady state. (Adapted from Charles et al. (2005a, 2008) with permission.)

These temperature relationships have been successfully applied to the model used in Figure 16.5 for predicting changes in gas composition in passive and active MAP of endives as a function of storage temperature (Charles et al., 2005a). Thus, modeling is a powerful and useful tool to assess optimal conditions for MAP of fresh products prior to experiments that might be time and money-consuming (Charoenchaitawornchit et al., 2003; Del Nobile et al., 2009; Guevara et al., 2003; Rai and Paul 2008; Salvador et al., 2002). However, it needs a complete database on fresh fruits and vegetables respiration parameters, films permeabilities, and scavengers absorption.

16.3.2 Passive Modified Atmosphere Packaging

Oriented polypropylene (OPP), low density polyethylene (LDPE), and poly(vinyl chloride) (PVC) are synthetic polymers used as trays, pouches, or overwraps for packing fresh fruits and vegetables. However, in most cases, they need to be perforated to allow sufficient gas and vapor transfer (Allan-Wojtas et al., 2008). Two kinds of perforated materials are available: (a) macroperforated films that slightly protect the produce against dehydration by slowing down water vapor transfers and (b) microperforated films that provide a large range of O_2 permeabilities (from 190 to 42,000 mL O_2 m^{-2} 24 hr^{-1}; Varoquaux and Ozdemir, 2005) and can be tailored to the O_2 requirement of most produce. However, they exhibit several physical limitations. CO_2 diffuses through the film at a similar rate to O_2. Therefore, it is impossible to achieve low O_2 (1–5%) without accumulating high CO_2 (15–20%) (Exama et al., 1993). Thus, these films are applicable only for products that can tolerate high CO_2 without experiencing injury. This is due to the permselectivity of materials, which is the ratio of the CO_2 permeability coefficient to the O_2 permeability coefficient; this ratio is 1 for perforated films. Commonly, in nonperforated synthetic films, the permselectivity is around 4–6 (Robertson, 2006; Yam and Lee, 1995). Another drawback is that detrimental changes in the steady state modified atmosphere cannot be avoided if the chill chain is disrupted since the activation energy (E_a) for gas permeation through synthetic films is twofold lower than the E_a for respiration rates of most of fresh produce (Exama et al., 1993; Varoquaux and Ozdemir, 2005). This means that O_2 consumption and CO_2 production by fresh produce are faster than diffusion of both gases through the packaging during temperature abuse. When synthetic films are perforated, the effect of temperature on permeation through pores (air) is much less compared to its effect on permeation through the polymer. As permeation through the pores accounts for most of the total permeation through a perforated film, the activation energies for perforated films are close to zero. Thus, perforated films are not able to compensate for the effect of temperature variations during storage on the respiration rates of fruits and vegetables.

16.3.2.1 Protein-Based Materials

The development of hydrophilic materials such as protein-based films appears to be of great interest to replace perforated materials in many applications. The film-forming ability of polymers such as proteins and polysaccharides from renewable resources is known and widely used for a long time, especially in the fruit and vegetable industry. Coating of fruits and vegetables is commonly employed to offer consumers a product with a shiny and glossy appearance; it also contributes to the establishment of internal modified atmospheres (Cisneros-Zevallos and Krochta, 2005) and limits physiological decay, dehydration, and microbial spoilage, as described in Table 16.1.

Although most of the commercial plastic films are relatively unaffected by relative humidity (RH), gas permeability of protein-based films sharply increases at high RH, such as in fresh fruit and vegetable packaging (Gontard et al., 1996; Mujica Paz and Gontard, 1997; Varoquaux and Ozdemir, 2005). For example, at 23°C, CO_2 permeability is 570 times lower for wheat gluten film stored at 0% RH than for LDPE stored under the same conditions, and it is 10 times higher for wheat gluten film stored at 95% RH than for LDPE stored under the same conditions (Gontard et al., 1996). The same effect was observed with O_2 permeability but to a lesser extent. Consequently, the permselectivity of wheat gluten films is also affected by RH and ranges from 5 in a dry atmosphere

TABLE 16.1
Overview of Potential Applications for Biobased Packaging and Coatings in MAP of Fresh or Minimally Processed Fruits and Vegetables

Biopolymer	Product	MAP Improvement	References
		Polysaccharides	
Xanthan gum	Baby carrots (coating)	Improve surface color, texture, and flavor	Mei et al., 2002
Starch	Strawberries (coating)	Reduce weight loss, maintain texture and surface color, delay maturation	Garcia et al.,1998; Mali and Grossmann, 2003
Cellulose	Mandarin (coating)	Reduce weight loss, improve visual appearance, and delay ripening	Togrul and Arslan, 2004
	Apples (coating)	Reduce weight loss	Togrul and Arslan, 2005
	Mini-peeled carrots (coating)	Prevent white surface discoloration	Howard and Dewi, 1995, 1996
HPC[a]	Guavas (coating)	Delay softening but surface blackening cannot be avoided	McGuire and Hallman, 1995
MC[b]	Avocados (coating)	Decrease respiration rate, maintain higher firmness and greener color, delay ripening	Maftoonazad and Ramaswamy, 2005
CMC[c]	Sweet cherries (coating)	Reduce weight loss, delay changes in firmness, skin color, ascorbic acid content, etc.	Yaman and Bayoindirli, 2002
	Tomatoes (coating)	Reduce weight loss, delay changes in firmness, ascorbic acid, lycopene, etc.	Tasdelen and Bayindirli, 1998
	Mangos (coating)	Reduce weight loss and delay changes in firmness and color	Carrillo-Lopez et al., 2000
		Polysaccharides/Proteins	
Cellulose/soy protein	Apples (coating)	Reduce water loss	Baldwin et al., 1996
Cellulose/wheat gluten	Mushrooms (packaging)	Delay cap opening	Guillaume et al., 2008
Starch/chitosan	Sliced carrots (coating)	Inhibit growth of total coliforms and lactic acid bacteria, slightly reduce counting of mesophilic aerobes, mold, and yeast	Durango et al., 2006
		Proteins	
Chitosan	Cucumber (coating)	Reduce water loss, respiration rate, discoloration, and fungal infection	Elghaouth et al., 1991
	Strawberries (coating)	Reduce decay caused by *Botrytis cinerea* and *Rhizopus stolonifer* Delay change in color and loss in firmness	Elghaouth et al., 1992 Hernandez-Munoz et al., 2006
	Tomatoes (packaging)	Reduce weight loss and formation of defective spots, and delay loss in firmness and color	Srinivasa et al., 2006
Corn zein	Broccoli (packaging)	Maintain original firmness and odor	Rakotonirainy et al., 2001
	Tomatoes (coating)	Delay color change, loss in firmness and weight	Park et al., 1994
Wheat gluten	Mushrooms (packaging)	Delay cap opening, but browning cannot be avoided	Barron et al., 2002

(Continued)

TABLE 16.1 ***(Continued)***

Biopolymer	Product	MAP Improvement	References
	Sliced mushrooms (coating)	Delay cap opening	Kim et al., 2006
	Strawberries (coating)	Retention of firmness and maintenance of visual acceptance and taste	Tanada-Palmu and Grosso, 2005

[a] HPC = hydroxypropyl cellulose.
[b] C = methyl cellulose.
[c] MC = carboxymethyl cellulose.

to 30 under 95% RH (Gontard et al., 1996). The increasing RH effect has been attributed to a modification of the wheat gluten network structure and polymeric chain mobility, related to a change from a glassy to a rubbery state (Cuq et al., 1997; Gontard et al., 1996; Gontard and Ring, 1996). The fact that CO_2 permeability is more affected than O_2 can be related to specific interactions between CO_2 and the water-plasticized protein matrix under high RH conditions. Water sorption could provide better accessibility to active sites for CO_2 sorption onto the protein matrix, mainly hydrogen binding with glutamine residue (Gontard, 1998). Thus generated modified atmospheres can be both low in O_2 and CO_2, as described for MAP of common mushrooms by Barron et al. (2002).

Temperature also affects gas permeability of protein-based films and this is more pronounced in films stored at high RH as observed for O_2 and CO_2 permeability (Mujica Paz and Gontard, 1997) as well as ethylene permeability (Mujica Paz et al., 2005) of wheat gluten films. When stored at high RH, ethylene permeability of wheat gluten is around 1.02×10^{-9} mL(STP) cm cm^{-2} s^{-1} (cm Hg)$^{-1}$ at 23°C, reaching 4.96×10^{-9} mL(STP) cm cm^{-2} s^{-1} (cm Hg)$^{-1}$ at 45°C. Such behavior is of great interest in delaying ripening of climacteric produce during shipping and distribution in cases of temperature abuse. The E_a for CO_2 permeation of these materials at high RH is close to that of CO_2 production for almost all fresh fruits and vegetables (Barron et al., 2002; Mujica Paz and Gontard, 1997; Varoquaux and Ozdemir 2005). This means that wheat gluten films should be able to self-adjust the CO_2 content in cases of temperature abuse. However, despite low cost and interesting functional properties, wheat-gluten-based films exhibit poor mechanical properties at both low and high RH. Combination with other materials or fillers appears to be the best way to overcome this defect, but problems of compatibility as well as loss of biodegradability have to be taken into consideration.

16.3.2.2 Fiber-Based and Protein (Nano)composites

Combinations of wheat gluten proteins with fiber-based materials such as paper and so-called composite materials have already been studied to overcome their poor mechanical properties (Gastaldi et al., 2007; Han and Krochta, 1999; Rhim et al., 2006). Recently, combinations of wheat gluten proteins with nanofillers such as montmorillonite (MMT) have been demonstrated to be an efficient way to reduce the drawback of moisture sensitivity (Hedenqvist et al., 2006; Olabarrieta et al., 2006; Tunc et al., 2007; Wang et al., 2005). Thus, combining wheat gluten proteins with paper and MMT to produce so-called nanocomposite materials could make the most of the functional properties of wheat gluten proteins while overcoming their drawbacks of insufficient mechanical properties and moisture sensitivity.

Paper coated with wheat gluten solution can no longer be considered as porous regarding O_2 and CO_2 permeability; the presence of MMT in the wheat gluten network did not significantly affect permeability, as indicated in Table 16.2. As permeability is known to be governed by two mechanisms (diffusion and sorption), it was assumed that introduction of MMT did not change

TABLE 16.2
Gas Permeability and Permselectivity of Paper, Paper Coated with Wheat Gluten and Paper Coated with Wheat Gluten Plus Nanoclay MMT at 25°C as a Function of Relative Humidity (RH %)

	RH (%)	PO_2[a]	PCO_2[a]	PCO_2/PO_2
Paper	80	OR[b]	OR[b]	1
	90	OR[b]	OR[b]	1
Paper + gluten	80	172	335	1.9
	90	281	2240	7.9
Paper + gluten + Nanoclay MMT	80	211	326	1.5
	90	290	2250	7.7

[a] O_2 and CO_2 permeability coefficients (PO_2 and PCO_2) expressed as 10^{-11} mL(STP) cm cm^{-2} sec^{-1} (cm Hg)$^{-1}$. Mean standard deviation is 10 and 14 × 10^{-11} mL(STP) cm cm^{-2} sec^{-1} (cm Hg)$^{-1}$ for PO_2 and PCO_2 respectively.

[b] OR: over range. Permeability values are higher than 5.7 × 10^{11} mL(STP) cm cm^{-2} sec^{-1} (cm Hg)$^{-1}$.

solubility nor diffusivity of O_2 and CO_2, as observed in a previous study (Tunc et al., 2007). It should be pointed out that O_2 and CO_2 permeability of both coated and nanocomposite material increased with increasing RH. As this phenomenon was also observed with pure wheat gluten films, it suggests that the wheat-gluten-based coating layer is the key element of gas barrier properties of the studied materials (composite and nanocomposite). As a consequence, gas permselectivity was highly affected by RH and rose from 1.9 to 7.9 and 1.5 to 7.5 for coated and nanocomposite materials, respectively. These results show that the unique gas permselectivity properties of proteins are preserved when combined with paper or MMT.

Passive MAP experiments were conducted on parsley with uncoated paper as the control and paper coated with wheat gluten containing MMT at 20°C. As expected for a highly porous material, O_2 and CO_2 partial pressures at the steady state obtained when using paper (Table 16.3) were close to those in air (21 and 0 kPa respectively). Such an atmosphere was detrimental to the quality of the product. After only 4 days of storage, more than 50% of the ascorbic acid and chlorophyll were lost, and the leaves were fully yellow. In comparison, nanocomposite material generated a headspace atmosphere containing lower O_2 (11 kPa) and higher CO_2 (4 kPa) content. This steady state atmosphere clearly improved the quality attributes of parsley during storage by maintaining high chlorophyll content (directly linked to the green color) and ascorbic acid during 8 days of storage.

For consumers the end of shelf life of parsley is mainly determined through its discoloration. Sensory analysis results were perfectly correlated with chlorophyll content. A critical chlorophyll level of 1.5 mg g^{-1} of fresh parsley (corresponding to 70% of the initial content) was reached after only 2 days of storage when packed at 20°C using uncoated paper. If a nanocomposite material was used, discoloration was delayed for more than 8 days. However, other deteriorative reactions occur simultaneously and must be taken into account, such as the decrease in ascorbic acid content, one of the major compounds of nutritional interest in parsley. If preserving at least 60% of the initial ascorbic acid content is considered as a critical level, it was reached after only 3 days of storage with uncoated paper against more than 8 days for composite material. The use of nanocomposite materials for MAP of parsley led to an equilibrium atmosphere favorable for maintaining the quality of parsley by slowing down oxidative and physiological reactions responsible for product degradation. In the context of sustainability, environmental protection, and health enhancement by increasing fresh fruits and vegetables consumption for their nutritional properties, MAP with nanocomposite materials could be a very interesting alternative to micro- or macroperforated conventional plastic films for storage of highly perishable products such as fresh fruits and vegetables.

TABLE 16.3
Atmosphere Composition of the Headspace and Remaining Percentage of Initial Chlorophyll and Ascorbic Acid Content in Parsley after 8 Days of Storage at 20°C, under MAP Using Paper or Paper Coated with Wheat Gluten Containing MMT

	Chlorophyll (%)[a]	Ascorbic Acid (%)[a]	O_2 (kPa)	CO_2 (kPa)
Paper	24 ± 2	30 ± 3	19 ± 1	0.5 ± 0.2
Nanocomposite	76 ± 4	61 ± 4	11 ± 1	4.0 ± 0.8

[a] Expressed as a percentage of the initial content (initial concentrations of chlorophyll and ascorbic acid were 2.1 ± 0.1 and 2.4 ± 0.1 mg g^{-1} of fresh parsley, respectively).

16.3.3 Active Modified Atmosphere Packaging

Another promising application of protein-based materials is their use as antimicrobial packaging. This relies on the ability of biobased polymers such as proteins to carry active compounds and release them in a controlled way, according to a moisture and temperature triggering effect (Carlin et al., 2001; Gennadios and Weller, 1991; Guilbert et al., 1997). The use of plant-derived antimicrobial compounds as natural preservatives has received increasing interest in recent years. The antimicrobial effect of essential oils has already been reported (Burt, 2004; Panizzi et al., 1993; Sivropoulou et al., 1996) and is often assigned to their major aroma compounds. The efficiency of thyme, clove, and oregano essential oils was related to the presence of the phenolic aroma compounds thymol, eugenol, and carvacrol, respectively (Farag et al., 1989; Kim et al., 1995; Lambert et al., 2001; Moleyar and Narasimham, 1992; Tsao and Zhou, 2000). The inhibitory effect of phenols could be explained by their interaction and accumulation in the cell membrane of microorganisms, correlated with the hydrophobicity of these compounds (Sikkema et al., 1995; Weber and de Bont, 1996).

Carvacrol was added to wheat gluten coating solution with or without MMT (2.5% w/w of wheat gluten) to produce nanocomposite or composite papers, respectively. Release of the volatile compound was assessed at 25°C on all materials as a function of time and using a two-step gradient of RH. This RH gradient was used to simulate, first, the average storage conditions of materials (60% RH for 20 days) and, second, the conditions when used for packaging food (100% RH for 15 days) such as fresh or minimally processed fruits and vegetables, generating an atmosphere close to 100% RH inside the package. Figure 16.6 shows that almost 100% of carvacrol was lost within 1 day when deposited onto paper, demonstrating the inadequacy of this material to control the release of a volatile active agent. Composite material lost more than 70% of carvacrol within 20 days of storage. This means that only 30% of the active agent would be available to be released to the food during storage. Once placed at 100% RH, this 30% was entirely released within 8 days (from day 22 to 30). In the presence of MMT added to the wheat gluten network (nanocomposite material), only 20% of the carvacrol was released during 20 days of storage at 60% RH. Consequently, 80% of the volatile active agent remained available to be released during the period in which food is packaged. Once placed at 100% RH, this 80% was entirely released within 13 days (from day 22 to 35).

It can be concluded from these results that the release of carvacrol from paper coated with wheat gluten is RH-dependent with or without MMT. It should be noted that only 2.5% of MMT was sufficient to (a) enable paper coated with wheat gluten to retain high amounts of volatile compounds at 60% RH and to (b) enhance the RH triggering effect for the release of the active volatile agent. Such behavior is extremely interesting for both limiting volatile active agent losses before using the material as food packaging and for triggering the active agent release in the presence of the food.

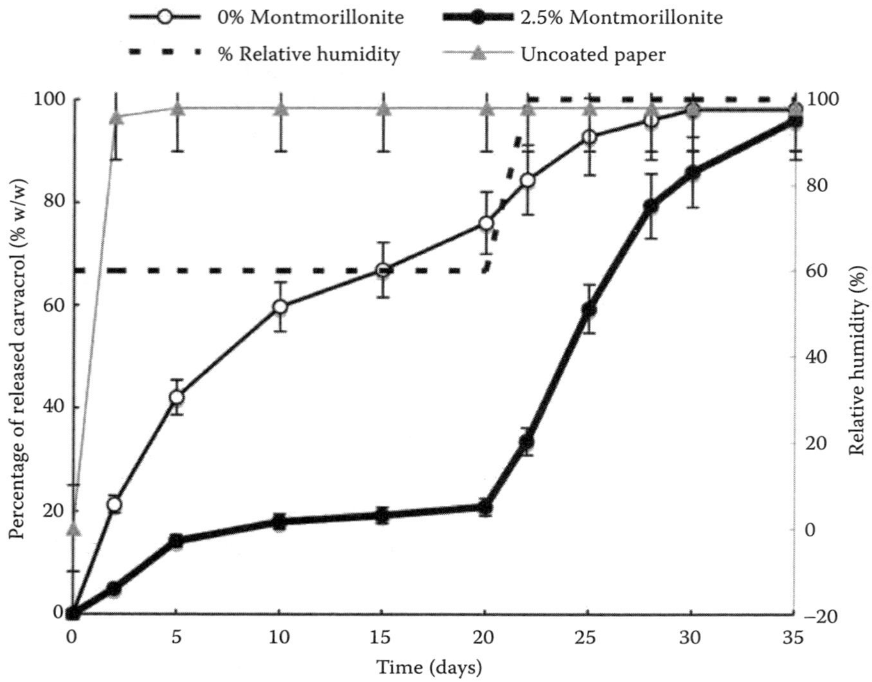

FIGURE 16.6 Release of carvacrol as a function of relative humidity (- - -) and material used: support paper (▲), wheat gluten coated paper coated with (●) or without (○) 2.5% of MMT.

16.4 FUTURE TRENDS

MAP has the potential to prolong shelf life and quality of fresh fruits and vegetables but the development of innovative materials is still required. To be effective, innovation should be conducted according to an integrated approach combining modeling and material science with plant physiology and food quality. Modeling tools could enable the definition of targeted functional requirements for designing materials that should maintain produce quality by using properly chosen mathematical models and input data. There is still a need to provide extensive data on the following:

- Physiological parameters for raw and fresh-cut produce and their temperature dependence
- Physiological tolerance of raw and fresh-cut produce and consumer perception toward common gases (O_2 and CO_2) but also natural preservatives such as aroma compounds and their combination
- Gas and vapor permeability values of materials under conditions of use (temperature and relative humidity)
- Absorption or release ability of active materials (kinetics and maximum absorption/release)

Thus, designing and processing of new MAP will rely on the smart use of materials tailored to the application. It could also be done in an environmentally friendly manner by using natural and biodegradable components that appear well adapted to the physiology of fresh fruits and vegetables. This means that investigations should focus on the buildup of biobased materials and fully understand the underlying mechanisms of the coating and thermomolding processes of such materials (proteins, layered silicate nanoclays, aroma compounds, etc.).

REFERENCES

Adato I., Gazit S. 1974. Water deficit stress, ethylene production and ripening in avocado fruits. *Plant Physiology* 53: 45–46.

Al-Ati T., Hotchkiss J.H. 2003. The role of packaging film permselectivity in modified atmosphere packaging. *Journal of Agricultural and Food Chemistry* 51: 4133–4138.

Allan-Wojtas P., Forney C.F., Moyls L., Moreau D.L. 2008. Structure and gas transmission characteristics of microperforations in plastic films. *Packaging Technology and Science* 21: 217–229.

Ayala-Zavala J.F., Oms-Oliu G., Odriozola-Serrano I., Gonzales-Aguilar G.A., Alvarez-Parrilla E., Martin-Belloso O. 2008. Bio-preservation of fresh-cut tomatoes using natural antimicrobials. *European Food Research and Technology* 226: 1047–1055.

Baldwin E.A., Nisperos M.O., Chen X., Hagenmaier R.D. 1996. Improving storage life of cut apple and potato with edible coating. *Postharvest Biology and Technology* 9: 151–163.

Barron C., Varoquaux P., Guilbert S., Gontard N., Gouble B. 2002. Modified atmosphere packaging of cultivated mushroom (*Agaricus bisporus* L.) with hydrophilic films. *Journal of Food Science* 67: 251–255.

Bishop C., Hanney S. 2008. Environmentally compatible packaging of fresh agricultural and horticultural produce. In: *Environmentally Compatible Food Packaging*. Chiellini E. (Ed). Boca Raton, Florida: CRC Press, pp. 459–477.

Brecht J.K. 1995. Physiology of lightly processed fruits and vegetables. *Hortscience* 30: 18–22.

Brecht J.K., Chau K.V., Fonseca S.C., Oliveira F.A.R., Silva F.M., Nunes M.C.N., Bender R.J. 2003. Maintaining optimal atmosphere conditions for fruits and vegetables throughout the postharvest handling chain. *Postharvest Biology and Technology* 27: 87–101.

Brody A.L., Strupinsky E.R., Kline L.R. 2001. *Active Packaging for Food Applications*. Lancaster, Pennsylvania: Technomic Publishing.

Burnett A.B., Beuchat L.R. 2001. Comparison of sample preparation methods for recovering *Salmonella* from raw fruits, vegetables and herbs. *Journal of Food Protection* 64: 1459–1465.

Burt S. 2004. Essential oils: their antibacterial properties and potential applications in foods—a review. *International Journal of Food Microbiology* 94: 223–253.

Cameron A.C., Boylanpett W., Lee J. 1989. Design of modified atmosphere packaging systems—modeling oxygen concentrations within sealed packages of tomato fruits. *Journal of Food Science* 54: 1413–1421.

Carlin F., Gontard N., Reich M., Nguyen-The C. 2001. Utilization of zein coating and sorbic acid to reduce *Listeria monocytogenes* growth on cooked sweet corn. *Journal of Food Science* 66: 1385–1389.

Carrillo-Lopez A., Ramirez-Bustamante F., Valdez-Torres J.B., Rojas-Villegas R., Yahia E.M. 2000. Ripening and quality changes in mango fruit as affected by coating with an edible film. *Journal of Food Quality* 23: 479–486.

Chalier P., Ben Arfa A., Guillard V., Gontard N. 2009. Moisture and temperature triggered release of a volatile active agent from soy protein coated paper: effect of glass transition phenomena on carvacrol diffusion coefficient. *Journal of Agricultural and Food Chemistry* 57: 658–665.

Charles F., Sanchez J., Gontard N. 2003. Active modified atmosphere packaging of fresh fruits and vegetables: modeling with tomatoes and oxygen absorber. *Journal of Food Science* 68: 1737–1742.

Charles F., Sanchez J., Gontard N. 2005a. Modeling of active modified packaging of endives exposed to several temperatures. *Journal of Food Science* 70: 443–449.

Charles F., Rugani N., Gontard N. 2005b. Influence of packaging conditions on natural microbial population growth of endive. *Journal of Food Protection* 68: 1020–1025.

Charles F., Sanchez J., Gontard N. 2006. Absorption kinetics of oxygen and carbon dioxide scavengers as part of active modified atmosphere packaging. *Journal of Food Engineering* 72: 1–7.

Charles F., Guillaume C., Gontard N. 2008. Effect of passive and active modified atmosphere packaging on quality changes of fresh endives. *Postharvest Biology and Technology* 48: 22–29.

Charoenchaitaworrnchit A., Kanlayanarat S., Tongta A. 2003. Modeling of respiration and modified atmosphere packaging of mango "Nam Dok Mai". Proceedings of the international conference postharvest unlimited, Leuven, Belgium. *Acta Horticulturae* 599: 489–494.

Cisneros-Zevallos L., Krochta J.M. 2005. Internal modified atmospheres of coated fresh fruits and vegetables: understanding relative humidity effects. In: *Innovations in Food Packaging*. Han J.H. (Ed). London, England: Elsevier Academic Press, pp. 173–184.

Crank J., Park G.S. 1968. *Diffusion in Polymers*. London, England: Academic Press.

Cuq B., Gontard N., Guilbert S. 1997. Thermoplastic properties of fish myofibrillar proteins: application to the biopackaging fabrication. *Polymer* 38: 4071–4078.

Del Nobile M.A., Conte A., Scrocco C., Brescia I., Speranza B., Sinigaglia M., Perniola R., Antonacci D. 2009. A study on quality loss of minimally processed grapes as affected by film packaging. *Postharvest Biology and Technology* 51: 21–26.
Durango A.M., Soares N.F.F., Andrade N.J. 2006. Microbiological evaluation of an edible antimicrobial coating on minimally processed carrots. *Food Control* 17: 336–341.
Edmond J.P., Castaigne F., Toupin C.J., Desilets D. 1991. Mathematical modeling of gas exchange in modified atmosphere packaging. *Transactions of the American Society of Agricultural Engineers* 34: 239–245.
Elghaouth A., Arul J., Grenier J., Asselin, A. 1992. Antifungal activity of chitosan on 2 postharvest pathogens of strawberry fruits. *Phytopathology* 82: 398–402.
Elghaouth A., Arul J., Ponnampalam R., Boulet M. 1991. Use of chitosan coating to reduce water-loss and maintain quality of cucumber and bell pepper fruits. *Journal of Food Processing and Preservation* 15: 359–368.
Exama A., Arul J., Lencki R.W., Lee L.Z., Toupin C. 1993. Suitability of plastic films for modified atmosphere packaging of fruits and vegetables. *Journal of Food Science* 58: 1365–1370.
Fan X., Argenta L., Mattheis J.P. 2000. Inhibition of ethylene action by 1-methylcyclopropene prolongs storage life of apricots. *Postharvest Biology and Technology* 20: 135–142.
Farag R.S., Daw Z.Y., Aboraya S.H. 1989. Influence of some spice essential oils on *Aspergillus parasiticus* growth and production of aflatoxins in a synthetic medium. *Journal of Food Science* 54: 74–76.
Fishman S., Rodov V., BenYehoshua S. 1996. Mathematical model for perforation effect on oxygen and water vapor dynamics in modified-atmosphere packages. *Journal of Food Science* 61: 956–961.
Floros J.D., Matsos K.I. 2005. Introduction to modified atmosphere packaging. In: *Innovations in Food Packaging*. Han J.H. (Ed). London, England: Elsevier Academic Press, pp. 159–172.
Fonseca S.C., Oliveira F.A.R., Brecht J.K. 2002. Modelling respiration rate of fresh fruits and vegetables for modified atmosphere packages: a review. *Journal of Food Engineering* 52: 99–119.
Garcia M.A., Martino M.N., Zaritzky N.E. 1998. Plasticized starch-based coatings to improve strawberry (*Fragaria x Ananassa*) quality and stability. *Journal of Agricultural and Food Chemistry* 46: 3758–3767.
Gastaldi E., Chalier P., Guillemin A., Gontard N. 2007. Microstructure of protein-coated paper as affected by physico-chemical properties of coating solutions. *Colloids and Surfaces A* 301: 301–310.
Gennadios A., Weller C.L. 1991. Edible films and coatings from soymilk and soy protein. *Cereal Foods World* 36: 1004–1009.
Gontard N., Thibault R., Cuq B., Guilbert S. 1996. Influence of relative humidity and film composition on oxygen and carbon dioxide permeabilities of edible films. *Journal of Agricultural and Food Chemistry* 44: 1064–1069.
Gontard N., Ring S. 1996. Edible wheat gluten film: influence of water content on glass transition temperature. *Journal of Agricultural and Food Chemistry* 44: 3474–3478.
Guevara J.C., Yahia E.M., Cedeno L. 2003. Modeling of O_2, CO_2 and H_2O vapor in modified atmosphere packaged prickly pear cactus stems (*Opuntia* spp.). Proceedings of the international conference on quality chains, Wageningen, Netherlands. *Acta Horticulturae* 604: 671–675.
Guilbert S., Cuq B., Gontard N. 1997. Recent innovations in edible and/or biodegradable packaging materials. *Food Additives and Contaminants* 14: 741–751.
Guillaume C., Chalier P., Gontard N. 2008. Modified atmosphere packaging using environmentally compatible and active food packaging materials. In: *Environmentally Compatible Food Packaging*. Chiellini E. (Ed). Boca Raton, Florida: CRC Press, pp. 396–418.
Han J.H., Krochta J.M. 1999. Wetting properties and water vapor permeability of whey-protein-coated paper. *Transactions of the ASAE* 42: 1375–1382.
Hedenqvist M.S., Backman A., Gallstedt M., Boyd R.H., Gedde U.W. 2006. Morphology and diffusion properties of whey/montmorillonite nanocomposites. *Composites Science and Technology* 66: 2350–2359.
Henig Y.S., Gilbert S.G. 1975. Computer analysis of variables affecting respiration and quality of produce packaged in polymeric films. *Journal of Food Science* 40: 1033–1035.
Hernandez-Munoz P., Almenar E., Ocio M.J., Gavara R. 2006. Effect of calcium dips and chitosan coatings on postharvest life of strawberries (*Fragaria x ananassa*). *Postharvest Biology and Technology* 39: 247–253.
Hershkovitz V., Saguy S.I., Pesis E. 2005. Postharvest application of 1-MCP to improve the quality of various avocado cultivars. *Postharvest Biology and Technology* 37: 252–264.
Howard L.R., Dewi T. 1995. Sensory, microbiological and chemical quality of mini-peeled carrots as affected by edible coating treatment. *Journal of Food Science* 60: 142–144.
Howard L.R., Dewi T. 1996. Minimal processing and edible coating effects on composition and sensory quality of mini-peeled carrots. *Journal of Food Science* 61: 643–645.

Jobling J., McGlasson W.B., Dilley D.R. 1991. Induction of ethylene synthesizing competency in Granny Smith apples by exposure to low temperature in air. *Postharvest Biology and Technology* 1: 111–118.

Kader A.A., Singh R.P., Mannapperuma J.D. 1998. Technologies to extend the refrigerated shelf life of fresh fruits. In: *Food Storage Stability*. Taub I.A., Singh R.P. (Eds). Boca Raton, Florida: CRC Press, chapter 16.

Kim J.M., Dersken F., Kolster P., Marshall M.R., Wei C.I. 1995. Antibacterial activity of some essential oil component against five foodborne pathogens. *Journal of Agricultural and Food Chemistry* 43: 2839–2845.

Kim K.M., Ko J.A., Lee J.S., Park H.J., Hanna M.A. 2006. Effect of modified atmosphere packaging on the shelf-life of coated, whole and sliced mushrooms. *LWT—Food Science and Technology* 39: 365–372.

Knee M., Looney N.E., Hatfield S.G.S., Smith S.M. 1983. Initiation of rapid ethylene synthesis by apple and pear fruits in relation to storage temperature. *Journal of Experimental Botany* 34: 1207–1212.

Lambert R.J.W., Skandamis P.N., Coote P.J., Nychas G.J.E. 2001. A study of minimum inhibitory concentration and mode of action of oregano essential oil, thymol and carvacrol. *Journal of Applied Microbiology* 91: 453–462.

Lee D.S., Haggar P.E., Lee J., Yam K.L. 1991. Model for fresh produce respiration in modified atmospheres based on principles of enzyme-kinetics. *Journal of Food Science* 56: 1580–1585.

Lelièvre J.M., Tichit L., Dao P., Fillion L., Nam Y.W., Pech J.C., Latche A. 1997. Effects of chilling on the expression of ethylene biosynthetic genes in Passe-Crassane pear (*Pyrus communis* L.) fruits. *Plant Molecular Biology* 33: 847–855.

Lopez Briones G., Varoquaux P., Chambroy Y., Bouquant J., Bureau G., Pascat B. 1992. Storage of common mushroom under controlled atmospheres. *International Journal of Food Science and Technology* 27: 493–505.

Maftoonazad N., Ramaswamy H.S. 2005. Postharvest shelf-life extension of avocados using methyl cellulose-based coating. *LWT—Food Science and Technology* 38: 617–624.

Maguire K.M., Sabarez H.T., Tanner D.J. 2006. Postharvest preservation and storage. In: *Handbook of Vegetable Preservation and Processing*. Hui Y.H., Ghazala S., Graham D., Murell K.D., Nip W.K. (Eds). New York: Marcel Dekker, pp. 39–65.

Mali S., Grossmann M.V.E. 2003. Effects of yam starch films on storability and quality of fresh strawberries (*Fragaria ananassa*). *Journal of Agricultural and Food Chemistry* 51: 7005–7011.

Mannapperuma J., Zagory D., Singh R.P., Kader A.A. 1989. Design of polymeric packages for modified atmosphere storage of fresh produce. *Proceeding of the 5th International Controlled Atmosphere Research Conference*, Wenatchee, Washington, pp. 225–233.

Marcellin P. 1975. Conservation des fruits et légumes frais. *Cahiers de nutrition et de diététique* 10: 49–52.

Marcellin P. 1992. Les maladies physiologiques du froid. In: *Les végétaux et le froid*. Côme D. (Ed). Paris, France: Herman, pp. 53–105.

Mathooko F.M. 1996. Regulation of respiratory metabolism in fruits and vegetables by carbon dioxide. *Postharvest Biology and Technology* 9: 247–264.

Mayer J. 1987. Polyphenol oxidases in plants. Recent progress. *Phytochemsitry* 26: 11–20.

McGuire R.G., Hallman G.J. 1995. Coating guavas with cellulose-based or carnauba-based emulsions interferes with postharvest ripening. *Hortscience* 30: 294–295.

Mei Y., Zhao Y., Yang J., Furr H.C. 2002. Using edible coating to enhance nutritional and sensory qualities of baby carrots. *Journal of Food Science* 67: 1964–1968.

Moleyar V., Narasimham P. 1992. Antibacterial activity of essential components. *International Journal of Food Microbiology* 16: 337–342.

Morin F., Rigault C., Hartmann C. 1985. Conséquences d'un séjour au froid sur le métabolisme de l'éthylène au cours de la maturation de la poire Passe-Crassane après récolte. *Physiologie Végétale* 23: 353–359.

Mujica Paz H., Gontard N. 1997. Oxygen and carbon dioxide permeability of wheat gluten film: effect of relative humidity and temperature. *Journal of Agricultural and Food Chemistry* 45: 4101–4105.

Mujica Paz H., Guillard V., Reynes M., Gontard N. 2005. Ethylene permeability of wheat gluten film as a function of temperature and relative humidity. *Journal of Membrane Science* 256: 108–115.

Nguyenthe C., Carlin F. 1994. The microbiology of minimally processed fresh fruits and vegetables. *Critical Reviews in Food Science and Nutrition* 34: 371–401.

Olabarrieta I., Gallstedt M., Ispizua I., Sarasua J.R., Hedenqvist M.S. 2006. Properties of aged montmorillionite-wheat gluten composite films. *Journal of Agricultural and Food Chemistry* 54: 1283–1288.

Panizzi L., Flamini G., Cioni P.L., Morelli I. 1993. Composition and antimicrobial properties of essential oils of 4 Mediterranean Lamiaceae. *Journal of Ethnopharmacology* 39: 167–170.

Park H.J., Chinnan M.S., Shewfelt R.L. 1994. Edible coating effects on storage life and quality of tomatoes. *Journal of Food Science* 59: 568–570.

Paull R.E. 1994. Tropical fruit physiology and storage potential. In: *Postharvest Handling of Tropical Fruits*. Champ B.R., Highley E., Johnson G.I. (Eds). Canberra, Australia: ACIAR Proceedings, pp. 198–204.

Paull R.E. 1999. Effect of temperature and relative humidity on fresh commodity quality. *Postharvest Biology and Technology* 15: 263–277.

Pech J.C., Balague C., Latche A., Bouzayen M. 1994. Postharvest physiology of climacteric fruits: recent developments in the biosynthesis and action of ethylene. *Sciences des aliments* 14: 3–15.

Rai D.R., Paul S. 2008. Water vapour and headspace temperature predictions for button mushrooms (*Agaricus bisporus*) under modified atmosphere packaging. *Journal of Food Science and Technology—Mysore* 45: 218–222.

Rakotonirainy A.M., Wang Q., Padua G.W. 2001. Evaluation of zein films as modified atmosphere packaging for fresh broccoli. *Journal of Food Science* 66: 1108–1111.

Rhim J.W., Lee J.H., Hong S.I. 2006. Water resistance and mechanical properties of biopolymer (alginate and soy protein) coated paperboards. *LWT—Food Science and Technology* 39: 806–813.

Robertson G.L. 2006. Packaging of horticultural products. In: *Food Packaging Principles and Practice*. Boca Raton, Florida: CRC Press, pp. 361–385.

Rothan C., Duret S., Chevalier C., Raymond P. 1997. Suppression of ripening-associated gene expression in tomato fruits subjected to a high CO_2 concentration. *Plant Physiology* 114: 255–263.

Rothan C., Nicolas J. 1994. High CO_2 levels reduce ethylene production in kiwi fruits. *Plant Physiology* 92: 1–8.

Roy S., Anantheswaran R.C., Beelman R.B. 1996. Modified atmosphere and modified humidity packaging of fresh mushrooms. *Journal of Food Science* 61: 391–397.

Salvador M.L., Jaime P., Oria R. 2002. Modeling of O_2 and CO_2 exchange dynamics in modified atmosphere packaging of Burlat cherries. *Journal of Food Science* 67: 231–235.

Sikkema J., de Bont J.A.M., Poolman B. 1995. Mechanisms of membrane toxicity of hydrocarbons. *Microbiological Review* 59: 201–222.

Shewfelt R.L. 1999. What is quality? *Postharvest Biology and Technology* 15: 197–200.

Sivropoulou A., Papanikolaou E., Nikolaou C., Kokkini C., Lanaras T., Arsenakis M. 1996. Antimicrobial and cytotoxic activities of *Origanum* essential oils. *Journal of Agricultural and Food Chemistry* 44: 1202–1205.

Soda I., Hasegawa T., Susuki T., Ogura N. 1987. Changes in polyuronides in kiwi-fruit during ripening. *Agricultural Biology and Chemistry* 2: 581–582.

Southon S., Faulks R. 2002. Health benefits of increased fruit and vegetable consumption. In: *Fruit and Vegetable Processing: Improving Quality*. Jongen W. (Ed). Cambridge, England: CRC Press, pp. 5–22.

Srinivasa P.C., Prashanth K.V.H., Susheelamma N.S., Ravi R., Tharanathan R.N. 2006. Storage studies of tomato and bell pepper using eco-friendly films. *Journal of the Science of Food and Agriculture* 86: 1216–1224.

Takeuchi K., Frank J.F. 2000. Penetration of *Escherichia coli* O157: H7 into lettuce tissues as affected by innoculum size and temperature and the effect of chlorine treatment on cell viability. *Journal of Food Protection* 63: 434–440.

Tanada-Palmu P.S., Grosso C.R.F. 2005. Effect of edible wheat gluten-based films and coatings on refrigerated strawberry (*Fragaria ananassa*) quality. *Postharvest Biology and Technology* 36: 199–208.

Tasdelen O., Bayindirli L. 1998. Controlled atmosphere storage and edible coating effects on storage life and quality of tomatoes. *Journal of Food Processing and Preservation* 22: 303–320.

Tewari G., Jayas D.S., Jeremiah L.E., Holley R.A. 2002. Absorption kinetics of oxygen scavengers. *International Journal of Food Science and Technology* 37: 209–217.

Togrul H., Arslan N. 2004. Carboxymethyl cellulose from sugar beet pulp cellulose as a hydrophilic polymer in coating of mandarin. *Journal of Food Engineering* 62: 271–279.

Togrul H., Arslan, N. 2005. Carboxymethyl cellulose from sugar beet pulp cellulose as a hydrophilic polymer in coating of apples. *Journal of Food Science and Technology—Mysore* 42: 139–144.

Tsao R., Zhou T. 2000. Antifungal activity of monoterpenoids against postharvest pathogens *Botrytis cinerea* and *Monilinia fructicola*. *Journal of Essential Oil Research* 12: 113–121.

Tunc S., Angellier H., Cahyana Y., Chalier P., Gontard N., Gastaldi E. 2007. Functional properties of wheat gluten/montmorillonite films processed by casting. *Journal of Membrane Science* 289: 159–168.

Utto W., Mawson A.J., Bronlund J.E. 2008. Hexanal reduces infection of tomatoes by *Botrytis cinerea* whilst maintaining quality. *Postharvest Biology and Technology* 47: 434–437.

Valero D., Valverde J.M., Martinez-Romero A., Guillen F., Castillo S., Serrano M. 2006. The combination of modified atmosphere packaging with eugenol or thymol to maintain quality, safety and functional properties of table grapes. *Postharvest Biology and Technology* 41: 317–327.

Valverde J.M., Guillen F., Martinez-Romero D., Castillo S., Serrano M., Valero D. 2005. Improvement of table grapes quality and safety by the combination of modified atmosphere packaging (MAP) and eugenol, menthol, or thymol. *Journal of Agricultural and Food Chemistry* 53: 7458–7464.

Varoquaux P., Ozdemir I. 2005. Packaging and produce degradation. In: *Produce Degradation: Pathways and Prevention*. Lamikanra O., Imam S.H., Ukuku D.O. (Eds). Boca Raton, Florida: CRC Press, pp. 117–153.

Wang S.F., Shen L., Tong Y.J., Chen L., Phang I.Y., Lim P.Q., Liu T.X. 2005. Biopolymer chitosan/montmorillonite nanocomposites: preparation and characterization. *Polymer Degradation and Stability* 90: 123–131.

Warriner K. 2005. Pathogens in vegetables. In: *Improving the Safety of Fresh Fruits and Vegetables*. Jongen W. (Ed). Cambridge, England: CRC Press, pp. 3–43.

Watada A.E., Qi L. 1999. Quality of fresh-cut produce. *Postharvest Biology and Technology* 15: 201–205.

Weber F.J., de Bont J.A.M. 1996. Adaptation mechanisms of microorganisms to the toxic effects of organic solvents on membranes. *Biochimica and Biophysica Acta* 1286: 225–245.

Yam K.L., Lee D.S. 1995. Design of modified atmosphere packaging for fresh produce. In: *Active food packaging*. Rooney M.L. (Ed). Glasgow, Scotland: Chapman and Hall, pp. 55–73.

Yaman O., Bayoindirli L. 2002. Effects of an edible coating and cold storage on shelf-life and quality of cherries. *LWT—Food Science and Technology* 35: 146–150.

Zagory D., Kader A.A. 1988. Modified atmosphere packaging of fresh produce. *Food Technology* 42(9): 70–77.

Zhang M., Zhan Z.G., Wang S.J., Tang J.M. 2008. Extending the shelf-life of asparagus spears with a compressed mix of argon and xenon gases. *LWT—Food Science and Technology* 41: 686–691.

Zhao Y. 2005. Pathogens in fruits. In: *Improving the Safety of Fresh Fruits and Vegetables*. Jongen W. (Ed). Cambridge, England: CRC Press, pp. 44–88.

17 Packaging and the Shelf Life of Vegetable Oils

Luciano Piergiovanni and Sara Limbo
Department of Food Science and Microbiology
University of Milan
Milan, Italy

CONTENTS

17.1 INTRODUCTION

The word oil is derived from the Latin word *oleum*, originally used for *olive oil*, but nowadays it means any of numerous combustible and unctuous substances that are liquid at room temperature (this distinguishes them from fats) and soluble in many organic solvents but not in water. Vegetable oils are derived from plants and chemically are composed of triglycerides and several other minor components, which may be very important for different aspects. This chapter deals exclusively with edible, vegetable oils. Lipids in general, and edible vegetable oils in particular, are very important in foods. They are, however, vulnerable to quality deterioration and must be adequately protected by packaging throughout their commercial life.

Sources of edible vegetable oils are many and varied, and their quality attributes such as nutritional properties, health benefits, lipid composition, odor, and color are very important. A precise knowledge of these attributes and their changes throughout the supply chain is required to guide shelf life testing and estimation. Oils are generally stable microbiologically due to very low moisture content. However, they are subject to important chemical and physical changes. Specific indices of failure (IoFs) of these products will be discussed, as will the role that different packaging materials and packages may have on their shelf lives.

17.1.1 Vegetable Oil Sources and Markets

Vegetable oils are derived from both annual (such as sunflower and soybean) and perennial (such as palm and olive) plants, and oil accumulates both in seeds (such as palm kernels and cottonseeds) and in fruits (such as olive, avocado, palm, and coconut). More detailed information on edible oils sources and their features can be found in specialized books such as those by Gunstone (2002) and Bockisch (1998).

Vegetables oils are produced and commercialized worldwide in very large quantities; global production of the major oilseeds is estimated at over 400 million tonnes per year. The two most widely produced edible oils are soybean and palm, with about 30 million tonnes of each produced annually worldwide; about 8–10 million tonnes of rapeseed and sunflower seed oils are produced annually, and 2.5–5 million tonnes each of oils such as olive, peanut, or cottonseed. Approximate world consumption of different vegetable oils is shown in Figure 17.1.

The greatest amounts of soybean and palm oils are used for cooking, as well as for margarine and soap production and even biofuel generation. The oils mainly or exclusively used for direct consumption and cooking or frying, such as olive, corn, and peanut oils, are produced in relatively

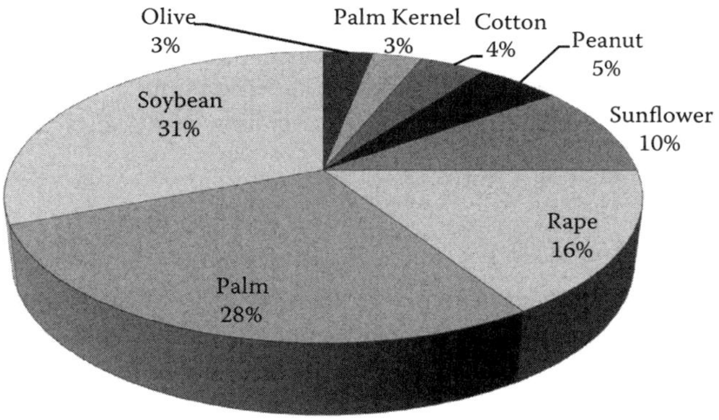

FIGURE 17.1 Approximate world market shares of different vegetable oils. (*Source*: USDA data, reported by SYNGENTA, http://www.soystats.com/2005/page_35.htm).

TABLE 17.1
Simplified Compositions of Various Vegetable Oils

Vegetable Oil	Saturated Fatty Acid (%)	Monounsaturated Fatty Acid (%)	Polyunsaturated Fatty Acid (%)	Vitamin E (mg/100 g)
Coconut oil	85.2	6.6	1.7	66
Cotton seed oil	25.5	21.3	48.1	42.77
Wheat germ oil	18.8	15.9	60.7	136.65
Soybean oil	14.5	23.2	56.5	16.29
Olive oil	14.0	69.7	11.2	5.10
Corn oil	12.7	24.7	57.8	17.24
Sunflower seed oil	11.9	20.2	63.0	49.0

small quantities commercially but are the ones for which package protection and strategies for shelf life extension are probably more crucial.

17.1.2 Vegetable Oil Chemical Composition

Several factors can affect the composition of a vegetable oil; many of them are related to the technological processes used for obtaining the oil but the majority are related to the vegetable oil source. Together with extrinsic factors coming from the packaging and the environment, the oil's compositional aspects primarily influence its shelf life, giving it more or less sensitivity to light, O_2, temperature, enzymes, and all the potential causes of shelf life reduction. The main components of vegetable oils (and of all edible lipids) are triglycerides, that is, esters of glycerin and free fatty acids, normally present as 98–99% of the total mass. The first and greatest variability in oil composition depends on which fatty acids are linked and with which of the three possible hydroxyls of glycerin. The simplified compositions of various vegetable oils are presented in Table 17.1.

The sensitivity to possible oxidation phenomena, that is, the level of unsaturated acids, and the degree of natural resistance to oxidation, that is, the level of antioxidant tocopherols (vitamin E), is very important. The antioxidant effects of several tocopherols, and vitamin E in particular, have been evaluated on the oxidative stability of virgin olive oils (Baldioli et al., 1996), as well as soybean oils (Jung and Min, 2006).

Vitamin E, in spite of its importance, may be considered a minor component of edible vegetable oils. Equally, other lipophilic constituents are present in low or very low concentrations, for example, vitamin K, carotenoids, phytosterols, hydrocarbons, polyphenols, alcohols, pigments, free fatty acids, flavoring, and aromatic substances. These are not directly related to oil stability over time, as is the case with unsaturated glycerides or antioxidant tocopherols, but they are important for their nutritional implications and sensory appreciation and, therefore, definitely pertinent to shelf life studies.

17.1.3 Vegetable Oil Processing

When fruits or seeds are separated from plants, the specificity of the cultivar, the degree of maturity, and climatic effects have already affected oil composition and its sensitivity to aging; however, these intrinsic aspects of the future product's shelf life are not definitely established yet. Other factors related to the handling operations and processing can still strongly influence oil stability.

Possible crushing or bruising of the vegetable sources during delivery to the extraction plants can directly modify the chemical composition and increase the product sensitivity to microbial attacks and enzymic degradation (Loew, 1973; Pritchard, 1983). Generally speaking, all possible measures that can reduce or avoid mechanical damage to seeds or fruits along the supply chain must be put into practice as a guarantee of longer shelf lives of the extracted oils.

The main technological problem in producing oil from a vegetable source, both fruits and seeds, is lipid extraction from the solid and heterogeneous material. Triglycerides reside in specialized locations in seeds and fruits (mainly in cotyledons and endosperm cells in seeds and in mesocarp vacuoles in fruits) and must be extracted in the most selective and safest way. The first unit operation, which follows preliminary cleaning steps, is always size reduction; that is, by crushing or milling the raw material, the oil extraction is made easier, but at the same time, the sensitive lipids are obviously exposed to light, heat, and O_2 and thus become susceptible to oxidation during their processing. After size reduction, the processes differ according to the vegetable source, the tradition, and the final results expected.

The traditional way of extracting oils (still largely used for olives) is physical extraction by pressure; an innovative form of physical extraction is the use of centrifuges, which have the great advantage, compared with the traditional way, of quickly separating the oil from the aqueous phase, thereby avoiding potential damage to the quality and stability of the oil.

Chemical extraction, which is faster, less expensive, and achieves higher yields, is the modern way of processing vegetable sources, particularly seeds. Hexane or similar petroleum-derived solvents are used. Chemical extraction makes it possible to recover about 99% of the oil contained in the seeds and avoids the overheating of the oil and meal that often occurs with mechanical extraction.

The crude oils must be refined before they can be considered edible. Filtration removes contaminants that could contribute to microbial, enzymic, and chemical deterioration. Distillation at low temperature is used to eliminate residual solvents. Winterizing (filtering the oil at near-freezing temperatures) is used in the preparation of some oils (particularly salad oils) to remove triglycerides that might cause them to become turbid. Some oils may be partially hydrogenated to produce cooking oils or various ingredient oils. Light hydrogenation makes the oil more resistant to rancidity, particularly those oils very rich in polyunsaturated triglycerides. Bleaching is one more possible operation in the refining process, particularly of seeds oils; it removes coloring matter and other minor components that are adsorbed on inorganic adsorbents such as montmorillonite clay or activated carbon. The last operation is the deodorizing step, which removes unwanted flavors and odors.

17.2 VEGETABLE OIL QUALITY ATTRIBUTES

For vegetable oils the expected final uses and destinations drive the definition of quality attributes. Vegetable oils come from very different raw materials and can be used as foods, as ingredients and for cooking. For these reasons their quality attributes are numerous and very different. A short overview of the quality attributes of vegetable oils will be given under three headings: sensory, nutritional, and technological characteristics.

17.2.1 Sensory Characteristics

17.2.1.1 Color

Color plays an important role in specifications for commercial oils, even if the expectations can be very different. For seed oils generally, a clear, almost colorless product is desired, whereas for olive and other common plant oils, more intense shades are well accepted. The color is deeply affected by the vegetable source (e.g., the fruit maturity) as well as by the extraction and refining processes, which can, in general, reduce the coloring substances originally present.

17.2.1.2 Odor

Volatiles in oils come straight from the vegetable source used, but they are also a consequence of the process applied to produce the oil. They are responsible for the odor of the oil that can be a positive and typical attribute of the product, but may also indicate a negative change in the oil, frequently related to oxidation and rancidity development. There are many different substances that are volatile and have sensory effects; some of them have extremely low perception thresholds in the range of ppb or even less, for example, hexanal, one of the main final products of the autoxidation of unsaturated fatty acids.

17.2.1.3 Flavor

Taste and flavor are sensory attributes particularly relevant for salad or ingredient oils (e.g., for mayonnaise production or fish canning). Criteria for the organoleptic assessment of virgin oils have been available in Europe since 1991 (EEC, 1991), making virgin oil the first food item in Europe for which a sensory panel analysis is mandatory for legal compliance.

17.2.2 Nutritional Characteristics

17.2.2.1 Polyunsaturated Fatty Acids

The polyunsaturated fatty acid (PUFA) content of oils is that fraction of fatty acids with more than one double bond within the molecule. Although general agreement on the role of PUFAs has not yet been achieved, omega-3 and omega-6 fatty acids appear to reduce the risk of cardiovascular disease and heart attacks. Omega-3 fatty acids can also reduce prostate cancer growth and slow histopathological progression. However, other studies show that the consumption of high amounts of PUFAs slightly increases low density lipoprotein (LDL) levels, and omega-6 fatty acids can contribute to allergies and inflammation, and may increase the risk of developing breast cancer in postmenopausal women and prostate cancer in men. All the supposed negative effects of PUFAs are related to their oxidation products, to free radical accumulation, and to the possible presence of *trans*-isomer fatty acids.

Partially hydrogenated vegetable oils can show a high proportion (up to 45%) of *trans*-fatty acids. This is a peculiarity of PUFAs, which can assume a *cis* or *trans* conformation depending on the geometry of the double bond. *Trans*-fatty acids give the triglycerides properties similar to saturated fatty acids in many respects, including increasing LDL and reducing high density lipoprotein (HDL) cholesterol. Ortega-Garcia et al. (2006) demonstrated that deodorization is the main step that increased the levels of *trans*-fatty acids, with the extent depending on temperature and heating time.

17.2.2.2 Essential Fatty Acids

Essential fatty acids (EFAs) are those fatty acids that cannot be synthesized by any known chemical pathways within humans but must be obtained from the diet. Many vegetable oils are excellent sources of EFAs, particularly soybean, rapeseed, safflower, and sunflower seed oils. Initially (1930–1950) only two PUFAs were defined as essential, namely arachidonic (20:4) and linolenic (18:3) acids; later studies demonstrated that any omega-3 and omega-6 fatty acid can remedy fatty acid deficiency. EFAs are particularly useful in the central nervous system. Nowadays it is recognized that some PUFAs are essential in some pathological states and in particular life phases such as lactation.

17.2.2.3 Fat-Soluble Vitamins

The fat-soluble vitamins A, D, E, and K are essential for the normal growth and development of humans, and their deficiency is associated with specific disease states. Recommended daily dietary allowances are in the range of 10–100 μg, except for vitamin E, of which at least 15 mg is required. They are present to different extents in vegetable oils in active forms or as precursors, with vitamin E always being the most abundant and important for its antioxidant activity.

17.2.2.4 Natural Antioxidants

For a long time the topic of antioxidants in oils has been approached only in terms of possible additives for protecting the product against rancidity. More recently, however, the presence of natural antioxidants in vegetable oils has been emphasized; their presence is both a nutritional attribute relevant to the consumer and a protective aid to the product. Antioxidants are compounds that extend the induction period of oxidation or slow down the oxidation rate.

Besides tocopherols and tocotrienols (both occur in alpha, beta, gamma, and delta forms, each form having slightly different vitamin E activity), other interesting antioxidants are always present in vegetable oils. Polyphenols (e.g., hydroxytyrosol, caffeic, protocatechuic, and syringic acids) are largely present in olive oil (Papadopoulos and Boskou, 1991). Phospholipids, which like lecithin, come from membrane constituents of vegetal cells, are widely diffused. Beta-carotene has been found both in seed and fruit oils. Lignans, which are polyphenols such as pinoresinol, podophyllotoxin, and staganacin, have been found in several oils coming from seed oils, particularly flax, sesame, and soybean. Among antioxidants, tocopherols are the most important group in various vegetable oils.

17.2.3 Technological Characteristics

There are few technological characteristics that are important for the possible use of oils as ingredients, or for cooking or frying. These attributes rarely change significantly unless very long or very bad storage conditions are assumed.

17.2.3.1 Heat Stability

The ability to withstand high temperatures such as those encountered in canning, frying, or cooking operations, without thermal decomposition or other negative side-effects, has been termed *heat stability* and is an important attribute of commercial oils. Frying and baking temperatures are often 200°C or higher, and at these temperatures polymers are formed via autoxidative pathways or thermal polymerization, which increases the viscosity and contributes to foam production.

17.2.3.2 Oil Crystallizability

The winterizing operation has been already mentioned as an important step in the preparation of salad oils, conducted in order to prevent the possible crystallization of the heaviest triglycerides that might cause turbidity. Crystallization of oils depends on their triglyceride composition, which determines their melting temperature (Sato, 1988). Crystallization is also a process used by the vegetable oil industry to obtain specific fractions with given properties. At subzero temperatures, a lipid is frequently a mixture of a liquid phase entrapped in a solid phase made of triglyceride crystals (Calligaris et al., 2004); this has consequences for the oxidative stability of oils as explained later.

17.2.3.3 Emulsification Ability

In several food preparations, oils have to be emulsified with an aqueous phase; manufacture of margarine, ice cream, and mayonnaise are typical examples. The approach to emulsification largely depends on oil composition (the fraction of polar lipids, in particular), which, in turn, affects surface energies, viscosities, and other rheological properties that influence the emulsification ability of different oils.

17.3 DETERIORATIVE REACTIONS AND INDICES OF FAILURE FOR VEGETABLE OILS

The stability of a vegetable oil must be regarded as the ability to maintain the original sensory and texture characteristics that are present immediately after manufacture for as long a time as possible despite the ongoing changes in its molecular structure (Kristott, 2000). The main deteriorative reactions that influence the quality of packaged vegetable oils during their shelf life are described in the following text and the IoFs to quantify the extent of deterioration are also defined.

17.3.1 Enzymic Reactions

Enzymes are involved in many different ways in deteriorative reactions of vegetable oils, and the final products of their activity may often be considered IoFs. The action of three main classes of enzymes will be discussed: lipases, lipoxygenases, and polyphenol oxidases.

17.3.1.1 Lipases

The cleavage of fatty acids from the triglyceride molecule in the presence of moisture is a reaction that leads to the formation of undesired acidity and, possibly, unpleasant off-flavors, particularly if short- and medium-chain length fatty acids are released. Lipases have both endogenous and microbiological origins. They can catalyze the hydrolysis of fatty acids at specific positions on triglyceride molecules, leading to the so-called lipolytic rancidity.

A special type of lipolytic rancidity development is the so-called ketonic rancidity. Molds of the genera *Penicillium*, *Aspergillus*, and *Citromyces* can release enzymes called desmolases, which catalyze the production of methyl ketones and alcohols from the liberated fatty acids. The methyl ketones formed by such enzymes have very characteristic sweet and fruity odors that resemble that of perfume (Kristott, 2000).

Because enzymes are usually inactivated at temperatures above 60°C and by the processes of refining and deodorization, lipolytic rancidity development can only occur in vegetable oils that have not been processed at high temperatures, such as the cold-pressed oils. The liberation of free fatty acids from triglycerides during storage makes an oil unpalatable and, therefore, shortens its shelf life. Spontaneous hydrolysis of triglycerides can also be triggered by heat in the presence of moisture, but this type of reaction is not of practical relevance to the stability of edible oils because these are usually stored at or below ambient temperature (Kristott, 2000).

17.3.1.2 Lipoxygenases

Lipoxygenases are nonheme iron-containing enzymes that catalyze the oxygenation of the 1,4-pentadiene sequence of PUFAs to produce their corresponding mainly 9 and 13 isomers, which are unstable and rapidly transformed into a variety of volatile and nonvolatile substances (Salas et al., 1999). The basic stoichiometry of the lipoxygenase oxidation reaction is the same as for autoxidation, but, in common with many enzyme reactions, it is both regiospecific and stereospecific about the substrate. The flavor components formed, such as aldehydes and alcohols, can be directly responsible for off-flavor genesis (Amirante et al., 2006). Virgin olive oil is a singular case because this enzymic pathway is also responsible for the genesis of its genuine and appreciated flavor. In addition, it has been suggested that lipoxygenase plays a role in the oxidation of pigments of the olive fruit (Georgalaki et al., 1998).

17.3.1.3 Polyphenoloxidases

The enzymic oxidation of polyphenols may be an important deteriorative reaction for some vegetable oils, in particular olive oils that have a high polyphenol content. The presence of such enzymic activities results in browning of olive fruits and may influence the oxidative stability of the oil during storage. Georgalaki et al. (1998) found that significant polyphenoloxidase and lipoxygenase catalytic activities were present in virgin olive oil samples, which could not be correlated with the moisture content of the oil samples. Filtration of oil samples resulted, in several cases, in up to a threefold increase in both activities, suggesting that the filtration process removed some inhibitory components.

17.3.2 Oxidative Rancidity

Oxidative stability is one of the most important indicators of the quality of edible oils. Oxidative rancidity is a complex of chemical changes that imply a series of reactions between unsaturated fatty acids or acylglycerols with O_2. The off-flavor compounds released make oil less acceptable or unacceptable to consumers or for industrial use as a food ingredient (Choe and Min, 2006). For this reason, oxidative stability is an important IoF to determine oil quality and shelf life because low molecular weight off-flavor compounds are produced during oxidation. Oxidation of oil also destroys EFAs and produces toxic compounds and oxidized polymers. Therefore, oxidation of oil is very important in terms of palatability, nutritional quality, and toxicity of edible oils. The process is complex because of the influence of multiple factors such as light, heat, composition of fatty acids, enzymes, metals, and antioxidants. Moreover, the type of O_2 influences the mechanism of oxidation of edible oils.

17.3.2.1 Autoxidation Pathway

The autoxidation pathway of oils proceeds through three steps: initiation, propagation, and termination. Heat, UV, and visible light can act as initiators and accelerate the formation of free radicals from fatty acids or acylglycerols. In the propagation step, the free radicals can react with triplet O_2 to form a reactive lipid peroxy radical, which can further react with another lipid molecule to generate hydroperoxide and another lipid alkyl radical (Angelo Allen, 1996). In the termination phase, two radicals react to give products that do not sustain the propagation phase; termination also occurs when antioxidants react with free radicals generated during propagation.

The hydroperoxides of unsaturated fatty acids (primary products) are intermediate products that, in the presence of metals or at high temperatures, are readily decomposed to alkoxy radicals that then form aldehydes, ketones, esters, acids, alcohols, and short-chain hydrocarbons. The presence of transition metals such as iron and copper in edible oils is due both to endogenous factors linked to plant metabolism and to exogenous factors such as fruit or seed contamination (Bendini et al., 2006). All the secondary products are much more volatile than the starting fatty acids and are responsible for the development of rancid off-flavors. The time for secondary product formation from the primary oxidation product varies with different oils. Secondary oxidation products are formed immediately after hydroperoxide formation in olive oil and rapeseed oils. However, in sunflower and safflower oils, secondary oxidation products are formed when the concentration of hydroperoxides is appreciable (Choe and Min, 2006).

17.3.2.2 Photo-oxidation Route

The exposure of vegetable oils to light accelerates oxidation through a mechanism called photo-oxidation. This is based on the generation of highly reactive singlet O_2 ($^1O_2$) from atmospheric triplet O_2. Chlorophylls and their degradation products (pheophytins and pheophorbides) act as sensitizers to produce $^1O_2$ in the presence of light and atmospheric O_2 and accelerate the oxidation of oil. Virgin olive oil and rapeseed oil contain chlorophyll at 10 ppm and 5–35 ppm, respectively (Salvador et al., 2001). Chlorophylls are generally removed during oil processing, especially by the bleaching process. Suzuki and Nishioka (1993) found that chlorophyll *a* in canola oil was 1.88 ppm and the concentration decreased to 0.22 ppm after refining.

Storage of virgin olive oil in the dark does not necessarily ensure the stability of the oil under conditions promoting photo-oxidation (Terao and Matsushita, 1977). Thus, the virgin olive oil photo-oxidation rate is not expected to depend on a high level of oleic acid to the extent observed in autoxidation but on the presence of photosensitizers and singlet O_2 quenchers. Although chlorophyll pigments act as photosensitizers, carotenoids are effective inhibitors of photo-oxidation by quenching singlet O_2 and excited triplet states of photosensitizers. The antioxidant activity of carotenoids is related to a light-filtering effect due to the extended conjugation system (Psomiadou and Tsimidou, 2002).

17.3.2.3 Oxidation-Derived Products

The primary oxidation products are (odorless and flavorless) monohydroperoxides that are precursors of unpleasant odors and flavors that diminish the quality of oils. The volatile aldehydes obtained from various unsaturated fatty acid monohydroperoxides and the vinyl ketones are mainly responsible for potent off-flavors because their threshold levels are very low. Other volatile oxidation products such as furan derivatives, vinyl alcohols, ketones, alcohols, alkynes, and short-chain fatty acids also contribute to undesirable flavors to varying extents (Kanavouras et al., 2006).

Morales et al. (1997) found that the volatile compounds identified in virgin olive oil flavor were quite different from those identified in virgin olive oil off-flavor. The explanation could be their different origins, mainly biochemical for flavors and chemical for off-flavors. The main differences that characterize off-flavors are the absence of C6 aldehydes and alcohols (produced from linolenic acid), which contribute to the green flavor of virgin olive oil, the absence of esters contributing to fruity flavor, and the presence of many aldehydes with low odor thresholds contributing to the typical rancid odor of oxidized oils. On the other hand, the amount of hexanal does not allow oxidized olive oils to be distinguished from virgin ones, as this compound can come from lipoxygenase cascade and oxidative pathways. The measurement of nonanal has clearly demonstrated its usefulness, as it does not appear in virgin olive oils but does appear in oxidized olive oils.

17.3.3 Loss of Natural Antioxidants

The presence and importance of natural antioxidants in vegetable oils have been mentioned previously; the reactions leading to loss of tocopherols and polyphenols are the main causes of quality deterioration of oils.

17.3.3.1 Tocopherols

In the case of cold-pressed oils and, in particular, of extra-virgin olive oils (EVOO), α-tocopherol is the major antioxidant, representing about 90% of the total tocopherols. These natural antioxidants, widely present in all vegetable oils, are not stable during oil refining and their content decreases during each processing step. Ortega-Garcia et al. (2006) found that the refining process removed 28.5% of the tocopherols of safflower oil. The major losses occur during deodorization where the tocopherols are removed with the stripping steam.

17.3.3.2 Polyphenols

Polyphenols in vegetable oils are always a complex mixture of compounds. Olive oil is particularly rich in these natural antioxidants, with aglycones derived from secoiridoid compounds present in olives being the most abundant phenolic compounds in virgin olive oil. Secoiridoid derivatives play an important role in oil stability and extend the shelf life of olive oil (Montedoro et al., 1992; Owen et al., 2000).

The presence of antioxidants in vegetable oils is also an important factor in the stabilization of free fatty acids. The higher oxidative stability of virgin olive oil compared to that of other vegetable oils is due to both the low PUFA content of the triacylglycerols and the level of natural phenolic components with antioxidant activity. Phenolic compounds in vegetable oils are strong free radical scavengers because they are able to donate a hydrogen atom to the lipid radical formed during the propagation phase of lipid oxidation. Similarly, polyphenols are effective stabilizers of α-tocopherol during olive oil heating, thus contributing to the nutritional value of cooked foods.

17.3.4 Oil Crystallization

The partial crystallization of triglycerides that causes a cloudy aspect to the oil has been already mentioned. This is a reversible defect that can be observed during winter storage and which progressively disappears with increasing temperature. Kristott (2000) reported that crystallization of oils and fats and subsequent transition of crystal type could promote oxidation reactions. In fact, the liquid phase surrounding fat crystals is expected to contain a high proportion of unsaturated fatty acids in addition to the concentrated dissolved O_2. Calligaris et al. (2006) highlighted that the relative concentration of reactants (unsaturated fatty acids and polyphenols) in the liquid phase surrounding fat crystals is, besides temperature, the main variable affecting the oxidation rate in partially crystallized EVOO.

17.3.5 Absorption of Flavor and Migration of Substances from Packaging

Scalping and migration are interactions that can occur between oil and packaging material and affect the quality and safety of oils (Kanavouras et al., 2006). In order to avoid alteration of the flavor profile or to reduce chemical contamination of oil during storage, knowledge of oil–package–environment interactions is required. Some mass transfer can occur between the environment and the food and some between the food and the packaging material.

17.3.5.1 Scalping

In recent years, plastics have been increasingly employed to package vegetable oils, due to their low weight, ease of handling, and competitive cost (Gambacorta et al., 2004; Kaya et al., 1993; Maloba et al., 1996). Plastic packaging materials can absorb different compounds from the food, a phenomenon called scalping (sorption). In particular, flavor scalping is a term used to describe the loss of quality of a packaged food due either to its volatile flavors being absorbed by the package or the food absorbing undesirable flavors from the packaging material. Sorption of food aromas, particularly by plastic packaging materials, is usually perceived as a major factor contributing to the quality alteration of most foods during storage. Also, nonvolatile compounds may be absorbed by packaging materials, but in this case, they primarily affect the packaging itself—its characteristics such as permeability and mechanical properties.

Kanavouras et al. (2004a, 2004b) studied the role of plastic materials as potential flavor sorbents for olive oil. Flavors dissolved in the oil were readily absorbed by low density polyethylene (LDPE), with the flavor concentration and storage temperature affecting the absorption of aroma compounds. Sorption of oil into packaging materials, especially olefins, causes swelling of the polymer, which in turn increases migration. The sorption of fatty acids increases with increased chain length due to increased van der Waals bonds between polymer and fatty acid. The sorption of olive oil flavor compounds by polymeric plastic materials during storage can result in a considerable decrease in oil quality due to losses of desirable organoleptic characteristics.

Several investigations have shown that considerable amounts of aroma compounds can be absorbed by plastic packaging materials, resulting in loss of aroma intensity or unbalanced flavor profile (e.g., van Willige et al., 2000a, 2000b). However, sorption may also indirectly affect food quality by causing delamination of multilayer packages (Olafsson and Hildingsson, 1995) or by altering the barrier and mechanical properties of plastic packaging materials (Tawfik et al., 1998). Oxygen permeability through the packaging can increase as a result of these interactions, but unfortunately, little information is available in the literature on this subject. Van Willige et al. (2002) studied the influence of flavor and off-flavor absorption on the O_2 permeability of LDPE, PP, PC, and poly(ethylene terephthalate) (PET). Absorption of some volatile substances (limonene, decanal, 2-nonanone, and hexyl-acetate) increased O_2 permeability of PP and LDPE; O_2 permeability of PET was not influenced by the presence of flavor compounds, meaning that PET remained a good O_2 barrier.

17.3.5.2 Migration

Migration is an important safety aspect to be considered when selecting food packaging materials. Plastic additives and residual monomers or oligomers are not chemically bound to the polymer molecules and can, therefore, move freely within the polymer matrix. Consequently, at the interface between the packaging material and food they can dissolve in the food product and thus adversely affect the flavor and acceptability of the food. The chemical nature of the packaging material has a notable influence on oil quality. A review by Kanavouras et al. (2006) suggested that edible oils should not be stored in PVC plastic materials as vinyl chloride monomer (VCM) and plasticizers can migrate into fatty foods, leading to the contamination of the oils. Castle et al. (1991) investigated the preferential migration by the size of oligomers from PVC into olive oil. The smaller oligomers migrated 90-fold more readily than the bulk of the plasticizer. PET is one of the most inert plastics and in recent years packing of oil into PET bottles has increased. Nevertheless, PET monomers, oligomers (cyclic trimers, pentamers, heptamers), plasticizers, colorants, stabilizers, and different additives used for flexibility purposes (adipic and phthalic acid esters) as well as degradation products are all prone to migration.

The migration of acetaldehyde from PET bottles is a major problem, as its presence may affect the organoleptic properties of oil (Tsimis and Karakasides, 2002). Migrating PET oligomers have been measured and the cyclic trimer was the most dominant. Little data concerning the influence of migration processes on the quality of olive oil are available; in general, PET bottles are usually considered suitable to contain not only seed oil but also olive oil (Cecchi et al., 2006; Kaya et al., 1993).

17.3.6 Indices of Failure of Vegetable Oils

Once the most relevant quality attributes and major deteriorative modes of vegetable oils are known, the subsequent task is to identify the IoFs that indicate that the product is no longer acceptable. To define shelf life, it is also necessary to establish a critical limit for each IoF beyond which the food product is no longer acceptable, and decide which one is the most relevant or reaches the critical limit first. The critical limit may vary from country to country as people from different locations have different preferences (Lee et al., 2008a). The presentation of the main IoFs for vegetable oils follows.

17.3.6.1 Triglyceride Hydrolysis

The level of free acidity during storage of oils measures the liberation of fatty acids as a result of hydrolytic rancidity development. Acidity values are used for classifying different categories of olive oil although, according to Kiritsakis et al. (2002), acidity is not the best criterion for evaluating olive oil quality, as one oil with relatively high acidity may have a good aroma while another one with low acidity may not. For EVOO the maximum acidity is 1%, increasing as the quality category of the oil changes. For seeds oils, the critical value of acidity is much less and in several countries the legal level is 0.5%.

17.3.6.2 Enzymic and Chemical Oxidation

Because lipid oxidation is one of the most undesirable yet common deteriorative reactions, several IoFs related to this phenomenon are known and regularly used. The peroxide value (PV) determines the quantity of hydroperoxides that are formed during the early stages of oxidative rancidity development. As hydroperoxides are decomposed in subsequent oxidation reactions, a low PV in an oil does not necessarily mean that it is fresh and IoFs related to the amount of secondary oxidation products are also necessary. The anisidine value (AnV) and thiobarbituric acid (TBA) test are common indices for detecting advanced lipid oxidation (Kristott, 2000).

The advance of oxidation processes in refined vegetable oils is indicated by the increase in total volatiles and the concentration of some specific volatile compounds such as hexanal. Morales et al. (1997) investigated the volatile components during the thermo-oxidation process and proposed the hexanal:nonanal ratio as an indicator of the level of oxidation of olive oil. In other words, as the amount of hexanal diminished in the olive oil headspace, the amount of nonanal increased and the oil moved toward higher oxidation levels and consequently lower acceptability. For quality control purposes, the presence of rancid off-flavors is also measured through sensory analysis. Basic flavor compounds, among other oxidation indicators, were investigated by Kanavouras et al. (2004a) as a useful tool to evaluate the tolerance of oil to oxidation or its oxidation level at a certain point, and to extrapolate the results for predicting the shelf life of the product under various conditions.

Another characteristics affected during oil oxidation is color. The color of olive oil, for example, is due to the solubilization of chlorophyll and carotenoid pigments present in the source fruit. The influence of extraction technology and the role of oxidative spoilage on chromatic characteristics of edible oils have been studied by Ranalli et al. (1994, 1996).

As oxidative deterioration proceeds, the level of important antioxidants such as tocopherols and polyphenols inevitably decreases. Therefore, the concentrations of natural antioxidants could theoretically be a valuable IoF to reveal early deterioration, but so far no significant literature is available on this subject.

17.3.6.3 Oil–Package Interactions

Both scalping and migration are time-related phenomena. In many cases legal limits to migration of packaging constituents already exist and analytical procedures for determining such potential migrants are known; diffusion equations to model the interactions are now readily available (CRL-FCM, 2008). Scientific knowledge on migration from packaging materials could provide some useful markers or IoFs for shelf life studies. For vegetable oils in plastic packaging, this seems to be a realistic perspective due to the strong chemical affinity that exists between the synthetic polymers used and the oils, making migration phenomena quite rapid and important.

17.4 HOW PACKAGING MIGHT IMPACT INDICES OF FAILURE

Many different kinds of packaging are used for vegetable oils: tinplate cans, glass bottles, PET, or HDPE plastic bottles, and paper-based cartons are most common. The selection of the kind of package to be used is generally done on the basis of marketing and economic criteria; however, proper packaging will in many cases provide conditions to assure adequate shelf life for distribution and marketing (Kanavouras et al., 2006). Even though oils are quite stable products, physicochemical characteristics of packaging materials may significantly affect oil quality during their shelf life. Furthermore, besides the specific properties of the materials, the packaging geometry (Del Nobile et al., 2003b) and the techniques of filling and closing the containers (de Oliveira et al., 2001) may also be very important.

17.4.1 Oxygen Permeability and Light Transmission

As far as physicochemical characteristics of the materials are concerned, the O_2 permeability and the UV/visible light transmission of the packaging walls are the major ones, due to the oxidative sensitivity of vegetable oils. Oxygen permeability is a property of plastic materials only, whereas light transmission is important both for glass and plastics. Different polymers may have very different permeabilities; for instance, the ratio of the highest O_2 permeability coefficient to the lowest can be greater than 5000 (Lee et al., 2008b). Light transparency is a less investigated property that depends mainly on the chemical nature of materials, their crystallinity, the presence of specific additives, color, and thickness. Thickness, however, is less relevant than molecular structure in comparisons between PET and glass. Only 25 μm of PET makes a better filter than 2.2 mm of glass (Lee et al., 2008c). El-Shattory et al. (1997) reported that the flavor of cottonseed and palm oils was preserved better in metal cans than in white plastic bottles, that is, in a container totally impermeable to gas and light. In recent times the effects of light on food stability and the role of packaging in protecting against light have been investigated (Bosset et al., 1994; Piergiovanni and Limbo, 2004; Wold et al., 2005). Many additives are available today for both plastics and glass to reduce UV transmission. It is also essential to take into account the light-emitting spectrum, which can be quite different in different circumstances as Torri et al. (2007) showed in a survey of the lighting conditions in large-scale food retail stores.

17.4.2 Packaging Geometry

The geometry of packaging can act in different ways in providing protection to the product. As discussed in Chapter 1, the size and shape of plastic packages can affect the ratio between permeable surface area and product volume, modulating the O_2 ingress per unit volume of product. For both plastic and glass or metal packages, shape and size can influence the headspace and, consequently, the amount of O_2 available. Del Nobile et al. (2003b) proposed a two-dimensional mathematical model to predict the time course of hydroperoxides and O_2 concentration profiles inside bottled virgin olive oil during storage. They showed that the quality decay kinetics of oil greatly depended on container geometry, material used, and initial value of the O_2 partial pressure in the bottle headspace (Figure 17.2).

17.4.3 Packaging Inertness

As already mentioned, oil–package interactions can affect product shelf life, reducing nutritional value and stability (by scalping) or increasing the level of chemical contamination (by migration). Therefore, the selection of packaging materials may also be done on the basis of their interaction with oils. Generally speaking, glass is the most inert material, followed by metals and plastics. However, large differences exist among the different materials, and consideration must be given to closures and their liners (often made of plastics even for metal and glass containers) when assessing the global inertness of a package.

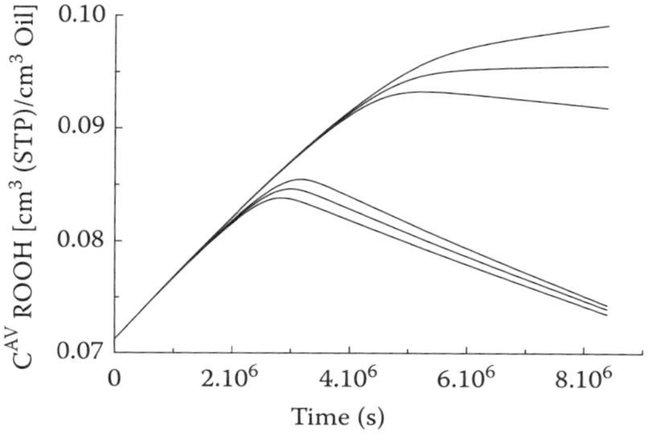

FIGURE 17.2 Predicted average hydroperoxide concentration versus time of virgin olive oil bottled in different PET containers. (From Del Nobile M.A., Bove S., La Notte E., Sacchi R. 2003b. Influence of packaging geometry and material properties on the oxidation kinetic of bottled virgin olive oil. *Journal of Food Engineering* 57: 189–197, with permission.)

17.4.4 Filling and Closing Technologies

The filling and capping steps are very relevant in the process of oil packaging, affecting the quality perceived by consumers. In order to reduce the O_2 residual inside the bottles, the oil is generally stripped with gaseous N_2, to lower the initial level of residual O_2 to below 0.5 ppm. Gaseous N_2 can be pressurized by injecting liquid N_2 into the headspace prior to closing (de Oliveira et al., 2001). In order to reduce O_2 ingress during shelf life, the effectiveness of the closures is also very important. The efficiency of closures is related to several factors: material used, design, and liner adopted. All these factors must guarantee, at the same time, hermeticity, easy opening, and the possibility of reclosing. As these goals are sometimes contradictory, efforts to develop new devices is ongoing, including the use of active and intelligent packaging.

17.5 SHELF LIFE OF VEGETABLE OILS IN DIFFERENT PACKAGES

As described previously, O_2, light, temperature, and the presence of heavy metals greatly influence the quality of oil. Therefore, all these factors must be taken into account when selecting packaging materials. The majority of shelf life research has focused on comparative evaluations using key IoFs of edible oil stored in different packaging materials. The conclusion is that oil stability can be enhanced by selection of a suitable package. Little information is available about light exposure of samples and the characteristics of the packaging materials in the UV and visible range. Moreover, most shelf life studies are conducted at constant temperature, relative humidity (RH), or lighting conditions; it would be of real interest to evaluate the shelf life of edible oils simulating the fluctuating temperature (or lighting conditions) during the whole distribution chain, that is, from producer to consumer. Ramirez et al. (2001), for instance, clearly showed the very negative effects on sunflower oil of exposing bottles to light for half a day; the shelf life was halved in comparison with bottles stored in the dark as shown in Figure 17.3A and 17.3B. Finally, packages used in end-point studies should have the same size and shape as the original containers for product distribution. This means that the contribution of the closure system or the volume of the headspace should always to be taken into account in the quality evolution of the oil.

A limited number of mathematical models have been presented in the literature concerning the role of packaging in influencing oil shelf life. In the majority of these studies, the main focus was to predict the shelf life of packaged oil in new package designs after taking into consideration the role of O_2, the geometrical and structural characteristics of the plastic container, and the volume of oil. Del Nobile et al. (2003a, 2003b) developed a mathematical model to assess the effectiveness of plastic containers to prolong the shelf life of virgin olive oil assuming average hydroperoxide concentration as a measure of oil quality. The mathematical model combined the mass balance equations of O_2 with those describing the rate of hydroperoxide formation and decomposition; it was validated by monitoring oil in glass and PET containers at 40°C. Couteliers and Kanavouras (2006) proposed a simple model based on the evolution of hexanal in order to estimate the reaction constants under various storage conditions of light, temperature, and O_2 availability. A mathematical predictive model was introduced to describe the mass transport from and to the oil phases through various packaging materials.

17.5.1 Metal Packaging

Metal containers for vegetable oils are manufactured using tinplate or aluminum and in two different ways:

a. Bottom and top closed at the metal fabrication plant with an orifice on the top for subsequent filling and capping.
b. Bottom seamed onto the body blank and the top closed only after filling at the bottling plant.

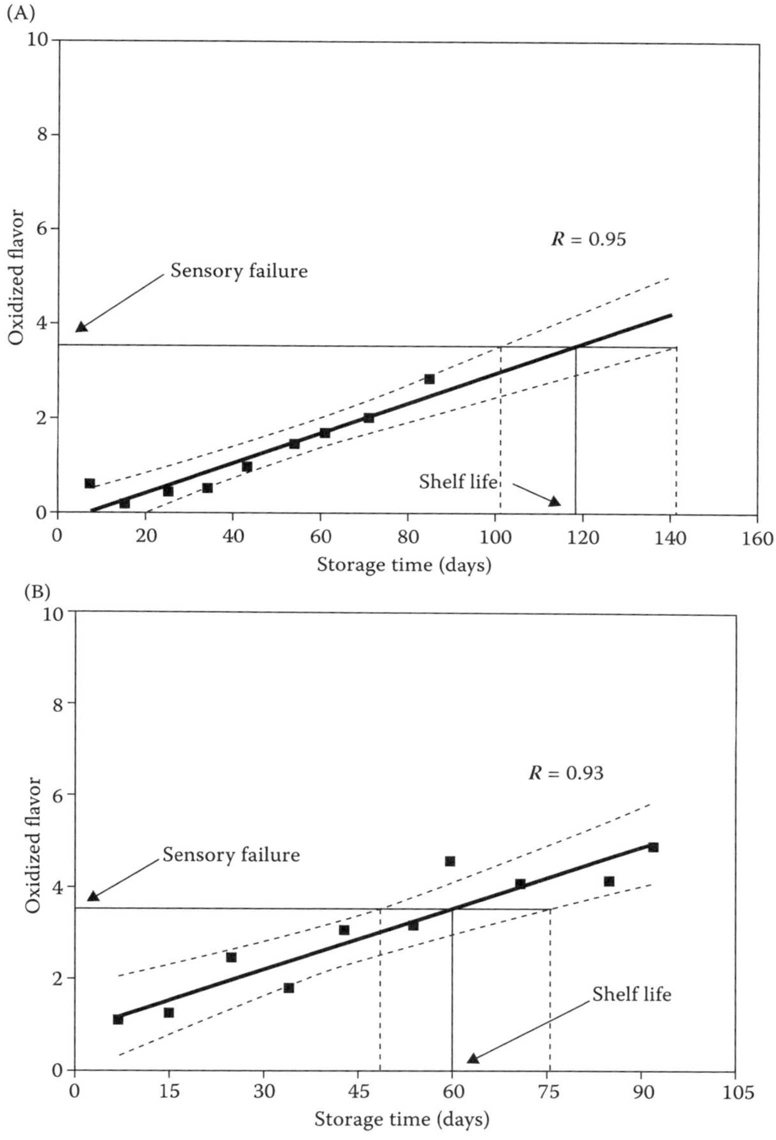

FIGURE 17.3 Estimated shelf lives for sunflower oil stored at 45°C kept in the dark (A) and exposed to light for 12 hours a day (B). (From Ramirez G., Hough G., Contarini A. 2001. Influence of temperature and light exposure on sensory shelf-life of a commercial sunflower oil. *Journal of Food Quality* 24: 195–204, with permission.)

Tinplate containers have been used for a long time for oil packaging and are still well appreciated because of their many advantages (Tsimis and Karakasides, 2002). They provide total protection against light, O_2, water vapor, and microorganisms, and are resistant to several types of mechanical abuses. In addition, the inside of the container is protected with food-approved special enamels (lacquers) that protect the metal from the corrosiveness of the product. Edible oils are generally packed in tinplate containers of different capacities, typically from 500 g to 15 kg. Grover (1982) studied the shelf life of several vegetable oils (mustard oil, groundnut oil, sesame oil, etc.) when packaged in new and reused tinplate containers. The quality of oil packed in new containers did not change during 1 year, whereas the quality of oils packed in containers reused several times remained intact for only 4–5 months. The reuse of containers, in fact, increases corrosion of the tin coating and the exposed steel base readily reacts with the free fatty acids in oil, leading to oxidative rancidity and organic tin salts with high toxicity (Tsimis and Karakasides, 2002).

Aluminum is also employed as a packaging material for edible oils as it is light and very resistant to corrosion. In order to increase its mechanical resistance, aluminum alloys with small amounts of Mg, Mn, and Si/Mg are recommended.

All these metallic containers are considered inert against oils, even though trace levels of metal ions such as Fe and Cu are known to have adverse effects on the oxidative stability of olive oil. In fact, transition metals catalyze the decomposition of hydroperoxides, contributing to off-flavor production (Sahan and Basoglu, 2008).

17.5.2 Glass Bottles

Glass containers are widely used for bottling olive oils and virgin olive oils in particular. This is due not only to marketing requirements but also because glass containers prevent the permeation of O_2 molecules into the bottle, slowing down the autoxidation rate of PUFAs. Transparent glass, however, leads to photo-oxidation of olive oil and reduction of its shelf life. The use of colored glass bottles prevents or slows down the oxidation process: green bottles, for instance, protect oil from wavelengths of 300–500 nm (Kiristakis et al., 2002). Rastrelli et al. (2002) found that α-tocopherol represents the first target of EVOO autoxidation stored in half-empty clear bottles after 12 months of storage at room temperature and under diffused lighting, with a reduction greater than 90%; in filled clear bottles, the vitamin E reduction was lower (about 25%). Results concerning the study of the effect of storage on secoiridoid and tocopherol contents and antioxidant activity of monovarietal EVOO showed that, despite antioxidant depletion, oils with high antioxidant content were still excellent after 240 days of storage at 40°C in closed dark bottles. These data led to the conclusion that the beneficial properties of extra-virgin olive oils due to antioxidant activity can be maintained throughout their commercial life if properly packaged and stored (Lavelli et al., 2006). In order to reduce light transmission, aluminum foil can be used to cover glass containers. This practice is used especially for EVOO that has high nutritional and sensory properties.

Metal and glass are the only packaging materials that provide a virtually total barrier to moisture and gases. The word "virtually" is used because such containers require a closure that incorporates other materials such as polymeric sealing compounds in cans and in closures, through which O_2 can easily permeate and promote oxidation (de Oliveira et al., 2001). The shelf life of edible oils packaged in metal containers or nontransparent glass bottles can be considered product-dependent because deteriorative reactions are driven by intrinsic stability of the product, not by environmental factors or packaging (Lee et al., 2008d). In other words, intrinsic stability is dictated by initial quality of the oil, processing conditions, and filling operations.

17.5.3 Plastic Bottles and Containers

Plastic containers are a relatively new means of edible oil packaging due to their comparatively low price and low weight. The polymers most frequently used are PET, HDPE, and PVC. Although they do not provide as long a shelf life as metal containers, they are economical compared to tinplate and therefore suitable for use where a very long shelf life is not required.

PET is one of the most used plastics in food packaging covering a wide range of packaging structures. PET satisfies many important requirements: good aesthetic aspect (brilliance and transparency); suitability for coloring; good mechanical, thermal, and chemical resistance; low production cost; good barrier properties against CO_2; suitability for prolonged storage, easy recyclability, and low weight with respect to glass bottles. The trend toward incorporating modifier compounds into PET packaging resins has grown in order to produce containers with a high degree of clarity, in a wide variety of custom shapes, and free from residual acetaldehyde (Kanavouras et al., 2006).

Also, the incorporation of antioxidant stabilizers in PET increases its application in the food area, particularly for vegetable oil storage.

The influence of PET bottle weight, closure performance, and filling technique on the O_2 content of soya cooking oil was investigated by Coltro et al. (2003). The experiment considered a PET bottle of standard weight (27 g) and two PET bottles of reduced weight (20 g), differently filled (flushed with gaseous N_2 or pressurized with liquid N_2). Considering the bottles without closures, results highlighted that PET bottles with reduced weight had a 20% increase in oxygen transmission rate (OTR). The differences in OTR decreased significantly (less than 14%) if closures were considered during the permeability measurements. Therefore, the closure system of PET containers for vegetable cooking oil is of critical importance to the overall barrier properties of the packaging system. This fact drastically diminishes the importance of the effect of weight reduction on the level of protection provided by the package to the oil.

Cecchi et al. (2006) assessed the role of clear PET bottles and dark-green glass bottles on the quality of EVOOs during 2 months storage. Samples stored in PET bottles developed a pungent and offensive off-flavor compared to samples stored in glass bottles. The PV of samples stored in glass bottles was always lower than those stored in PET bottles. Kucuk and Caner (2005) studied sunflower oil packaged in both PET and glass bottles, both with and without headspace and stored in conditions of light and darkness for 9 months. Oil stored in glass and PET in the dark showed very little oxidation and maintained its original profile for a long period (Table 17.2). De Oliveira et al. (2001) referred to an experiment on soya oil stored in biaxially oriented PVC. In the dark, the shelf life was 360 days at 23°C and 135 days at 35°C; stored under light, shelf life was reduced to less than 60 days at 23°C. The shorter shelf life of the oil in PVC compared to metal and glass packages, when stored at 23°C and 35°C in the presence of light, was attributed to the higher OTR of PVC and the type of closure used.

Kakuda et al. (2008) studied the effect of packaging and light exposure on vitamin A stability in fortified vegetable oil. Three packaging materials (PET, PVC, and HDPE) and exposure to light and dark conditions were variables in the experimental design. The major factors affecting vitamin A stability was exposure to light and the type of packaging material. After 37 days of storage in clear PET bottles, the levels of vitamin A dropped to 17–33% when exposed to light. Samples stored in brown PET bottles and exposed to light retained 72–88% vitamin A. When completely protected from light, vitamin A in the clear and brown PET bottles was very stable (81–91% remained after 6 months). Similar results were obtained with clear PVC, opaque HDPE, and brown HDPE bottles. The clear and opaque bottles showed rapid loss of vitamin A (13–36% remained) when exposed

TABLE 17.2
Effect of Packaging Materials and Storage Time on Mean Peroxide Values

Package	With–Without Headspace (Air)	Dark/Light	Storage Time (Months)				
			0	3	6	9	Mean
PET	Air	Light	0.230	0.800	3.306	14.85	4.79[c]
		Dark	0.230	0.400	0.800	1.250	0.67[d]
	Without air	Light	0.230	0.800	3.000	13.90	4.48[f]
		Dark	0.230	0.400	0.500	1.100	0.55[ad]
Glass	Air	Light	0.230	0.990	4.740	10.60	4.14[h]
		Dark	0.230	0.310	0.430	1.010	0.49[a]
	Without air	Light	0.230	0.776	3.133	8.670	3.20[c]
		Dark	0.230	0.300	0.530	0.770	0.44[a]

Source: From Kukuk M., Caner C. 2005. Effect of packaging materials and storage conditions on sunflower oil quality. *Journal of Food Lipids* 12: 222–231, with permission.

Note: Means with different letters are significantly different ($p < 0.01$); standard error 0.0764.

to light for 49 days, whereas brown HDPE bottles retained 79–88% vitamin A after 3 months. Vegetable oil is an effective carrier for vitamin A and may retain high potency if protected from light or packaged in nontransparent plastic containers.

HDPE is largely used as a packaging material because of its tensile strength and hardness and good chemical resistance. Blow-molded HDPE containers in the form of bottles, jars, and jerry cans are used for packaging edible oils. Coltro et al. (2003) investigated the quality deterioration of soya oil in 1-L plastic bottles made of HDPE, coextruded with a layer containing black pigment, during storage at 23°C. Soya oil stored in a metal container was used as a control. No differences were found between the organoleptic properties of the oil contained in the metal and plastic containers throughout the total storage time investigated (113 days). Over the same period, the chemical quality of the oil remained within the limits of stability.

PVC is a popular packaging material for edible oils in many countries, mainly due to its transparency, adaptability to all types of closures, total compatibility with existing packaging lines, and potential for personalized design features (Kanavouras and Coutelieris, 2005). Mainly driven by issues such as the protection of the environment, PET has been supplanting PVC in the edible-oil market. As with other transparent plastic materials, PVC increases light exposure of the oil, enhancing oxidation. UV absorbers can be added to plastic materials in order to reduce their light transmission. Azeredo et al. (2004) studied the oxidation of refined soybean oil during 6 months storage at 25°C under a constant illumination (1720 lux) as affected by combining different primary antioxidants to oil or PVC resin. In particular, tert-butylhydroquinone (TBHQ), β-carotene, and citric acid were added to the oil while an UV absorber (Tinuvin P) was in the PVC bottles. The results highlighted that, among the compounds used to reduce oxidation rates of soybean oil, TBHQ was the most effective, followed by Tinuvin P. Moreover, the latter solution allowed the protection of oil without modification of oil color.

17.5.4 Multilayer Pouches and Paper-Based Cartons

In recent years, the adoption of multilayer pouches for oil storage has increased due to consumer preference for unit packages. Generally, limited quantities of edible oil are packed in flexible pouches (up to 500 g). Flexible pouches may be manufactured from laminates or multilayered films of different compositions and the pouches may be in the form of a pillow or stand-up pouch. The selection of a laminate or multilayer film is governed primarily by the compatibility of the contact layer, heat sealability, heat seal strength, and shelf life required, together with machinability and physical strength parameters.

Mahadevaiah et al. (1992) studied the storage of double-filtered groundnut oil in different multilayer film pouches. The materials were based on polyolefins, nylon, and PET layers, and after filling, the pouches were stored at 27°C/65% RH and 38°C/90% RH. The groundnut oil kept better in multilayer film pouches consisting of nylon and PET films than in non-nylon-based film pouches under both conditions. Packaging and storage studies of palm oil in seven different flexible pouches were carried out in order to design a suitable package for oil (Narasimhan et al., 2001). Pouches containing 200 g of oil were stored under three conditions representative of different climatic conditions. The quality deterioration was comparatively less in films containing polyamide (PA) as one of the layers, and leakage rates were minimal in films containing ethylene-acrylic acid copolymer (EAA) as the sealant layer.

Recently, new packaging formats have been introduced in the market including bag-in-box systems, lined cartons, and paperboard laminate cartons. In particular, Tetra Brik® cartons have been used in Spain, Brazil, and other countries and are considered more suitable for packaging olive oil (Kiritsakis et al., 2002).

17.5.5 Active Packaging

In order to reduce the diffusion of O_2 into bottled oil, various solutions have been tried; the most popular involve the use of "oxygen scavengers" (OS), which remove O_2 dissolved in the oil and

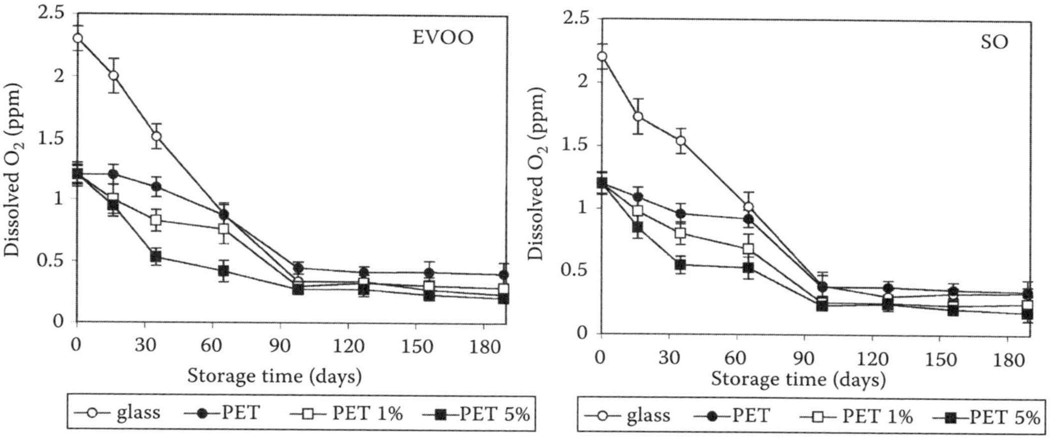

FIGURE 17.4 Evolution of dissolved oxygen (ppm) in extra-virgin olive oil (EVOO) and sunflower oil (SO) bottled in different containers (glass; PET; PET + 1% OS; PET + 5% OS where OS is oxygen scavenger) during shelf life test of 6 months at 25 ± 4°C under 400 lux. (From Sacchi R., Savarese M., Del Regno A., Paduano A., Terminiello R. Ambrosino M.L. 2008. Shelf life of vegetable oils bottled in different scavenging PET containers. *Packaging Technology and Science* 21: 269–277, with permission.)

provide a barrier to O_2 diffusion from the atmosphere. These scavengers can be easily incorporated into the packing material without altering its other properties. Sacchi et al. (2008) studied the oxidation of EVOO and sunflower oil (SO) stored in PET bottles with two different OS concentrations (1% and 5%). The shelf life test was carried out for 6 months at 25°C under a constant illumination of 400 lux. During the first 3 months of storage, the effect of scavengers was evident: oils bottled in PET loaded with 5% of OS showed a dissolved O_2 (DO) content lower than oils bottled in PET with 1% OS and in standard PET. Between 3 and 6 months, the level of DO remained almost constant in all packages, indicating that the O_2 consumed during storage was nearly limited to the initial content in the oil (Figure 17.4).

Maloba et al. (1996) measured the oxidative stability of sunflower oil during storage at two temperatures (23°C and 37°C) in the presence of a novel OS film that contained polyfuryloxirane (PFO). Commercially refined and deodorized sunflower oil was stored in a lighted room in sealed transparent packages containing either PFO film or the antioxidant butylated hydroxytoluene (BHT) at 0.02%. Sunflower oil stored in the presence of the OS film was more stable than oil stored without the film, or than film stored with 0.02% BHT. The PFO film scavenges O_2 through energy transfer sensitization of singlet O_2.

Gambacorta et al. (2004) studied the shelf life of EVOO stored in five different materials: PET, PET including 1% and 3% OS, PET coated with high-barrier resin, and PET coated with high-barrier resin including an O_2 scavenger. Glass was used as a reference material. The packaged oil was stored in the dark at room temperature and at 37°C for 12 months. A significant influence of the package on the quality decay kinetics was found, suggesting that PET bottles coated with a high-barrier resin and those including an OS could replace traditional glass bottles.

REFERENCES

Amirante P., Clodoveo M.L., Dugo G., Leone A., Tamborrino A. 2006. Advanced technology in virgin olive oil production from traditional and de-stoned pastes: influence of the introduction of a heat exchanger on oil quality. *Food Chemistry* 98: 797–805.

Angelo Allen J. St. 1996. Lipid oxidation in foods. *Critical Reviews in Food Science and Nutrition* 36: 175–224.

Azeredo H.M.C., Faria J.A.F., Aparecida M., da Silva A.P. 2004. Minimization of peroxide formation rate in soybean oil by antioxidant combinations. *Food Research International* 37: 689–694.

Baldioli M., Servili M., Perretti G., Montedoro G.F. 1996. Antioxidant activity of tocopherols and phenolic compounds of virgin olive oil. *Journal of the American Oil Chemists' Society* 76: 1589–1593.

Bendini A., Cerretani L., Vecchi S., Carrasco-Pancorbo A., Lercker G. 2006. Protective effects of extra virgin olive oil phenolics on oxidative stability in the presence or absence of copper ions. *Journal of Agricultural and Food Chemistry* 54: 4880–4887.

Bockisch M. (Ed). 1998. *Fats and Oils Handbook.* Urbana, Illinois: AOCS Press.

Bosset J.O., Gallmann P.U., Sieber R. 1994. Influence of light transmittance of packaging materials on the shelf life of milk and dairy products—review. In: *Food Packaging and Preservation.* Mathlouthi M. (Ed). London, England: Blackie Academic & Professional, pp. 222–268.

Calligaris S., Manzocco L., Conte L.S., Nicoli M.C. 2004. Application of a modified Arrhenius equation for the evaluation of oxidation rate of sunflower oil at subzero temperatures. *Journal of Food Science* 69: E361–E366.

Calligaris S., Sovrano S., Manzocco L., Nicoli M.C. 2006. Influence of crystallization on the oxidative stability of extra virgin olive oil. *Journal of Agricultural and Food Chemistry* 54: 529–535.

Castle L., Mercer A.J., Gilbert J. 1991. Migration from plasticized films into foods. 5. Identification of individual species in a polymeric plasticizer and their migration into foods. *Food Additives and Contaminants* 8: 565–576.

Cecchi T., De Marco C., Passamonti P., Pucciarelli F. 2006. Analytical definition of the quality of extra-virgin olive oil stored in polyethylene terephthalate bottles. *Journal of Food Lipids* 13: 251–258.

Choe E., Min D.B. 2006. Mechanisms and factors for edible oil oxidation. *Comprehensive Reviews in Food Science and Technology* 5: 169–186.

Coltro L.B., Padula M., Segantini Saron E., Borghetti J., Penteado Buratin A.E. 2003. Evaluation of a UV absorber added to PET bottles for edible oil packaging. *Packaging Technology and Science* 16: 15–20.

Coutelieris F.A., Kanavouras A. 2006. Experimental and theoretical investigation of packaged olive oil: development of a quality indicator based on mathematical predictions. *Journal of Food Engineering* 73: 85–92.

CRL-FCM. 2008. *Community Reference Laboratory for Food Contact Materials. Public Documents.* http://crl-fcm.jrc.it/. Last accessed 13th May, 2009.

De Oliveira L.M., Sarantopulos C., Bordin M., Nakandakari Y. 2001. Influence of PET bottle weight, closure performance and filling technique on the oxygen content of soya cooking oil. *Packaging Technology and Science* 14: 203–208.

Del Nobile M.A., Ambrosino M.L., Sacchi R., Masi P. 2003a. Design of plastic bottles for packaging of virgin olive oil. *Journal of Food Science* 68: 170–175.

Del Nobile M.A., Bove S., La Notte E., Sacchi R. 2003b. Influence of packaging geometry and material properties on the oxidation kinetics of bottled virgin olive oil. *Journal of Food Engineering* 57: 189–197.

EEC. 1991. Regulation (EEC) No 2568/91 on the characteristics of olive oil and olive-residue oil and on the relevant methods of analysis. *Official Journal of the European Communities* 248, 5.9.1991, p. 1.

El-Shattory Y., Saadia M.A., Said F.H. 1997. Flavour changes due to effect of different packaging materials on storage of cottonseed oil, hydrogenated oil and margarine. *Grasas Aceites* 48(2): 61–67.

Gambacorta G., Del Nobile M.A., Tamagnone P., Leonardi M., Faccia M., La Notte E. 2004. Shelf-life of extra virgin olive oil stored in packages with different oxygen barrier properties. *Italian Journal of Food Science* 16: 417–428.

Georgalaki M.D., Sotiroudis T.G., Xenakis A. 1998. The presence of oxidizing enzyme activities in virgin olive oil. *Journal of the Oil Chemists' Society* 75: 155–159.

Grover M.R. 1982. Studies on shelf life of vegetable oils packed in tin containers. *Journal of Food Science and Technology* 19: 268–270.

Gunstone F.D. (Ed). 2002. *Vegetable Oils in Food Technology: Composition, Properties and Uses.* Boca Raton, Florida: CRC Press.

Jung M.Y., Min D.B. 2006. Effects of α-, γ-, and δ-tocopherols on oxidative stability of soybean oil. *Journal of Food Science* 55: 1464–1465.

Kakuda Y., Smadi H., Gao K., Abraham V., Laleye L. 2008. Effect of packaging and light exposure on vitamin A stability in fortified vegetable oil. http://ift.confex.com/ift/2001/techprogram/paper_8889.htm. Last accessed 13th May 2009.

Kanavouras A., Coutelieris F.A. 2005. Shelf life predictions for packaged olive oil based on simulations. *Food Chemistry* 96: 48–55.

Kanavouras A., Hernandez-Munoz P., Frank A. 2006. Packaging of olive oil: quality issues and shelf life predictions. *Food Reviews International* 22: 381–404.

Kanavouras A., Hernandez-Munoz P., Coutelieris F.A. 2004a. Shelf life predictions for packaged olive oil using flavor compounds as markers. *European Journal of Lipid Science and Technology* 219: 190–198.
Kanavouras A., Hernandez-Münoz P., Coutelieris F., Selke S. 2004b. Oxidation-derived flavor compounds as quality indicators for packaged olive oil. *Journal of the American Oil Chemists' Society* 81: 251–257.
Kaya A., Tekin A.R., Oner D. 1993. Oxidative stability of sunflower and olive oils: comparison between a modified active oxygen method and long-term storage. *LWT—Food Science and Technology* 26: 464–468.
Kiristakis A., Kanavouras A., Kiristakis K. 2002. Chemical analysis, quality control and packaging issues of olive oil. *European Journal of Lipid Science and Technology* 104: 628–638.
Kristott J. 2000. Fats and oils. In: *The Stability and Shelf Life of Food.* Kilcast D., Subramaniam P. (Eds). Cambridge, England: Woodhead Publishing, pp. 279–310.
Kukuk M., Caner C. 2005. Effect of packaging materials and storage conditions on sunflower oil quality. *Journal of Food Lipids* 12: 222–231.
Lavelli V., Fregapane G., Desamparados Salvador D.M. 2006. Effect of storage on secoiridoid and tocopherol contents and antioxidant activity of monovarietal extra virgin olive oils. *Journal of Agricultural and Food Chemistry* 54: 3002–3007.
Lee D.S., Yam K.L., Piergiovanni L. 2008a. Shelf life of packaged food products. In: *Food Packaging Science and Technology*. Boca Raton, Florida: CRC Press, pp. 479–542.
Lee D.S., Yam K.L., Piergiovanni L. 2008b. Permeation of gas and vapor. In: *Food Packaging Science and Technology*. Boca Raton, Florida: CRC Press, pp. 79–108.
Lee D.S., Yam K.L., Piergiovanni L. 2008c. Physical properties of packaging materials. In: *Food Packaging Science and Technology*. Boca Raton, Florida: CRC Press, pp. 43–78.
Lee D.S., Yam K.L., Piergiovanni L. 2008d. Metal packaging. In: *Food Packaging Science and Technology*. Boca Raton, Florida: CRC Press, pp. 197–242.
Loew G. 1973. Precautions in treatment of oilseeds designed to prevent quality deterioration during production of edible oils and grits. *Seifen-Oele-Fette-Wachse* 99(21): 625–628.
Mahadevaiah B., Indiramma A.R., Balasubrahmanyam N. 1993. Packaging and storage studies on double-filtered groundnut oil. *European Food Research and Technology* 196: 252–254.
Maloba F.W., Rooney M.L., Nguyen M. 1996. Improved oxidative stability of sunflower oil in the presence of an O_2-scavenging film. *Journal of the American Oil Chemists' Society* 73: 181–185.
Montedoro G.F., Servili M., Baldioli M., Miniati E. 1992. Simple and hydrolysable phenolic compounds in virgin olive oil. I. Their extraction, separation, and quantitative and semiquantitative evaluation by HPLC. *Journal of Agricultural and Food Chemistry* 40: 1571–1576.
Morales M.T., Rios J.J., Aparicio R. 1997. Changes of the volatile compounds of virgin olive oil during oxidation: flavors and off-flavors. *Journal of Agricultural and Food Chemistry* 45: 2666–2673.
Narasimhan S., Rajalakshmi D., Nagin C., Mahadeviah B., Indiramm A.R. 2001. Palm oil quality in different packaging materials: sensory and physicochemical parameters. *Journal of the American Oil Chemists' Society* 78: 257–265.
Olafsson G., Hildingsson I. 1995. Sorption of fatty acids into LDPE and its effect on adhesion with aluminum foil in laminated packaging material. *Journal of Agricultural and Food Chemistry* 43: 306–312.
Ortega-García J., Gámez-Meza N., Noriega-Rodriguez J., Dennis-Quiñonez O., García-Galindo H., Angulo-Guerrero J., Medina-Juárez L. 2006. Refining of high oleic safflower oil: effect on the sterols and tocopherols content. *European Food Research and Technology* 223: 775–779.
Owen R.W., Mier W., Giacosa A., Hull W.E., Spiegelhalder B., Bartsch H. 2000. Phenolic compounds and squalene in olive oils: the concentration and antioxidant potential of total phenols, simple phenols, secoiridoids lignans and squalene. *Food Chemistry and Toxicology* 38: 647–659.
Papadopoulos G., Boskou D. 1991. Antioxidant effect of natural phenols on olive oil. *Journal of the American Oil Chemists' Society* 68: 669–671.
Piergiovanni L., Limbo S. 2004. The protective effect of film metallization against oxidative deterioration and discoloration of sensitive foods. *Packaging Technology and Science* 17: 155–164.
Pritchard J.R. 1983. Oilseed quality requirements for processing. *Journal of the American Oil Chemists' Society* 60: 322–332.
Psomiadou E., Tsimidou M. 2002. Stability of virgin olive oil. 2. Photo-oxidation studies. *Journal of Agricultural and Food Chemistry* 50: 722–727.
Ramirez G., Hough G., Contarini A. 2001. Influence of temperature and light exposure on sensory shelf-life of a commercial sunflower oil. *Journal of Food Quality* 24: 195–204.
Ranalli A., Angerosa F. 1996. Integral centrifuges for olive oil extraction: the qualitative characteristics of products. *Journal of the American Oil Chemists' Society* 73: 417–422.

Ranalli A., Martinelli N. 1994. Tecniche Industriali Biologiche e non Convenzionali per il Recupero dell'Olio dalla Pasta di Olive. *Industrie Alimentari* 33: 1073–1083.

Rastrelli L., Passi S., Ippolito F., Vacca G., De Simone F. 2002. Rate of degradation of α-tocopherol, squalene, phenolics, and polyunsaturated fatty acids in olive oil during different storage conditions. *Journal of Agricultural and Food Chemistry* 50: 5566–5570.

Sacchi R., Savarese M., Del Regno A., Paduano A., Terminiello R. Ambrosino M.L. 2008. Shelf life of vegetable oils bottled in different scavenging polyethyleneterephthalate (PET) containers. *Packaging Technology and Science* 21: 269–277.

Sahan Y., Basoglu F. 2008. Trace metals in olive oil. *ISHS Acta Horticulturae* 791: 719–723. www.actahort.org/books/791/791_109.htm. Last accessed 13th May 2009.

Salas J.J., Williams M., Harwood J.L., Sánchez J. 1999. Lipoxygenase activity in olive (*Olea europaea*) fruit. *Journal of the Oil Chemists' Society* 76: 1163–1168.

Salvador M.D., Aranda F., Gomez-Alonso S., Fragapane G. 2001. Cornicabra virgin olive oil: a study off five crop season. Composition, quality and oxidative stability. *Food Chemistry* 74: 267–274.

Sato K. 1988. Crystallisation of fats and fatty acids. In: *Crystallisation and Polymorphism of Fats and Fatty Acids*. Garti N., Sato K. (Eds). New York: Marcel Dekker, pp. 227–266.

Suzuki K., Nishioka A. 1993. Behavior of chlorophyll derivatives in canola oil processing. *Journal of the American Oil Chemists' Society* 70: 837–841.

Tawfik M.S., Devlieghere F., Huyghebaert A. 1998. Influence of d-limonene absorption on the physical properties of refillable PET. *Food Chemistry* 61: 157–162.

Terao J., Matsushita S. 1977. Products formed by photosensitized oxidation of unsaturated fatty acid esters. *Journal of the American Oil Chemists' Society* 54: 234–238.

Torri L., Limbo S., Piergiovanni L. 2007. A survey of the food lighting conditions in the large-scale retail stores. *Industrie Alimentari* 46(5): 1–7.

Tsimis D.A., Karakasides N.G. 2002. How the choice of container affects olive oil quality. *Packaging Technology and Science* 15: 147–154.

Van Willige R.W.G., Linssen J.P.H., Voragen A.G.J. 2000a. Influence of food matrix on absorption of flavour compounds by linear low density polyethylene: proteins and carbohydrates. *Journal of the Science of Food and Agriculture* 80: 1779–1789.

Van Willige R.W.G., Linssen J.P.H., Voragen A.G.J. 2000b. Influence of food matrix on absorption of flavour compounds by linear low density polyethylene: oil and real food products. *Journal of the Science of Food and Agriculture* 80: 1790–1797.

Van Willige R.W., Linssen J.P.H., Meinders M.B.J., van der Stege H.J., Voragen A.G.J. 2002. Influence of flavour absorption on oxygen permeation through LDPE, PP, PC and PET plastics food packaging. *Food Additives and Contaminants* 19: 303–313.

Wold J.P., Veberg A., Nilsen A., Iani V., Juzenas P., Moan J. 2005. The role of naturally occurring chlorophyll and porphyrins in light-induced oxidation of dairy products. A study based on fluorescence spectroscopy and sensory analysis. *International Dairy Journal* 15: 343–353.

18 Packaging and the Shelf Life of Cereals and Snack Foods

Sea C. Min
Division of Food Science
Seoul Women's University
Seoul, Korea

Young T. Kim
Department of Packaging Science
Clemson University
Clemson, South Carolina

Jung H. Han
PepsiCo Fruit and Vegetable Research Center
Frito-Lay Inc.
Plano, Texas

CONTENTS

18.1 INTRODUCTION

18.1.1 Cereals and Snack Foods

18.1.1.1 Cereals

Cereals (cereal crops or grains) are mostly grasses cultivated for their edible grains or fruit seeds and represent a major component of the human diet. Wheat, rice, barley, maize, oats, rye, sorghum, and millets are the primary cereals. They are accepted by nearly all consumers as safe and healthy ingredients in food products. Cereal products such as breakfast cereals are an important part of the diet in many countries. Cereal products comprise extruded, puffed, flaked, shredded, and granulated products, generally made from wheat, rice, barley, or maize with added sugar, honey, or malt extract for sweetness (Robertson, 2006). Grains are cooked to gelatinize the starch and the gelatinized grains are shredded; the shreds are baked (e.g., 20 min at 260°C), dried (e.g., to 1% moisture content), and then packaged (Robertson, 2006). The marketing of cereal products has been growing, resulting in both improvement of their quality and developments in their production, including packaging.

18.1.1.2 Snack Foods

A snack food can be defined as a type of food not meant to be eaten as a main meal of the day, but one consumed to get a brief supply of energy for the body or consumed between meals purely for the enjoyment of its taste (Wikipedia, 2008). Ingredients for snack foods include cereal crops, fats, oils, emulsifiers, antioxidants, sweeteners, salt, sugar, eggs, milk, nuts, vegetables, fruits, spices, colorants, and flavorants (Matz, 1993).

Further growth of snack products is expected because of the following reasons:

a. Materials for snack products are relatively inexpensive sources of food energy and nutrition.
b. Cereals and snack products are convenience meals consumers enjoy easily at many places.
c. Changes in formulas, reducing salt and adding nutrients, and functional compounds can be accomplished.

18.1.1.2.1 Savory Snacks

Savory snack foods are multipurpose foods that can be consumed with a meal or on the go and are often associated with picnics and sports events. Many savory snacks are based on the adaptation of a yeast fermentation process and, to some extent, the formulas used for bread. Savory snacks are microbiologically and chemically safe, do not require any refrigeration for product stability and safety, and are usually seasoned with salt and additional flavorings (Lusas, 2001).

18.1.1.2.2 Chips

Potato chips make up about 25% of the snack food market, and lead snack food sales (McCarthy, 2001). Potato chips are produced by frying and seasoning sliced potatoes or frying dough containing dehydrated potatoes or potato puree (Matz, 1993). The unit operations involved in manufacturing potato chips are presented in Figure 18.1A. Washed potato slices are dried before being passed into a very hot bath of vegetable oil (~120–130°C). During frying in hot oil, the potatoes are dehydrated and cooked. The excess oil is drained or centrifuged and the chips are salted and flavored before being packed in moisture-proof containers (Huber and Rokey, 1990; Robertson, 2006). The moisture content of the potato is reduced from about 80% to 5% and the final fat content ranges from 35% to 40% on a dry basis (Robertson, 2006). Fabricated potato snacks may be defined as snacks made from dried potatoes as opposed to those processed directly from fresh potatoes, such as regular potato chips. The most common potato materials used in fabricated potato snacks are potato starch, potato flakes, potato flour, and dried ground potatoes (Hix, 2001).

Two major kinds of fried chips made from corn are tortilla chips and corn chips, which command about 20% of the snack food market (Matz, 1993; McCarthy, 2001). Both are made from ground, alkalized corn dough known as "masa." Tortilla chips have become popular in many other countries in Europe and Asia (Mehta, 2001). Tortilla chips were originally made without extrusion by pressing a thin sheet of masa, baking it on a hot surface, cutting shapes, and frying

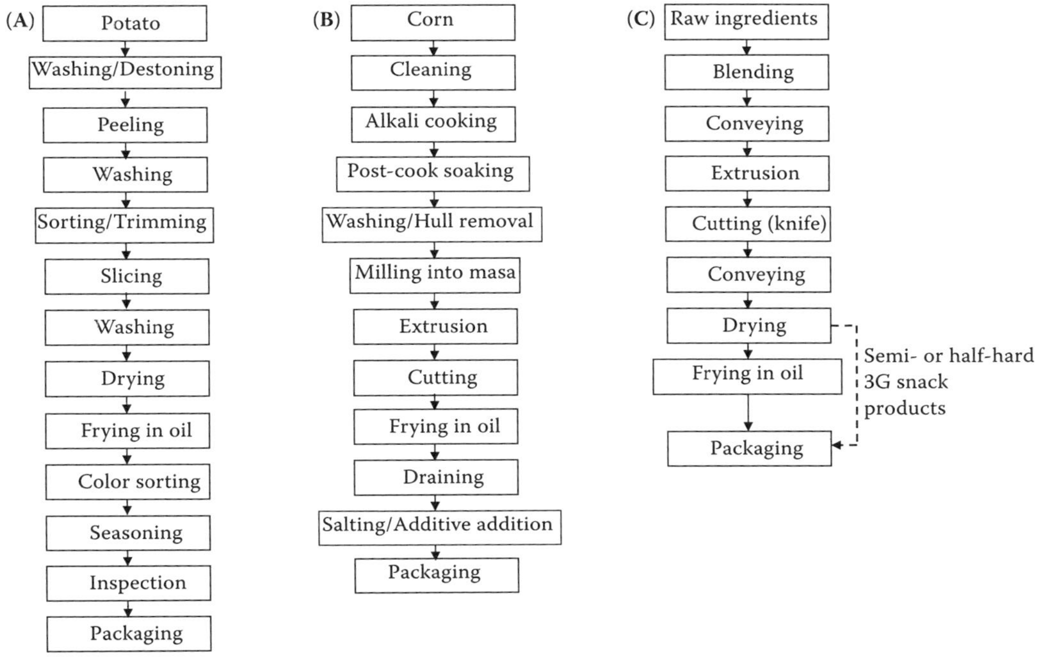

FIGURE 18.1 Flowcharts for the manufacture of (A) potato chips, (B) corn chips, and (C) general extruded snack products.

the chips in hot oil, but now most commercial tortilla chips are made by extrusion methods (Matz, 1993). Masa is converted to a thick sheet by an extruder or a masa presheeter and cut into discs (typically triangular shapes) of appropriate size. The masa is thoroughly mixed, gelatinized, and expanded or puffed during extrusion. The discs are dried in an oven, tempered, and then deep-fried. After cooling, the flavors are applied before packing (Mehta, 2001). Traditional masa tortillas are being replaced to a considerable extent by wheat flour tortillas. They are generally made from bleached, enriched winter wheat flour (Matz, 1993). Corn chips are another product made from masa. Corn chips are fried without drying and tempering, whereas tortilla chips are dried, tempered, and then fried (Matz, 1993). The flowchart for the manufacture of corn chips is shown in Figure 18.1B. Chips are also made from other cereal grains; the alkalizing step is not included in most of these cases.

18.1.1.2.3 Pretzels

The pretzel had its origins sometime after 610 AD in southern France, where monks baked scraps of dough in the image of arms folded in prayer to reward children for learning their prayers (McCarthy, 2001). The development of extruders marked a major breakthrough in the high-volume production of pretzels. The array of die patterns are very easily interchangeable, which allows extruders to make pretzels of all shapes (Groff, 2001). After the dough is extruded, rolled, and twisted, the pretzels are passed through a spray of hot soda solution, sprinkled with salt, and baked. Pretzels are often dried further after they are baked in the oven (Matz, 1993).

18.1.1.2.4 Rice-Based Snacks

Rice-based snacks are becoming more available and variable. The mixtures in rice-based snacks include vegetables; tofu; tahini (ground sesame seeds); cereals such as oats, wheat flour, or seaweed; nuts; herbs; or seeds (Rice, 1990). Traditional Korean and Japanese cuisines include a range of deep-fried snacks consisting of a carbohydrate-based glutinous batter containing vegetables. Puffing rice by frying is popular in Asia. Rice is first cooked by either boiling or steaming. The cooked rice is compacted, shaped/cut, dried to 12–15% moisture, and fried in oil at 220°C for several seconds in a deep fryer (Lu and Lin, 2001).

18.1.1.2.5 Semi- or Half-Hard Third-Generation Snack Products

Snack products or pellets that are ready to be consumed after frying in hot oil, puffing with hot air, or secondary expansion extrusion by either the snack processor or the consumers are called semi- or half-hard or third-generation (3G) snack products (Huber and Rokey, 1990) (Figure 18.1C). They can be sold as a consumer item to be fried at home and in the restaurant for immediate consumption. Drying is a very critical step in the production of these products. Proper drying will reduce the moisture content to approximately 10–15%. The drying temperature and drying time for the semihard products range from 70°C to 95°C and from 1 to 3 hr, respectively, depending on the product properties (Huber and Rokey, 1990). Many 3G snack products and pellets are produced by extruding the cooked dough into the form of a sheet, with the raw materials used mostly being starch-based (Huber, 2001).

18.1.2 Dried Fruits and Nuts

18.1.2.1 Dried Fruits for Cereals and Snack Foods

Popular types of dried fruits are apricots, raisins, dates, figs, apples, pears, peaches, prunes, cranberries, blueberries, cherries, currants, pears, and nectarines. Infusion of dried fruit is becoming popular for cereal applications. The fruit is typically infused with a combination of sweeteners and acids, and sometimes humectants and flavors, and then air- or freeze-dried. Fruit bars are healthy snacks as they are low in fat, high in fiber, and generally do not contain much sugar or salt. Binders and fillers such as wheat flour, potato flour, apple fiber, rice flour, or oat flakes are all used in such products of one sort or another (Rice, 1990).

18.1.2.2 Nuts

Nuts are consumed on their own or with dried fruits. They are also used as components in various food products such as chocolate confectionery items (Booth, 1990). These nuts include almonds, Brazil nuts, cashew nuts, chestnuts, coconut, hazelnuts, macadamia, peanuts, pecans, pine nuts, pistachio nuts, sunflower seeds, and walnuts (Booth, 1990).

Peanuts are almost always fry- or dry-roasted and salted for use as snacks. Coconut is a main ingredient in many snacks that also contain sugar, eggs, and chocolate because of its unique flavor and texture. Chestnuts are readily cooked by boiling or roasting and are not always salted like most other nuts. Pine nuts are not edible when raw; they are roasted to be eaten as snacks and used in confectionery recipes. Their delicious flavor and relative scarcity make them a high-priced commodity. Pistachio nuts are typically salted while in the shell in brine and roasted for retail sale as a snack. Those shelled and unsalted are also consumed as components in a variety of confections including ice cream and baked goods (Booth, 1990).

A close watch has to be maintained when storing and transporting nuts due to their susceptibility to mold growth and insect, animal, and bird attention. Some nuts become tooth-breakingly hard when they are too dry (Booth, 1990). Allergies caused by the consumption of nuts have been of concern. More studies are needed on nut-induced allergic reactions to educate consumers and to provide accurate information on allergies (Allen, 2008).

The role of nuts in reducing the risk of cardiovascular disease, diabetes, and cancer has been emphasized (King et al., 2008a, 2008b). This risk reduction by nuts has been considered to be related to their bioactivity in reducing oxidative stress and inflammation. Nuts are an excellent source of vitamin E, phytochemicals (phytosterols, phenolic acids, flavonoids, and carotenoids), and magnesium. The total antioxidant capacity of nuts is comparable to broccoli and tomatoes (King et al., 2008a). A great variety of flavors, textures, and nutritional values of nuts allows the development of a wide range of nut-based snacks and snack foods.

18.2 MANUFACTURING

18.2.1 Frying

All cereals contain starch at high concentrations and so they need to be cooked to be digestible and acceptable for consumption. Frying is not a complicated process but its chemistry is complex and involves extensive changes in both the frying oil and the food being fried (Banks and Lusas, 2001). Fat is used in most fried snack foods as a processing aid to dehydrate the products and develop characteristic flavors (Robertson, 2006). The primary change in frying is moisture reduction or dehydration (Banks, 1996). When starch-based material is immersed in hot oil, the starch on the surface is rapidly gelatinized and the product is uniformly covered with small steam bubbles as the surface moisture begins to vaporize. The outer layer of the product dehydrates and forms a veneer-like structure (case hardening). As surface moisture diminishes by frying, the internal moisture becomes steam, rupturing channels through the structure of the product (Banks, 1996). Continued frying keeps lowering the moisture content of the product. Low moisture content and high temperature support flavor-producing reactions involving amino acids, proteins, and carbohydrates (Banks and Lusas, 2001).

Industrial fryers consist of a heating oil, conveying systems, and a hood to exhaust steam and vapors (Banks and Lusas, 2001). Fryers directly heated by gas burners under the fryer pan (direct-fired fryers) are commonly used, but they are not as efficient as fryers with heat exchangers although they are more economical to purchase. Fryers usually operate at 149–218°C at atmospheric pressure (Banks and Lusas, 2001). Vacuum fryers are sometimes used to fry heat-sensitive products. Vacuum frying at lower temperature results in lighter color in the products than those processed at atmospheric conditions (Banks and Lusas, 2001).

During oil frying, oils can undergo oxidation, cyclicization, polymerization, degradation to volatile compounds, and hydrolysis. These chemical changes cause off-flavors, rancid aromas, greasy mouthfeel, and impaired nutritional value in fried foods. Thus, maintaining the quality of frying oil under high-temperature conditions is critical to the fried foods (Ory et al., 1985).

18.2.2 Extrusion

The majority of expanded snacks on the market is manufactured by extrusion and puffing. Extrusion has revolutionized many conventional snack-producing processes and has provided a means of producing new products (Huber and Rokey, 1990).

Extrusion offers many advantages over traditional food unit operations such as minimizing time, energy, and cost inputs while adding versatility and flexibility to the manufacturing process (Huber, 2001). A wide variety of different products can be produced by changing ingredients, operating conditions, and/or minor components of the extruder. The functions of an extruder include feeding, heating, conveying, mixing, kneading, compressing, shearing, cooking, and shaping (forming). When a pressurized, heated, and sheared mixture of ingredients exits the die of an extruder, the moisture in the mixture vaporizes instantaneously, which puffs and expands the mixture, producing a snack product. The shape and size of the product are determined by the design of the die and the cutting components. The shaped extrudates are dried or baked and coated with oils, flavors, and/or seasonings (Huber and Rokey, 1990; Matz, 1993). The important variables in extrusion are extrusion temperature, screw speed, feed rate, barrel/screw designs (configuration), extrusion rate (residence time), and the composition of the feed, including moisture, sugar, salt, and fat content.

Crispiness of expanded snacks is controlled by expansion. More expansion (more crispiness) is expected at a higher extrusion pressure and a higher content of starch in the extruded materials. Fat usually reduces expansion. Salt helps expansion and results in a tender texture. Emulsifiers generally weaken starch structure, resulting in less expansion.

The moisture content of extruded snack products is normally between 8% and 10% (wet basis). Additional drying or frying is applied to the extruded products to reduce the moisture content to 1–2%, providing desired crispness (Robertson, 2006).

Coextrusion has been used to produce snacks that typically have an extrusion-cooked outer shell with a pumpable (but not free flowing) filling (Huber and Rokey, 1990). The extrusion-cooked portion of the snack will flow through the die parallel to the flow direction in the extruder barrel. The filling is pumped into the die and it flows within the shell extrudate.

18.2.3 Puffing

The base of puffed cereal products contains conditioned whole grain wheat, rice, barley, maize, or dough made from corn meal or oat flour and is mixed with other ingredients, such as sugar and salt, and cooked under pressure, dried to 14–16% moisture content, and pelleted by extrusion (Robertson, 2006). Popcorn is used as the basis for many snack foods due to its pleasing textural effects at low cost, with the basic popcorn snack simply being popcorn coated with butter and salt (Matz, 1993). Blending the puffed products with different flavors provides enormous opportunities for increasing acceptance and the usage of puffed products (Nath et al., 2007). Puffing is also a commercially applied method to produce breakfast cereals (Matz, 1993). The expansion of water by puffing causes a several-fold increase in volume. Puffing is useful in producing cereal products that need to be reconstituted later.

The raw material is heated in an enclosed vessel so that its vapors create a high pressure atmosphere around the product. Under these conditions, the product temperature is above the glass transition temperature, so that it is in a highly deformable, liquid-like state. The pressure is suddenly released, so that the product "explodes" due to evaporation of moisture from within. This evaporation cools the product and its temperature drops below the glass transition, to possess solid-like behavior. The main process parameters affecting puffing are puffing temperature, puffing duration, initial moisture content of the material to be puffed, and starch content of the raw materials. The degree of puffing of the cooked grain is mainly dependent on the suddenness of change in temperature or pressure (Nath et al., 2007). The advantage of puffing is that different cereal materials can be processed together to produce cereal products with different formulations. The puffed product is usually dried to 3% moisture content by toasting and then cooled and packaged (Robertson, 2006).

Two types of popping equipment, wet (oil) and dry, are commercially used to produce popcorn (Cretors, 2001). Popping corn in oil (wet-popping) is simple and enables the end product to be made at the point of purchase. Dry-popping corns without oil relies on heating. A commercial version of dry-popping uses a motorized rotating wire drum over an open flame or electric heating elements (Cretors, 2001).

18.2.4 Flaking

Various flakes, such as corn flakes, wheat flakes, and rice flakes, can be produced from intact whole grains, grain mills, or individual or mixtures of ground grains (Lawes, 1990). The grain materials are cooked at elevated pressures and then passed between the steel rollers of a drum dryer (15–20% moisture content). The partially dried flakes are scraped off from the rollers and then usually dried on a traveling wire mesh belt or conditioned for 1–3 days (Lawes, 1990; Robertson, 2006). The flakes are toasted or browned by radiant heat before packing in moisture-proof materials. Flaked grains are sometimes coated with sugar or candy to provide a hard and transparent coating that does not become sticky (Robertson, 2006).

18.2.5 Baking

Virtually all snack food production needs a baking, cooking, frying, or drying step, generally near the end of the production line (Matz, 1993). Ovens and driers transfer heat by conduction, convection, and radiation. For example, in ovens, heat transfer is by conduction from the pan, convection from hot oven gases, and radiation from flames and hot oven parts.

Baking is carried out to produce biscuits, cookies, and crackers and this is usually done in tunnel ovens. The product from the oven cools and loses traces of moisture. Some of those products are directly packaged, but many other products require additional processing to add nonbaked ingredients or flavor enrichments (Robertson, 2006).

18.2.6 Drying

If the drying is not properly done, various microorganisms such as foodborne pathogens or spoilage microorganisms can grow (Robertson, 2006). Potato chips fried in oil have enriched flavor. Due to the increased market share of low-fat and fat-free chips, drying or partial drying prior to frying is being used, which, however, results in significant reduction in flavor (Gould, 2001).

A high-temperature short-time (HTST) fluidized bed drying technique was developed for the production of quick cooking products from starch vegetables (e.g., potato, sweet potato, green peas, carrot) (Jayaraman et al., 1982). HTST drying can result in a considerable reduction in the drying time with improvement in the texture of the cooked starch vegetables (Nath et al., 2007).

Freeze-drying where heat generation is not involved during processing is often used when dried fruits need to maintain their nutritional values. Generally, less loss of α- and β-carotenes and ascorbic acid is found in freeze-dried fruit products compared to those dried by conventional convection drying methods.

18.3 DETERIORATIVE REACTIONS AND INDICES OF FAILURE

Many cereal and snack products are dry and contain lipids. They are stable against microbial growth due to their low water activity ($a_w < 0.6$), but not against chemical and enzymic reactions that result in product deterioration (Labuza, 1980). The most important modes of deterioration in cereal products and snack foods are loss of crispness (i.e., moisture uptake) and lipid oxidation causing rancidity/off-flavors.

18.3.1 Loss of Crispness

Dry food systems can lose their desired crispness during storage or upon opening of the package. Loss of crispness due to moisture uptake is a major cause of snack food rejection by consumers (Robertson, 2006). Puffed rice cakes were reported to lose their crispness and become tough as the a_w increases through moisture absorption; rice cakes with a_w between 0.2 and 0.4 have the best crispness and low hardness (Hsieh et al., 1990). A critical a_w for potato chips and corn chips was reported to be 0.4; potato chips became organoleptically unacceptable at a_w of 0.5 (Katz and Labuza, 1981; Quast and Karel, 1972; Robertson, 2006). The values for the critical a_w for puffed corn curl and extruded rice snacks have been reported as 0.36 and 0.43, respectively (Chauhan and Bains, 1990; Katz and Labuza, 1981).

18.3.2 Lipid Oxidation

The primary mode of chemical deterioration in cereals and snack foods is lipid oxidation (rancidity development) since fat (lipid) is used in most snacks as a processing aid or flavor inducer (Robertson, 2006). When cereals and snack foods are dried at low moisture content (e.g., <3%), they become highly susceptible to oxidation (Labuza, 1980). In some systems (e.g., peanut flakes), the water activity corresponding to a Brunauer–Emmett–Teller (BET) or Guggenheim–Anderson–de Boer (GAB) monolayer may be optimal with respect to minimizing lipid oxidation (Jensen and Risbo, 2007).

The content and quality of lipids in breakfast cereals are important factors determining the shelf life of the products. Generally, products made from wheat, barley, rice, and maize (lipid content 1.5–2.0%) have a longer shelf life than those made from oats (lipid content 7%) (Robertson, 2006). The susceptibility of cereals and snack foods to lipid oxidation is associated with the concentration and type (quality) of fat used and the number of unsaturated bonds in the fatty acids. These lipid-containing foods need to be protected from O_2, light, and pro-oxidants (e.g., metal ions) to minimize lipid oxidation.

18.3.3 Vitamin Degradation

A substantial loss of fortified vitamins in cereal products is not observed during normal shelf life at room temperature (Robertson, 2006). The major factor influencing vitamin loss in packaged cereal products is storage temperature. The loss of aroma/flavor can be a problem with certain cereal products to which fruit flavors have been added (Robertson, 2006).

18.3.4 Mechanical Damage

Physical breakage of most cereals and snack foods, especially chips, is unacceptable to consumers. Many snack foods are very fragile because they are usually fried, extruded, or dried from natural products. The packaging material must be physically strong to withstand the processes of vacuumizing and gas flushing. Many cereals and snack foods are protected by paperboard or plastic cases. The rigidity of printed fiberboard (PFB) for cereal packaging and the compression resistance of the finished package seem to resolve the problem of physical breakage (Robertson, 2006). More delicate packages (inert fluffing N_2 gas for pressure buffering or loop array packaging) have been developed for cereals and snack foods.

The shelf life of cereals and snack foods can be extended by (a) the removal of residual O_2 in packages, (b) the use of antioxidants or O_2 scavengers in food and/or packaging material, (c) the use of selective gas barrier packaging material, and (d) the initial substitution of gas composition inside the package.

18.4 PACKAGING AND SHELF LIFE OF CEREALS AND SNACK FOODS

Packaging should protect cereals and snack foods during the period from manufacture until ultimate consumption by the consumer. Generally, a good package will protect the food from various physical and/or chemical deteriorative reactions.

18.4.1 Packaging Materials for Cereals and Snack Foods

18.4.1.1 Paper, Paperboard, and Printed Fiberboard

Most cereals and snack foods are packaged with paper-based materials made from wood fibers. Microflute corrugated paperboards have unique characteristics including good strength properties, excellent shock absorbing ability, good aesthetic appearance, environmental advantages, and distinctive print properties. White board is suitable for contact with food and is often coated with low density polyethylene (LDPE), poly(vinyl chloride) (PVC), or wax. It is used for snack, chocolate, and frozen food cartons.

18.4.1.2 Plastic Films

Flexible plastic films have been used for cereals in single packaging or multiserving size packages with other packaging materials. Typically, the majority of snacks are in flexible bags. Biaxially oriented films are mostly widely used for snack foods; biaxially oriented polypropylene (BOPP) has qualities of toughness (against puncture and abrasion) and clarity, and is rendered heat sealable by coextrusion or coating with polyolefin copolymers. Films are also coated with other polymers or aluminum to improve the barrier properties or to impart heat sealability.

18.4.1.3 Metals

Metal containers have been rarely used for cereals and snack foods due to their cost, despite their perfect gas barrier properties, convenience, and extreme strength. However, composite containers are used for molded chips and nuts. The body of the container is made of LDPE-coated foil on spirally wound paperboard. The top and bottom ends of the containers may be made of metal or plastic. An aluminum pull-tab top and a reclosable plastic lid on the container form a reclosable canister.

18.4.2 Requirements of Packaging for Shelf Life Extension

18.4.2.1 Oxygen Gas Barrier

Low O_2 content may be achieved by vacuum packaging, flushing with N_2 or CO_2, or using O_2 absorbers in combination with low oxygen transmission rate (OTR) materials (Holaday et al., 1979; Jensen et al., 2005). This system is effective for special foods possessing high quantities of lipid or fat (Jensen et al., 2003, 2005; Lima et al., 1999).

18.4.2.2 Water Vapor Barrier

Generally, many cereals and snacks are dry products, which imply that crispness is the major consideration for their quality and shelf lives. This requires the use of high-water-vapor barrier packaging materials. Sometimes, sachets containing desiccants are added to snack food packages to scavenge moisture. Generally, the BET monolayer moisture content provides optimal stability with respect to oxidative degradation (Hill and Rizvi, 1982). As an example, peanuts had the best stability with the BET monolayer moisture content, whereas the monolayer value was lower than the a_w at which the oxidation was minimal (Evranuz, 1993). In some cases, however, the packaging humidity should be controlled to extend the shelf lives of products. For example, oatmeal and muesli have a high content of unsaturated fatty acids and, therefore, a high susceptibility to lipid oxidation. It has been reported that humidity (a_w) is a very important packaging parameter for oatmeal and muesli because the optimal humidity with respect to oxidative stability does not coincide with the BET monolayer value (Jensen and Risbo, 2007).

18.4.2.3 Light Barrier

Depending on the light transmission through plastic films, the shelf life of cereal and snack foods could be seriously affected through various chemical reactions. Generally, the increased exposure to light results in accelerated lipid oxidation. Light accounts for the greatest variation of the kinetics of free radical formation. For example, the shelf life of oatmeal and muesli packaged with poly(ethylene terephthalate) (PET) with 0–60% transmission of visible light resulted in better quality than those packaged with more transparent PET (Firman, 1973). In other studies, no rancid odors or flavors developed in extruded oats during 3 months of storage in a N_2 atmosphere with low-OTR packages in darkness at 23°C (Larsen et al., 2005). Extruded oats exposed to light and stored at 38°C in packages with medium and high OTR developed the highest degree of rancidity (Larsen et al., 2005). The effectiveness of the package as a light barrier is improved by the addition of ink pigment into packaging materials.

18.4.3 Packaging Requirements for Cereals

Choosing packaging materials for cereal products should be related to the major indices of failure (IoFs) of cereals, which include loss of crispness, lipid oxidation, and nutrient loss.

18.4.3.1 Loss of Crispness

Crispness is an essential characteristic of dried cereal foods. It is affected by the moisture content of products, which is strongly related to the shelf life of products. Packaging materials for cereals may be combined with PFBs for outer packaging and plastic films for inner packaging. Plastic film layers [e.g., polypropylene (PP), laminated PP, and aluminum-metalized polyester films] have been used to improve both moisture barrier and seal performance. The glassine paper liner has been largely replaced by various plastic materials, in particular thin-gauge high density polyethylene (HDPE) coextruded with ethylene-vinyl acetate copolymer (EVA) (Robertson, 2006). Under the low water activity and low O_2 availability in vacuum or gas-flushed packaging, bacterial growth is significantly restrained in cereals. These conditions are also very helpful to prevent the oxidative discoloration, destruction of nutrients, and growth of insects (Arvanitoyannis and Traikou, 2005). The aroma of baked cereal products is one of the most important factors appealing consumers. The aroma of dried foods is generated during cooking or baking through chemical, enzymic, and thermal reactions. To keep the desirable aroma and flavor in dry foods, and also to prevent external odor absorption, gas barrier materials and gas substitution have been utilized effectively (Lozano et al., 2007; Meiron and Saguy, 2007; Safa et al., 2008).

18.4.3.2 Lipid Oxidation

Lipid oxidation is actively affected by light, a_w, and O_2. To minimize oxidative rancidity, it is important to exclude light, control the a_w of packaged cereals, and use high-O_2-barrier packaging material. It was reported that the storage stability of a flaked oat cereal packaged in high-O_2-barrier materials with the addition of iron-based O_2 absorbers was improved by retarding lipid oxidation. Although the use of an antioxidant in the package liner has been shown to be successful in extending shelf life, it is not generally permitted in most countries (Robertson, 2006). In high-OTR systems, rancid odors, which develop in the packaged product, can escape as soon as they are formed (Labuza, 1982). This was well tested by Matz et al. (1955), who reported that rancid odors accumulated if wheat flakes were stored in airtight containers. Therefore, O_2-sensitive products are usually sold in breathable packaging without inner or outer linings (Jensen et al.,

2005). However, such a package system may have the disadvantage of moisture absorption and desorption.

18.4.3.3 Others

In a study of the effects of processing and storage conditions on micronutrients in breakfast cereals, it was concluded that micronutrient loss would not be a major factor in determining the shelf life of dry cereals. Flavor scalping can occur when the glassine liners are substituted by HDPE (Robertson, 2006).

18.4.4 Packaging Requirements for Snack Foods

18.4.4.1 Fried Snacks

Because of the nature of fried snack foods, packaging materials should block the factors accelerating lipid oxidation. The oil is spread over a large surface area and exposed to the O_2 in the atmosphere. Unsaturated fatty acids in the oil are prone to oxidative rancidity in the presence of air (Anon., 2008). Fried snack foods are typically packaged in multilayer structures. Spiral-wound, paperboard cans lined with aluminum foil or a barrier polymer are used for some specialty products that require mechanical protection. In addition, the use of metal cans for fried nuts is popular for premium products, where the container is usually N_2 gas flushed (Robertson, 2006). Requirements for fried snack packaging material are as follows:

a. Grease-proofness: An oil barrier film is required to prevent unsightly staining of the package, smudging of the printing and the feeling of a greasy package.
b. Prevention of rancidity can be achieved by an O_2 absorber and/or high O_2 and light barrier films.
c. Loss of crispness can be prevented with high water-vapor barrier films.
d. Package should be strong enough to withstand the process of N_2 or CO_2 flushing.

18.4.4.2 Extruded and Puffed Snacks

Many extruded and puffed snack foods are packaged in the same packaging materials used for fried snack foods. However, because the major mode of quality deterioration of extruded/puffed snacks is loss of crispness, a package that provides a good water vapor barrier is the primary requirement. Some extruded and puffed snack are comparatively less sensitive to O_2 than are fried snack foods, and, therefore, the O_2 barrier requirements of these package are consequently less important than those of fried snacks (Robertson, 2006).

18.4.4.3 Fruit-Based Snacks

Many cereal products contain dried fruits. Packaging materials for these products must prevent the development of stickiness as a result of moisture uptake. The a_w should be maintained at <0.5. The moisture equilibrium between dried cereals and fruits in the package should also be considered (Risbo, 2003). Most dried fruits in cereals and snacks are intermediate moisture foods that have relatively higher moisture contents than other cereal-based parts of the products. Therefore, it is very important to study moisture transfer between the fruits and cereal portions based on water activity and moisture equilibrium.

18.4.4.4 Nuts

Lipid oxidation is the most important quality parameter in nut products. Lipid oxidation creates undesirable rancid taste and odor. External factors such as storage temperature, O_2 concentration, light, relative humidity (RH), and gas atmosphere have a great influence on this oxidative quality deterioration (Jensen et al., 2003). For internal factors, shelf life is influenced by maturity, fatty acid composition, and variety (Lee and Krochta, 2002). The shelf life of nut products can be extended by various special packaging techniques, such as N_2 and CO_2 flushing, vacuum packaging, use of high-O_2-barrier material, and O_2-absorbing sachets or film ingredients.

Packaging peanuts in cans with a gas mixture of 0–1% O_2 and 100–99% N_2 guarantees the longest shelf life (Ucherek, 2004). Comparison tests of storage at 21% and <2.5% O_2 revealed lower peroxide values and hexanal content for shelled peanuts and walnuts (Mate et al., 1996) and a better sensory quality of walnuts (Jan et al., 1998) when they were packed with N_2. Coating the peanut with whey-protein film significantly delayed hexanal production (Lee and Krochta, 2002). Exchanging the atmospheric air with CO_2 extended the shelf life of pecans and raw peanuts from 2 to 27 weeks (Holaday et al., 1979). CO_2 flushing is also economical for packaging peanuts and pecans for long-term storage. CO_2 is adsorbed into the pores of nuts resulting in the formation of a vacuum inside the pouches (Robertson, 2006). In another study, the optimal storage condition for walnuts were obtained with an O_2 absorber at low temperature (<11°C). Also, without chill storage and O_2 absorbers, it was possible to achieve an acceptable quality of walnuts with a high-O_2-barrier packaging material combined with N_2 flushing (Jensen et al., 2003). The use of combined technologies (e.g., O_2 absorber combined with high-O_2-barrier packaging material) is more effective against lipid oxidation of nut products (Dull et al., 1988; Jensen et al., 2003; Ribeiro et al., 1993). High-O_2-barrier packaging material without gas flushing could develop lower sensory quality, probably due to fermentation of the nuts (Dull et al., 1988). Some nuts (e.g., almonds, peanuts, pine nuts, and walnuts) had lowered sensory quality after storage under light than samples stored in the dark (Sattar et al., 1990). Walnuts and peanuts stored at 54% RH have higher peroxide values than nuts stored at 21% RH (Mate et al., 1996). The optimum storage conditions for shelled walnuts are 0–3.5°C and 55–65% RH (Matz, 1993). Generally, low-temperature conditions are good for shelf life extension of nut products.

REFERENCES

Allen L.H. 2008. Priority areas for research on the intake, composition and health effects of tree nuts and peanuts. *The Journal of Nutrition* 138: 1763S–1765S.

Anon. 2008. Packaging of Snack Food. http://www.iip-in.com/foodservice/22_snackfood.pdf. Accessed October 2008.

Arvanitoyannis I.S., Traikou A. 2005. A comprehensive review of the implementation of hazard analysis critical control point (HACCP) to the production of flour and flour-based products. *Critical Reviews in Food Science and Nutrition* 45: 327–370.

Banks D. 1996. Industrial frying. In: *Deep Frying: Chemistry, Nutrition, and Practical Applications*. Perkins E.G., Erickson M.D. (Eds). Champaign, Illinois: AOCS Press, pp. 258–270.

Banks D.E., Lusas E.W. 2001. Oils and industrial frying. In: *Snack Foods Processing*. Lusas E.W., Rooney L.W. (Eds). Lancaster, Pennsylvania: Technomic Publishing, pp. 137–204.

Booth R.G. 1990. Nuts. In: *Snack Food*. Booth R.G. (Ed). New York: Van Nostrand Reinhold, pp. 247–263.

Chauhan G.S., Bains G.S. 1990. Equilibrium moisture content, BET monolayer moisture and crispness of extruded rice-legume snacks. *International Journal of Food Science & Technology* 25: 360–363.

Cretors C. 2001. Popcorn products. In: *Snack Foods Processing*. Lusas E.W., Rooney L.W. (Eds). Lancaster, Pennsylvania: Technomic Publishing, pp. 385–420.

Dull G.G., Kays S.J. 1988. Quality and mechanical stability of pecan kernels with different packaging protocols. *Journal of Food Science* 53: 565–567.

Evranuz E.O. 1993. The effects of temperature and moisture content on lipid peroxidation during storage of unblanched salted roasted peanuts: shelf life studies for unblanched salted roasted peanuts. *International Journal of Food Science & Technology* 28: 193–199.

Firman E.F. 1973. Effects of light on snack food packaging. *Snack Food* 62: 70–72.

Gould W.A. 2001. Potatoes and potato chips. In: *Snack Foods Processing*. Lusas E.W., Rooney L.W. (Eds). Lancaster, Pennsylvania: Technomic Publishing, pp. 227–246.

Groff. 2001. Perfect pretzel production. In: *Snack Foods Processing*. Lusas E.W., Rooney L.W. (Eds). Lancaster, Pennsylvania: Technomic Publishing, pp. 369–384.

Hill P.E., Rizvi S.S.H. 1982. Thermodynamic parameters and storage stability of drum dried peanut flakes. *LWT—Food Science and Technology* 15: 185–190.

Hix V.M. 2001. Use of dried potatoes in snack foods. In: *Snack Foods Processing*. Lusas E.W., Rooney L.W. (Eds). Lancaster, Pennsylvania: Technomic Publishing, pp. 247–260.

Holaday C.E., Pearson J.L., Slay W.O. 1979. New packaging method for peanuts and pecans. *Journal of Food Science* 44: 1530–1533.

Hsieh F., Hu L., Huff H.E., Peng I.C. 1990. Effects of water activity on texture characteristics of puffed rice cake. *LWT—Food Science and Technology* 23: 471–473.

Huber G. 2001. Snack foods from cooking extruders. In: *Snack Foods Processing*. Lusas E.W., Rooney L.W. (Eds). Lancaster, Pennsylvania: Technomic Publishing, pp. 315–367.

Huber G.R., Rokey G.J. 1990. Extruded snacks. In: *Snack Food*. Booth R.G. (Ed). New York: Van Nostrand Reinhold, pp. 107–138.

Jan M., Langerak D.I., Wolters T.G., Farkas J., Kamp H.J.V.D., Muuse B.G. 1988. The effect of packaging and storage conditions on the keeping quality of walnuts treated with disinfestation doses of gamma rays. *Acta Alimentaria* 17: 13–31.

Jayaraman K.S., Gopinath V.K., Pitchamuthu P., Vijayaraghavan P.K. 1982. The preparation of quick cooking dehydrated vegetables by high temperature short time pneumatic drying. *Journal of Food Technology* 17: 19–26.

Jensen P.N., Danielsen B., Bertelsen G., Skibsted L.H., Andersen M.L. 2005. Storage stabilities of pork scratchings, peanuts, oatmeal and muesli: comparison of ESR spectroscopy, headspace-GC and sensory evaluation for detection of oxidation in dry foods. *Food Chemistry* 91: 25–38.

Jensen P.N., Sørensen G., Brockhoff P., Bertelsen G. 2003. Investigation of packaging systems for shelled walnuts based on oxygen absorbers. *Journal of Agricultural and Food Chemistry* 51: 4941–4947.

Jensen P.N., Risbo J. 2007. Oxidative stability of snack and cereal products in relation to moisture sorption. *Food Chemistry* 103: 717–724.

Katz E.E., Labuza T.P. 1981. Effect of water activity on the sensory crispness and mechanical deformation of snack food products. *Journal of Food Science* 46: 403–409.

King J.C., Blumberg J., Ingwersen L., Jenab M., Tucker K.L. 2008a. Tree nuts and peanuts as components of a healthy diet. *The Journal of Nutrition* 138: 1736S–1740S.

King J.C., Rechkemmer G., Geiger C.J. 2008b. Second international nuts and health symposium, 2007: introduction. *The Journal of Nutrition* 138: 1734S–1735S.

Labuza T.P. 1980. The effect of water activity on reaction kinetics of food deterioration. *Food Technology* 34(4): 36–41, 59.

Labuza T.P. 1982. Shelf-life of breakfast cereals In: *Shelf-life Dating of Foods*. Westport, Connecticut: Food & Nutrition Press, pp. 98–118.

Larsen H., Lea P., Rodbotten M. 2005. Sensory changes in extruded oat stored under different packaging, light and temperature conditions. *Food Quality and Preference* 16: 573–584.

Lawes M.J. 1990. Potato-based textured snacks. In: *Snack Food*. Booth R.G. (Ed). New York: Van Nostrand Reinhold, pp. 265–284.

Lee S.Y., Krochta J.M. 2002. Accelerated shelf life testing of whey-protein-coated peanuts analyzed by static headspace gas chromatography. *Journal of Agricultural and Food Chemistry* 50: 2022–2028.

Lima J.R., da Silva M.A.A.P., Goncalves L.A.G. 1999. Sensory characterization of cashew nut kernels. *Cienciae e Tecnologis de Alimentos* 19: 123–126.

Lozano P.R., Miracle R.E., Krause A.J., Cadwallader K.R., Drake M.A. 2007. Effect of cold storage and packaging material on butter flavor. *Journal of Animal Science* 85: 482–482.

Lu S., Lin T.-C. 2001. Rice-based snack foods. In: *Snack Foods Processing*. Lusas E.W., Rooney L.W. (Eds). Lancaster, Pennsylvania: Technomic Publishing, pp. 439–455.

Lusas E.W. 2001. Overview. In: *Snack Foods Processing*. Lusas E.W., Rooney L.W. (Eds). Lancaster, Pennsylvania: Technomic Publishing, pp. 3–28.

Mate J.I., Saltveit M.E., Krocha J.M. 1996. Peanut and walnut rancidity: effect of oxygen concentration and relative humidity. *Journal of Food Science* 61: 465–472.

Matz S.A. 1993. *Snack Food Technology*. 3rd edn. New York: Van Nostrand Reinhold.

Matz S.A., McWilliams C.S., Larsen R.A., Mitchell J.H., McMullen J., Layman B. 1955. The effect of variations in moisture content on the storage deterioration rate of cake mixes. *Food Technology* 6: 276–285.

McCarthy J.A. 2001. The snack industry: history, domestic and global status. In: *Snack Foods Processing*. Lusas E.W., Rooney L.W. (Eds). Lancaster, Pennsylvania: Technomic Publishing, pp. 29–35.

Mehta S.P. 2001. Tortilla chip processing. In: *Snack Foods Processing*. Lusas E.W., Rooney L.W. (Eds). Lancaster, Pennsylvania: Technomic Publishing, pp. 261–280.

Meiron T.S., Saguy I.S. 2007. Wetting properties of food packaging. *Food Research International* 40: 653–659.

Nath A., Chattopadhyay P.K., Majumdar G.C. 2007. High temperature short time air puffed ready-to-eat (RTE) potato snacks: process parameter optimization. *Journal of Food Engineering* 80: 770–780.

Ory R.L., St Angelo A.J., Gwo Y.-Y., Flick G.J., Mod R.R. 1985. Oxidation-induced changes in foods. In: *Chemical Changes in Food during Processing*. Richardson T., Finley J.W. (Eds). Westport, Connecticut: AVI Publishing, pp. 205–217.

Quast D.G., Karel M. 1972. Effects of environmental factors on the oxidation of potato chips. *Journal of Food Science* 37: 584–588.

Ribeiro M.A.A., Regitano-d'Arce M.A.B., Lima U.A., Nogueira M.C.S. 1993. Storage of canned shelled brazil nuts; effects on the quality. *Acta Alimentaria* 22: 295–303.

Rice R. 1990. Health food snacks. In: *Snack Food*. Booth R.G. (Ed). New York: Van Nostrand Reinhold, pp. 285–300.

Risbo J. 2003. The dynamics of moisture migration in packaged multi-component food systems I: shelf life predictions for a cereal-raisin system. *Journal of Food Engineering* 58: 239–246.

Robertson G.L. 2006. *Food Packaging Principles and Practice*. Boca Raton, Florida: CRC Press, pp. 61, 417–446.

Safa L., Zaki O., Leprince Y., Feigenbaum A. 2008. Evaluation of model compounds—polypropylene film interactions by Fourier transform infrared spectroscopy (FTIR) method. *Packaging Technology and Science* 21: 149–157.

Sattar A., Mohammad J., Saleem A., Jan M., Ahmad A. 1990. Effect of fluorescent light, gamma radiation and packages on oxidative deterioration of dry nuts. *Sarhad Journal of Agriculture*. 6: 235–240.

Ucherek M. 2004. An integrated approach to factors affecting the shelf life of products in MAP. *Food Reviews International* 20: 297–307.

Wikipedia. 2008. Olestra. Wikimedia Foundation, Inc. http://en.wikipedia.org/wiki/Olestra. Accessed April 2008.

19 Shelf Life of Foods in Biobased Packaging

Vibeke Kistrup Holm
Danish Technological Institute
Kolding, Denmark

CONTENTS

19.1 INTRODUCTION AND CLASSIFICATION

Polymers originating from renewable resources are attracting more interest in a society where a large part of the garbage and waste consists of polymers derived from petroleum sources (Robertson, 2008). The most widely used renewable packaging materials are paper and board, which are based on cellulose, the most abundant renewable polymer worldwide. However, alternative biopolymers have been identified for the production of food packaging materials and research is taking place in order to develop materials that meet the requirements of the foods in order to obtain the desired shelf life and maintain quality.

Several materials have demonstrated favorable properties that may prove useful for packaging of specific foods. Some of the materials are already competitive alternatives to conventional food packaging [polylactate (PLA) being one], whereas other materials still need further optimization in order to be suitable for the packaging of foods.

This chapter provides an overview of biobased food packaging. The relevant food packaging properties and the impact on shelf life of particular foods are reviewed. Finally, suggestions for future developments and use of biobased packaging are given.

Biobased materials are here defined as materials derived primarily from annually renewable sources. These materials may be compostable. Some biobased packaging materials still contain nonrenewable substances due to inferior properties of the biobased materials (e.g., starch-based materials). The definition also includes edible films and coatings and, to some extent, cellulose-based materials, but they are not discussed in this chapter. However, it is interesting to note that the first step in the development of biobased packaging for food applications was in the field of edible films and coatings.

Depending on the production process and on the source, biopolymers can be divided into three groups: polyesters, starch-based, and others.

The polyesters can be:

1. Polymers directly extracted from biomass such as proteins, lipids, and polysaccharides
2. Polymeric materials synthesized by a classical polymerization procedure such as PLA
3. Polymeric materials produced by microorganisms and bacteria like poly(hydroxyalkanoates) (PHAs)

PLA has a high potential for use in food packaging and offers numerous opportunities for tailoring the properties of the final packaging solution. PLA is polymerized from lactic acid monomers. Lactic acid has two enantiomeric forms: L- and D-lactic acid. Depending on which of the forms are present and in what proportions, it is possible to produce the semicrystalline poly-L-lactic acid (PLLA) and poly-D-lactic acid (PDLA) (Frits et al., 1994; Ikada and Tsuji, 2000) and the amorphous mixture of poly-L-lactic acid and poly-D-lactic acid (PDLLA) (Li and Vert, 1994).

High molecular weight PLA exhibits the attractive properties of styrenes, enabling thermoforming without displaying the brittleness inherent in styrene polymers (Fang et al., 2005) as well as several of the relevant properties for the packaging applications of PLA such as optical, physical, and mechanical properties that are between those of poly(ethylene terephthalate) (PET) and polystyrene (PS) (Auras et al., 2005, 2006).

PHAs are thermoplastics with promising potential for food packaging applications as they possess properties close to those of conventional packaging materials. PHAs are accumulated by a large number of bacteria as energy and carbon reserves. Poly(hydroxybutyrate) (PHB) is one of the well-known PHAs. PHB is a natural thermoplastic polyester and has many mechanical properties comparable to synthetically produced degradable polyesters (Freier et al., 2002).

Polymers primarily extracted from agricultural or forest plants and trees include cellulose, starch, pectin, and protein. The polysaccharide-based polymers are hydrophilic and are somewhat crystalline, with both factors causing processing and performance problems, especially in relation to packaging of moist foods. Therefore, many of these polymers are blended with more hydrophobic polymers. An example is the starch-based packaging materials. Commercial starch-based polymers can be mixed with oil-based monomers with different percentages of starch used as additives. Depending on starch percentages and other materials, the properties of these materials can be varied a lot (Siracusa et al., 2008).

19.2 PROPERTIES OF BIOBASED PACKAGING

As with conventional packaging, biobased packaging aspects are important when evaluating the shelf life of packaged foods. Among other parameters, the shelf life of a packaged food is determined by the properties of the package, which depend on the chemical and physical nature of the polymers used. Parameters such as molecular structure, crystallinity, and chemical composition of polymers subsequently determine the protection that can be provided by a particular food packaging material. Due to the biobased and biodegradable nature of some packaging, material safety and stability are also important issues to consider.

19.2.1 Safety

Due to the natural origin of biobased polymers and the fact that some of the raw materials are present in the foods, that is, starch and lactic acid (dimers, trimers, etc.), many of the biobased packaging materials are considered safe for food packaging purposes. Thus, the categories lactic acid, edible and hydrolyzed starch, and PHB are approved for use in the manufacture of materials and articles (Commission Directive, 2002/72/EC). However, it is imperative that the final food packaging material is safe.

PLA is approved by the US Food and Drug Administration for contact with food. The overall migration from different PLA polymers into water, acetic acid, isooctane, olive oil, and a solid simulant modified phenylene oxide (MPPO) was well below the 10 mg dm^{-2} limit (Conn et al., 1995; Plackett et al., 2006; Selin, 1997). Migration into ethanol, however, has been reported to be less than 1 mg dm^{-2} in 15% ethanol (Selin, 1997) but from 13 to more than 200 mg dm^{-2} in 95% ethanol (Plackett et al., 2006). Hence, the studied PLA should not be used for packaging products containing a high percentage of alcohol.

A sensory panel found a strong odor in mineral water packaged in PHB jars, which was due to a high level of residual solvent. Hence, the packaging materials could not be approved for mineral water or aqueous products (Bucci et al., 2005).

The conformity of a starch–clay nanocomposite film with EU directives was verified in a study by migration tests and by putting the films in contact with vegetables and simulants (Avella et al., 2005).

The use of biobased materials as primary food packaging is challenging in relation to microbial growth as the packaging material itself may serve as a source of energy for the microorganisms. Such growth may present a potential risk of growth of undesirable microorganisms on the packaging, and should the packaging be degraded during storage of the foods, exterior microbial migration may occur, thereby contaminating the food (Jakobsen et al., 2008).

Bergenholtz and Nielsen (2002) and Plackett et al. (2006) showed that packaging materials based on starch/polycaprolactone (PCL) supported growth of undesirable food-related fungi. The authors suggested a modification: incorporation of antimicrobial compounds into the starch-based packaging materials. These studies also showed that PLA and PHB did not serve as a growth substrate for undesirable food-related fungi.

19.2.2 Stability

Due to the degradable nature of some biobased packaging materials, it is important to consider the stability of the package when evaluating and determining the shelf life of packaged foods. Parameters such as moisture, temperature, and ultraviolet (UV) light affect the stability of particular biobased materials. These parameters are also important when biobased packages are composted after use. During the biodegradation process, temperature, moisture, and enzymes from microorganisms degrade the polymers. In addition, pH, available nutrients, oxygen, storage time, and temperature are important to the biodegradation process. It is therefore important to control these parameters during product shelf life in order to protect the packaging material from degradation and also protect the packaged products.

PLA is degraded on contact with moisture due to hydrolytic cleavage of the ester bonds in the PLA polymer, which impairs polymer stability and subsequently barrier and mechanical properties of the package (see Figure 19.1) (Holm et al., 2006a). As moisture is often present in the entire food packaging system, including the food, the package headspace, and the surroundings, the stability aspects become important when matching the shelf life of PLA with the shelf life of the food. Despite the well-known hydrolytic effect of water on PLA (Cairncross et al., 2005), it is used commercially for packaging water, with the end of shelf life being when the amount of water remaining in the bottle is less than the quantity declared on the label.

Reports on the temperature stability of PLA cups indicate that the cups will remain stable only up to 55°C (Anon., 1997, 1998). Commercial usage of PLA cups for hot drinks, however, indicates

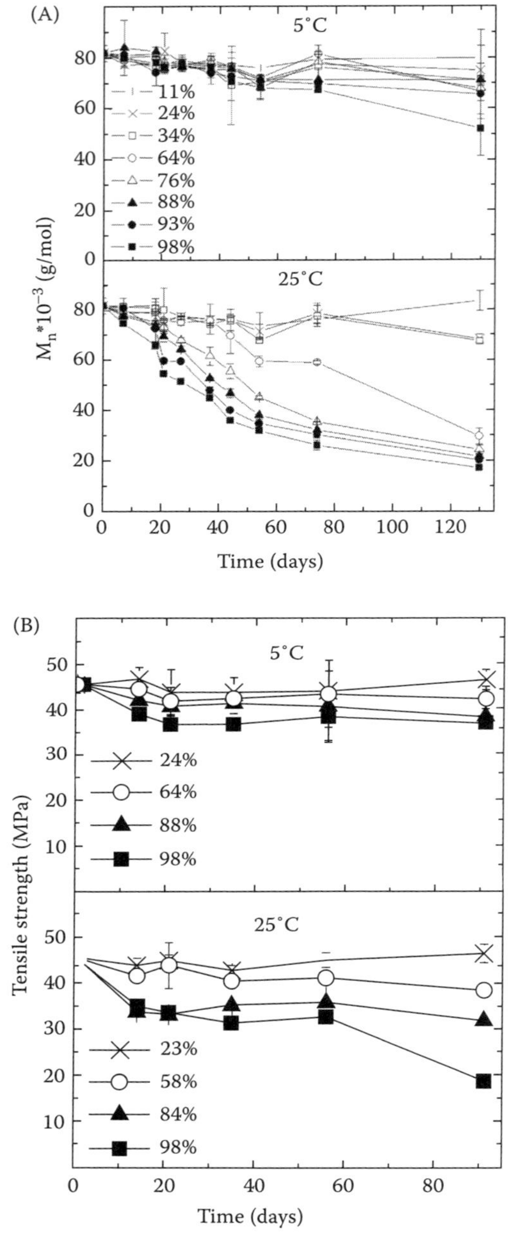

FIGURE 19.1 (A) Changes in M_n (number–average molecular weight) and (B) tensile strength at break of PLLA film exposed to different relative humidities at 5°C and 25°C. (From Holm V.K., Ndoni S., Risbo J. 2006. The stability of poly(lactic acid) packaging films as influenced by humidity and temperature. *Journal of Food Science* 71: E40–E44. With permission from Wiley-Blackwell.)

that this particular material may be used for some types of hot foods. Thermal stability of PLA may be enhanced by, for example, nanoclay addition (Plackett et al., 2006).

With respect to the stability toward UV light, a study has shown that the resulting decrease in physical integrity and degradation of the polymer were much lower for PLA than for low density polyethylene (LDPE) (Ho and Pometto, 1999).

19.2.3 Barrier Properties

Light barrier properties are important in order to prevent photo-oxidation of proteins, lipids, and nutrients. The light transmission of a packaging material can be altered by different means. As transparency of plastics is a function of crystallinity, the light barrier can, to a certain extent, be tailored by the crystallinity. Amorphous polymers free from fillers and impurities are transparent unless chemical groups capable of absorbing visible light are present. On the other hand, crystalline polymers are generally translucent. In addition, modification of the light barrier can be achieved by incorporation of dyes or application of coatings that absorb light at specific wavelengths (Robertson, 2006).

The water vapor barrier of PLA is generally regarded as being moderate (Ikada and Tsuji, 2000) and subsequently comparable to certain conventional food packaging materials. Water vapor transmission rates (WVTRs) close to that of a polyamide (PA6) film and a PS cup have been reported (van Tuil et al., 2000; Petersen et al., 2001). Other studies showed water vapor barriers of PLA materials to be 4 times greater than poly(vinylidene chloride) (PVdC) copolymer and LDPE, 2 times greater than PS cups, and 40–60 times higher than polypropylene (PP) and high density polyethylene (HDPE) cups (Petersen et al., 2001). Rhim et al. (2007) found that coating PLA films with soy protein isolate improved the water vapor barrier compared to those of cellulose acetate films. Suyatma et al. (2004) improved the water vapor barrier properties of PLA by coating it with chitosan. PHAs have a low WVTR resembling that of LDPE materials (van Tuil et al., 2000).

The literature reports WVTRs of starch-based cups to be 100–300 times greater than those of PP and HDPE cups, and 5–9 times greater than those of PS cups. WVTRs of blends of starch, and oil-based films were 4–6 times greater than those of LDPE and HDPE films (Petersen et al., 2001).

Martin et al. (2001) successfully prepared multilayer films based on plasticized wheat starch and various biodegradable aliphatic polyesters in order to improve the water vapor barrier properties of the starch. Garcia et al. (2006) improved the water vapor barrier of starch films markedly by addition of chitosan.

Barrier properties of polar packaging materials may be affected by moist environments, for example, cellophane (Del Nobile et al., 2002) and ethylene-vinyl alcohol (EVOH) copolymer. Thus, as biobased polymers are polar, moisture may affect their water vapor barrier properties. However, the water vapor permeability (WVP) of PLA did not depend on the steady state moisture content as shown by subjecting PLA films to different internal and external relative humidities simulating different food packaging environments. A negative activation energy of the temperature dependence on WVP of PLA has been reported (Auras et al., 2003), which means that WVP increases along with a decrease in temperature. Subsequently, the behavior of PLA is similar to many synthetic polymeric materials where moisture transport is described by Fickian diffusion, making the WVP constant dependent only on temperature (Robertson, 2006).

The oxygen transmission rate (OTR) of PLA is high (Ikada and Tsuji, 2000). The OTR of PLA films has been reported to be higher than that of the high-oxygen-barrier materials such as EVOH, PVdC, and PA6, and lower than that of the oxygen-permeable LDPE film (Petersen et al., 2001; van Tuil et al., 2000). When comparing cups, the OTR of PLA was lower than that of HDPE, PP, and PS, which are all regarded as low-oxygen-barrier materials (Petersen et al., 2001). Auras et al. (2005) reported the OTR of oriented PLA to be approximately 10 times lower than that of oriented PS, but approximately 6 times higher than that of PET.

No major effects of changes in OTR at normal food storage conditions on PLA and PHA have been observed when the relative humidity (RH) is increased. The OTR was reported to decrease slightly when the RH increased at 40°C, but no major effect was noted at 5°C and 23°C (Auras et al., 2004).

In general, starch-based polymers constitute materials with medium to excellent gas barriers (van Tuil et al., 2000). Specific studies showed OTRs of cups based on PHB to be lower than those of PP, PS, and HDPE (Petersen et al., 2001). Furthermore, packaging films based on blends of wheat starch and PCL, as well as blends of corn starch and PCL, had OTRs considerably lower than LDPE and HDPE films (Petersen et al., 2001).

The gas barriers of biobased packaging materials may be improved by combining them with a layer of chitosan, protein, or modified starch film, all of which have high-oxygen-barrier properties (Kittur et al., 1998; van Tuil et al., 2000). The materials may provide alternatives to the gas barrier materials currently used, such as EVOH and PA.

19.2.4 Mechanical Properties

Mechanical properties can to some extent be tailored to meet the desired properties by means of plasticizing, blending with other polymers or fillers, adding fibers (van Tuil et al., 2000), orienting polymers during processing (Sinclair, 1996), and by controlling the crystallinity. Subsequently, there are different processing parameters that may be varied to obtain biobased materials with specific mechanical properties.

PLA has mechanical properties similar to conventional packaging materials such as HDPE, PP, and PET (Auras et al., 2003; Kharas et al., 1994). Studies showed that compression of PLA cups was in the range of that for PP, PS, and HDPE cups (Krochta and De Mulder-Johnston, 1996; Petersen et al., 2001). However, mechanical properties of PLA films have been reported to be lower than those of LDPE and HDPE films (Ikada and Tsuji, 2000; Petersen et al., 2001). Tensile strength, percent elongation, and tear strength of a PLA, and a starch/PCL film were lower than that for LDPE and HDPE films (Ikada and Tsuji, 2000; Petersen et al., 2001), whereas soy-protein-coated PLA films exhibited mechanical properties similar to LDPE (Rhim et al., 2007). Blending of PLA with other polymers is a useful strategy to impart flexibility and toughness (Lim et al., 2008).

Mechanical properties of PHB resemble those of isotactic PP (van Tuil et al., 2000), although PHB is somewhat stiffer and more brittle (Siracusa et al., 2008).

Native starch alone does not meet the requirements for food packaging applications as its mechanical strength and stability are too low (Ahvenainen et al., 1997).

19.3 BIOBASED MATERIALS AND SHELF LIFE

As both the WVTR and OTR of biobased packaging are high, changes in the moisture content and the oxidative, and microbial changes are important indices of failure when evaluating the shelf life of foods packaged in biobased materials.

19.3.1 Moisture Loss or Gain

Many biobased packaging materials do not provide a high water vapor barrier (see Section 19.2.3), and moisture loss or gain is therefore a critical parameter for foods packaged in such materials, as has been shown in several studies involving the packaging of fruits and vegetables. The water activity (a_w) of fruits and vegetables is typically very high, and thus moisture will be released from the packages due to a higher internal RH generated by a high a_w of the products compared to the external RH, resulting in drying of the foods.

Weight loss of tomatoes packaged in PLA- and PHB-coated paperboard trays overwrapped with perforated starch-based bags was studied for 22 days and compared with LDPE packages (Kantola and Helén, 2001). Although the biobased materials offered the same protection against quality changes as did the conventional ones, the tomatoes lost more of their weight in the biobased packages than in the LDPE packages, which was identified as the shelf life-limiting factor.

PLA did not provide proper protection of mushrooms. The WVTR of a specific PLA film studied for packaging of mushrooms was approximately four times higher than the conventional HDPE film. This difference was reflected in the moisture loss rate, which was up to five times higher for PLA than for HDPE (Holm and Mortensen, 2004). After 3 weeks of storage, the moisture

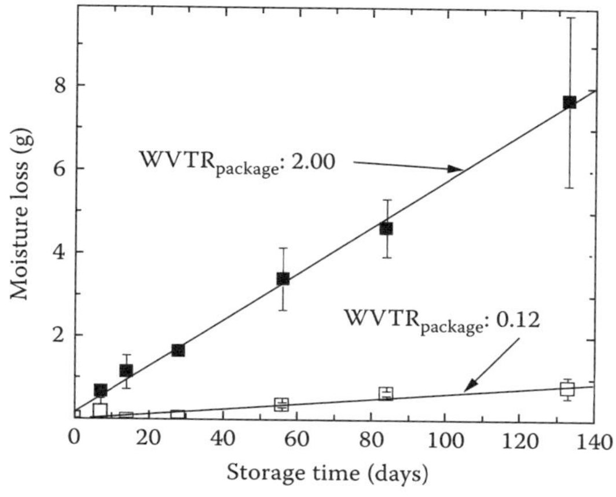

FIGURE 19.2 Moisture loss from Danbo cheese packages relative to weight at time of packaging at 4°C. (Modified from Holm V.K., Risbo J., Mortensen G. 2006. Quality changes in semi-hard cheese packaged in a poly(lactic acid) material. *Food Chemistry* 97: 401–410. With permission from Elsevier.)

loss amounted to more than 2.5% from the PLA packages, resulting in considerable product shrinkage.

Another study showed that weight loss of mangoes stored in chitosan-covered cellulose-based boxes was considerably higher compared to the reference packages, where the boxes were covered with HDPE, due to poor water vapor characteristics (Srinivasa et al., 2002).

The water vapor barrier has also been demonstrated to be a shelf life-limiting factor when using biobased materials for packaging of cheese. Semihard cheese was packaged in PLA and compared with a conventional cheese-packaging material consisting of amorphous PET/LDPE (APET/LDPE). In PLA-packaged cheeses, moisture loss was the prevalent process due to a higher internal RH, generated by the high a_w of the cheese, compared to the external RH. The particular PLA material did not provide sufficient protection of cheeses against moisture loss, and surface drying was observed after 56 days of storage. For cheese packaged in conventional materials, a shelf life of 84 days is achieved. The PLA package provided a 10-times poorer protection than the reference package, which is due to the much higher WVTR of the PLA (see Figure 19.2) (Holm et al., 2006b). In another cheese study, the moisture loss from PLA packages was six times higher than that from APET packages (Holm et al., 2006c).

19.3.2 Lipid Oxidation

Lipid oxidation is an important index of failure of fat-containing foods. As both oxygen and light initiate lipid oxidation, the light transmission and OTRs are very important in protecting the packaged foods against oxidation of the lipids. As discussed in Section 19.2.3., biobased packaging materials do not provide a high oxygen barrier. Therefore, lipid oxidation must be studied when evaluating the shelf life of foods packaged in biobased materials.

Lipid oxidation of different fatty foods such as semihard cheeses and salad dressings packaged in biobased materials has been documented in shelf life studies simulating realistic storage conditions.

Cheeses are very prone to lipid oxidation. Therefore, an oxygen-free modified atmosphere (MA) is often applied to the packages. Lipid oxidation of semihard cheese has been evaluated in systems where cheeses were packaged in a MA with a maximum allowed oxygen concentration of 0.5% at the time of packaging in either a PLA or a PET material (Holm et al., 2006b, 2006c). In both packaging systems, light transmission was almost identical and unlimited, but the OTR of the PLA packages was 1.6 (Holm et al., 2006b) and 5.4 times higher than the references (Holm et al., 2006c). Lipid oxidation was most pronounced in the PLA systems when exposed to light (Holm et al., 2006b, 2006c). The two types of materials are not expected to provide different protection against light-initiated lipid oxidation as the important wavelengths for this phenomenon are situated in the part of the spectrum where the transmittance of the two films is identical. Hence, the difference in lipid oxidation cannot only be ascribed to different light transmission spectra, but rather to a combined action of light and oxygen, with oxygen being present in much higher levels in the PLA than the reference packages due to the more permeable nature of the PLA. This was consistent with the fact that lipid oxidation was limited when the packages were stored in the dark, where PLA provided almost the same protection as the reference (Holm et al., 2006b, 2006c).

Reduction of lipid oxidation in cheeses packaged in PLA has been attempted by including oxygen scavengers in the packages. A sixfold reduction in the content of the secondary lipid oxidation product (hexanal) was obtained, but dark storage still provided a better protection even without oxygen scavengers applied to the packages (Holm et al., 2006c).

An evaluation of Canestrato Pugliese cheese (Italian semihard cheese made from sheep's milk) packaged in starch-based materials (Di Marzo et al., 2006) reported that biodegradable films could be used for packaging of cheese although their performance proved lower than that of high-barrier conventional films.

Development of secondary lipid oxidation products in yoghurt (Frederiksen et al., 2003), sour cream (Holm and Mortensen, 2004), and salad dressing packaged in PLA has been studied (Haugaard et al., 2003). Such products are conventionally packaged in PS or HDPE, both of which have high OTRs. The OTR of the packaging material is not critical for such products as the products are packaged in atmospheric air and thus sufficient oxygen will be available for lipid oxidation to occur. Transmission of light at wavelengths below 500 nm was lower for the PLA package than the PS and HDPE packages (Frederiksen et al., 2003; Haugaard et al., 2003). This difference resulted in a superior protection against lipid oxidation of products packaged in PLA than in PS and HDPE, as studied by development of the secondary lipid oxidation product, hexanal, in sour cream (see Figure 19.3). Light of the critical wavelengths 405–435 nm (Lennersten and Lingnert, 2000) was transmitted through PS. This was not the case for PLA. Furthermore, hexanal was not detected in any of the samples stored in the dark. Hence, the different levels of protection provided by the two materials are due to the different light transmissions in the critical wavelength regions.

19.3.3 Oxidation of Vitamins

The light transmission and OTR of the packaging material also affect the degradation of vitamins as found for different food systems such as riboflavin in yogurt and cheese (Frederiksen et al., 2003; Holm et al., 2006b, 2006c; Kristensen et al., 2000), ascorbic acid in orange juice (Haugaard et al., 2002, 2003), β-carotene in yogurt (Frederiksen et al., 2003), and α-tocopherol in salad dressing (Haugaard et al., 2003). Light-induced oxidation of vitamins is due to the absorption of light at specific wavelengths. For instance, light at wavelengths below 465 nm initiates the oxidation of β-carotene (Sattar et al., 1977), whereas wavelengths at, especially, 415 nm and 455 nm initiate oxidation of riboflavin (Sattar et al., 1977; Tagliaferri, 1989).

The effect of PLA on the oxidation of vitamins in particular food systems revealed a superior protection of ascorbic acid in orange juice (Haugaard et al., 2002) and α-tocopherol in salad dressing (Haugaard et al., 2003) by PLA compared to PS or HDPE references, respectively. As these products were packaged in atmospheric air, the specific OTR values of the packaging materials were of little importance. The PLA packages had, in general, a lower light transmission than the respective reference packages (Haugaard et al., 2003), which may explain the superior protection against vitamin loss. Similarly, a more pronounced degradation of riboflavin was observed in light-exposed

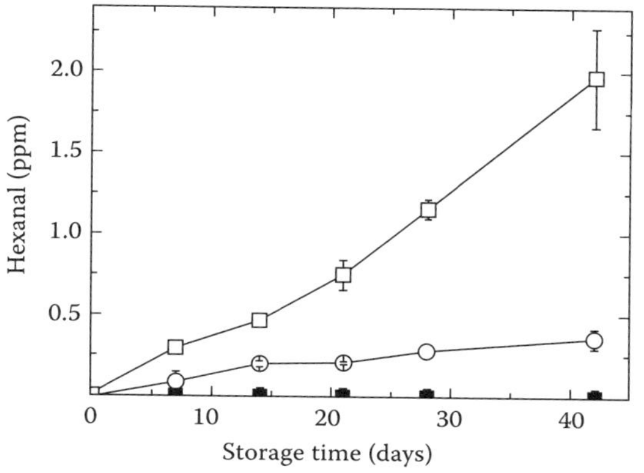

FIGURE 19.3 Concentration of hexanal in sour cream packaged in PLA (o) and PS (□) stored under fluorescent light (open symbols) or in the dark (dark symbols). (Modified from Holm V.K., Mortensen G. 2004. Food packaging performance of polylactate (PLA). In: *14th IAPRI World Conference on Packaging*. Stockholm, June 13–16, with permission.)

yogurts packaged in atmospheric air in PS than in PLA. The effect was ascribed to differences in light transmission below 455 nm (Frederiksen et al., 2003), as wavelengths below 455 nm initiate the oxidation of riboflavin (Sattar et al., 1977; Tagliaferri, 1989).

Photo-oxidative degradation of riboflavin has also been observed in other dairy products (Hansen and Skibsted, 2000; Holm et al., 2006c; Kristensen et al., 2000). In MA-packaged cheese, the PLA package did not provide sufficient protection against diffusion of oxygen into the package headspaces as compared to the gastight reference package. The maximum oxygen concentration in the PLA and reference packages during storage was 10% and 2%, respectively, although the oxidation of riboflavin was not affected by the package type, and hence the different OTRs. However, a distinct effect of light was noted: all light-exposed cheeses lost up to 80% of their initial riboflavin content, whereas loss of riboflavin in samples stored in the dark was close to zero. With respect to the oxidation of riboflavin, performance of PLA equaled that of the reference packages. Hence, it was the presence of oxygen rather than the actual concentration that was important for the oxidation of riboflavin in cheese. These findings suggest that the oxygen concentration in the packages must be very low or zero in order to prevent riboflavin degradation in cheeses exposed to light. This could not be reached even when incorporating oxygen scavengers in the gastight reference packages (Holm et al., 2006c).

In addition, both PLA and PHB have provided similar protection to conventional HDPE for another type of fatty food (salad dressing) with respect to loss of α-tocopherols, lipid oxidation, and color changes (Haugaard et al., 2003).

19.3.4 Color Changes

The color of many foods is due to the presence of natural pigments such as carotenoids in orange juice and riboflavin in cheese. As such pigments are oxidized at rates that are dependent on light (Robertson, 2006), the OTR and light transmission of the packaging material obviously become crucial parameters in preventing discoloration of the packaged foods. It has been demonstrated that the oxidation of natural food pigments impairs the color of certain foods. An examples is the loss of yellowness, which correlates with the degradation of riboflavin and β-carotene in yogurt (Bosset and Flückinger, 1986, 1989; Frederiksen et al., 2003) and citrus juices (Mannheim et al., 1987) and the degradation of α- and β-carotene in vegetable juices (Pesek and Warthesen, 1987).

The ability of PLA to protect orange juices (Haugaard et al., 2002, 2003), yogurt (Frederiksen et al., 2003), and salad dressing (Haugaard et al., 2003) from discoloration was evaluated by objective color measurements and compared with measurements of products packaged in PS or HDPE. PLA provided better protection against color changes than did the references for the lightness parameter of salad dressing although the differences were small.

Wavelengths in the range of 410–450 nm have the greatest impact on the discoloration of mayonnaise (Lennersten and Lingnert, 2000), which may also be the case for salad dressing. As light at these wavelengths was not transmitted through PLA but through HDPE, and as no changes in samples stored in the dark were observed, the better color stability provided by PLA must be due to the different light transmission spectra at 410–450 nm (Haugaard et al., 2003). A relationship between color stability and light exposure has also been demonstrated by Weichold et al. (1988). They found that cheese exposed to light developed a more reddish color compared to cheese kept in darkness, which may be ascribed to the transmission of light at wavelengths of especially 405 and 436 nm (Mortensen et al., 2003).

19.3.5 Microbiology

Koide and Shi (2007) reported on the effects of a PLA film as compared with an LDPE film on the microbial quality of green peppers. Lower coliform bacteria counts were observed on the peppers in PLA than in LDPE packaging. Another study, however, revealed that blueberries packaged in PLA containers showed higher fungal development than those in a PET container (Almenar et al., 2008).

Use of chitosan has been examined in a storage study of mangoes packaged in cellulose-based boxes with top surfaces covered with either chitosan film or LDPE as a reference material (Srinivasa et al., 2002). The results revealed that mangoes stored in chitosan-covered boxes showed an extension of shelf life of up to 18 days, without microbial growth. Thus, the authors concluded that chitosan films are a useful alternative to synthetic films for storage of mangoes.

Kim and Pometto (1994) and Strantz and Zottola (1992) found that the addition of starch to LDPE packaging films did not enhance the growth of bacteria in the meat products or on the surface of the packaging materials.

19.4 PERSPECTIVES

Due to their renewable origin, biobased packaging is indeed a material for future food packaging applications. However, such materials are not expected to replace conventional food packaging materials in the short term as they will be applicable only for products with limited barrier requirements. At the moment, PLA is the biobased polymer of choice when it comes to the packaging of many types of foods. As PLA properties approach those of PS and PET (Auras et al., 2005, 2006), it is expected that PLA will be suitable for short-shelf-life products such as fruits and vegetables packaged in atmospheric air, and dairy products such as yogurt, milk, and sour cream, as these products do not require any specific gas barrier.

In the longer term, biobased materials are expected to compete with conventional materials when their properties are further developed to resemble such materials. It is obvious that more work is required to reach such a goal. Continuous developments make it possible to improve the barrier properties of biobased materials substantially by plasma deposition of glass-like silicon oxide coatings, by applying nanocomposites from natural polymers and modified clays (Fischer et al., 2000; Johannson, 2000; Ray et al., 2006; Sorrentino et al., 2007), or by using biofillers derived from renewable resources (e.g., natural fibers, starches, and proteins) (Lim et al., 2008). Hence, in the future, biobased packaging materials may also be applicable for MAP, which will enhance their usage range considerably.

REFERENCES

Ahvenainen R., Myllärinen P., Poutanen K. 1997. Prospects of using edible and biodegradable protective films for foods. *The European Food & Drink Review* Summer: 73–80.

Almenar E., Samsudin H., Auras R., Harte B., Rubino M. 2008. Postharvest shelf life extension of blueberries using a biodegradable package. *Food Chemistry* 110: 120–127.

Anon. 1997. Danone setzt auf Öko-Verpackung. *Welt Der Milch* 51: 882.

Anon. 1998. Focus on natural fibre. *Food Process* 58: 38.

Auras R., Harte B., Selke S. 2004. Effect of water on the oxygen barrier properties of poly(ethylene terephthalate) and polylactide films. *Journal of Applied Polymer Science* 92: 1790–1803.

Auras R., Harte B., Selke S., Hernandez R. 2003. Mechanical, physical, and barrier properties of poly(lactide) films. *Journal of Plastic Film and Sheeting* 19: 123–135.

Auras R.A., Singh S.P., Singh J.J. 2005. Evaluation of oriented poly(lactide) polymers vs. existing PET and oriented PS for fresh food service containers. *Packaging Technology and Science* 18: 207–216.

Auras R., Singh S.P., Singh J. 2006. Performance evaluations of PLA against existing PET and PS containers. *Journal of Testing and Evaluations* 34: 530–536.

Avella M., De Vlieger J.J., Errico M.E., Fischer S., Vacca P., Volpe M.G. 2005. Biodegradable starch/clay nanocomposite films for food packaging applications. *Food Chemistry* 93: 467–474.

Bergenholtz K.P., Nielsen P.V. 2002. New improved method for evaluation of growth by food related fungi on biologically derived materials. *Journal of Food Science* 67: 2745–2749.

Bosset J.O., Flückinger E. 1986. Einfluβ der Licht- und Gasdurchlässigkeit verschiedener Packungsarten auf die Qualitätserhaltung von Naturjoghurt. *Deutsche Milchwirtschaft* 29: 908–914.

Bosset J.O., Flückinger E. 1989. Die Verpackung als Mittel zur Qualitätserhaltung von Lebensmitteln, dargestellt am Beispiel der Lichtschutzbedürfigkeit verschiedener Joghurtsorten. *Lebensmittel Wissenschaft und Technologie* 22: 292–300.

Bucci D.Z., Tavares L.B.B., Sell I. 2005. PHB packaging for the storage of food products. *Polymer Testing* 24: 564–571.

Cairncross R.A., Becker J.G., Ramaswamy S., O'Connoer R. 2005. Moisture sorption, transport, and hydrolytic degradation in polylactide. *Applied Biochemistry and Biotechnology* 129–132: 774–785.

Commission Directive. 2002/72/EC relating to plastics materials and articles intended to come into contact with foodstuffs as amended by 2004/19/EC.

Conn R.E., Kolstad J.J., Borzelleca J.F., Dixler D.S., Filer Jr. L.J., LaDu B.N., Pariza M.W. 1995. Safety assessment of polylactide (PLA) for use as a food-contact polymer. *Food and Chemical Toxicology* 33: 273–283.

Del Nobile M.A., Buonocore G.C., Dainelli D., Battaglia G., Nicolais L. 2002. A new approach to predict the water-transport properties of multilayer films intended for food packaging applications. *Journal of Food Science* 69: 85–90.

Di Marzo S., Di Monaco R., Cavella S., Borriello I., Masi P. 2006. Correlation between sensory and instrumental properties of *Canestrato Pugliese* slices packed in biodegradable films. *Trends in Food Science & Technology* 17: 169–176.

Fang J.M., Fowler P.A., Escrig C., Gonzalez R., Costa J.A., Chamudis L. 2005. Development of biodegradable laminate films derived from naturally occurring carbohydrate polymers. *Carbohydrate Polymers* 60: 39–42.

Fischer S., de Vlieger J., Kock T., Gilberts J., Fischer H., Batenburg L. 2000. Green composites—the materials of the future—a combination of natural polymers and inorganic particles. In: *Biobased Packaging Materials for the Food Industry. Status and Perspectives*. Weber C.J. (Ed). Frederiksberg: KVL Department of Dairy and Food Science, p. 109.

Frederiksen C.S., Haugaard V.K., Poll L., Miquel Becker E. 2003. Light-induced quality changes in plain yoghurt packed in polylactate and polystyrene. *European Food Research and Technology* 217: 61–69.

Freier T., Kunze C., Nischan C., Kramer S., Sternberg K. 2002. *In vitro* and *in vivo* degradation studies for development of a biodegradable patch based on poly(3-hydroxybutyrate). *Biomaterials* 23: 2649–2657.

Fritz H.G., Seidenstücker T., Bölz U., 1994. Use of whole cells and enzymes for the production of polyesters. In: *Production of Thermo-Bioplastics and Fibres based mainly on Biological Materials*. Agro-Industrial Research Division, Directorate-General XII Science, Research and Development, pp. 109–139.

Garcia M.A., Pinotti A., Zaritzky N.E. 2006. Physicochemical, water vapor barrier and mechanical properties of corn starch and chitosan composite films. *Starch/Stärke* 58: 453–463.

Hansen E., Skibsted L.H. 2000. Light-induced oxidative changes in a model dairy spread. Wavelength dependence of quantum yields. *Journal of Agricultural and Food Chemistry* 48: 3090–3094.

Haugaard V.K., Danielsen B., Bertelsen G. 2002. Quality changes in orange juice packed in materials based on polylactate. *European Food Research and Technology* 214: 423–428.

Haugaard V.K., Danielsen B., Bertelsen G. 2003. Impact of polylactate and poly(hydroxybutyrate) on food quality. *European Food Research and Technology* 216: 233–240.

Ho K.-L.G., Pometto A.L. 1999. Effects of electron-beam irradiation and ultraviolet light (365 nm) on polylactic acid plastic films. *Journal of Environmental Polymer Degradation* 7: 93–100.

Holm V.K., Mortensen G. 2004. Food packaging performance of polylactate (PLA). In: *14th IAPRI World Conference on Packaging*. Stockholm, June 13–16.

Holm V.K., Ndoni S., Risbo J. 2006a. The stability of poly(lactic acid) packaging films as influenced by humidity and temperature. *Journal of Food Science* 71: E40–E44.

Holm V.K., Risbo J., Mortensen G. 2006b. Quality changes in semi-hard cheese packaged in a poly(lactic acid) material. *Food Chemistry* 97: 401–410.

Holm V.K., Mortensen G., Vishart M., Agerlin Petersen M. 2006c. Impact of poly-lactic acid packaging material on semi-hard cheese. *International Dairy Journal* 16: 931–939.

Ikada Y., Tsuji H. 2000. Biodegradable polyesters for medical and ecological applications. *Macromolecular Rapid Communications* 21: 117–132.

Jakobsen M., Holm V., Mortensen G. 2008. Biobased packaging of dairy products. In: *Environmentally Compatible Food Packaging*. Chiellini E. (Ed). Cambridge: Woodhead Publishing Limited, pp. 478–495.

Johannson K.S. 2000. Improved barrier properties of renewable and biodegradable polymers by means of plasma deposition of glass-like SiOx coatings. In: *Conference Proceedings, The Food Biopack Conference*. Weber C.J. (Ed). Copenhagen, August 27–29, p. 110.

Kantola M., Helén H. 2001. Quality changes in organic tomatoes packaged in biodegradable plastic films. *Journal of Food Quality* 24: 167–176.

Kharas H., Sanchez-Riera F., Severson D.K. 1994. Polymers of lactic acid. In: *Plastics from Microbes. Microbial Synthesis of Polymers and Polymer Precursors*. Mobley D.P. (Ed). Munich: Carl Hanser Verlag, pp. 93–137.

Kim M., Pometto III A.-L. 1994. Food packaging potential of some novel degradable starch-polyethylene plastics. *Journal of Food Protection* 57: 1007–1012.

Kittur F., Kumar K.R., Tharanathan N. 1998. Functional packaging properties of chitosan films. *Zeitschrift für Lebensmittel Untersuchung und Forschung A* 206: 44–47.

Koide S., Shi J. 2007. Microbial and quality evaluation of green peppers stored in biodegradable film packaging. *Food Control* 18: 1121–1125.

Kristensen D., Orlien V., Mortensen G., Brockhoff P., Skibsted L. 2000. Light-induced oxidation in sliced Havarti cheese packaged in modified atmosphere. *International Dairy Journal* 10: 95–103.

Krochta J.M., De Mulder-Johnston C. 1996. Biodegradable polymers from agricultural products. In: *Agricultural Materials as Renewable Resources: Nonfood and Industrial Applications*. ACS Symposium Series #647, Fuller G., McKeon T.A., Bills D.D. (Eds). Washington, DC: American Chemical Society, pp. 121–140.

Lennersten M., Lingnert H. 2000. Influence on wavelength and packaging material on lipid oxidation and color changes in low-fat mayonnaise. *Lebensmittel Wissenschaft und Technologie* 33: 253–260.

Li S., Vert M. 1994. Morphological changes resulting from the hydrolytic degradation of stereocopolymers derived from l- and dl-lactides. *Macromolecules* 27: 3107–3110.

Lim L.-T., Auras R., Rubino M. 2008. Processing technologies for poly(lactic acid). *Progress in Polymer Science* 33: 820–852.

Mannheim C.H., Miltz J., Letzter A. 1987. Interaction between polyethylene laminated cartons and aseptically packed citrus juices. *Journal of Food Chemistry* 52: 737–740.

Martin O., Schwach E., Avérous L., Couturier Y. 2001. Properties of biodegradable multilayer films based on plasticized wheat starch. *Starch/Stärke* 53: 372–380.

Mortensen G., Sørensen J., Danielsen B., Stapelfeldt H. 2003. Effect of specific wavelengths on light-induced quality changes in Havarti cheese. *Journal of Dairy Research* 70: 413–421.

Pesek C.A., Warthesen J.J. 1987. Photodegradation of carotenoids in a vegetable juice system. *Journal of Food Science* 52: 744–746.

Petersen K., Nielsen P.V., Olsen M.B. 2001. Physical and mechanical properties of biobased materials. *Starch/Stärke* 53: 356–361.

Plackett D.V., Holm V.K., Johansen P., Ndoni S., Nielsen P.V., Sipilainen-Malm T., Södergård A., Verstichel S. 2006. Characterization of l-polylactide and l-polylactide-polycaprolactone co-polymer films for use in cheese-packaging applications. *Packaging Technology and Science* 19: 1–24.

Ray S., Quek S.Y., Easteal A., Chen X.D. 2006. The potential use of polymer-clay nanocomposites in food packaging. *International Journal of Food Engineering* 2, 4, article 5: 1–11.

Rhim J.-W., Lee J.H., Ng P.K.W. 2007. Mechanical and barrier properties of biodegradable soy protein isolate-based films coated with polylactic acid. *LWT—Food Science and Technology* 40: 232–238.

Robertson G.L. 2006. *Food Packaging Principles and Practice*, 2nd edn. Boca Raton, Florida: CRC Press.

Robertson G.L. 2008. State-of-the-art biobased food packaging materials. In: *Environmentally Compatible Food Packaging*. Chiellini E. (Ed). Cambridge, England: Woodhead Publishing, pp. 3–28.

Sattar A., deMan J., Alexander J.C. 1977. Light-induced degradation of vitamins II. Kinetic studies on ascorbic acid decomposition. *Journal of the Canadian Institute of Food Science and Technology* 10: 65–68.

Selin J.F. 1997. Polylactides and their applications. In: *Technology Programme Report 13/9*. Helsinki: Technology Development Centre Tekes, pp. 111–127.

Sinclair R.G. 1996. The case for polylactic acid as a commodity packaging plastic. *JMS—Pure Applied Chemistry* A33: 585–597.

Siracusa V., Rocculi P., Romani S., Dalla Rosa M. 2008. Biodegradable polymers for food packaging: a review. *Trends in Food Science & Technology* 19: 634–643.

Sorrentino A., Gorassi G., Vittoria V. 2007. Potential perspectives of bio-nanocomposites for food packaging applications. *Trends in Food Science & Technology* 18: 84–95.

Srinivasa P.C., Baskaran R., Ramesh M.N., Harish Prashanth K.V., Tharanathan R.N. 2002. Storage studies of mango packed using biodegradable chitosan film. *European Food Research and Technology* 215: 504–508.

Strantz A.A., Zottola E.A. 1992. Bacterial survival on cornstarch-containing polyethylene film held under food storage conditions. *Journal of Food Protection* 55: 782–786.

Suyatma N.C., Copinet A., Tighzert L., Coma V. 2004. Mechanical and barrier properties of biodegradable films made from chitosan and poly(lactic acid) blends. *Journal of Polymers and Environment* 12: 1–6.

Tagliaferri E. 1989. Effect protecteur de l'emballage contre la photo-oxydation, IV. Etude de la stabilité des vitamins A et B_2 dans divers yoghourts en cours de stockage. *Traveaus de Chimie Alimentaire et d'Hygiene* 80: 77–86.

van Tuil R., Fowler P., Lawther M., Weber C.J. 2000. Properties of biobased packaging materials. In: *Biobased Packaging Materials for the Food Industry. Status and Perspectives*. Weber C.J. (Ed). Frederiksberg: KVL Department of Dairy and Food Science, pp. 13–44.

Weichold U., Seiler H., Busse M., Klostermeyer H. 1988. Rotfärbungen bei Käse und deren Ursachen. *Deutsch Milchwirtschaft* 39: 1671–1675.

20 Active Packaging and the Shelf Life of Foods

Kay Cooksey
Department of Packaging Science
Clemson University
Poole Agricultural Center
Clemson, South Carolina

CONTENTS

20.1 INTRODUCTION

Active packaging has been defined as "packaging in which subsidiary constituents have been deliberately included in or on either the packaging material or the package headspace to enhance the performance of the package system" (Robertson, 2006). Table 20.1 lists active packaging systems or components and their common applications. Much of the work on active packaging has remained at the research stage but some have been commercialized. Testing active packaging systems using foods has begun to increase but few researchers have actually performed true shelf life studies. Most data provided relate to microbial challenges, quality changes, and sensory

TABLE 20.1
Examples of Some Active Packaging Systems and Their Applications

Active Component	Commercially Available	Applications
Moisture absorbers/emitters	Yes	Fresh meat purge pads Cheese ripening
Oxygen absorbers/emitters	Yes	Modified atmosphere packaging Prevention of oxidation in bakery and meat products
Carbon dioxide absorbers/emitters	Yes	Modified atmosphere packaging Antimicrobials
Volatile odor absorbers	Yes	Quality in fresh fruit
Ethylene absorbers and adsorbers	Yes	Control ripening of fruit
Antimicrobials		Microbial control
Ethanol emitters	Yes (sachets)	
Bacteriocins	Yes (pediocin)	
Chlorine dioxide	Yes (sachets)	
Organic acid	No	
Chitosan	No	
Natural extracts and essential oils	No	
Allyl isothiocyanate	Yes (labels)	
Silver ions	Yes (nanocomposite)	
Glucose oxidase	Yes (sachets)	

analysis, and are reported as days of acceptable storage compared to a package without an active component.

20.2 BAKERY PRODUCTS

The use of active components to preserve bakery products has been mainly performed for military rations and retail products in Europe and Asia. The main focus has been to prevent mold growth or oxidation if the product contains enough fat and is of a high enough water activity for lipid oxidation. One problem presented with bakery products is their ability to pick up odors from some of the active components.

Pastry dough was packaged in a cellulose-based film (25 or 70 μm thick) containing either 7% or 3% sorbic acid. Dough stored in low density polyethylene (LDPE) was used as the control. All samples were stored at 8°C for 40 days. Films with sorbic acid were more effective in reducing mesophilic, psychotropic, and *Staphylococcus* spp. throughout the 40-day storage period compared to control films. For example, dough with 25-mm-thick film with 7% sorbic acid and 70-mm-thick film with 3% sorbic acid had a 2 log reduction compared to a 1.5 log increase for mesophilic bacteria after 40 days of storage. It is of interest to note that the authors also measured the diffusion of sorbic acid into the dough and found that it was below the limit allowed by legislation (Silveira et al., 2007).

A variety of essential oils and oleoresins from spices and herbs were used to inhibit common molds found on hot dog buns contained in a high-barrier modified atmosphere package (MAP). The only essential oil found to inhibit several different types of molds tested was allyl isothiocyanate (AITC), from mustard oil. The researchers found that the incorporation of AITC into the package allowed hot dog buns to meet their required shelf life but an off-flavor and odor was detected by a sensory panel (Nielsen and Rios, 2000).

Franke et al. (2002) studied the effects of an ethanol-releasing sachet on the shelf life of prebaked buns. Without the ethanol sachet, the buns had a shelf life of 2–3 days. When the ethanol sachet was incorporated into the packaging, the shelf life was extended to 13 days. Although the ethanol is completely absorbed by the buns, the off-flavor can be reduced by heating the buns prior to serving, which vaporizes the ethanol prior to consumption.

Sponge cakes (0.8–0.9 a_w, which is the water activity) were packed in MAP with O_2 absorber sachets of two different absorption capacities (100 and 210 mL). The cakes were analyzed for mold growth over a period of 28 days at 25°C storage. MAP alone provided some benefits regarding mold prevention; however, combining O_2 scavengers with modified atmosphere (30% CO_2) prevented mold growth entirely during the 28 days of storage. The results also indicated that there was a greater benefit for cakes with a higher a_w (0.9) compared to a lower a_w (0.8) (Guynot et al., 2003).

In another study (Berenzon and Saguy, 1998), O_2 absorbers were added to wheat crackers formulated with high levels of oil for storage in hermetically sealed cans used as military rations. The study included storage at 15°C, 25°C, and 35°C. Shelf life was assessed using sensory panels, as well as hexane concentration and headspace O_2 measurements. As storage temperature increased, headspace O_2 decreased within the can. Overall, cans of crackers without O_2 sachets reached unacceptable levels of rancidity within 24 weeks at 25°C and 35°C. Cans of crackers with O_2 sachets did not have rancid odors after 44 weeks of storage, regardless of the storage temperature. Thus, the shelf life of canned crackers was extended for 20 weeks with O_2 absorbers added to the can.

20.3 DAIRY PRODUCTS

Active packaging, which affects the shelf life of dairy products, has mainly focused on mold prevention and moisture control in cheese during ripening. Although some of the active components included are also used as additives in formulations of the products, some are not typically used with dairy products and may add flavors or odors that could be detected by the consumer. In such cases, work is ongoing to control and reduce the effects of active components that can effectively increase the shelf life of the dairy products but not alter the sensory properties of the product.

20.3.1 CHEESE

The moisture content of natural cheeses is important to maintain their desirable properties. Often, a permeable paper or perforated regenerated cellulose film (RCF) is used in a multilayer film, but sometimes these options do not work well for certain types of cheese. In a study by Pantaleão et al. (2007), a regional Portuguese cheese called Saloio was packaged using Humidipaks (see Figure 20.1). Previously, the cheese had been sold unpackaged (which made the cheese too hard) or vacuum packaged (which made the cheese too soft). Therefore, humidity control was important for maintaining the characteristics of the cheese during storage. The Humidipaks maintained two different levels (78% and 84%) of relative humidity. Cheese in the Humidipaks were contained within a poly(vinyl chloride) (PVC) shell with a polystyrene (PS) base and were stored for 60 days at 8°C. Cheese in the perforated film was stored for 84 days at 8°C. It was determined that the cheese contained within the package with 84% RH Humidipak provided the best results with regard to hardness, color, and sensory acceptability during the 60 days of storage. The authors stated that the

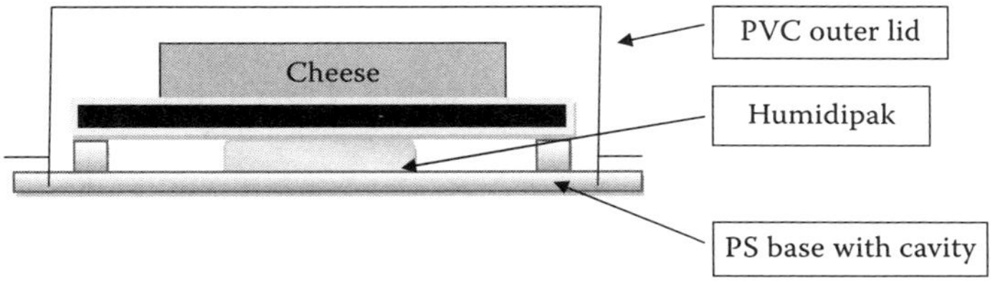

FIGURE 20.1 Design of packaging using Humidipak. (From Pantaleão I., Pintado M.M.E., Poças M.F.F. 2007. Evaluation of two packaging systems for regional cheese. *Food Chemistry* 102: 481–487, with permission from ACS.)

shelf life was significantly extended with either type of Humidipak compared to previous packaging methods (Pantaleão et al., 2007).

Mozzarella cheese was packaged with an "active gel" composed of agar, dissolved saline, and lemon extract at three different levels (500, 1000, and 1500 ppm). The samples were stored at 15°C to simulate temperature abuse and shelf life was calculated using the Gompertz equation. In addition, microbial inhibition with dairy cultures as well as spoilage organisms were used in the study. It was concluded that the lemon extract provided inhibitory action toward the spoilage organisms at temperature-abuse conditions, thus extending the shelf life of the mozzarella cheese (Conte et al., 2007).

Winter and Nielsen (2006) studied the effect of active packaging of cheese with AITC-releasing labels on cheese packaged in atmospheric conditions and modified atmosphere conditions. Danish Danbo cheese was inoculated with a mix of *Penicillium* spp., *Geotrichum* spp., *Aspergillus* spp., and *Debaryomyces* spp. spores and stored at 5°C for up to 28 weeks. Atmospheric packaging extended the shelf life (sensory and volatile detection) from 4.5 to 13 weeks and 4.5 to 28 weeks for 1 and 2 labels, respectively. When combining the labels with MAP, the shelf life was extended from 18 to 28 weeks regardless of whether 1 or 2 labels were used. Cheese stored with labels for up to 12 weeks did have an unacceptable mustard flavor (based on sensory results), but this decreased to acceptable levels between weeks 12 and 28.

A cellulose-based film containing natamycin at 2% and 4% was used to package Gorgonzola cheese inoculated with *Penicillium roqueforti* and stored for 45 days in the ripening chamber. The film was also coated onto the surface of an aluminum foil to form a laminate film, which was also used to wrap the cheese. Both types of films were effective in reducing *P. roqueforti*, but the foil-coated film was considered more effective. In addition, the amount of natamycin released into the cheese over the 45-day ripening period was below the level permitted by legislation (Oliveria et al., 2007).

Another study was performed using a cellulose-based film coated with natamycin and also included nisin for the inhibition of bacteria and mold in mozzarella cheese. Films containing nisin (50%) combined with natamycin (8%) improved the shelf life by 6 days compared to films without the antimicrobial agents. Nisin was found to be less effective against yeast and mold when used alone due to lack of diffusion from the polymer structure (Pires et al., 2008).

20.4 MEAT, FISH, AND POULTRY PRODUCTS

The purpose of much active packaging technology for use in the meat, fish, and poultry market is to improve color, off-flavors, and spoilage. Although MAP might be considered a form of active packaging, it will only be discussed in combination with other active packaging components. A great deal of work has been done on antimicrobial packaging for ready-to-eat (RTE) meat products due to several outbreaks caused by post-process contamination. These outbreaks have been of great concern to consumers and industry alike. Useful review papers on active packaging in meat products include Aymerich et al. (2008), Coma (2008), Kerry et al. (2006), and Cagri et al. (2004).

20.4.1 OXYGEN SCAVENGERS

Much of the research on O_2 scavengers for red meat products has focused on the effect of the scavenger on the color of the meat. Studies found that packages with O_2 scavengers had less discoloration (i.e., lower concentrations of metmyoglobin) than those without sachets, as well as lower levels of *Pseudomonas* and less oxidative changes to flavor (Gill and McGinnis, 1995; Tewari et al., 2001; Vermeiren et al., 1999). Martinez et al. (2006) reported that the use of O_2 scavengers combined with 20% CO_2:80% N_2 extended the shelf life of fresh pork sausages for up to 20 days, with improvements focusing on reduced psychrotrophic aerobic counts and better color and lipid stability.

Oxygen scavengers are also incorporated into labels for RTE meat products such as ham, bologna, and turkey. Labels are manufactured that absorb 10–20 mL of O_2, and larger labels are beginning to become available that scavenge 100–200 mL O_2 (Kerry et al., 2006). Besides labels, O_2

scavengers can also be incorporated into the polymer film. The film is capable of reducing O_2 in the headspace to less than 1 ppm in 4–10 days for products such as dried, smoked meat, as well as processed meat products (Butler, 2002).

Rosemary and oregano were studied separately as antioxidant films for extending the display shelf life of fresh lamb steaks based on color and odor (Camo et al., 2008). Two sets of packaged lamb samples were packaged under modified atmosphere, without active film; a third set was packaged using 4% rosemary impregnated film; a fourth set used a 4% oregano extract; and the last set of lamb was sprayed with 2 mL rosemary extract solution per 50 g meat. The lamb using active packaging and the spray treatment were also packaged under high-oxygen modified atmospheres. Based on color, sensory evaluation, and lipid oxidation analysis, the lamb packaged in either active film and the spray-treated lamb effectively increased shelf life compared to the control from 8 to 11 days. However, active films with oregano were more effective than the rosemary-containing film, extending shelf life from 8 to 13 days compared to the samples without antioxidant film or treatment.

20.4.2 Carbon Dioxide Emitters

The use of carbon dioxide emitters appears to be somewhat controversial for use in active packaging of fresh meat products. According to Coma (2008), moderate levels of CO_2 (10–20%) are known to inhibit spoilage bacteria such as *Pseudomonas*, whereas the growth of lactic acid bacteria can be stimulated. Further, pathogens such as *Clostridium perfringens*, *Clostridium botulinum*, and *Listeria monocytogenes* are not significantly affected by levels of 50% or less CO_2. The concern is that conditions may be created where pathogenic bacteria may thrive in an environment where normal spoilage bacteria cannot grow and provide the normal indicators that consumers associate with a spoiled meat product. This was confirmed by Lovenklev et al. (2004), who reported higher production of *C. botulinum* toxin with high-CO_2 environments. However, Vermeiren et al. (1999) suggested that for most applications of meat and poultry, high levels of CO_2 would be desirable to prevent surface microbial growth and would be useful for extending the shelf life of such products.

20.4.3 Antimicrobials

A tremendous effort has been made over the last decade to develop and test films with antimicrobial properties to improve food safety and shelf life. A comprehensive review (Joerger, 2007) catalogued and analyzed the outcome of these research efforts. The methodologies for measuring antimicrobial activity varied considerably among the studies. Results, defined as the difference in the log colony-forming units (cfu) of a test organism exposed to a control film and the log cfu of the organism exposed to the antimicrobial film, ranged from 0 to 9 for many of the antimicrobials tested. The majority of results centered around 2 log reductions, suggesting that antimicrobial films still face limitations and are perhaps best viewed as part of a hurdle strategy to provide safe foods.

20.4.3.1 Nisin

Natrajan and Sheldon (2000) found that nisin-coated (100 $\mu g\ mL^{-1}$) PVC increased the shelf life of fresh broiler skin and drumsticks. Log reductions of *Salmonella typhimurium*^{NAr} (nalidixic acid-resistant strain) inoculated onto the surface of the chicken ranged from 0.4 to 2.1 depending on the level of citric acid added to the nisin-containing film. Overall, shelf life was extended from 0.6 to 2.2 days following a 3-min immersion in a nisin-containing treatment solution followed by storage in a foam tray pack containing a nisin-treated PVC overwrap and a nisin-treated absorbent tray pad.

Listeria monocytogenes was completely inhibited on the surface of ham, turkey, and beef using a cellulose casing spray-coated with pediocin and nisin (Ming et al., 1997). Viskase Co., Inc., produced the film and received a patent for the antimicrobial casing (U.S. Patent, 5,573,797); however, it is not widely used at this time.

The objective of a study by Franklin et al. (2004) was to determine the effectiveness of packaging films coated with a methylcellulose/hydroxypropyl methylcellulose (MC/HPMC)-based solution containing nisin (10,000, 7500, 2500, or 156.3 IU mL^{-1}) for controlling *L. monocytogenes* on the surface of vacuum-packaged hot dogs. Hot dogs were placed in control and nisin-containing pouches, and inoculated with a five-strain *L. monocytogenes* cocktail (~5 log cfu/package), vacuum sealed, and stored for up to 60 days at 4°C. The *L. monocytogenes* counts on hot dogs packaged with 156.3 IU mL^{-1} nisin level films decreased slightly (~0.5 log reduction) through day 15 of refrigerated storage but were statistically the same ($p > 0.05$) as that for hot dogs packaged in films without nisin after 60 days of storage. Packaging films coated with a cellulose-based solution containing 10,000 and 7500 IU mL^{-1} of nisin significantly decreased ($p < 0.05$) *L. monocytogenes* populations on the surface of hot dogs by greater than 2 log cfu/package throughout the 60-day study.

The shelf life of hams was extended using immobilized nisin and lacticin with MAP (60% N_2:40% CO_2). The film was a polyethylene:nylon (70:30) material, which formed a bond with the antimicrobial agents and remained active for 3 months at room temperature and refrigerated storage (Scannell et al., 2000).

Inhibition of *L. monocytogenes* in cooked ham using natural antimicrobials delivered through an active packaging system combined with high pressure processing was studied by Jofre et al. (2007). The antimicrobial agents (enterocins A and B, sakacin K, and nisin) were incorporated into interleavers (11 × 11 cm) of perforated polypropylene (PP), polyamide, and nonperforated PP. The antimicrobials were spread onto the perforated PP layer and allowed to dry overnight. After high pressure processing (400 MPa) the ham samples were compared to the interleavers without antimicrobials. The nisin-plus-lactate interleavers were found to be most effective with an almost 2 log lower population compared to the nonpressurized ham without antimicrobials stored for 3 months at 6°C. The combination of high pressure processing and antimicrobials (nisin-plus-lactate combination) had a significantly lower population of *L. monocytogenes* than the nonpressurized ham samples (1.5 log cfu g^{-1} population compared to 6.5 log cfu g^{-1} for ham without antimicrobials) after 90 days of storage. The same study was repeated for cooked ham inoculated with *Salmonella* spp., with similar results. Using 400 MPa high pressure processing and nisin-containing interleavers, *Salmonella* spp. counts reduced from 4 log cfu g^{-1} to <10 cfu g^{-1}, a level that was maintained for 3 months of storage at 6°C (Jofre et al., 2008). Overall, it may be concluded that the combination of high pressure processing and bacteriocin-containing packaging could extend the shelf life of cooked ham with regard to microbial safety.

Dawson et al. (2002) used biocide-impregnated films for reducing *L. monocytogenes* on turkey bologna stored at refrigerated temperatures. Two biocidal additives were used (8% lauric acid and 4% nisin) and they were either incorporated singly or combined into a soy film used to wrap the bologna. Bologna in films without biocidal agents showed a 0.5 log increase from an initial 6 log population after 21 days of refrigerated storage. In films with one or both biocidal agents, *L. monocytogenes* population decreased by 1 log during the 21-day storage period. Both biocidal agents (and particularly lauric acid) showed greater promise in liquid media.

Calcium alginate was used as a carrier for two different types of lysozyme (oyster and hen) combined with nisin. Lysozyme is commonly used in combination with nisin to improve its efficacy against gram-negative bacteria. Smoked salmon was inoculated with *L. monocytogenes* and *Salmonella anatum*, coated with an antimicrobial film containing 1000 IU g^{-1} nisin, hen lysozyme (160 μg mL^{-1}), or oyster lysozyme (160 μg mL^{-1}) and stored at 4°C for up to 35 days. Growth of *L. monocytogenes* and *S. anatum* was reduced by 2.2–2.8 log cfu g^{-1} with the containing calcim alginate film nisin and oyster lysozyme and the containing calcim alginate film nisin and hen lysozyme, repectively. There was no difference with regard to type of lysozyme used in the study (Datta et al., 2008).

A nisin-containing coating solution was applied to LDPE film and used to package cold-smoked salmon. Normal (4°C) and temperature abuse (10°C) storage temperatures as well as low (2 log) and high (5 log) inoculation levels were studied. Films containing 2000 IU cm^{-2} nisin reduced *L. monocytogenes* by 3.9 log (at low innoculation level) at either storage temperature, compared to

on appearance, it could be concluded that the shelf life, as it relates to purchasing characteristics, was increased but extension of shelf life related to food safety was not proven (Nerin et al., 2006).

Ku et al. (2008) produced an edible film using *Gelidium corneum* (a type of red algae harvested from Jeju Island, South Korea) and 0.01% cinnamonaldehyde and glycerol. This film was used as a carrier of catechin, which was the active component of the study. Three different levels of catechin (50, 100, and 150 mg) were added to the film and used to wrap sausages, which were further wrapped by a PP film. Sausages inoculated with *Escherichia coli* O157:H7 and *L. monocytogenes* showed a decrease in population by 1.81 and 1.44 log cfu g^{-1}, respectively, after 5 days of storage. Sausages in the catechin-containing film also showed decreased oxidation compared to those without catechin. It was concluded that the film could be used to extend the shelf life of sausages.

20.4.3.5 Grapefruit Seed Extract

Ha et al. (2001) produced a solution-coated multilayer polyethylene film and coextruded film containing grapefruit seed extract at 0.5% or 1% levels. Both films were used to package fresh minced beef that was tested for microbial population, color, and oxidation. The solution-coated multilayer film showed greater antimicrobial activity compared to the coextruded film; it was also effective against a wider range of bacteria than the coextruded film. Further testing was done with the multilayer film and the 1% level of grapefruit seed extract was more effective than 0.5% for inhibiting bacterial growth in minced beef stored at 3°C for 18 days. However, the level of grapefruit seed extract did not affect the beef color or lipid oxidation significantly during storage.

20.4.3.6 Triclosan

Vermeiren et al. (2002) reported that 1% triclosan films were effective in inhibiting psychrotrophic bacteria when tested *in vitro*, but when tested *in vivo*, it did not reduce *L. monocytogenes* in vacuum-packaged chicken breasts stored at 7°C. Intimate contact of the product with the coated material was discussed as necessary for the effectiveness of the triclosan-containing material. Little work has continued on the use of triclosan for food packaging, although a recent paper (Pinto et al., 2008) reported its effectiveness against phytopathogenic microorganisms when incorporated into plastic crates used for shipping fruits and vegetables.

20.4.4 Odor Removal

Franzetti et al. (2001) extended the shelf life of fillets of sole, steaks of cod, and whole cuttlefish by the use of a patented tray (European Patent Application 830018.8-2308) that absorbed trimethylamine from the headspace and liquids in the package. The trays were also gas-flushed with 40% CO_2:60% N_2 and kept under strict temperature control (3±0.5°C). The shelf life was extended up to 10 days and delayed microbial growth, especially of gram-negative and hydrogen-sulfide-producing bacteria. Another interesting effect of the active packaging was that the package conditions favored growth of bacterial strains such as *Moraxella phenylpiruvic*, which did not produce volatiles associated with spoilage.

20.5 FRESH PRODUCE

The key to increasing the shelf life of fresh produce (generally after reducing temperature) has typically been the balancing of oxygen and carbon dioxide concentrations in a MAP to reduce respiration. The goal is to maintain the fresh quality of the produce for as long as possible from the point of shipment to the point of consumption. Another method of extending shelf life has been the use of ethylene scavengers or emitters. Such methods have been effectively used in distribution packaging and have had limited use in films and overwraps for primary or retail packaging. In addition to maintaining the quality of the produce, prevention of mold growth has also been a goal. Bacteria associated with foodborne outbreaks resulting from postharvest handling contamination have been of concern. Recently, several outbreaks involving *E. coli* and *Salmonella* in fresh produce have heightened the concern for pathogenic bacteria in produce.

20.5.1 Tomatoes

A study by Bailen et al. (2006) found that combining MAP with activated carbon improved the sensory properties, reduced spoilage, and improved the quality attributes of fresh tomatoes compared to MA packages without activated carbon. The observed effect was attributed to the ability of activated carbon to function as an ethylene adsorber. The effects were further improved with the use of activated carbon combined with palladium.

20.5.2 Grapes

The shelf life of table grapes was improved by adding eugenol and thymol to a MA package stored at refrigerated temperatures for up to 56 days (Valero et al., 2006). For example, grapes in the MA package without eugenol or thymol had a 50% decay rate, whereas those contained within the MA package with 150 μL of eugenol (saturated onto a gauze pad placed inside the package) had a 12% decay rate. Other factors such as weight loss, color, texture, and sensory properties were also significantly improved in the MA packages containing the essential oils.

A shipping container layered with a paper containing an active component has been used to release sulfur dioxide (SO_2) as a preservative. The packaging system is designed to react with humidity generated in the package by grapes and sodium metabisulfite, which can be incorporated into a paper layer, plastic, or sachet. It has been found to be effective for reducing a common mold *Botyris cinerea* (Scully and Horsham, 2007). A commercially available SO_2-releasing paper pad is called Grape Guard. According to Mustonen (1992), table grapes stored at 0°C using a grape guard pad could provide a shelf life of 8–10 weeks. Some problems with this technology are sudden high releases of SO_2 with changes in temperature and humidity, causing bleaching of the grapes. Another problem is that the U.S. Environmental Protection Agency (EPA) has set a maximum level of SO_2 at 10 ppm due to severe asthmatic reactions in people sensitive to SO_2 (Scully and Horsham, 2007).

20.5.3 Strawberries

Wild strawberries are particularly susceptible to spoilage. A study performed by Almenar et al. (2007) incorporated an antifungal component (nonanone) in a sachet on the inner surface of the lid of a PP container (see Figure 20.2). The strawberries (20 per container) were placed in equilibrium

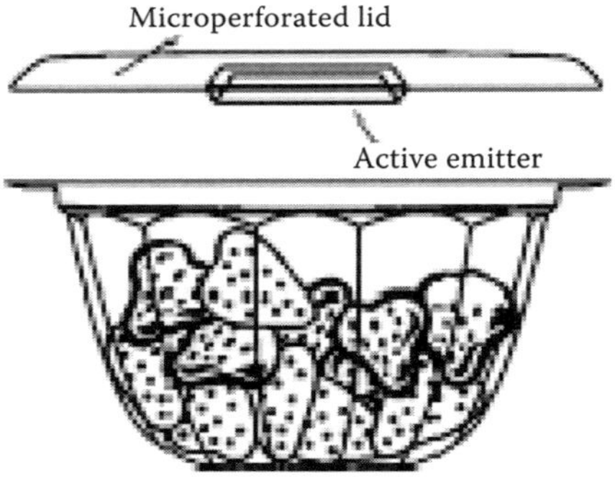

FIGURE 20.2 Diagram of package designed to contain strawberries with nonanone-emitting sachet in lidding. (From Almenar E., Valle V., Catala R., Gavara R. 2007. Active package for wild strawberry fruit (*Fragaria vesca* L.). *Journal of Agricultural and Food Chemistry* 55: 2240–2245, with permission from ACS.)

MA packages containing nonanone-emitting sachets with 2.5, 5, or 10 μL nonanone. The largest amount of nonanone absorbed by the fruit occurred within the first 24 hr, after which the researchers conjectured that there was loss of nonanone through the permeable sachet into the outer atmosphere. Decay was reduced but was not attributed to nonanone alone; the synergistic effect of the MA, storage temperature, and the antifungal compound was also important. Other quality measurements such as weight loss, acidity, anthocyanin concentration, and soluble solids were improved with increasing concentration of nonanone during 3 or 10 days of storage at 22°C or 10°C, respectively.

20.5.4 Onions

Howard et al. (1994) studied the effects of potassium permanganate and activated alumina on diced onions. They used potassium permanganate to absorb sulfur volatiles and reduce levels of CO_2 inside the package. It was determined that the onions were acceptable after 10 days of storage at 2°C with the addition of potassium permanganate to the package (Ahvenainen, 2003).

20.5.5 Fresh-Cut Produce

Mahmoud and Linton (2008) spot-inoculated lettuce leaves with *E. coli* O157:H7 and *Salmonella enterica*. The inoculated lettuce was treated with 0.5, 1.0, 1.5, 3.0, or 5.0 mg L^{-1} ClO_2 gas for 10 min at 22°C/90–95% RH and stored in PVC containers at 4°C for 7 days. Samples were removed after 0, 1, 2, 3, 5, and 7 days and tested for microbial load and color. It was determined that a 5 log reduction could be achieved with 5.0 mg L^{-1} ClO_2 gas exposure for 14.8 and 19 min for *E. coli* and *S. enterica*, respectively. However, color was significantly affected by exposure to ClO_2 gas; it turned the lettuce leaves white with brown spots due to oxidation by ClO_2.

To date, no one has been able to test ClO_2 in a packaging system but some attempts have been made using sachets. The problem has been that, with high-humidity products, ClO_2 is highly reactive and too much ClO_2 is used up initially so that it becomes a one-time high-level treatment rather than a desired slow-release and low-level treatment. To overcome this problem attempts are being made to incorporate ClO_2 into nanoparticles or to encapsulate it. Barrier Safe Solutions International (BSSI) acquired Bernard Technologies Incorporated, which owned and licensed a technology referred to as Microsphere Technology. This technology involved an encapsulation technique that could be used for release of ClO_2. It could be incorporated into film, coatings, adhesives, powders, and received a food contact notification no-objection letter. Early attempts to commercialize the technology failed, but the technology still exists and may warrant further examination for use in packaging applications for RTE produce to help solve the release rate problem associated with ClO_2.

Nanoparticles containing TiO_2 were powder-coated onto oriented PP. TiO_2 is activated by black light and the study included a comparison of nanoparticle size as well as time of exposure to light. *E. coli* was reduced by 3 log compared to only a 1 log reduction with the uncoated oriented PP. When *E. coli* was inoculated onto lettuce, stored in bags containing nanoparticles, and exposed to light for 180 min, *E. coli* was reduced from 6.4 log cfu g^{-1} on day 0 to 4.9 log cfu g^{-1} after 1 day of storage compared to 6.1 log cfu g^{-1} for uncoated bags of lettuce. Nanoparticle size did not affect the results (Chawengkijwanich and Hayata, 2008).

Grapefruit seed extract (0.1% and 1%) was coated onto LDPE to measure the inhibition of several microbial species on agar and for packaging curled lettuce and soybean sprouts stored for 15 days at refrigerated temperature (Lee et al., 1998). Films containing grapefruit seed extract at 1.0% inhibited *E. coli* and *S. aureus* in a disc diffusion test. There was no inhibition of total aerobic bacteria at both concentrations of grapefruit seed extract. Growth of lactic acid bacteria, yeast, and aerobic bacteria were decreased in lettuce and soybean sprouts, particularly if the microbial load was less than 2 log cfu g^{-1}.

An et al. (1998) blended *Rheum palmatum* extract, *Coptis chinensis* extract, sorbic acid, and silver-substituted inorganic zirconium matrix at 1% into an LDPE film. The film was used to package curled lettuce and cucumber inoculated with a variety of bacteria. The film was not effective in inhibiting most bacteria except for *E. coli*, *S. aureus*, and *L. mesenteroides*. When curled lettuce and cucumber were stored at 5°C or 10°C, total aerobic bacteria was reduced and attributes such as hardness and ascorbic acid content were not affected.

Packaging companies have been interested in research on antimicrobial packaging, but for obvious reasons, their internal testing is not published. However, some information can be gained from patents. A patent was granted for the development of an edible coating containing 3% (by weight) chitosan or modified chitosan with 0.1% organic acid (acetic acid was preferred) and 0.02–0.1% surfactant to assist with coating wettability. Methods of application cited included dipping, spraying, and curtain coating. The film was tested against lime, avocados, papayas, tomatoes, and leafy vegetables (Yang et al., 2001). Bell (2004) also filed a patent for an antimicrobial film for use with fresh produce such as bananas and pineapples. The coating was designed to be used with gas-permeable, microperforated, or macroperforated film. The coating is aqueous and uses methyl paraben as the antimicrobial agent. Triclosan and imazalil were also suggested as possible antimicrobial agents.

20.5.6 Ethylene Control

For many years, ethylene control has been used extensively in the bulk shipping of climacteric fruit to delay or control ripening until the products reach the point of use. Films and sachets have been developed for primary packaging with some success. The sachets contain potassium permanganate, which not only scavenges ethylene but can also remove undesirable aromas from the products (Scully and Horsham, 2007). A number of films have been developed that utilize clays, zeolites, and carbon to remove ethylene. Examples include Evert-Fresh (now marketed as Debbie Meyer Greenbags), Peakfresh, Orega, and Bio-fresh. According to a study by Jacobsson et al. (2004), the shelf life of broccoli stored in a plain LDPE bag compared to an ethylene-scavenging LDPE bag stored at 4°C were nearly the same (11 and 12 days, respectively). However, when stored at 10°C, the ethylene scavenging bag resulted in a longer shelf life of 9 days compared to 7 days for the broccoli in a plain bag. Shelf life was determined based on weight loss, visual inspection, total chlorophyll, and texture.

20.6 ISSUES TO CONSIDER FOR SHELF LIFE EXTENSION

Although active packaging may provide many benefits to shelf life extension, there are several issues to consider before implementing such a packaging system. First, of course, is the regulatory status of the active packaging system. The US and EU regulatory systems are somewhat different and, therefore, both must be understood to implement a packaging system for international use. Both regulatory systems are directed toward protection and safety of consumers and are quite complicated. Active packaging, under US regulation, may be considered under the Food Additive Petition or Food Contact Notification process depending upon whether it may intentionally or unintentionally become part of the food, and if so, at what level. Other factors such as functional barriers and nonmigration may apply. The EU directives are somewhat similar but have some important differences regarding food simulants and migration levels. An excellent review of EU regulations as they apply to active packaging has been published by de Kruijf et al. (2002) with an update by van Dongen and de Kruijf (2007).

Other issues include the cost-to-benefit ratio, production capability, consumer acceptance, and sensory effects on the food. Some of the active packaging systems are commercially viable but do not provide enough benefit to justify the cost. In addition, many of the methods developed in laboratories have not been run on a commercial-scale equipment and a procedure such as drying overnight at room temperature is simply not practical. Technologies such as oxygen scavenging films are not very obvious to consumers and, therefore, do not cause any concern, but other systems such as antimicrobial-releasing materials may be noticeable to consumers and cause confusion or mistrust, particularly if the food is altered in any way. In addition, if the active device does not work as designed, liability regarding the cause of failure may be an issue.

Generally, the shelf life has clearly been extended through implementation of active packaging. Combinations of systems along with new technologies to be further developed will continue to improve the quality and safety of foods around the world.

REFERENCES

Ahvenainen R. 2003. *Novel Food Packaging Techniques.* Boca Raton, Florida: Woodhead Publishing.

Almenar E., Valle V., Catala R., Gavara R. 2007. Active package for wild strawberry fruit (*Fragaria vesca* L.). *Journal of Agricultural and Food Chemistry* 55: 2240–2245.

An D.S., Hwang Y.I, Cho S.H., Lee D.S. 1998. Packaging of fresh curled lettuce and cucumber by using low density polyethylene films impregnated with antimicrobial agents. *Journal of the Korean Society of Food Science and Nutrition* 27: 675–681.

Aymerich T., Picouet P.A., Monfort J.M. 2008. Decontamination technologies for meat products. *Meat Science* 78: 114–129.

Bailen G., Guillen F., Castillo S., Serrano M., Balero D., Martinez-Romero D. 2006. Use of activated carbon inside modified atmosphere packages to maintain tomato fruit quality during cold storage. *Journal of Agricultural and Food Chemistry* 54: 2229–2235.

Bell L.D. 2004. Method of Preserving Fresh Perishables. In: *U.S. Patent Application Publication*, edited by U.S. Patent Office, USA. Pub. No.WO/2004/066758.

Berenzon S., Saguy I.S. 1998. Oxygen absorbers for extension of crackers shelf-life. *LWT—Food Science and Technology* 31: 1–5.

Butler B. 2002. Cryovac OS2000 Polymeric Oxygen Scavenging Systems. Paper presented at Worldpak 2002, East Lansing, Michigan.

Cagri A., Ustunol Z., Ryzer E.T. 2004. Antimicrobial edible films and coatings. *Journal of Food Protection* 67: 833–848.

Camo J., Beltran J.A., Roncales J. 2008. Extension of the display life of lamb with an antioxidant active packaging. *Meat Science* 80: 1086–1091.

Chawengkijwanich C., Hayata Y. 2008. Development of TiO_2 powder-coated food packaging film and its ability to inactivate *Escherichia coli* in vitro and in actual tests. *International Journal of Food Microbiology* 123: 288–292.

Coma V. 2008. Bioactive packaging technologies for extended shelf life of meat-based products. *Meat Science* 78: 90–103.

Conte C.S., Scorocco C., Sinigaglia M., Del Nobile M.A. 2007. Innovative active packaging systems to prolong the shelf life of Mozzarella cheese. *Journal of Dairy Science* 90: 2126–2131.

Datta S., Janes M.E., Xue Q.-G., Losso J., La Peyre J.E. 2008. Control of *Listeria monocytogenes* and *Salmonella anatum* on the surface of smoked salmon coated with calcium alginate coating containing oyster lysozyme and nisin. *Journal of Food Science* 73: M67–M71.

Dawson P.L., Carl G.D., Acton J.C., Han I.Y. 2002. Effect of lauric acid and nisin-impregnated soy-based films on the growth of *Listeria monocytogenes* on Turkey bologna. *Poultry Science* 81: 721–726.

De Kruijf N., van Beest M., Rijk R. Sipilainen-Malm T., Paserio Losada P., De Meulenaer B. 2002. Active and intelligent packaging: applications and regulatory aspects. *Food Additives and Contaminants* 19: 144–162.

Ellis M., Cooksey K., Dawson P., Vergano P., Han I. 2006. Quality of fresh chicken breasts using a combination of modified atmosphere packaging and chlorine dioxide sachets. *Journal of Food Protection* 69: 1991–1996.

Franke I., Wijma E., Bouma K. 2002. Shelf life extension of pre-baked buns by an active packaging ethanol emitter. *Food Additives and Contaminants* 19:314–322.

Franklin N., Cooksey K., Getty K.J.K. 2004. Inhibition of *Listeria monocytogenes* on the surface of individually packaged hot dogs using a packaging film coating containing nisin. *Journal of Food Protection* 67: 480–485.

Franzetti L., Martinoli S., Piergiovanni L., Galli A. 2001. Influence of active packaging on the shelf life of minimally processed fish products in a modified atmosphere. *Packaging Technology and Science* 14: 267–274.

Gill C.O., McGinnis J.C. 1995. The use of oxygen scavengers to prevent the transient discolouration of ground beef packaged under controlled, oxygen depleted atmospheres. *Meat Science* 41: 19–27.

Guynot M.E., Sanchis V., Ramos A.J., Marín S. 2003. Mold-free shelf-life extension of bakery products by active packaging. *Journal of Food Science* 68: 2547–2552.

Ha J.U., Kim Y.M., Lee D.S. 2001. Multilayered antimicrobial polyethylene films applied to the packaging of ground beef. *Packaging Technology and Science* 15: 55–62.

Howard L.R., Yoo K.S., Pike L.M., Miller G.H. 1994. Quality changes in diced onions stored in film packages. *Journal of Food Science* 59: 110–117.

Jacobsson A., Nielsen T., Sjohom I. 2004. Effects of type of packaging material on shelf life of fresh broccoli by means of changes in weight, colour and texture. *European Food Research Technology* 218: 157–163.

Joerger R.D. 2007. Antimicrobial films for food applications: a quantitative analysis of their effectiveness. *Packaging Technology and Science* 20: 231–274.

Joerger R.D., Sabesan S., Visioli D., Urian D., Joerger M. 2009. Antimicrobial activity of chitosan attached to ethylene copolymer films. *Packaging Technology and Science* 22: 125–138.

Jofre A., Aymerich T., Garriga M. 2008. Assessment of the effectiveness of antimicrobial packaging combined with high pressure to control *Salmonella* sp. in cooked ham. *Food Control* 19: 634–638.

Jofre A., Garriga M., Aymerich T. 2007. Inhibition of *Listeria monocytogenes* in cooked ham through active packaging with natural antimicrobials and high-pressure processing. *Journal of Food Protection* 70: 2498–2502.

Kerry J.P., O'Grady M.N., Hogan S.A. 2006. Past, current and potential utilisation of active and intelligent packaging systems for meat and muscle-based products: a review. *Meat Science* 74: 113–130.

Ku K.-J., Hong Y.-H., Song K.B. 2008. Mechanical properties of *Gelidium corneum* edible film containing catechin and its application in sausages. *Journal of Food Science* 73: C217–C221.

Lee D.S., Hwang Y.I., Cho S.H. 1998. Developing antimicrobial packaging film for curled lettuce and soybean sprouts. *Food Science and Biotechnology* 7: 117–121.

Lovenklev M., Artin I., Hagberg O., Borch E., Holst E., Radstrom P. 2004. Quantitative interaction effects of carbon dioxide, sodium chloride and sodium nitrite on neurotoxin gene expression in nonproteolytic *Clostridium botulinum* Type B. *Applied and Environmental Microbiology* 70: 2928–2934.

Mahmoud B.S.M., Linton R.H. 2008. Inactivation kinetics of inoculated *Escherichia coli* O157:H7 and *Salmonella enterica* on lettuce by chlorine dioxide gas. *Food Microbiology* 25: 244–252.

Martinez L., Djenane D., Cilla I., Beltran J.A., Roncales P. 2006. Effect of varying oxygen concentrations on the shelf-life of fresh pork sausages packaged in modified atmosphere. *Food Chemistry* 94: 219–225.

Ming X., Weber G.H., Ayres J.W., Sandine W.E. 1997. Bacteriocins applied to food packaging materials to inhibit *Listeria monocytogenes* on meats. *Journal of Food Science* 62:413–415.

Mustonen H.M. 1992. The efficacy of a range of sulfur dioxide generating pads against botyris cinerea infection and on out-turn quality of Calmeria table grapes. *Australian Journal of Experimental Agriculture* 32: 389–394.

Natajan N., Sheldon B.W. 2000. Efficacy of nisin-coated polymer films to inactivate *Salmonella typhimurium* on fresh broiler skin. *Journal of Food Protection* 63: 1189–1196.

Neetoo H., Ye M., Chen H., Joerger R., Hicks D.T., Hoover, D.G. 2008a. Use of nisin-coated plastic films to control *Listeria monocytogenes* on vacuum-packaged cold-smoked salmon. *International Journal of Food Microbiology* 122: 8–15.

Neetoo H., Ye M., Chen H. 2008b. Potential antimicrobials to control *Listeria monocytogenes* in vacuum-packaged cold-smoked salmon pate and fillets. *International Journal of Food Microbiology* 123: 220–227.

Nerin C., Tovar L., Djenane D., Camo J., Salafranca J., Beltran J.A., Roncales P. 2006. Stabilization of beef meat by a new active packaging containing natural antioxidants. *Journal of Agricultural and Food Chemistry* 54: 7840–7846.

Nielsen P.V., Rios R. 2000. Inhibition of fungal growth on bread by volatile components of spices and herbs, and the possible application in active packaging, with special emphasis on mustard essential oil. *International Journal of Food Microbiology* 60: 219–229.

Oliveira T.M., Soares N.F.F., Pereira R.M., Fraga K.F. 2007. Development and evaluation of antimicrobial natamycin-incorporated film in Gorgonzola cheese conservation. *Packaging Technology and Science* 20: 147–153.

Ouattara B., Simard R.E., Piette G., Begin A., Holley R.A. 2000. Inhibition of surface spoilage bacteria in processed meats by application of antimicrobial film prepared with chitosan. *International Journal of Food Microbiology* 62: 139–148.

Pantaleão I., Pintado M.M.E., Poças M.F.F. 2007. Evaluation of two packaging systems for regional cheese. *Food Chemistry* 102: 481–487.

Pinto M.S., Soares N.F.F., Fontes E.A.F., Geraldine R.M., Júnior J.C.B, Melo N.R. 2008. Development of active plastic crate and evaluation of phytopathogenic and pathogenic food spoilage micro-organism inhibition. *Journal of the Science of Food and Agriculture* 88: 1939–1942.

Pires A.C.S., Soares N.F.F., Andrade N.J., Silva L.H.M., Camilloto G.P., Bernardes P.C. 2008. Development and evaluation of active packaging for sliced mozzarella preservation. *Packaging Technology and Science* 21:375–383.

Robertson G.L. 2006. *Food Packaging Principles and Practice*, 2nd edn. Boca Raton, Florida: CRC Press, p. 291.

Scannell A.G.M., Hill C., Ross R.P., Marx S., Hartmeier W., Arendt E.K. 2000. Development of bioactive food packaging materials using immobilised bacteriocins Lacticin 3147 and Nisaplin®. *International Journal of Food Microbiology* 60: 241–49.

Scully A.D., Horsham M.A. 2007. Active packaging for fruits and vegetables. In: *Intelligent and Active Packaging for Fruits and Vegetables*. Wilson C.L. (Ed). Boca Raton, Florida: CRC Press, pp. 57–73.

Silveira M.F.A., Soares N.F.F., Geraldine R.M., Andrade N.J., Goncalves M.P.J. 2007. Antimicrobial efficacy and sorbic acid migration from active films into pastry dough. *Packaging Technology and Science* 20: 287–292.

Skandamis P.N., Nychas E.G.-J. 2002. Preservation of fresh meat with active and modified atmosphere packaging conditions. *International Journal of Food Microbiology* 79: 35–45.

Tewari G., Jayas D.S., Jeremia L.E., Holley R.A. 2001. Prevention of transient discoloration of beef. *Journal of Food Science* 66: 506–510.

Valero D., Valverde J.M., Martýnez-Romero D., Guillén F., Castillo S., Serrano M. 2006. The combination of modified atmosphere packaging with eugenol or thymol to maintain quality, safety and functional properties of table grapes. *Postharvest Biology and Technology* 41: 317–327.

Van Dongen W.D., de Kruijf N. 2007. ACTIPAK in Europe. In: *Intelligent and Active Packaging for Fruits and Vegetables*. Wilson C.L. (Ed). Boca Raton, Florida: CRC Press, pp. 315–328.

Vermeiren L., Devlieghere F., Van Beest M., De Kruijf N., Debevere J. 1999. Developments in the active packaging of foods. *Trends in Food Science and Technology* 10: 77–86.

Vermeiren L., Devlieghere F., Debevere J. 2002. Effectiveness of some recent antimicrobial packaging concepts. *Food Additives and Contaminants* 19 (Suppl 1): 163–171.

Winter M., Nielsen P.V. 2006. Active packaging of cheese with allyl isothiocyanate, an alternative to modified atmosphere packaging. *Journal of Food Protection* 69: 2430–2435.

Yang L., Yang X., Petcavich R., Mao L. 2001. Process and Coating Composition for Extending the Shelf Life of Post Harvest Produce. United States Patent Office (Ed.). US Patent No. 20080063762.

Ye M., Neetoo H., Chen H. 2008a. Effectiveness of chitosan-coated plastic film incorporating antimicrobials in inhibition of *Listeria monocytogenes* on cold-smoked salmon. *International Journal of Food Microbiology* 127: 235–240.

Ye M., Neetoo H., Chen H. 2008b. Control of *Listeria monocytogenes* on ham steaks by antimicrobials incorporated into chitosan-coated plastic films. *Food Microbiology* 25: 260–268.

Yingyuad S., Ruamsin S., Reekprkhon D., Douglas S., Pongamphai S., Siripatrawan U. 2006. Effect of chitosan coating and vacuum packaging on the quality of refrigerated grilled pork. *Packaging Technology and Science* 19: 149–157.

Index

C

D

E

F

Q

R

S

T